PROCEEDINGS OF SYMPOSIA IN PURE MATHEMATICS
VOLUME XIV

# GLOBAL ANALYSIS

AMERICAN MATHEMATICAL SOCIETY
Providence, Rhode Island
1970

Proceedings of the Symposium in Pure Mathematics
of the American Mathematical Society

Held at the University of California
Berkeley, California
July 1–26, 1968

*Prepared by the American Mathematical Society*
*under National Science Foundation Grant GP–8410*

SHIING-SHEN CHERN
STEPHEN SMALE
Editors

Standard Book Number 8218–1414–1
Library of Congress Catalog Number 70–95271
Copyright © 1970 by the American Mathematical Society
AMS 1968 Primary Subject Classification 3465
Printed in the United States of America

1387848

# CONTENTS

Preface . . . . . . . . . . . . v

Bumpy Metrics . . . . . . . . . . . 1
   By R. Abraham

Nongenericity of $\Omega$–Stability . . . . . . . 5
   By R. Abraham and S. Smale

On Certain Automorphisms of Nilpotent Lie Groups . . . 9
   By Louis Auslander and John Scheuneman

Periodic Points of Anosov Diffeomorphisms . . . . . 17
   By Thomas F. Banchoff and Michael I. Rosen

Topological Entropy and Axiom A . . . . . 23
   By Robert Bowen

Zeta Functions of Restrictions of the Shift Transformation . 43
   By R. Bowen and O. E. Lanford III

On the Generic Nature of Property H1 for Hamiltonian Vector Fields . 51
   By Michael A. Buchner

Locating Invariant Sets . . . . . . . . 55
   By R. W. Easton

Anosov Diffeomorphisms . . . . . . . . 61
   By John Franks

Endomorphisms of the Riemann Sphere . . . . . 95
   By John Guckenheimer

Expanding Maps and Transformation Groups . . . . 125
   By Morris W. Hirsch

Stable Manifolds and Hyperbolic Sets . . . . . 133
   By Morris W. Hirsch and Charles C. Pugh

Commuting Diffeomorphisms . . . . . . . 165
   By Nancy Kopell

A Generic Phenomenon in Conservative Hamiltonian Systems . . 185
   By Kenneth R. Meyer and Julian Palmore

Nondensity of Axiom A(a) on $S^2$ . . . . . . 191
   By Sheldon E. Newhouse

Nonsingular Endomorphisms of the Circle . . . . 203
   By Zbigniew Nitecki

A Note on $\Omega$–Stability . . . . . . . 221
   By J. Palis

Structural Stability Theorems . . . . . . 223
   By J. Palis and S. Smale

A Global Approximation Theorem for Hamiltonian Systems . . 233
   By R. Clark Robinson

Strongly Mixing Transformations . . . . . . 245
   By Richard Sacksteder

The Depth of the Center of 2–Manifolds . . . . . . 253
    By A. J. Schwartz and E. S. Thomas

Second Order Ordinary Differential Equations on Differentiable Manifolds  265
    By S. Shahshahani

Expanding Maps . . . . . . . . . 273
    By Michael Shub

Notes on Differentiable Dynamical Systems . . . . . 277
    By Stephen Smale

The Ω–Stability Theorem . . . . . . . . 289
    By S. Smale

Anosov Flows on Infra-Homogeneous Spaces . . . . . 299
    By Per Tomter

The "DA" Maps of Smale and Structural Stability . . . . 329
    By R. F. Williams

The Zeta Function in Global Analysis . . . . . . 335
    By R. F. Williams

Classification of One Dimensional Attractors . . . . . 341
    By R. F. Williams

Author Index . . . . . . . . . 363

Subject Index . . . . . . . . . 365

# PREFACE

The papers in these Proceedings grew out of lectures given at the fifteenth Summer Mathematical Institute of the American Mathematical Society, whose topic was global analysis. The Institute was held at the University of California at Berkeley from July 1 to July 26, 1968, and was partially financed by the National Science Foundation.

Notes of lectures were distributed at the time of the conference and some of the papers here are just as in those notes. These volumes, however, can be distinguished from the notes in the sense that in general the papers here are not just expositions of material that has or will appear elsewhere; most of the articles could just as well have appeared in Journals.

The unity given by the subject matter makes it desirable to collect them here. It is hoped that the volumes will provide an important start to the scientist who wishes to learn what is going on in that part of mathematics called global analysis.

The organizing committee for the institute consisted of: F. Browder, S.–S. Chern, L. Hörmander, I. Singer, and S. Smale, with the co–editors serving as co–chairmen.

Seminar organizers were: F. Browder, E. Calabi, H. Goldschmidt, R. Hermann, C. Morrey, R. Palais, C. Pugh, I. Singer, and D. Spencer.

Finally the editors would like to thank the many people who made the institute and volumes possible. Of especially direct help to ourselves were Celeste Andrade, Ann Harrington, Gordon and Jacqueline Walker.

S.–S. Chern

December 1968

S. Smale

# BUMPY METRICS

R. ABRAHAM

On a compact Riemannian manifold, $M$, there ought to be infinitely many closed geodesics (a classical conjecture). This is obvious if the isometry group of $M$ has dimension greater than zero, so we should examine the "generic case" of minimal symmetry. For example, suppose $M$ is a 2-sphere embedded in 3-space, with the induced metric. In the case of the standard embedding, every point is in a 1-parameter family of closed geodesics. But if the embedding is perturbed to an ellipsoid with unequal axes, most of these geodesics disappear. Three short ones remain, and there are arbitrarily long spiraling ones as well. Perturbed further by bumps or undulations, this is an example of a *bumpy metric*, with a countable set of closed orbits (finitely many of bounded length), all stable in some sense, and a 0-dimensional isometry group. The definition is stated later, and the main theorem: *on a compact manifold, almost all metrics are bumpy*. We conjecture that every bumpy metric on a compact manifold has infinitely many distinct closed geodesics, and this has been proved for some manifolds by Gromoll and W. Meyer [4].

1. **The definition of bumpy.** We consider closed geodesics from the point of view of Marston Morse. Let $H = H^1(S^1, M)$ be the Hilbert manifold of absolutely continuous maps $c : S^1 \to M$, and $J_g : H \to R$ the energy function for a Riemannian metric $g$ on $M$ (see Palais [6]). Then a critical point $c \in H$ of $J_g$ is a closed geodesic parameterized proportionately to arclength (we identify $S^1 \approx [0, 1]/\{0, 1\}$). As the group $S^1$ acts continuously on $H$ by $\theta(c)(t) = c(\theta + t)$ for $(\theta, c, t) \in S^1 \times H \times S^1$ and $J_g$ is invariant under this action, the orbit $S^1(c)$ of a critical point $c \in H$ consists entirely of critical points. In fact if $c \in H$ is a $C^\infty$ curve, we may prove that $S^1(c)$ is a $C^\infty$ submanifold of $H$, so this always occurs for critical points (by the usual regularity theorem). Then $S^1(c)$ is a critical manifold of $J_g$ corresponding to a single closed geodesic in $M$.

DEFINITION. A Riemannian metric $g$ on a manifold $M$ is *bumpy* iff for every nonconstant critical point $c \in H$ of $J_g$, $S^1(c)$ is a nondegenerate critical manifold of $J_g$. That is, the index form (or Hessian $d^2 J_g(c)$) has a 1-dimensional null space, the tangent space of $S^1(c)$ at $c$.

The constant curves of $H$ are excluded because they comprise a submanifold of $H$ diffeomorphic to $M$, which is a nondegenerate critical manifold of $J_g$ (at least if $M$ is compact) of higher dimension. Because of Bott's formula relating the nullity of an iterated closed geodesic to its Poincaré rotation numbers [3], this definition is equivalent to: every closed geodesic has irrational rotation numbers.

2. **Properties of bumpy metrics.** If $g$ is a $C^{r+4}$ metric on $M$, then $dJ_g$ is a $C^r$ section of the cotangent bundle $T^*H$. Let $H_0 \subset H$ denote the closure of the constant

curves, $X = H/H_0$ the complement, and $\rho_0(g) = dJ_g|X$. If $\mathcal{M}^{r+4}$ is the Banach manifold of $C^{r+4}$ metrics on $M$, and $C^r(T^*X)$ the space of $C^r$ sections of the cotangent bundle of $X$, the map

$$\rho_0 : \mathcal{M}^{r+4} \to \mathscr{C}^r(T^*X)$$

is a $C^r$ representation. That is,

$$ev_{\rho_0} : \mathcal{M}^{r+4} \times X \mapsto T^*X : (g, x) \mapsto \rho_0(g)(x)$$

is $C^r$. It is easy to see that in this context, $g$ is bumpy if $\rho_0(g)$ is as transversal as possible to the zero-section of $T^*X$. As in equivariant Morse Theory, this would be easily made precise if the action $S^1 : X$ were smooth. Unfortunately it is not. Nevertheless, it is possible to construct a vectorbundle $\pi : E \to X$ and a $C^\infty$ representation

$$\rho_1 : \mathcal{M}^{r+4} \to \mathscr{C}^r(\pi)$$

such that $g$ is bumpy iff $\rho_1(g)$ is transversal to the zero-section of $\pi$. Application of an appropriate transversal density theorem then yields the following.

THEOREM 1. *If $M$ is compact and $r \geq 1$, the bumpy metrics in $\mathcal{M}^{r+4}$ comprise a residual subset.*

Some of the essential lemmas in the proof yield these properties.

THEOREM 2. *If $g$ is a bumpy $C^5$ metric on $M$, then*

(a) *the Poincaré rotation numbers of all closed nonconstant geodesics of $g$ are irrational,*

(b) *every closed nonconstant geodesic $c$ of $g$ is geometrically isolated, that is, for all $\varepsilon > J_g(c)$, there exists a neighborhood $S$ of $c \in H$ (or $\Delta$ of $c(S^1) \subset M$) such that if $c'$ is a closed nonconstant geodesic of $g$ with $c' \in \delta$ (or $c'(S^1) \subset \Delta$), then $J_g(c') > \varepsilon$,*

(c) *if $B > 0$, and $g$ is a bumpy $C^5$ metric, then $g$ has at most a finite number of closed nonconstant geodesics of energy (or length) less than $B$, and*

(d) *the set $\overline{\Gamma}(g) = closure \{c(S^1)|c \text{ nonconstant geodesic}\}$ is lower semicontinuous in $g$, that is, the function $\overline{\Gamma} : \mathcal{M}^5 \to 2_0^M$ is lower semicontinuous at $g$, where $2_0^M$ is the set of closed subsets of $M$ with the Hausdorf topology.*

Of these, only the last is nontrivial, and requires condition (C) of Palais-Smale. The appropriate transversality theorem is a mildly strengthened form of the usual one [2].

THEOREM 3. *Let $\mathcal{M}$, $X$, and $E$ be $C^r$ Banach manifolds, $W \subset Y$ a $C^r$ submanifold, and $\rho : \mathcal{M} \to C^r(X, E)$ a $C^r$ representation. Then $\{m \in M | \rho(m) \pitchfork W\} \subset \mathcal{M}$ is residual if*

(i) *for all $m \in M$, $\dim(\rho(m) \cap W) \leq k_1$,*

(ii) *for all $m \in M$ and $x \in X$ such that $\rho(m)(x) \in W$, $T_x\rho(m)$ has closed split range and finite-dimensional kernel, with $\dim \ker T_x\rho(m) \leq k_2$,*

(iii) *$r > \max\{0, k_2 - k_1\}$, and*

(iv) *$ev_\rho \pitchfork W$.*

The proof of Theorem 1 is analogous to the proof of the Kupka-Smale Theorem in [2]. Because condition (iv) of Theorem 3 fails at iterated closed geodesics in the representation $\rho_1$, a device similar to the Peixoto induction argument in [2] is used. For the induction step, a local reduction in $\mathcal{M}$ is made, using properties (c) and (d) of Theorem 2. Then Theorem 3 is applied to the reduced representation to complete the induction. In this application, only hypothesis (iv) is difficult to establish, as in the Kupka-Smale case. It follows from direct construction of a perturbing metric, using Fermi coordinates at a fixed closed geodesic of the original metric.

3. **Remarks.** These generic properties of Riemannian metrics are interesting not only because of geometric questions about closed geodesics [4], [8], but also as candidates for generic qualitative features of Hamiltonian systems. Analogous properties have recently been proved to be generic in the Hamiltonian case by K. Meyer [5] and Robinson [7].

If there were a closing lemma for geodesic flows, then Theorem 2(d) would imply that for almost every metric, closed geodesics are dense in the manifold.

It is a pleasure to thank W. Klingenberg for suggesting the bumpy question, and W. Meyer for several helpful discussions.

REFERENCES

1. R. Abraham and J. Marsden, *Foundations of mechanics,* Benjamin, New York, 1967.
2. R. Abraham and J. Robbin, *Transversal mappings and flows,* Benjamin, New York, 1967.
3. R. Bott, *On the iteration of closed geodesics and the Sturm intersection theory,* Comm. Pure Appl. Math. **9** (1956), 171–206.
4. D. Gromoll and W. Meyer, *Periodic geodesics on compact Riemannian manifolds* (to appear).
5. K. Meyer, *Genericity of periodic solutions for Hamiltonian systems* (to appear).
6. R. Palais, *Morse theory on Hilbert manifolds,* Topology **2** (1963), 299–340.
7. R. C. Robinson, *A Kupka-Smale theorem for Hamiltonian vector fields,* Lecture Notes, Summer Inst. Global Analysis (Univ. of California, Berkeley 1968).
8. A. Weinstein, *The generic conjugate locus,* these Proceedings, vol. 15.

UNIVERSITY OF CALIFORNIA, SANTA CRUZ

# NONGENERICITY OF Ω-STABILITY

R. ABRAHAM AND S. SMALE

We prove here that in general, $\Omega$-stable diffeomorphisms are not dense in Diff($M$), the space of $C^r$ diffeomorphisms on a $C^\infty$ manifold $M$, with the uniform $C^r$ topology, $1 \leq r \leq \infty$. Recall from [1] that if $f \in$ Diff($M$), then $x \in M$ is a *non-wandering point* of $f$ if and only if for every neighborhood $U$ of $x \in M$ there is a nonzero integer $m \in Z$ such that $f^m(U) \cap U \neq \emptyset$. The set $\Omega = \Omega(f)$ of all nonwandering points of $f$ is a closed invariant set. If $\Lambda \subset M$ is a closed invariant set, $\Lambda$ has a *hyperbolic structure* if and only if the tangent bundle of $M$ restricted to $\Lambda$, $T(M)$, splits into a sum of $C^0$ subbundles $E^s$ and $E^u$, invariant under the tangent of $f$, $Tf: T_\Lambda(M) \to T_\Lambda(M)$ such that $Tf$ is expanding on $E^u$ and contracting on $E^s$ (see [1] for complete definitions). Then $f$ satisfies *Axiom* A if and only if:

(Aa)  $\Omega(f)$ has a hyperbolic structure, and

(Ab)  The periodic points of $f$ are dense in $\Omega(f)$.

If $f, g \in$ Diff($M$), they are $\Omega$-*conjugate* if and only if there exists a homeomorphism $h: \Omega(f) \to \Omega(g)$ such that $gh = hf$, and $f$ is $\Omega$-*stable* if and only if there is a neighborhood $N(f)$ of $f \in$ Diff($M$) such that every $g \in N(f)$ is $\Omega$-conjugate to $f$.

In this paper we construct an open set $N \subset$ Diff($T^2 \times S^2$) such that every $g \in N$ violates both (Aa) and $\Omega$-stability. The basic idea is to construct $f \in$ Diff($M$) with disjoint closed invariant sets $\Lambda_1$ and $\Lambda_2$, having hyperbolic structures of different dimensions, such that an orbit goes from $\Lambda_1$ to $\Lambda_2$, and another goes from $\Lambda_2$ to $\Lambda_1$. This implies that $\Lambda_1, \Lambda_2$, and the two orbits are contained in $\Omega(f)$, which therefore cannot have a hyperbolic structure. Further, this "pathology" is stable under perturbations of $f$ in the $C^1$ topology.

In §1 we establish a criterion for the behavior described above, and in §2 we construct a diffeomorphism satisfying the criterion. §3 establishes $\Omega$-instability for this example.

1.  We begin by recalling some aspects of the Stable Manifold Theorem [1, §7.3], or [2], or [3].

If $\Lambda$ is a compact invariant set of $f \in$ Diff($M$) with hyperbolic structure, $T(M) = E^s + E^u$, then there is defined for each $x \in \Lambda$, a stable manifold $W^s(x)$ which is a one-to-one immersed cell in $M$, and consists of points $y \in M$ with the property that $d(f^m(x), f^m(y)) \to 0$ as $m \to \infty$. Then $W^u(x)$ is defined as the stable manifold at $x \in \Lambda$ for $f^{-1}$. Then define $W^s(\Lambda) = \bigcup_{x \in \Lambda} W^s(x)$. Finally, $W^s(x)$ varies smoothly on compact sets as $x$ varies in $\Lambda$.

The *type* of $\Lambda$ is the pair $(a, b)$ where $a =$ fiber dim $E^s$ and $b =$ fiber dim $E^u$.

DEFINITION. *A subbasic set* for $f \in$ Diff($M$) is a compact invariant set $\Lambda \subset M$ with hyperbolic structure such that $f/\Lambda$ is topologically transitive and the periodic points are dense in $\Lambda$.

If $\Lambda_1$ and $\Lambda_2$ are disjoint subbasic sets of $f$, we write $\Lambda_1 < \Lambda_2$ when $W^s(\Lambda_1)$ $\cap W^u(\Lambda_2) \neq \varnothing$, and $\Lambda_1 \ll \Lambda_2$ when there are points $p_i \in \Lambda_i$ such that $W^s(p_1)$ and $W^u(p_2)$ have a point of transversal intersection.

THEOREM. *If $\Lambda_1$ and $\Lambda_2$ are disjoint subbasic sets of $f \in \mathrm{Diff}(M)$ and $\Lambda_2 \ll \Lambda_1 < \Lambda_2$, then*

$$W^s(\Lambda_1) \cap W^u(\Lambda_2) \subset \Omega = \Omega(f).$$

COROLLARY. *If $\Lambda_1$ and $\Lambda_2$ are disjoint subbasic sets of $f \in \mathrm{Diff}(M)$, $\Lambda_2 \ll \Lambda_1 < \Lambda_2$, and type $(\Lambda_1) \neq$ type $(\Lambda_2)$, then $f$ does not satisfy Axiom A.*

PROOF. Suppose $f$ satisfies Axiom A. If $\Lambda_2 \ll \Lambda_1 < \Lambda_2$ and $x \in W^s(\Lambda_1) \cap W^u(\Lambda_2)$, then $x \in \Omega(f)$ by the theorem above, while $f^m(x) \to \Lambda_1$ as $m \to \infty$, and $f^m(x) \to \Lambda_2$ as $m \to -\infty$. As the orbit closure $\overline{O(x)} \subset \Omega(f)$, $\Lambda_1$ and $\Lambda_2$ must be in the same basic set $\Omega_i$ of $f$ (that is, the same indecomposable piece of $\Omega(f)$, see [1, 6.2]). As $\Omega_i$ has a hyperbolic structure and is indecomposable, $\dim E_x^s$ is constant for all $x \in \Omega_i$, a contradiction.

The proof of the theorem requires the following

LEMMA. *Let $f: \Lambda \to \Lambda$ be a topologically transitive homeomorphism of a compact metric space with periodic points dense in $\Lambda$. Then given nonempty open sets $V_1$, $V_2$ in $\Lambda$, there is a periodic point $p \in V_1$ such that $f^m(p) \in V_2$ for some $m$.*

PROOF. From the topological transitivity $f^m(V_1) \cap V_2 \neq \varnothing$ for some $m$. Let $q$ be a periodic point in this intersection and $p = f^{-m}(q)$.

PROOF OF THE THEOREM. Let $x \in W^s(\Lambda_1) \cap W^u(\Lambda_2)$ and let $U$ be a neighborhood of $x$.

Now by the hypothesis $\Lambda_2 \ll \Lambda_1$, $W^u(p_1)$ and $W^s(p_2)$ have a point of transversal intersection for $p_1 \in \Lambda_1$, $p_2 \in \Lambda_2$. Let $V_1$ be a neighborhood of $p_1$ in $\Lambda_1$, $U_1$ a neighborhood of $p_2$ in $\Lambda_2$ such that for every $p \in V_1$, $q \in U_1$, $W^u(p)$ and $W^s(q)$ have a point of transversal intersection. Choose an open set $V_2$ in $\Lambda_1$, $U_2$ in $\Lambda_2$ such that if $q' \in V_2$, $p' \in U_2$, then $W^s(q')$ and $W^u(p')$ intersect $U$.

Now apply the previous lemma to obtain periodic orbits $\gamma_1$ meeting $V_1$, $V_2$ and $\gamma_2$ meeting $U_1$, $U_2$. Thus $W^u(\gamma_1)$ and $W^s(\gamma_2)$ have a point of transversal intersection while $W^s(\gamma_1)$ and $W^u(\gamma_2)$ both meet $U$. Now apply the argument of [1, 7.2] to obtain that $f^m(U) \cap U \neq \varnothing$ for some $m$. This proves the theorem.

2. We now construct a diffeomorphism $f \in \mathrm{Diff}(T^2 \times S^2)$ having a subbasic set configuration $\Lambda_1 \ll \Lambda_2 < \Lambda_1$. Let $g \in \mathrm{Diff}(T^2)$ be the Thom diffeomorphism, defined by the linear isomorphism of $R^2$ having matrix

$$\begin{pmatrix} 1 & 2 \\ 1 & 1 \end{pmatrix}$$

(see [I, §1–3]), and $p \in T^2$ the fixed point corresponding to the origin, which is of type $(1, 1)$, that is, $\dim E_p^s = 1$, $\dim E_p^u = 1$. Let $h \in \mathrm{Diff}(S^2)$ be the horseshoe diffeomorphism [I, §1–5], which has two fixed points, $q_1$ and $q_2$, of type $(1, 1)$, at each of which the map $h$ is linear in a $C^\infty$ coordinate chart. In addition, recall that $\{q_1\} \ll \{q_2\} \ll \{q_1\}$ with respect to $h$. Let $f_0 = g \times h \in \text{Diff}(T^2 \times S^2)$. Then $\Lambda_1 = T^2 \times \{q_1\}$ is a 2-dimensional subbasic set of type $(2, 2)$ and $\Lambda_2 = \{(p, q_2)\}$ is a one point subbasic set of type $(2, 2)$. Using local coordinate charts on $T^2$ and $S^2$, at $p$ and $q_2$ respectively, with respect to which $g$ and $h$ have linear local representatives, we modify $f_0$ through a curve of linear maps so that the fixed point $\Lambda_2$ becomes fixed of type $(1, 3)$. We may do this in such a way that the new local diffeomorphism at $\Lambda_2$ can be extended so as to agree with $f_0$ outside a small neighborhood of $\Lambda_2$, and so that the new diffeomorphism $f$ satisfies:

$$W^u_{f_0}(\Lambda_2) \subset W^u_f(\Lambda_2)$$

and $W^s_f(\Lambda_2)$ contains a connected part of the 1-dimensional stable manifold $W^s(q_2)$ in $S^2$ which contains $q_2$ and a point $y \in S^2$ of transversal intersection, $y \in W^s(q_2) \cap W^u(q_1)$.

Thus with respect to $f$, $\Lambda_1 \ll \Lambda_2 < \Lambda_1$, completing the construction.

Finally we claim there exists a neighborhood $N$ of $f \in \text{Diff}(T^2 \times S^2)$ such that for all $g \in N$, there are subbasic sets $\Lambda_i(g)$, homeomorphic to $\Lambda_i$, $i = 1, 2$, such that

$$\Lambda_1(g) \ll \Lambda_2(g) < \Lambda_1(g).$$

First, we observe that the local Ω-stability Theorem [3, Proposition 3.1] applies to yield the following: There is a neighborhood $N_0$ of $f \in \text{Diff}(T^2 \times S^2)$ and for every $g \in N_0$, subbasic sets $\Lambda_i(g)$ and a conjugating homeomorphism $h(g)$: $\Lambda_1(g) \cup \Lambda_2(g) \to \Lambda_1 \cup \Lambda_2$, $fh(g) = h(g)g$. Furthermore, the stable manifolds $W^s(\Lambda_1(g))$ depend continuously ($C^r$ topology on compact subsets in the fibers, $C^0$ topology on $\Lambda_1(g)$) on $g$ ($C^r$ topology, $r \geq 1$). This follows from the continuity conclusion of the Stable Manifold Theorem [2]. Thus $\Lambda_1 \ll \Lambda_2$ is always an open condition. The relation $\Lambda_2 < \Lambda_1$ can in general be destroyed by arbitrarily small $C^1$ perturbations, but for this particular diffeomorphism $f \in \text{Diff}(T^2 \times S^2)$ both $\Lambda_1$ and $\Lambda_2$ are submanifolds, so $W^u(\Lambda_1)$ and $W^s(\Lambda_2)$ are smoothly immersed, with $W^u(\Lambda_1)$ transversal to $W^s(\Lambda_2)$. But nonempty transversal intersections are preserved even by $C^0$ perturbations of the submanifolds (that is, intersection but not transversality is preserved), so in this case $\Lambda_2 < \Lambda_1$ is an open condition also. Thus there is a neighborhood $N$ of $f \in \text{Diff}(T^2 \times S^2)$ contained in $N_0$ such that every $g \in N$ has a configuration of subbasic sets of the form $\Lambda_1(g) \ll \Lambda_2(g) < \Lambda_1(g)$, and $\Lambda_1(g)$ has type $(2, 2)$, while $\Lambda_2(g)$ has type $(1, 3)$. Thus every $g \in N$ violates condition (Aa) by the corollary of §1, and thus violates Axiom A.

3. In this section we show that the neighborhood $N$ of $f$ in $\text{Diff}(T^2 \times S^2)$ has the property that every diffeomorphism $g \in N$ is not Ω-stable.

We argue that if $g \in N = N(f)$, then either

(a) $W^s(\Lambda_2) \cap W^u(\Lambda_1)$ contains at least one point in some $W^s(\Lambda_2) \cap W^u(z)$ where $z \in \Lambda_1$ is periodic or

(b) not (a).

Nongenericity of $\Omega$-stability now follows from the following facts:

(A) A diffeomorphism $g \in N$ satisfying (a) may be approximated by one satisfying (b).

(B) A diffeomorphism $g \in N$ satisfying (b) may be approximated by one satisfying (a).

(C) If $g, g' \in N$ satisfy (a), (b) respectively, then $g, g'$ are not $\Omega$-conjugate.

Now (A) is a consequence of a general approximation theorem (see for example [4, p. 100]). Fact (B) is proved easily since the periodic points are dense in $\Lambda_1$ and the $W^u(z)$ for $z$ periodic are dense in $W^u(\Lambda_1)$.

Finally we see the truth of (C) by following the orbit of $W^s(\Lambda_2) \cap W^u(z)$ and its image under a possible conjugacy.

## REFERENCES

1. S. Smale, *Differentiable dynamical systems*, Bull. Amer. Math. Soc. **73** (1967), 747–817.
2. M. Hirsch and C. Pugh, *Stable manifolds and hyperbolic sets*, these Proceedings, vol. 14.
3. S. Smale, *The $\Omega$-stability theorem*, these Proceedings, vol. 14.
4. R. Abraham and J. Robbin, *Transversal mappings and flows*, Benjamin, New York, 1967.

UNIVERSITY OF CALIFORNIA, SANTA CRUZ AND

UNIVERSITY OF CALIFORNIA, BERKELEY

# ON CERTAIN AUTOMORPHISMS OF NILPOTENT LIE GROUPS

LOUIS AUSLANDER AND JOHN SCHEUNEMAN

1. **Introduction.** Let $\mathcal{N}$ be a Lie group and let $\Gamma$ be a discrete subgroup such that $\mathcal{N}/\Gamma$ is compact. We seek information about automorphisms of $\mathcal{N}$ whose restriction to $\Gamma$ is an automorphism of $\Gamma$ and whose differential at the identity has no eigenvalues of absolute value 1. Such automorphisms, which arise in connection with the work of Anosov on dynamical systems, and which exist only for certain nilpotent Lie groups, will be called Anosov automorphisms of $\mathcal{N}$ with respect to $\Gamma$ (see [1, p. 760, ff.]).

This paper is a study of the question of existence of Anosov automorphisms of nilpotent Lie groups. In §2, we reduce this problem to that of existence of semi-simple Anosov automorphisms of nilpotent Lie algebras. In §3, we give a description of all possible semisimple Anosov automorphisms. §4 is devoted to giving examples of Anosov automorphisms, §5 to a test for nonexistence of such automorphisms, and §6 to a certain generalization of a result of §4.

2. **Reduction of the problem.** A connected simply connected nilpotent Lie group may be identified with its Lie algebra. The group operation is then given in terms of the operations in the Lie algebra by the Campbell-Hausdorff formula, and the exponential mapping is the identity. This identification having been made, automorphisms of a connected simply connected nilpotent Lie group and automorphisms of its Lie algebra are one and the same thing, so from here on we consider automorphisms of nilpotent Lie algebras over $R$, the field of real numbers.

Let $N$ be a nilpotent Lie algebra over $R$. The set $C$ of all integer linear combinations of a basis of $N$ having integer constants of structure is called a $Z$-subalgebra of $N$ (not all nilpotent Lie algebras have $Z$-subalgebras; see [2]). An automorphism of $N$ whose restriction to $C$ is an automorphism of $C$ none of whose eigenvalues has absolute value 1 will be called an Anosov automorphism of $N$ with respect to $C$. Given an Anosov automorphism $A$ of $N$ with respect to $C$, one obtains an Anosov automorphism of the associated Lie group $\mathcal{N}$ as follows. There exists a $Z$-subalgebra $C_0$ of $N$ which is contained in $C$ and has the properties that $A|_{C_0}$ is an automorphism and $\Gamma = \exp C_0$ is a subgroup of $\mathcal{N}$. (If $C = Zu_1 + \ldots + Zu_n$, take $C_0 = Zmu_1 + \ldots + Zmu_n$, where $m$ is chosen so that the denominators in the Campbell-Hausdorff formula divide the products of constants of structure of $\{mu_1, \ldots, mu_n\}$ that arise.) Then $A$ is an Anosov automorphism of $\mathcal{N}$ with respect to $\Gamma$. Conversely, if $A$ is an Anosov automorphism of $\mathcal{N}$ with respect to $\Gamma$, there exists a subgroup $\Gamma^* \supseteq \Gamma$ such that $\Gamma^* = \exp C$, where $C$ is a $Z$-subalgebra of $N$, and $A\Gamma^* = \Gamma^*$ (see [4, Theorem 1]).

9

Another reduction of our problem is possible. The group of automorphisms of a nilpotent Lie algebra is a linear algebraic group. Thus, if $A$ is an automorphism, $A$ can be factored $A = US = SU$, where $S$ is semisimple, $U$ is unipotent, and both $S$ and $U$ are themselves automorphisms. As far as eigenvalue properties of automorphisms are concerned, it suffices to consider only semisimple automorphisms. The following observation shows that as far as lattice properties are concerned, it again suffices to consider semisimple automorphisms.

PROPOSITION. *Let $N$ be a nilpotent Lie algebra over $R$ and let $C$ be a $Z$-subalgebra of $N$. If there exists an Anosov automorphism of $N$ with respect to $C$, then there exists a semisimple Anosov automorphism of $N$ with respect to $C$.*

PROOF. Let $A$ be such an automorphism and let $A = SU = US$ be the splitting of $A$ into its semisimple and unipotent parts. Then with respect to a $Z$-basis of $C$, the matrix of $A$ is an integer unimodular matrix, and those of $U$ and $S$ are rational matrices of determinant 1. Since $U = I + N$, where $N$ is nilpotent, it is easy to see that there exists an integer $r$ such that $U^r$ has an integer matrix, so $S^r = U^{-r}A^r$ has an integer matrix with respect to a $Z$-basis of $C$, has determinant 1, and is a semisimple automorphism of $N$ having no eigenvalues of absolute value 1.

3. **On the classification of Anosov automorphisms.** This section is devoted to a study of the role of the free nilpotent Lie algebras in problems of semisimple Anosov automorphisms of arbitrary nilpotent Lie algebras. We begin with a description of the free nilpotent Lie algebras and some of their properties.

Let $V$ be a vector space with basis $\{x_1, ..., x_n\}$. The free Lie algebra $N(V)$ generated by $V$ may be described as the algebra spanned by all nonassociative words $x_1, ..., x_n, [x_1 x_2], ..., [[x_1 x_2][x_3 x_4]]$, etc., subject only to the Jacobi and skew symmetry relations. The (finite dimensional) free $k$-step nilpotent Lie algebra $N_k(V)$ is obtained by including the condition that every word of length $k+1$ is zero.

Given any nonsingular linear transformation $A_0: V \to V$, there is a unique extension $\tilde{A}: N_k(V) \to N_k(V)$ of $A_0$ to an automorphism of $N_k(V)$. Also, given any $k$-step nilpotent Lie algebra $N$ and a complement $V$ to the commutator ideal $[N, N]$ of $N$, there is a unique homomorphism $h: N_k(V) \to N$ such that $h|_V = $ identity. Note that if $A: N \to N$ is any semisimple automorphism then $A$ preserves a complement to $[N, N]$.

Suppose now that we are given a $k$-step nilpotent Lie algebra $N$, a complement $V$ to $[N, N]$ and a semisimple linear transformation $A_0: V \to V$. It is our purpose to give necessary and sufficient conditions for $A_0$ to be extendible to an Anosov automorphism of $N$, these conditions being stated in terms of $A_0$ and the mappings $h$ and $\tilde{A}$ above.

LEMMA 1. *Let $\{x_1, ..., x_n\}$ be a basis for $V$ and let $V_r$ be the subspace of $N_k(V)$ spanned by words in $x_1, ..., x_n$ of length $r$, so $N_k(V) = V_1 \oplus ... \oplus V_k$. Let $\{w_1, ..., w_N\}$ be a basis for $V_r$ consisting of such words. Then the total number of times $x_t$ occurs in all $w_1, ..., w_N$ is the same for each $t = 1, 2, ..., n$.*

PROOF. If $r = 1, 2$ then the total number is $1$, $n - 1$, respectively. If $k \geq 3$, the Jacobi identities make it difficult to actually pick out bases, so we use other means to prove our result. Let $A$ be the automorphism of $N_k(V)$ obtained by extending the linear transformation of $V$ defined by $x_i \to \alpha_i x_i$, where the $\alpha$'s are arbitrary. The matrix of $A$ with respect to a basis consisting of words in $x_1, ..., x_n$ is diagonal. Let $M_r$ be the matrix of $A|_{V_r}$ with respect to $\{w_1, ..., w_N\}$. The diagonal entries of $M_r$ are of the form $\alpha_1^{m_1} ... \alpha_n^{m_n}$, and $\det M_r = \alpha_1^{p_1} ... \alpha_n^{p_n}$, where $p_i$ is the total number of times $x_i$ occurs among all the words $w_1, ..., w_N$. If a different basis for $V_r$ is chosen from among the words of length $r$, and if $M_r'$ is the matrix of $A|_{V_r}$ with respect to this new basis, we have $\det M_r = \det M_r' = \alpha_1^{p_1} ... \alpha_n^{p_n} = \alpha_1^{p_1'} ... \alpha_n^{p_n'}$. Since the $\alpha$'s are arbitrary, we must have $p_i = p_i'$ for $i = 1, 2, ..., n$. Now suppose $p_i \neq p_j$ for some $i \neq j$. By replacing $x_i$ by $x_j$ and $x_j$ by $x_i$ in each of the words $w_1, ..., w_N$, we get a new basis $\{w_1', ..., w_N'\}$ for $V_r$. Again taking determinants, we get a contradiction.

COROLLARY. *If $A_0 : V \to V$ is semisimple and $\tilde{A} : N_k(V) \to N_k(V)$ is the above-mentioned extension of $A_0$ to an automorphism of $N_k(V)$, then $\det \tilde{A}$ is a power of $\det A_0$.*

THEOREM 1. *Suppose that $N$ is a $k$-step nilpotent Lie algebra, that $V$ is a complement to $[N, N]$ with basis $\{x_1, ..., x_n\}$, and that the image under $h$ of the $Z$-subalgebra of $N_k(V)$ spanned by words in $x_1, ..., x_n$ is a $Z$-subalgebra $C$ of $N$. Let $A_0 : V \to V$ be a semisimple linear transformation whose matrix with respect to $\{x_1, ..., x_n\}$ is integer unimodular. Then $A_0$ can be extended to an Anosov automorphism of $N$ with respect to $C$ such that $Ah = h\tilde{A}$ if and only if the subspaces of $N_k(V)$ belonging to the eigenvalues of $\tilde{A}$ of absolute value 1 are contained in $\ker h$, $\tilde{A}$ preserves $\ker h$, and $\tilde{A}$ is unimodular on $\ker h$.*

PROOF. Suppose that $A$ is an extension of $A_0$ to an Anosov automorphism of $N$ with respect to $C$ such that $Ah = h\tilde{A}$. If $v \in N_k(V)$ and $\tilde{A}(v) = \alpha v$ where $|\alpha| = 1$, we get $Ah(v) = \alpha h(v)$, so $h(v) = 0$ since $A$ is Anosov. Obviously, $\tilde{A}$ preserves $\ker h$, and is unimodular on $\ker h$ because $A$ is unimodular by hypothesis and $\tilde{A}$ is unimodular on $N_k(V)$ by the corollary to Lemma 1.

Conversely, if $x \in N$, then $x = h(\tilde{x})$ and we may define $A(x) = h\tilde{A}(x)$ since $\tilde{A}$ preserves $\ker h$. Obviously, $A$ is an automorphism of $N$ and $Ah = h\tilde{A}$. It remains to show that $A$ is an Anosov automorphism. First of all, $A$ has an integer matrix with respect to a basis consisting of images under $h$ of words in $x_1, ..., x_n$, since the matrix of $A$ with respect to a basis of $N_k(V)$ consisting of words in $x_1, ..., x_n$ is an integer matrix. Next, since $\tilde{A}$ is unimodular by the corollary to Lemma 1, and since $\tilde{A}$ is unimodular on $\ker h$, the matrix of $A$ is unimodular. Finally, $A$ has no eigenvalues of absolute value 1, since $\tilde{A}$ is semisimple and all eigenvectors belonging to such eigenvalues of $\tilde{A}$ lie in $\ker h$.

We turn now to more specialized considerations.

4. **Examples of Anosov automorphisms.** The effect of Theorem 1 is to reduce questions of semisimple Anosov automorphisms to the study of semisimple

automorphisms of $N_k(V)$. One gets all semisimple Anosov automorphisms as follows[*]: Let $A_0 : V \to V$ be nonsingular, semisimple, and integer unimodular with respect to the basis $\{x_1, \dots, x_n\}$; find an ideal $J$ of $N_k(V)$ which contains all eigenvectors of $\tilde{A}$ of absolute value 1 which is invariant under $\tilde{A}$, on which $\tilde{A}$ is unimodular, and has the property that modulo $J$, the words in $x_1, \dots, x_n$ form a $Z$-subalgebra. Then $\tilde{A}$ induces an Anosov automorphism of $N_k(V)/J$.

In this section we review one of the examples of Anosov automorphisms in [1] (due to Borel) from the above point of view, and then prove a theorem about Anosov automorphisms of $N_k(V)$ itself ($J = 0$).

Let $V$ be a vector space with basis $\{x_1, x_2, y_1, y_2\}$. Define $A_0 : V \to V$ by

$$A_0 x_1 = \lambda x_1, \quad A_0 x_2 = \lambda^{-1} x_2, \quad A_0 y_1 = \lambda^2 y_1, \quad A_0 y_2 = \lambda^{-2} y_2,$$

where $|\lambda| \neq 1$ is a root of $x^2 + 2ax + 1 \in Z[x]$. Then $A_0$ is semisimple and unimodular. The vectors $[x_1, x_2], [y_1, y_2] \in N_2(V)$ are the eigenvectors of $\tilde{A}$ belonging to eigenvalues of absolute value 1. Let $J$ be the ideal of $N_2(V)$ with basis $\{[x_1 x_2], [y_1 y_2], [x_1 y_2], [x_2 y_1]\}$, so $\tilde{A}$ preserves $J$ and is unimodular on $J$. The set

$$\{x_1 + x_2, \sqrt{a^2 - 1}(x_1 - x_2), y_1 + y_2, \sqrt{a^2 - 1}(y_1 - y_2),$$
$$[x_1 y_1] + [x_2 y_2], \sqrt{a^2 - 1}([x_1 y_1] - [x_2 y_2])\}$$

is a basis for a complement to $J$ in $N_2(V)$, and this complement is a $Z$-subalgebra of $N_2(V)$ modulo $J$. Finally, notice that

$$A_0(x_1 + x_2) = \lambda x_1 + \lambda^{-1} x_2 = -a(x_1 + x_2) + (\sqrt{a^2 - 1}(x_1 - x_2))$$
$$A_0(\sqrt{a^2 - 1}(x_1 - x_2)) = \sqrt{a^2 - 1}(\lambda x_1 - \lambda^{-1} x_2) = (a^2 - 1)(x_1 + x_2)$$
$$- a(\sqrt{a^2 - 1}(x_1 - x_2)),$$

etc., so that $\tilde{A}$ has an integer matrix with respect to the above basis. Hence, by Theorem 1, $\tilde{A}$ induces an Anosov automorphism of $N_2(V)/J$. This is example 1 of [1, page 762].

To conclude this section, we prove the following theorem about Anosov automorphisms of $N_k(V)$ itself.

THEOREM 2. *Let $A_0$ be a nonsingular linear transformation of $V$ which has an integer unimodular matrix with respect to the basis $\{x_1, \dots, x_n\}$. Assume that the eigenvalues $\alpha_1, \dots, \alpha_n$ of $A_0$ have the property that $|\alpha_{i_1} \dots \alpha_{i_r}| \neq 1$ for all $r < n$ (repeated factors allowed). Then if $k < n$, the extension $\tilde{A}$ of $A_0$ to an automorphism of $N_k(V)$ is a semisimple Anosov automorphism with respect to the $Z$-subalgebra spanned by words in $x_1, \dots, x_n$.*

PROOF. First notice that the algebraic integers $\alpha_1, \dots, \alpha_n$ are distinct, because our hypotheses imply that the characteristic polynomial of $A_0$ must be irreducible in $Q[x]$. Hence $A_0$ is semisimple, so $\tilde{A}$ is semisimple. The corollary to Lemma 1 says that $\tilde{A}$ is unimodular. The matrix of $\tilde{A}$ with respect to a basis consisting of words in $x_1, \dots, x_n$ is an integer matrix, and our hypotheses imply that $\tilde{A}$ has no eigenvalues of absolute value 1. Hence $\tilde{A}$ is as required.

5. **On nonexistence of Anosov automorphisms in certain cases.** Attached to a nilpotent Lie algebra $N$ is the 2-step nilpotent Lie algebra obtained by factoring by the ideal $[N, [N, N]]$. Nonexistence of Anosov automorphisms of this 2-step nilpotent algebra implies nonexistence of Anosov automorphisms of $N$, and from here on we restrict attention to the case where $N$ is already 2-step nilpotent.

Let $V$ be a complement to $[N, N]$. We shall be interested in linear transformations of $V$ that can be extended to Anosov automorphisms of $N$. As noted in the introduction, it is sufficient to study this case.

Given a basis $\{x_1, ..., x_n\}$ of $V$ and a basis $\{y_1, ..., y_k\}$ of $[N, N]$, we obtain the skew-symmetric matrix $([x_ix_j])$, which may be considered as a matrix of linear forms in $y_1, ..., y_k$ or as a linear form $y_1C_1 + ... + y_kC_k$ with skew symmetric matrix coefficients.

If $A_0$ is a linear transformation of $V$ and $x'_i = A_0x_i$ then

$$([x'_ix'_j]) = A_0([x_ix_j])A_0^t = y_1A_0C_1A_0^t + ... + y_kA_0C_kA_0^t.$$

From this we see that $A_0$ can be extended to an automorphism of $N$ if and only if each matrix $A_0C_iA_0^t$ is a linear combination of $\{C_1, ..., C_k\}$. Moreover, if $A_0 \oplus A_1$ is an automorphism of $N$, where $A_0 : V \to V$, then the matrix $([x_ix_j])$ expressed in terms of $y_1, ..., y_k$, is the same as the matrix $([A_0x_i, A_0x_j])$ expressed in terms of $A_1y_1, ..., A_1y_k$.

Let $B = \{x_1, ..., x_n, y_1, ..., y_k\}$ be the basis for $N$ given above, and define the polynomial $I_B$ as follows:

$$I_B(y_1, ..., y_k) = (\det([x_ix_j]))^{1/2}.$$

In case $\dim N - \dim[N, N]$ is even, $I_B$ is the invariant described in [3]. $I_B$ is zero if the center of $N$ is larger than $[N, N]$ or if $\dim N - \dim[N, N]$ is odd, but if $I_B$ is not zero, it contains some useful information. To see this, we will need the following proposition, whose proof follows easily from the preceding discussion.

LEMMA 2. *Let $A_0$ be a linear transformation of $V$ of determinant $\pm 1$ which extends to an automorphism $A_0 \oplus A_1$ of $N$. If $I_B(u_1, ..., u_k) = r$ for $(u_1, ..., u_k) \in R^k$, then $I_B(u'_1, ..., u'_k) = r$, also, where $(u'_1, ..., u'_k) = A_1(u_1, ..., u_k)$.*

Now suppose that $A = A_0 \oplus A$, is an automorphism of $N$ that has an integral unimodular matrix with respect to the basis $B = \{x_1, ..., x_n, y_1, ..., y_k\}$ of $N$.

THEOREM 3. *If there exists $r > 0$ such that the region $U_r = \{(u_1, ..., u_k) \in R^k |$ $|I_B(u_1, ..., u_r| \le r\}$ contains a basis of lattice points, but only finitely many lattice points, then $A_1$ has roots of unity as eigenvalues, and hence $A$ is not an Anosov automorphism.*

PROOF. By the preceding lemma, $A_1$ preserves $U_r$. Since $A_1$ has an integral matrix, it preserves the lattice points in $U_r$. Since there are sufficiently many, but not too many lattice points in $U_r$, some power of $A_1$ is the identity, so the eigenvalues of $A_1$ are roots of unity.

EXAMPLE. Define the 2-step nilpotent Lie algebra $N$ with basis

$$B = \{x_1, x_2, x_3, x_4, y_1, y_2\}$$

by $[x_1 x_2] = [x_3 x_4] = y_1$, $[x_1 x_3] = [x_4 x_2] = y_2$, $[x_1 x_4] = [x_2 \dot{x}_3] = [x_i x_j] = 0$.
Then $N$ admits no Anosov automorphisms.

PROOF. It suffices to consider $A = A_0 \oplus A_1$ which preserve $V = (x_1, x_2, x_3, x_4)$.
$I_B(y_1, y_2) = y_1^2 + y_2^2$, so for any basis $B'$, $\{|I_{B'}(y_1', y_2')| \leq r\}$ is an elliptical region,
so Theorem 2 applies.

6. **On dual automorphisms of two-step nilpotent Lie algebras.** In [3], there is
described a process of attaching a "dual algebra" $N^*$ to any 2-step nilpotent Lie
algebra $N$. The underlying vector space of $N^*$ is the subspace $H^1(N) \oplus (H^1(N)$
$\cup H^1(N))$ of the usual cohomology ring of $N$, and the operation is the obvious one
(using $\cup$) which makes $N^*$ a 2-step Lie algebra. Now, automorphisms of $N$ induce
automorphisms of the cohomology ring of $N$ and hence of $N^*$. It is our purpose in
this section to investigate when an Anosov automorphism of $N$ induces an Anosov
automorphism of $N^*$.

In order to prove our results, we need information about the "dual basis" of
$N^*$. Choose a basis $B = \{x_1, ..., x_n, y_{i_1 j_1}, ..., y_{i_r j_r}\}$ of $N$ as follows. Let $\{x_1, ..., x_n\}$
be a basis for a complement $V$ of $[N, N]$. Then from among the vectors $\{[x_i x_j]$
$|1 \leq i < j \leq n\}$ in $[N, N]$, select a basis $\{[x_{i_1} x_{j_1}] = y_{i_1 j_1}, ..., [x_{i_k} x_{j_r}] = y_{k_r j_r}\}$ of
$[N, N]$. We denote by $\{(i_t, j_t)|t = 1, 2, ..., r\}$ the pairs involved in the definition of
this basis for $[N, N]$ and by $\{(a_k, b_k)|k = 1, 2, ..., \frac{1}{2}n(n-1) - r\}$ the rest of the
pairs $(i, j)$ with $1 \leq i < j \leq n$. The essential constants of structure of the basis
$B$ of $N$ are given by

$$[x_{a_k}, x_{b_k}] = \sum_{t=1}^{r} C_{a_k, b_k}^{i_t j_t} y_{i_t j_t}.$$

Now, $H^1(N)$ can be identified with the dual space of $V$, and if $\{f_1, ..., f_n\}$ is the dual
basis of $\{x_1, ..., x_n\}$, it is proved in [3] that

$$(1) \qquad B^* = \{f_1, ..., f_n, \overline{f_{a_1} \cup f_{b_1}}, ..., \overline{f_{a_s} \cup f_{b_s}}\}$$

is a basis for $N^*$, and the essential constants of structure of $B^*$ are given by

$$[f_{i_t}, f_{j_t}] = -\sum_{k=1}^{(1/2)n(n-1)-r} C_{a_k, b_k}^{i_t j_t} \overline{f_{a_k} \cup f_{b_k}}.$$

(Here the bar denotes cohomology class, and $(i_t j_t)(a_k, b_k)$, and $C_{a_k, b_k}^{i_t j_t}$ have the
previously explained significance.)

From these facts, several observations are immediately apparent. First, if the
set of integer linear combinations of $B$ is a $Z$-subalgebra of $N$, then the set of
integer linear combination of $B^*$ is a $Z$-subalgebra of $N^*$. Next, suppose $A_0: V \to V$
can be extended to an automorphism $A$ of $N$. Then $A_0^t: H^1(N) \to H^1(N)$ can be

extended to the automorphism $A^*$ of $N^*$ induced by $A$. If $A_0$ is semisimple with eigenvalues $\alpha_1, \ldots, \alpha_n$ and eigenvectors $x_1, \ldots, x_m$, the eigenvalues of $A$ are $\alpha_1, \alpha_2, \ldots, \alpha_n, \alpha_{i_1}\alpha_{j_1}, \ldots, \alpha_{i_r}\alpha_{j_r}$. The eigenvalues of $A^*$ are $\alpha_1, \ldots, \alpha_n, \alpha_{a_1}\alpha_{b_1}, \ldots, \alpha_a \alpha_b$. Finally, if $A_0 : V \to V$ is a semisimple linear transformation which can be extended to an automorphism of $N$ whose matrix with respect to $B$ is an integer unimodular matrix, it is easy to see that the induced automorphism of $N^*$ has an integer unimodular matrix with respect to $B^*$.

These observations prove the following theorem, which in the 2-step nilpotent case is a generalization of Theorem 1.

THEOREM 4. *Let $N$ be a 2-step nilpotent Lie algebra, and let $V$ be a complement to $[NN]$. If $A_0 : V \to V$ can be extended to a semisimple Anosov automorphism $A$ of $\dot{N}$ with respect to the $Z$-subalgebra of $N$ spanned by $B$, and if no product $\alpha_i\alpha_j$ of eigenvalues of $A_0$ has absolute value 1, then the automorphism $A^*$ of $N^*$ induced by $A$ is an Anosov automorphism of $N^*$ with respect to the $Z$-subalgebra spanned by $B^*$.*

REFERENCES

1. S. Smale, *Differentiable dynamical systems,* Bull. Amer. Math. Soc. **73** (1967, 747–817.
2. A. Malcev, *On a class of homogeneous spaces,* Amer. Math. Soc. Transl. **39** (1951).
3. J. Scheuneman, *Two-step nilpotent Lie algebras,* J. Algebra **7** (1967), 152–159.
4. L. Auslander, *On a problem of Philip Hall,* Ann. of Math. **86** (1967), 112–116.

CITY UNIVERSITY OF NEW YORK, GRADUATE CENTER
INDIANA UNIVERSITY

# PERIODIC POINTS OF ANOSOV DIFFEOMORPHISMS

THOMAS F. BANCHOFF AND MICHAEL I. ROSEN

In his recent article on differentiable dynamical systems [1], Smale considers in detail two types of Anosov diffeomorphisms—the toral diffeomorphisms introduced by Thom and nontoral Anosov diffeomorphisms defined on nilmanifolds. For a proof of the fact that the periodic points of toral diffeomorphisms are dense, Smale presents two arguments, one based on the generalized Birkhoff theorem for dynamical systems, and the other based on an algebraic argument (involving one-parameter subgroups of Lie groups). The purpose of this article is to present a direct and elementary proof of a slightly stronger result:

THEOREM 1. *Every point with all coordinates rational is a periodic point of any toral diffeomorphism. Moreover, for almost all toral diffeomorphisms, all periodic points are rational.*

We then extend this result in part 2 to one of the examples of a nontoral diffeomorphism in Smale's article, to show the analogy with the toral case and in the third, more formal part of the paper we give a proof of the theorem in full generality for this class of Anosov diffeomorphisms on nilmanifolds.

1. **The toral case.** Recall the definition of toral diffeomorphism: Let $f_0$ be an element of $M(n, Z)$ the ring of $(n \times n)$ matrices with integer coefficients, and let $\det f_0 = \pm 1$ so $f_0 \in GL(n, Z)$. Then, $f_0$ acting on $R^n$ restricts to an automorphism of $Z^n$, the integer points, and induces a *toral diffeomorphism* $f: T^n \to T^n$ on the quotient space $T^n = R^n/Z^n$. The equivalence class $[x]$ of a point $x$ in $R^n$ is a periodic point of $f$ if, for some $m > 0$, $f^m[x] = [x]$, i.e., $f_0^m(x) - x = (f_0^m - I)(x)$ is in $Z^n$.

Consider a point $x$ in $Q^n$, and let $r = 1$cm of denominators of components of $x$. Let $\phi_r: M(n, Z) \to M(n, Z_r)$ be defined by reducing all the entries of a matrix modulo $r$. This is a ring homomorphism, so $\phi_r$ maps $GL(n, Z)$ into $GL(n, Z_r)$. Since $GL(n, Z_r)$ is a finite multiplicative group, there is an $m$ such that $\phi_r(f_0)^m = \phi_r(I)$, so $\phi_r(f_0^m - I) = \phi_r(f_0^m) - \phi_r(I) = \phi_r(f_0)^m - \phi_r(I) = 0_r$, and all entries of $f_0^m - I$ are divisible by $r$. But then, $(f_0^m - I)(x)$ is in $Z^n$.

Moreover, if no eigenvalue of $f_0$ is a root of unity, then, for any $m$, $\det(f_0^m - I) \neq 0$, so if $(f_0^m - I)x \in Z^n$, then $x \in Q^n$.

The above method allows us to give a very easy proof of the well-known fact that any integer $r$ divides some Fibonacci number.

The Fibonacci numbers are given by $f_n$ where

$$F^n = \begin{pmatrix} f_{n-1} & f_n \\ f_n & f_{n+1} \end{pmatrix}$$

17

for the matrix

$$F = \begin{pmatrix} 0 & 1 \\ 1 & 1 \end{pmatrix}$$

in GL(2, Z). Thus, $\phi_r(F^m - I) = 0$, for some $m$.

2. **A nontoral example.** We now recall Smale's first example of a nontoral Anosov diffeomorphism on a nilmanifold. We consider a 6-dimensional real Lie group $G$ represented by matrices of the form

$$A = \left( \begin{array}{c|c} A_1 & 0 \\ \hline 0 & A_2 \end{array} \right)$$

where

$$A_i = \begin{pmatrix} 1 & x_i & z_i \\ 0 & 1 & y_i \\ 0 & 0 & 1 \end{pmatrix}$$

The Lie algebra $\mathfrak{g}$ of $G$ is represented by the space of matrices

$$B = \left( \begin{array}{c|c} B_1 & 0 \\ \hline 0 & B_2 \end{array} \right)$$

with

$$B_i = \begin{pmatrix} 0 & a_i & c_i \\ 0 & 0 & b_i \\ 0 & 0 & 0 \end{pmatrix}$$

In $\mathfrak{g}$ we find a lattice $\Gamma_0$ of matrices of the form

$$\left( \begin{array}{c|c} B & \\ \hline & \bar{B} \end{array} \right)$$

where the entries of $B$ lie in the ring $Z[\sqrt{3}]$ and

$$\bar{B} = \begin{pmatrix} 0 & \bar{a}_1 & \bar{c}_1 \\ 0 & 0 & \bar{b}_1 \\ 0 & 0 & 0 \end{pmatrix}$$

where the conjugation $Q(\sqrt{3}) \to Q(\sqrt{3})$ is defined by $\overline{p + q\sqrt{3}} = p - q\sqrt{3}$.

Under the exponential map $\Gamma_0$ corresponds to the uniform discrete subgroup $\Gamma$ of matrices with $x_1 = \bar{x}_2 \in Z[\sqrt{3}]$, $y_1 = \bar{y}_2 \in Z[\sqrt{3}]$, and $z_1 + \frac{1}{2}x_1 y_1 = z_2 + \frac{1}{2}x_2 y_2 \in Z[\sqrt{3}]$.

Let $\lambda = 2 + \sqrt{3}$. We now consider the automorphism $f: G \to G$ induced by $x_1 \to \lambda x_1,\ y_1 \to \lambda^2 y_1,\ z_1 \to \lambda^3 z_1,\ x_2 \to \lambda^{-1} x_2,\ y_2 \to \lambda^{-2} y_2,\ z_2 \to \lambda^{-3} z_2$. This leaves $\Gamma$ invariant, and we let $\bar{f}: G/\Gamma \to G/\Gamma$ be the induced map on the quotient space.

A point $[A]$ in $G/\Gamma$ is a periodic point for $\bar{f}$ if for some $m$, $f^m(A) - A \in \Gamma$, so $(\lambda^m - 1)x_1 = (\lambda^{-m} - 1)x_2$, $(\lambda^{2m} - 1)y_1 = (\lambda^{-2m} - 1)y_2$, and

$$(\lambda^{3m} - 1)z_1 + \tfrac{1}{2}(\lambda^m - 1)x_1(\lambda^{2m} - 1)y_1 = (\lambda^{-3m} - 1)z_2 + \tfrac{1}{2}(\lambda^{-m} - 1)x_2(\lambda^{-2m} - 1)y_2$$

are elements of $Z(\sqrt{3})$. Since for any $r$, $(\lambda^{-r} - 1) = (\overline{\lambda}^r - 1) = (\overline{\lambda^r - 1}) \neq 0$, this implies that $x_1 = \bar{x}_2$, $y_1 = \bar{y}_2$, and $z_1 = \bar{z}_2$, and all are elements of $Q(\sqrt{3})$. Call such a point a "rational" point.

We shall show that any rational point $[A]$ is a periodic point for $f$. To bring out the analogy with the previous case, consider the monomorphism

$$\psi: Q(\sqrt{3}) \to M(2, Q), \quad \psi(a + b\sqrt{3}) \to \begin{pmatrix} a & b \\ 3b & a \end{pmatrix}$$

Under this mapping

$$\psi(\lambda) = \begin{pmatrix} 2 & 1 \\ 3 & 2 \end{pmatrix} \in \mathrm{GL}(2, Z),$$

so by the previous result we may find an $n$ so large that $\lambda^n - 1 = r + s\sqrt{3}$ where $r$ and $s$ are rational integers divisible by twice the lowest common multiple of the denominators of $x_1, y_1$, and $z_1$. Since $\lambda^n - 1$ divides $\lambda^{kn} - 1$ for any $k$, it follows that $\bar{f}^n([A]) = [A]$ so that $[A]$ is a periodic point of $\bar{f}$.

3. **The general nontoral case.** Let $T_n$ denote the group of upper triangular, real, $n \times n$ matrices with ones on the diagonal. $T_n$ is a nilpotent Lie group of dimension $n(n - 1)/2$. Let $S_n$ be the set of upper triangular real matrices with zero on the diagonal. Under the operation $[B_1, B_2] = B_1 B_2 - B_2 B_1$, $S_n$ is a Lie algebra. It can be identified with the Lie algebra of $T_n$.

For $B \in S_n$ define $\exp B = \sum_{m=1}^{\infty} B^m/m!$. This series has only finitely many terms since $B^m = 0$ for $m \geq n$. $\exp B \in T_n$.

For $A \in T_n$ define $\ln A = \sum_{m=1}^{\infty} (-1)^m (A - I)^m/m$. Since $A - I \in S_n$, this series is also finite. Notice $\ln A \in S_n$. $\exp$ defines a map from $S_n$ to $T_n$ and $\ln$ is the inverse map. Thus $\exp: S_n \to T_n$ is $1 - 1$ and onto.

A vector space basis for $S_n$ over $R$, the real numbers, is given by the matrices $E_{ij}$, $i < j$, where $E_{ij}$ is the matrix with a 1 in the $ij$th place, and zeros elsewhere. We have the following rules of composition: $E_{ij} E_{lm} = \delta_{jl} E_{im}$ and

$$\begin{aligned} [E_{ij}, E_{lm}] &= E_{im} \quad \text{if } j = l, \\ &= -E_{l_j} \text{ if } i = m, \\ &= 0 \quad \text{otherwise.} \end{aligned}$$

We are now going to generalize the example given in §2.

Let $\lambda \neq 0$ be a real number and $\{r(i, j) | i < j\}$ a set of nonzero integers such that $r(i, j) = r(i, k) + r(k, j)$ whenever $i < k < j$. Let $\alpha_0(E_{ij}) = \lambda^{r(i,j)} E_{ij}$ and extend by linearity to obtain a map $\alpha_0: S_n \to S_n$. If $B_1, B_2 \in S_n$, one easily checks that $\alpha_0(B_1, B_2) = \alpha_0(B_1)\alpha_0(B_2)$ and $\alpha_0([B_1, B_2]) = [\alpha_0(B_1), \alpha_0(B_2)]$.

Since $\alpha_0$ is a Lie algebra automorphism of $S_n$, and $\exp$ is onto, there is an induced group homomorphism of $T_n$ given by $\alpha(\exp B) = \exp \alpha_0(B)$.

There is a simpler description of $\alpha$. Notice $\exp \alpha_0(B) = I + \sum_{m=1}^{\infty} \alpha_0(B)^m/m! = I + \alpha_0(\sum_{m=1}^{\infty} B^m/m!)$. Thus $\alpha(A) = I + \alpha_0(A - I)$ for all $A \in T_n$.

Let $K \subset R$ be a totally real number field, $[K : Q] = t$, $\lambda \in K$ a unit, $\lambda \neq \pm 1$, and $\sigma_1, \sigma_2, \ldots, \sigma_t$ the isomorphisms of $K$ into $R$. We take $\sigma_1$ to be the identity.

Let $G_1, G_2, ..., G_t$ be copies of $T_n$ and set $G = G_1 \times G_2 \times ... \times G_t$. The Lie algebra of $G$ is $\mathfrak{g} = \mathfrak{g}_1 \oplus \mathfrak{g}_2 \oplus ... \oplus \mathfrak{g}_t$ where each $\mathfrak{g}_i$ is a copy of $S_n$. Let $E_{i\,j}^s = (0, 0, ..., E_{ij}, ..., 0)$ where $E_{ij}$ occurs at the $s$th component. Define an automorphism $\alpha_0$ of $\mathfrak{g}$ by $\alpha_0(E_{ij}^s) = (\lambda^{\sigma_s})^{r(i,j)} E_{ij}^s$. As above, this leads to a group automorphism $\alpha$ of $G$.

Let $\mathfrak{g}_i(\mathcal{O}_i) \subset \mathfrak{g}_i$ be the matrices in $\mathfrak{g}_i$ with coefficients in $\mathcal{O}_i$, the ring of integers in $K_i = K^{\sigma_i}$. Define

$$\Gamma_0 = \{\gamma \in \Sigma \mathfrak{g}_i(\mathcal{O}_i) | \gamma = \Sigma \gamma_i \text{ and } \gamma_i = \gamma_1^{\sigma_i}\}.$$

$\Gamma_0$ is a discrete subalgebra of $\mathfrak{g}$ of $Z$ rank $tn(n-1)/2$, which is also the dimension of $\mathfrak{g}$ over $R$. Notice that $\alpha_0(\Gamma_0) = \Gamma_0$.

Let $\Gamma = \exp \Gamma_0$. From the properties of $\Gamma_0$ it is not hard to show that $\Gamma$ is a uniform, discrete subgroup of $G$ and $\alpha(\Gamma) = \Gamma$. Thus $\alpha$ induces a diffeomorphism on $G/\Gamma$. We denote this diffeomorphism by $\bar{\alpha}$.

If $A \in G$, denote its image in $G/\Gamma$ by $[A]$. $[A]$ is a periodic point of $\bar{\alpha}$ if $\bar{\alpha}^m([A]) = [A]$ for some positive integer $m$. This is equivalent to requiring $A^{-1}\alpha^m(A) \in \Gamma$, or $\ln A^{-1}\alpha^m(A) \in \Gamma_0$.

We are interested in discovering the nature of the periodic points. It is possible to give an exact description.

Let $\mathfrak{g}_i(K_i) \in \mathfrak{g}_i$ be the matrices in $\mathfrak{g}_i$ with coefficients in $K_i$. Let

$$\Gamma_0(K) = \{\gamma \in \sum \mathfrak{g}_i(K_i) | \gamma = \sum \gamma_i \text{ and } \gamma_i = \gamma_1^{\sigma_i}\},$$

and $\Gamma(K) = \exp \Gamma_0(K)$. $\Gamma(K)$ can be described in another way; $\Gamma(K)$ is the set of all $A = \prod A_i \in G$, $A_i \in G_i$, such that $A_1$ has coefficients in $K$ and $A_i = A_1^{\sigma_i}$. This follows from the fact that the series defining exp has rational coefficients.

THEOREM. $\Gamma(K)/\Gamma$ *is the precise set of periodic points for the diffeomorphism* $\bar{\alpha}$.

Before proving this we need an elementary lemma of an arithmetic nature.

LEMMA. *Let* $K$ *be an algebraic number field, and* $\lambda \in K$ *a unit. If* $a_1, a_2, ..., a_l \in K$, *there is a positive integer* $m$ *such that* $(\lambda^m - 1)a_i$ *is an algebraic integer for* $i = 1, 2, ..., l$.

PROOF. Let $\mathcal{O}$ be the ring of integers of $K$. There is a rational integer $n$ such that $na_i \in \mathcal{O}$ for $i = 1, 2, ..., l$. If we can find an $m$ such that $\lambda^m - 1$ is divisible by $n$, we will be done. Consider $\mathcal{O}/n\mathcal{O}$. This is a finite ring and $\bar{\lambda}$, the residue class of $\lambda$, is a unit. Consequently, $\bar{\lambda}^m = \bar{1}$ for some $m$. Thus $\lambda^m - 1$ is divisible by $n$.

We now turn to the proof of the theorem. By previous remarks it is sufficient to prove the following:

(i) If $A \in G$ and $A^{-1}\alpha^m(A) \in \Gamma$ for some $m > 0$, then $A \in \Gamma(K)$.

(ii) If $A \in \Gamma(K)$, then there is an integer $m > 0$ such that $\ln A^{-1}\alpha^m(A) \in \Gamma_0$.

To prove (i) we compute the matrix elements of $A^{-1}\alpha^m(A)$. Let $A = I + B$. Then $B$ is nilpotent and $A^{-1} = \sum_{k=0}^{\infty}(-1)^k B^k$. Also, notice that $\alpha^m(A) = I + \alpha_0^m(B)$. Thus $A^{-1}\alpha^m(A) = I + \sum_{k=1}^{\infty}(-1)^k(B^k - B^{k-1}\alpha_0^m(B)) = I + C$.

If $B = \sum_{s=1}^{t} \sum_{i<j} b_{ij}^s, E_{ij}^s$ then the matrix coefficient $c_{ij}^s$ of $C$ is given by

$$c_{ij}^s = (\lambda_s^{mr(i,j)} - 1)b_{ij}^s - \sum_{i<k<j} (\lambda_s^{mr(k,j)} - 1)b_{ik}^s b_{kj}^s$$

(*)
$$+ \sum_{i<k<l<j} (\lambda_s^{mr(l,j)} - 1)b_{ik}^s b_{kl}^s b_{lj}^s - \cdots$$

In this formula we have set $\lambda_s = \lambda^{\sigma_s}$.

If $j - i = 1$ and $s = 1$, we see $c_{ij}^1 = (\lambda^{mr(i,j)} - 1)b_{ij}^1 \in K$, and thus $b_{ij}^1 \in K$. Also, since $c_{ij}^s = (c_{ij}^1)^{\sigma_s}$, we deduce from (*) that $b_{ij}^s = (b_{ij}^1)^{\sigma_s}$. Proceeding inductively (induction on the number $j - i$) we obtain our result.

To prove (ii) we note that the formula (*) shows that the matrix $C$ has entries which are linear forms in the quantities $\lambda_s^{mr(i,j)} - 1$ with coefficients in $K_s$, $s = 1$, ..., $t$. $\ln A^{-1}\alpha^m(A) = \ln(I + C) = \sum_{k=1}^{\infty} (-1)^k C^k/k = D$. The entries of $D$ are thus polynomials in the quantities $\lambda_s^{mr(i,j)} - 1$ with coefficients in $K_s$, $s = 1, \ldots, t$. These polynomials are all without constant term. Apply the lemma to the set of coefficients of the polynomials $d_{ij}^1$ (notice that these coefficients do not depend on $m$). The fact that $\lambda^m - 1$ divides $\lambda^{mr(i,j)} - 1$ for all $r(i,j)$ shows that $D_1 \in \mathfrak{g}(\mathcal{O}_1)$ and consequently that $D \in \Gamma_0$. Thus $[A]$ is a fixed point $\alpha^m$.

## REFERENCE

1. S. Smale, *Differentiable dynamical systems*, Bull. Amer. Math. Soc. **73** (1967), 747–817.

BROWN UNIVERSITY

# TOPOLOGICAL ENTROPY AND AXIOM A

ROBERT BOWEN

1. **Introduction.** In the recent Russian studies of dynamical systems consider-able attention has been given to entropy, an invariant of automorphisms of finite measure spaces (see V. A. Rokhlin [9] for an introduction). In this paper we shall study a topological entropy analogous to the measure theoretic one. This invariant was defined by Adler, Konheim and McAndrew in [2].

Our principal result is that the topological entropy of a diffeomorphism $f$ satisfying Smale's Axiom A [13] is given by

$$\text{ent} f = \overline{\lim_n} \frac{1}{n} \log N_n(f)$$

where $N_n(f)$ is the number of fixed points of $f^n$. Such an $f$ has positive entropy unless its nonwandering set is finite. These results are based on a method of constructing periodic points in the presence of canonical coordinates (Proposition 3.5) which is motivated by Bernoulli shifts. Later we characterize zero-dimensional sets having canonical coordinates as subshifts of finite type.

In the last section of this paper we show that our expression for entropy holds for a certain class of one-dimensional maps. We do not know just how generally it does hold. H. Furstenburg [16] has an example of a subshift which has positive entropy but no periodic points; so the expression does not work for all expansive homeomorphisms.

A significant problem now seems to be to calculate the entropy of the Abraham-Smale examples of [1].

The paper [12] of Sinai seems to include some of our results in the case of an Anosov diffeomorphism.

Several conversations with Stephen Smale have been very helpful during the preparation of this paper.

2. We review the definition of topological entropy given in [2]. Suppose $M$ is a compact topological space and $f: M \to M$ is continuous. For $\mathscr{A}$ an open cover of $M$ let $N(\mathscr{A})$ denote the minimum number of members of a subcover of $\mathscr{A}$. If $\mathscr{A}$, $\mathscr{B}$ are two open covers, let $\mathscr{A} \vee \mathscr{B} = \{A \cap B : A \in \mathscr{A}, B \in \mathscr{B}\}$. We set $\mathscr{A}_{f,n} = \mathscr{A} \vee f^{-1}\mathscr{A} \vee \ldots \vee f^{-n}\mathscr{A}$ and $M_n(f, \mathscr{A}) = N(\mathscr{A}_{f,n})$; $h(f, \mathscr{A}) = \lim_n (1/n) \log M_n(f, \mathscr{A})$ (elementary analysis shows this limit exists). Finally we define the topological entropy of $f$

$$\text{ent} f = \sup h(f, \mathscr{A})$$

where the supremum is over all open covers $\mathscr{A}$ of $M$. One can easily show that if $M$ is a metric space and $\mathscr{A}_m$ is a sequence of open covers with diam$(\mathscr{A}_m) \to 0$, then $h(f, \mathscr{A}_m) \to$ ent $f$.

Given a finite cover $\mathscr{A}$ we think of $M_n(f, \mathscr{A})$ as the minimum cardinality of a set $E_n$ of $n + 1 -$ strings $(A_{i_0}, \ldots, A_{i_n})$ of members of $\mathscr{A}$ such that for every $x \in M$ there is some $(A_{i_0}, \ldots, A_{i_n}) \in E_n$ with $f^r(x) \in A_{i_r}$ for $r = 0, 1, \ldots, n$. With this viewpoint the following fundamental fact is useful.

(2.1) LEMMA. *Suppose we are given for each positive integer $r$ a set $D_r$ of $r$-strings of symbols from some fixed finite set and that $K$ is a fixed integer. Let $F_n$ be the set of all $n$-strings which can be decomposed into disjoint substrings belonging to various $D_r$'s, using at most $K$ substrings. Then*

$$\overline{\lim_n} \frac{1}{n} \log \overline{\overline{F}}_n \leqslant \overline{\lim_r} \frac{1}{r} \log \overline{\overline{D}}_r.$$

PROOF. Let $a = \overline{\lim}_r (1/r) \log \overline{\overline{D}}_r$ and $d > 0$. Then $(1/r) \log D_r \leqslant a + d$ for $r$ sufficiently large; so for $c$ large enough we have $\log D_r \leqslant r(a + d) + c$ for all $r$. There are $P(n) = \sum_{s=1}^{K} \binom{n}{s}$ ways of specifying the positions in an $n$-string where the substrings of the decomposition are to begin. Such a specification $j_1, \ldots, j_s$ gives rise to $A_{j_1, \cdots, j_s} = \overline{\overline{D}}_{j_2 - j_1} \overline{\overline{D}}_{j_3 - j_2} \cdots \overline{\overline{D}}_{n - j_s + 1}$ elements of $F_n$. Setting $j_{s+1} = n + 1$ we have

$$\log A_{j_1, \ldots, j_s} = \sum_{i=1}^{s} \log \overline{\overline{D}}_{j_{i+1} - j_i} \leqslant \sum_{i=1}^{s} ((j_{i+1} - j_i)(a + d) + c)$$

$$\leqslant sc + n(a + d) \leqslant Kc + n(a + d).$$

From this follows $(1/n) \log \overline{\overline{F}}_n \leqslant (1/n) \log P(n) + Kc/n + a + d$. As $P(n)$ is a polynomial, we get $\overline{\lim}_n (1/n) \log F_n \leqslant a + d$. As $d$ was arbitrary we are done.

We quote a result of Adler, Konheim and McAndrew [2].

(2.2) PROPOSITION [2]. *If $M = M_1 \cup M_2$ with the $M_i$ closed and $f(M_i) \subseteq M_i$, then*

$$\text{ent } f = \max_{i=1,2} \text{ ent } f|M_i.$$

(2.3) COROLLARY. *If $A$ is a closed, $f$-invariant subset of $M$, then*

$$\text{ent } f \geqslant \text{ent } f|A.$$

PROOF. Consider $M = M \cup A$.

A point $x$ in $M$ is a *wandering* point of $f$ if it has a neighborhood $U$ in $M$ such that $U \cap \bigcup_{n \in Z; n \neq 0} f^n(U) = \varnothing$; otherwise $x$ is called *nonwandering*. The set $\Omega(f)$ of all nonwandering points of $f$ is easily seen to be closed and invariant under $f$. One idea of Smale's program [13] is to classify maps according to their action on their nonwandering sets. The following theorem shows that topological entropy fits in with such a viewpoint.

(2.4) THEOREM. *Let* $f: M \to M$ *be a continuous map on a compact metric space. If* $\Omega$ *is the nonwandering set of* $f$, *then*

$$\text{ent } f = \text{ent } f | \Omega.$$

PROOF. Let $\mathscr{A}_\varepsilon$ be a finite cover of $\Omega$ by open sets of diameter less than $\varepsilon > 0$. Then $h(f|\Omega, \mathscr{A}_\varepsilon) \to \text{ent} f$ as $\varepsilon \to 0$. We will construct finite coverings $\mathscr{B}_\varepsilon$ of $M$ by open sets of diameter less than $3\varepsilon$ such that $h(f, \mathscr{B}_\varepsilon) \leqslant h(f|\Omega, \mathscr{A}_\varepsilon) + \varepsilon$. Then, as $\varepsilon \to 0$, $\text{ent} f = \lim h(f, \mathscr{B}_\varepsilon) \leqslant \lim h(f|\Omega, \mathscr{A}_\varepsilon) = \text{ent} f|\Omega$. The reverse inequality needed for the theorem is a case of (2.3).

We proceed with the construction of $\mathscr{B}_\varepsilon$. Let $\mathscr{A}_\varepsilon$ be $\{A_1, ..., A_s\}$ and $c_n = M_n(f, \mathscr{A}_\varepsilon) = N(\mathscr{A}_\varepsilon \vee ... \vee f^{-n}\mathscr{A}_\varepsilon)$. Choose $N$ large enough so that $(1/N) \log c_{N-1} \leqslant h(f|\Omega, \mathscr{A}_\varepsilon) + \varepsilon$. By uniform continuity of $f$ on $M$ choose $\alpha > 0$ such that $d(f^k(x), f^k(y)) < \varepsilon$ for $k = 0, 1, ..., N-1$ whenever $x, y \in M$ and $d(x, y) < \alpha$. Let $U = \{y \in M: d(y, \Omega) < \alpha\}$ and $B_j = \{y \in U: d(y, A_j) < \varepsilon\}$ for $1 \leqslant j \leqslant s$. Let $E = \{(i_0, ..., i_{N-1})\}$ have $c_{N-1}$ elements with $\{\bigcap_{k=0}^{N-1} f^{-k}A_{i_k}: (i_0, ..., i_{N-1}) \in E\}$ a subcover of $\mathscr{A}_\varepsilon \vee f^{-1}\mathscr{A}_\varepsilon \vee ... \vee f^{-N+1}\mathscr{A}_\varepsilon$. For $x$ a wandering point let $N_x$ be an open neighborhood of $x$ of diameter less than $3\varepsilon$ such that $N_x \cap \bigcup_{m \neq 0} f^m(N_x) = \varnothing$; let $\mathscr{B}_\varepsilon = \{B_1, ..., B_s, N_{x_1}, ..., N_{x_t}\}$ where $\{N_{x_1}, ..., N_{x_t}\}$ is a finite subcover of the open cover $\{N_x : x \in X \backslash U\}$ of $X \backslash U$.

We now define a set $D_r$ of $r$-strings of members of $\mathscr{B}_\varepsilon$ for $r > 0$. Set $D_1 = \{(X) : X \in \mathscr{B}_\varepsilon\}$. For $r > 1$ and $r = pN + q$, $0 \leqslant q < N$, let $D_r$ consist of all $r$-strings $(B_{j_1}, ..., B_{j_r})$ such that

$$(j_{mN+1}, ..., j_{(m+1)N}) \in E \text{ for } m = 0, 1, ..., p-1.$$

Then $D_r = c_{N-1}^p s^q \leqslant c_{N-1}^p s^N$ and so

$$\varlimsup_r (1/r) \log \overline{\overline{D}}_r \leqslant \varlimsup_r (1/pN)(\log s^N + p \log c_{N-1})$$
$$\leqslant (1/N) \log c_{N-1} \leqslant h(f|\Omega, \mathscr{A}_\varepsilon) + \varepsilon.$$

Let $x \in M$ and $n \geqslant 0$. We will show that $x$ is in some $\bigcap_{i=0}^n f^{-i}C_i$ with $(C_0, ..., C_n)$ in $F_{n+1}$, where $F_n$ is defined as in the Basic Lemma with $K = 2t + 1$. Then $\{\bigcap_{i=0}^n f^{-i}C_i : (C_0, ..., C_n) \in F_{n+1}\}$ will be a subcover of $\mathscr{B}_\varepsilon \vee ... \vee f^{-n}\mathscr{B}_\varepsilon$ and the Basic Lemma will give us the desired inequality: $h(f, \mathscr{B}_\varepsilon) \leqslant \varlimsup_n (1/n) \log \overline{\overline{F}}_n \leqslant h(f|\Omega, \mathscr{A}_\varepsilon) + \varepsilon$.

Given $x$ we now define the $C_k$. For $f^k(x) \notin U$ let $C_k$ be an $N_{x_i}$ containing $f^k(x)$. If $f^{k_1}(x)$, $f^{k_2}(x)$ were both in $N_{x_i}$ with $k_1 < k_2$, we would have $f^{k_2-k_1}(f^{k_1}(x)) \in N_{x_i} \cap f^{k_2-k_1}(N_{x_i})$—contradicting the definition of $N_{x_i}$. As there are only $tN_{x_i}$'s, $f^k(x) \in U$ except for at most $t$ values of $k$. Thus, omitting $f^k$ with $f^k(x) \notin U$, we are left with at most $t + 1$ blocks of successive iterates of $f$ whose value at $x$ is in $U$. Suppose $f^{k_1}, ..., f^{k_2}$ is such a block. Let $r = k_2 - k_1 + 1, r = pN + q, 0 \leqslant q < N$. For $0 \leqslant m < p$ let $y_m \in \Omega$ with $d(y_m, f^{k_1+mN}(x)) < \alpha$. Then $d(f^i(y_m), f^{k_1+mN+i}(x)) < \varepsilon$ for $0 \leqslant i \leqslant N-1$. Suppose $y_m \in \bigcap_{i=0}^{N-1} f^{-i}A_{j_i}$ with $(j_0, ..., j_{N-1}) \in E$; then $f^{k_1+mN+i}(x) \in B_{j_i}$ for $0 \leqslant i \leqslant N-1$. Set $C_{k_1+mN+i} = B_{j_i}$. We then have $(C_{k_1}, ..., C_{k_1+pN-1})$ in $D_{pN}$. Letting $C_k$ be an arbitrary $B_i$ containing $f^k(x)$ for

$k_1 + pN \leqslant k \leqslant k_2$, we get $(C_{k_1}, \ldots, C_{k_2}) \in D_r$ and $f^k(x) \in C_k$ for $k_1 \leqslant k \leqslant k_2$. We have defined $(C_0, \ldots, C_n)$ and by its construction it decomposes into at most $2t + 1$ substrings in various $D_r$'s. We are done.

(2.5) COROLLARY. *If $f$ and $g$ are $\Omega$-conjugate, then* ent $f =$ ent $g$.

(2.6) COROLLARY. *If $\Omega(f)$ is finite, then* ent $f = 0$.

PROOF. It is easy to see that any endomorphism of a finite space has entropy 0.

We now suppose that $M$ is a compact differentiable manifold and Diff$(M)$ is the space of all $C^r$ diffeomorphisms $(r \geqslant 1)$ of $M$ with the $C^r$ topology. Let $f \in$ Diff$(M)$. A closed $f$-invariant subset $\Lambda$ $M$ is hyperbolic under $f$ if the tangent bundle of $M$ restricted to $\Lambda$, $T_\Lambda(M)$, has a continuous splitting $T_\Lambda(M) = E^s + E^u$ which is invariant under $Df$ such that $Df : E^s \to E^s$ is contracting and $Df : E^u \to E^u$ is expanding (see [13] for definitions). $f$ satisfies Axiom A if

(Aa) $\Omega(f)$ is hyperbolic, and

(Ab) the periodic points of $f$ are dense in $\Omega(f)$.

We now have a theorem which formally resembles a "second law of thermodynamics."

(2.7) THEOREM. *If $f \in$ Diff$(M)$ satisfies Axiom A, then $f$ has a neighborhood $N_f$ in* Diff$(M)$ *such that* ent $f \leq$ ent $g$ *for all $g \in N_f$.*

PROOF. By Smale's Spectral Decomposition Theorem [13], $\Omega(f) = \Omega_1 \cup \ldots \cup \Omega_k$ where the $\Omega_i$ are certain "basic" sets. By (2.2) and (2.4) we have ent $f = \max_i$ ent $f | \Omega_i$. In [19] Smale shows that $f$ has a neighborhood $N_f$ such that for $g$ in $N_f$ $f | \Omega_i$ is topologically conjugate to $g | \Lambda_i$ for some closed $g$-invariant subset $\Lambda_i$ of $M$. Hence, using (2.3),

$$\text{ent } f | \Omega_i = \text{ent } g | \Lambda_i \leqslant \text{ent } g.$$

Thus ent $f \leqslant$ ent $g$.

We now assume $M$ is a compact metric space and $f$ is a homeomorphism. Then $f$ is expansive if there is an $\varepsilon > 0$ (called an expansive constant) such that for any $x, y$ in $M$ either $x = y$ or $d(f^n(x), f^n(y)) > \varepsilon$ for some integer $n$. If $f$ is a diffeomorphism satisfying Axiom A, then $f | \Omega$ is expansive (3.2).

(2.8) PROPOSITION. *If $f$ is an expansive homeomorphism of a compact metric space, then*

$$\text{ent } f \geqslant \overline{\lim_n} \frac{1}{n} \log N_n(f) < \infty$$

*where $N_n(f)$ is the number of fixed points of $f^n$.*

PROOF. Let $\varepsilon > 0$ be an expansive constant and $\mathscr{A} = \{A_1, \ldots, A_r\}$ be a finite open cover of $M$ of diameter less than $\varepsilon$. Suppose $x$ and $y$ have period $n$ and $x, y \in \bigcap_{j=0}^n A_{i_j}$. Then $f^j(x), f^j(y) \in A_{i_j}$ for $j = 0, 1, \ldots, n$; hence $f^m(x), f^m(y) \in A_{i_j}$ for $m \equiv j \pmod n$. It follows that $d(f^m(x), f^m(y)) < \varepsilon$ for all $m$ and so $x = y$. Thus each member of $\mathscr{A}_{f,n}$ contains at most one periodic point of order $n$, and so $N_n(f) \leqslant N(\mathscr{A}_{f,n})$. Also $\mathscr{A}_{f,n} \leqslant r^{n+1}$, so $N(\mathscr{A}_{f,n}) \leqslant r^{n+1}$. It follows that

$$\overline{\lim_n} \frac{1}{n} \log N_n(f) \leqslant h(f, \mathscr{A}) \leqslant r.$$

We are done as ent $f \geqslant h(f, \mathscr{A})$.

REMARK. The above proof resembles a proof by M. Shub of the finiteness statement of the proposition. P. Walters also has a proof of the proposition. Keynes and Robertson [6] show that ent $f = h(f, \mathscr{A})$ in the above situation. J. P. Conze [18] also proves 2.8.

3. **Canonical coordinates.** Let $(X, d)$ be a compact metric space and $f: X \to X$ be a homeomorphism. For $\delta > 0$ and $x \in X$ define

$$W^s(x, \delta) = \{y \in X : d(f^n(x), f^n(y)) \leqslant \delta \text{ for all } n \geqslant 0\}$$

and

$$W^u(x, \delta) = \{y \in X : d(f^n(x), f^n(y)) \leqslant \delta \text{ for all } n \leqslant 0\}.$$

$f$ has *canonical coordinates* if for every $\delta > 0$ there is an $\varepsilon(\delta) > 0$ such that $W^s(x, \delta) \cap W^u(y, \delta) \neq \varnothing$ whenever $x, y \in X$ and $d(x, y) \leqslant \varepsilon(\delta)$. By the uniform equivalence of metrics for a compact space this property is seen to be independent of the metric used; topologically conjugate maps have this property or not together. We say $f$ has *hyperbolic* canonical coordinates if in addition there are constants $\delta^* > 0$, $0 < \lambda < 1$, and $c \geqslant 1$ such that if $x \in X$ and $y \in W^s(x, \delta^*)$ (or $y \in W^u(x, \delta^*)$), then $d(f^n(x), f^n(y)) \leqslant c\lambda^n d(x, y)$ for all $n \geqslant 0$ (or $n \leqslant 0$). One notices that the condition then holds for all sufficiently small $\delta^*$. This condition depends on the metric used.

The above conditions arise out of Smale's study of diffeomorphisms. Specifically, from S. Smale [13, I.7] and Hirsch and Pugh [5]:

(3.1) THEOREM. *Let $M$ be a compact Riemannian manifold, $f \in \mathrm{Diff}(M)$ satisfy Axiom A, and $\Omega_i$ be a basic set in the spectral decomposition of $\Omega(f)$. Then $f|\Omega_i$ has hyperbolic canonical coordinates with respect to some metric.*

During the remainder of this section we assume that $f$ is a homeomorphism of a compact metric space $(X, d)$ which has hyperbolic canonical coordinates. The following fact was first proved by M. Shub in [11].

(3.2) LEMMA. $\delta^*$ *is an expansive constant for $f$.*

PROOF. Suppose $x, y \in X$ and $d(f^n(x), f^n(y)) \leqslant \delta^*$ for all $n$. Then $f^n(y) \in W^s(f^n(x), \delta^*)$ for all $n$; so $d(f^{n+r}(x), f^{n+r}(y)) \leqslant c\lambda^r \delta^*$ for all $n$ and nonnegative $r$. Letting $r \to +\infty$ and setting $n = -r$, we get $d(x, y) \leqslant c\lambda^r \delta^* \to 0$. Thus $x = y$.

The function $\varepsilon$ in the definition of canonical coordinates may be chosen so that $\varepsilon(\delta) < \delta$ and $\varepsilon(\delta_1) \geqslant \varepsilon(\delta_2)$ when $\delta_1 \geqslant \delta_2$. For small positive $\delta$ (say $\delta \geqslant \frac{1}{2}\varepsilon(\delta^*)$) define $b(\delta) = \delta + \inf\{\delta' : \varepsilon(\delta') \geqslant 2\delta\}$. One easily checks that $b(\delta) \to 0$ monotonely as $\delta \to 0$. Let $\omega > 0$ be such that $b(\delta) \leqslant \frac{1}{12}\delta^*$ for $0 < \delta \leqslant \omega$. Set $\gamma = \frac{1}{2}\delta^*$.

(3.3) LEMMA. *If* $0 < \delta \leqslant \omega$ *and* $d(z, y) \leqslant 2\delta$, *then*

$$W^s(z, b(\delta)) \cap W^u(y, b(\delta)) = W^s(z, \gamma) \cap W^u(y, \gamma)$$

*and consists of a single point.*

PROOF. As $b(\delta) \leqslant \gamma$, we clearly have

$$W^s(z, b(\delta)) \cap W^u(y, b(\delta)) \subseteq W^s(z, \gamma) \cap W^u(y, \gamma).$$

As $b(\delta) > \inf\{\delta' : \varepsilon(\delta') \geqslant 2\delta\}$ and $\varepsilon$ is monotone, $\varepsilon(b(\delta)) \geqslant 2\delta$. Hence $W^s(z, b(\delta))$ $\cap W^u(y, b(\delta)) \neq \varnothing$. We will complete the proof by showing that $W^s(z, \gamma) \cap W^u(y, \gamma)$ can have at most one point. Suppose it had two, $v$ and $w$. Then for $n \geqslant 0$, $d(f^n(v), f^n(z)) \leqslant \gamma$ and $d(f^n(w), f^n(z)) \leqslant \gamma$; so $d(f^n(v), f^n(w)) \leqslant \delta^*$. For $n \leqslant 0$ we have $d(f^n(v), f^n(y)) \leqslant \gamma$ and $d(f^n(w), f^n(y)) \leqslant \gamma$; so $d(f^n(v), f^n(w)) \leqslant \delta^*$. Thus $d(f^n(v), f^n(w)) \leqslant \delta^*$ for all $n$, contradicting $\delta^*$ an expansive constant.

Let $x \in X$ and $0 < \delta \leqslant \omega$. We define the canonical coordinate neighborhood

$$A(\delta, x) = \{w \in X : w \in W^s(z, \delta) \cap W^u(y, \gamma) \text{ for some } z \in W^u(x, \delta) \text{ and some}$$
$$y \in W^s(x, \delta)\}.$$

For $w \in A = A(\delta, x)$ we define $W^u(w, A) = W^u(w, \gamma) \cap A$ and $W^s(w, A) = W^s(w, \gamma) \cap A$. We collect a few facts about these sets which will be useful later.

(3.4) LEMMA. (a) $A(\delta, x) \supseteq B_{\varepsilon(\delta)}(x)$ *where* $B_r(x)$ *is the open ball of radius* $r$ *about* $x$.

(b) *If* $y \in W^s(x, \delta)$, *then* $W^u(y, A) = W^u(y, b(\delta)) \cap A$; *if* $z \in W^u(x, \delta)$, *then* $W^s(z, A) = W^s(z, b(\delta)) \cap A$.

(c) *If* $y \in W^u(z, \delta_1)$, *then* $W^u(y, \delta_2) \subseteq W^u(z, \delta_1 + \delta_2)$. *A similar statement holds for* $W^s$'s.

(d) *If* $v, w \in A$, *then* $W^s(v, A) \cap W^u(w, A) = W^s(v, 2b(\delta)) \cap W^u(w, 2b(\delta)) = W^s(v, \gamma)$ $\cap W^u(w, \gamma)$ *and consists of a single point of* $A$.

(e) *If* $w \in A$, *then* $W^u(w, A) = W^u(w, 2b(\delta)) \cap A$ *and* $W^s(w, A) = W^s(w, 2b(\delta)) \cap A$.

(f) *If* $v, w \in A$, *then* $W^u(v, A)$ *and* $W^u(w, A)$ *are either disjoint or equal; if* $w \in W^u(v, A)$, *then* $W^u(w, A) = W^u(v, A)$. (*Similarly for* $W^s$'s.)

(g) *If* $t \in B_{\varepsilon(\delta)}(x)$ *and* $0 < \eta \leqslant \delta^*$, *there is an* $I_0$ *such that* $f^I W^u(y, A) \supseteq W^u(f^I(y), \mathscr{A})$ *for* $I \geqslant I_0$.

(h) *Suppose* $V \subseteq B_{\varepsilon(\delta)}(x)$ *is compact. Then there is an* $I_0$ *such that if* $I \geqslant I_0$ *and* $y \in V \cap f^{-m}(V)$ *then* $f^I W^u(y, A) \supseteq W^u(f^I(y), A)$ *and* $f^I W^s(y, A) \subseteq W^s(f^I(y), A)$.

PROOF. (a) Let $w \in B_{\varepsilon(\delta)}(x)$. By the very definition of $\varepsilon(\delta)$ let $z \in W^u(x, \delta) \cap W^s(w, \delta)$ and $y \in W^s(x, \delta) \cap W^u(w, \delta)$. Then $w \in W^s(z, \delta) \cap W^u(y, \delta) \subseteq W^s(z, \gamma) \cap W^u(y, \gamma)$ and so $w \in A = A(\delta, x)$.

(b) Let $w \in W^u(y, A) = W^u(y, \gamma) \cap A$. Then $w \in W^s(z, \gamma)$ for some $z \in W^u(x, \delta)$. Now $d(y, x) \leqslant \delta$ and $d(z, x) \leqslant \delta$; so $d(z, y) \leqslant 2\delta$ and by Lemma 3.3 we have $w \in W^u(y, \gamma) \cap W^s(z, \gamma) \subseteq W^u(y, b(\delta))$. The second statement is proved similarly.

(c) If $w \in W^u(y, \delta_2)$, then for $n \leqslant 0$ we have $d(f^n(w), f^n(y)) \leqslant \delta_2$ and $d(f^n(y), f^n(z)) \leqslant \delta_1$. Apply the triangle inequality.

(d) Let $v = W^s(z_1, \gamma) \cap W^u(y_1, \gamma)$ and $w = W^s(z_2, \gamma) \cap W^u(y_2, \gamma)$ with $z_i \in W^u(x, \delta)$ and $y_i \in W^s(x, \delta)$. By (b) we have $v \in W^s(z_1, b(\delta))$ and $w \in W^u(y_2, b(\delta))$. As

$d(z_1, y_2) \leqslant 2\delta$, Lemma 3.3 gives us a point $t \in W^s(z_1, b(\delta)) \cap W^u(y_2, b(\delta)) \subseteq A$. Now $t, v \in W^s(z_1, b(\delta))$; using (c), $t \in W^s(v, 2b(\delta))$. Similarly $t \in W^u(w, 2b(\delta))$. As

$$t \in W^s(v, 2b(\delta)) \cap W^u(w, 2b(\delta)) \subseteq W^s(v, \gamma) \cap W^u(w, \gamma)$$

and the right set has at most one point (proof of Lemma 3.3), each of the two sets is $\{t\}$. (d) now follows from the fact that $t \in A$.

(e) Suppose $v \in W^u(w, A)$. Then $v \in W^s(v, A) \cap W^u(w, A) \subseteq W^u(w, 2b(\delta))$ by (d). The second statement is proved similarly.

(f) Suppose $t \in W^u(v, A) \cap W^u(w, A) \subseteq W^u(v, 2b(\delta)) \cap W^u(w, 2b(\delta))$. Then, via (c), we have $v \in W^u(w, 4b(\delta))$. Hence $W^u(v, A) = W^u(v, 2b(\delta)) \cap A \subseteq W^u(w, 6b(\delta)) \cap A \subseteq W^u(w, \delta) \cap A = W^u(w, A)$, using (c) and the fact that $6b(\delta) \leqslant \gamma$. The reverse inclusion follows symmetrically. The second statement of (f) follows from the first. For $W^s$'s the proof is similar.

(g) Set $\beta = \varepsilon(\delta) - d(x, y)$. Then $W^u(y, \beta) \subseteq B_{\varepsilon(\delta)}(x) \subseteq A$. Pick $I_0$ so that $\beta \geqslant c\lambda^{I_0}\eta$. Let $z \in W^u(f^I(y), \eta)$, $I \geqslant I_0$. For $J \geqslant 0$, $d(f^{-J}(f^{-I}(z)), f^{-J}(y)) = d(f^{-(I+J)}(z), f^{-(I+J)}(f^I(y))) \leqslant c\lambda^{I+J}\eta \leqslant \beta$. So $f^{-I}(z) \in W^u(y, \beta) \subseteq W^u(y, A)$.

(h) Let $\beta = \varepsilon(\delta) - d(x, V)$ and $\eta = 2b(\delta)$. The proof of (g) gives $f^I W^u(y, A) \supseteq W^u(f^I(y), 2b(\delta)) \supseteq W^u(f^I(y), A)$. For the second statement, apply the first to $f^{-1}$.

We shall now present a way to use canonical coordinates to construct periodic points. For motivation we look at the Cantor shift $X_N$ on $N$ symbols. Here $X_N$ consists of all two-sided infinite sequences $x = (x_i)$ of integers between 0 and $N - 1$; $f: X_N \to X_N$ is defined by $f(x)_i = x_{i+1}$. We have stable and unstable sets for $x = (x_i)$:

$$W^s(x) = \{y \in X_N : f^r(x) \text{ and } f^r(y) \text{ agree in the zeroth place for all } r \geqslant 0\}$$
$$= \{y \in X_N : x_r = y_r \text{ for all } r \geqslant 0\}.$$

Suppose we want to construct a periodic element of $X_N$ with period $a_0 a_1 \dots a_{n-1}$ and we have a point $y$ such that $x_i = a_i$ for $i = 0, \dots, n$ (where $a_n = a_0$). Let $w = f^n(y)$.

$$y = \qquad \dots a_0 a_1 \dots a_0 \dots$$
$$w = \dots a_0 \dots a_{n-1} a_0 \dots$$
$$\text{Set } w^1 = W^s(y) \cap W^u(w) = \dots a_0 \dots a_{n-1} a_0 a_1 \dots a_0 \dots$$
$$y^1 = f^{-n}(w^1) = \qquad \dots a_0 a_1 \dots a_{n-1} a_0 a_1 \dots a_{n-1} a_0 \dots.$$

Define inductively $w^{k+1} = W^s(y^k) \cap W^u(w)$ and $y^{k+1} = f^{-n}(w^{k+1})$. One sees that $y^k$ is a point with $a_0 \dots a_{n-1}$ occurring $k + 1$ successive times starting at the 0th place. $z_k = f^{kn}(y^{2k})$ has $2k$ successive $a_0 \dots a_{n-1}$'s, centered at the 0th place. As $k \to \infty$, $z_k \to z$, our periodic point. This seemingly farfetched construction works in our general situation.

(3.5) PROPOSITION. *Suppose* $0 < \delta \leqslant \omega$, $e = 2b(\delta)c(1 + \sum_{j=0}^{\infty} \lambda^j) \leqslant \gamma$, $A = A(\delta, x)$, $y \in A \cap f^{-m}(A)$ *and* $f^m(A) \supseteq W^u(f^m(y), A)$ *where* $m$ *is some positive integer. Then there is a point* $z$ *of period* $m$ *under* $f$ *such that* $d(f^k(z), f^k(y)) \leqslant e$ *for* $0 \leqslant k \leqslant m$.

PROOF. Set $y_0 = y$, $w_0 = f^m(y)$ and define recursively

$$w_{k+1} = W^s(y_k, A) \cap W^u(f^m(y), A),$$
$$y_{k+1} = f^{-m}(w_{k+1}).$$

Inductively, $w_k$ and $y_k$ are in $A$. This is true for $k = 0$; if it is true for $k$, then $W^s(y_k, A)$ and $W^u(f^m(y), A)$ intersect in a point of $A$ by 3.4d. This point $w_{k+1}$ is in $f^m(A)$ as it is in $W^u(f^m(y), A)$; hence $y_{k+1} = f^{-m}(w_{k+1})$ is in $A$.

By 3.4d we also have $w_{k+1} \in W^s(y_k, 2b(\delta)) \cap W^u(f^m(y), 2b(\delta))$. By hyperbolicity we have, for $r \geq 0$,

(i) $d(f^{m+r}(y_{k+1}), f^r(y_k)) = d(f^r(w_{k+1}), f^r(y_k)) \leq 2b(\delta)c\lambda^r$,

(ii) $d(f^{m-r}(y_k), f^{m-r}(y)) = d(f^{-r}(w_k), f^{-r}(f^m(y))) \leq 2b(\delta)c\lambda^r$.

We show inductively that

(iii) $d(f^p(y_k), f^s(y)) \leq 2b(\delta)c(1 + \sum_{j=0}^{t-1} \lambda^{mj})$

for $0 \leq p \leq mk$, $p = mt + s$, and $0 \leq s \leq m$. By (ii) this is true for $t = 0$; this covers the case $k = 0$ also. Thus suppose $t, k \geq 1$ and that (iii) holds for smaller $k$'s and $t$'s. Then by (i)

$$d(f^p(y_k), f^{p-m}(y_{k-1})) \leq 2b(\delta)c\lambda^{p-m} \leq 2b(\delta)c\lambda^{m(t-1)}.$$

By our inductive hypothesis

$$d(f^{p-m}(y_{k-1}), f^s(y)) \leq 2b(\delta)c(1 + \sum_{j=0}^{t-2} \lambda^{mj}).$$

(iii) now follows from the above two inequalities and the triangle inequality.

(iii) gives us immediately $d(f^p(y_k), f^s(y)) \leq e$ under the same conditions on $p$ and $s$. Set $z_t = f^{mt}(y_{2t})$. Then we have

$$z_t \in A_t = \{v: d(f^r(v), f^s(y)) \leq e \text{ for } |r| \leq mt, r \equiv s \text{ (mod } m), 0 \leq s \leq m\}.$$

Now $A_{t+1} \subseteq A_t$ and the $A_t$ are compact. By the finite intersection property let $z \in \bigcap_{t>0} A_t = \{v: d(f^r(v), f^r(y)) \leq e \text{ for } r \equiv s \text{ (mod } m) \text{ and } 0 \leq s \leq m\}$. Then we have, for any $r$ and the right $s$,

$$d(f^r(f^m(z)), f^r(z)) \leq d(f^{m+r}(z), f^s(y)) + d(f^r(z), f^s(y)) \leq 2e.$$

As $2e \leq 2\gamma = \delta^*$ is an expansive constant, $f^m(z) = z$. The rest of the lemma follows from $z \in A_1$.

4. In this section we compute the topological entropy of the homeomorphisms introduced in the last section and confirm a conjecture of Smale that they have positive entropy (unless $\Omega$ is finite). It turns out that the entropy is just the lower bound given by Proposition 2.8 for expansive homeomorphisms.

(4.1) THEOREM. *Let $f$ be a homeomorphism of a compact metric space $X$ having hyperbolic canonical coordinates. Then*

$$\text{ent } f = \overline{\lim_n} \frac{1}{n} \log N_n(f).$$

PROOF. Because of 2.8 and 3.2 we need only show that $T = \varlimsup_n (1/n) \log N_n(f)$ $\geq$ ent $f$. Let $\mathscr{A}$ be a finite open cover of $X$. Choose $\delta$ small enough that $e' = 2b(\delta)c(1 + \sum_{j=0}^{\infty} \lambda^j) \leq \frac{1}{3}\gamma$ and every $e'$-ball of $X$ lies in some member of $\mathscr{A}$. Choose $\delta' < \delta$ so small that $2b(\delta')c(1 + \sum_{j=0}^{\infty} \lambda^j) \leq \frac{1}{4}\varepsilon(\frac{1}{2}\varepsilon(\delta))$ and set

$$e^* = \min\{\tfrac{1}{2}\varepsilon(\delta'), \tfrac{1}{4}\varepsilon(\tfrac{1}{2}\varepsilon(\delta))\}.$$

Let $\mathscr{B} = \{B_{e^*}(x_1), \ldots, B_{e^*}(x_q)\}$ be a cover of $X$.

(4.2) LEMMA. Suppose for $m \geq 0$ and $V \in \mathscr{B}$ there are integers $D(m, V)$ and sets $C_i^j(m, V)$, $0 \leq i \leq m$, $1 \leq j \leq D(m, V)$, defined such that

$$V \cap f^{-m}(V) \subseteq \bigcup_{j=1}^{D(m,V)} \bigcap_{i=0}^{m} f^{-i}(C_i^j(m, V)).$$

If $\varlimsup_m (1/m) \log D(m, V) \leq T$ for each $V \in \mathscr{B}$, then $h(f, \mathscr{A}) \leq T$.

PROOF. We will apply our basic counting Lemma 2.1. Let $D_{m+1}$ be the totality of $(C_0^j(m, V), \ldots, C_m^j(m, V))$ with $V \in \mathscr{B}$ and $1 \leq j \leq D(m, V)$. Then $\overline{\overline{D}}_{m+1} \leq \sum D(m, V)$ and so $\varlimsup_m (1/m) \log \overline{\overline{D}}_m \leq \max_V \varlimsup_m (1/m) \log D(m - 1, V) \leq T$. Consider a string $(V_0, \ldots, V_n)$ of elements of $\mathscr{B}$ with $\bigcap_{i=0}^{n} f^{-i}V_i \neq \varnothing$. Then $(V_0, \ldots, V_n)$ may be decomposed into at most $K = q$ substrings, each with first and last elements the same. (Easily proved by induction on $q$: let $V_s$ be the last appearance of $V_0$. Then in $(V_{s+1}, \ldots, V_n)$ appear at most $q - 1$ distinct elements of $\mathscr{B}$; apply induction.) Suppose $(V_r, \ldots, V_s)$ is a substring with $V_r = V_s$. Then $V_r \cap f^{r-s}(V_s)$ is covered by the $\bigcap_{i=0}^{s-r} f^{-i}(C_i^j(s - r, V_r))$, $j = 1, \ldots, D(s - r, V_r)$. It follows that $\bigcap_{i=0}^{n} f^{-i}V_i$ is covered by various $\bigcap_{i=0}^{n} f^{-i}C_i$ with $(C_0, \ldots, C_n)$ in $F_{n+1}$, where $F_n$ is constructed from the $D_m$ as in Lemma 2.1. Thus $\mathscr{A} \vee \ldots \vee f^{-n}\mathscr{A}$ has a subcover with $F_{n+1}$ elements and $h(f, \mathscr{A}) \leq \varlimsup_n (1/n) \log F_{n+1}$. By Lemma 2.1 we have $h(f, \mathscr{A}) \leq T$.

We proceed now to show that the hypotheses of the above lemma actually hold. Consider a particular $V \in \mathscr{B}$. If $V \cap f^{-I}(V) \neq \varnothing$ for only finitely many $I \geq 0$, then we may take $D(m, V) = 0$ for $m$ sufficiently large, thus trivially satisfying the hypotheses. We assume $V \cap f^{-I}(V) \neq \varnothing$ for arbitrarily large $I$. Now $\overline{V} \subseteq B_{\varepsilon(\delta')}(x_s)$ where $V = B_{e^*}(x_s)$. Applying Lemma 3.4h, for large enough $I$ and $y^* \in V \cap f^{-I}(V)$ we have

$$f^I(A) \supseteq f^I W^u(y^*, A) \supseteq W^u(f^I(y^*), A)$$

where $A = A(\delta', x_s)$. By 3.5 we have a point $p$ of period $I$ with

$$d(p, y^*) \leq 2b(\delta')c(1 + \sum_{j=0}^{\infty} \lambda^j) \leq \tfrac{1}{4}\varepsilon(\tfrac{1}{2}\varepsilon(\delta)).$$

As $y^* \in V$, $d(y^*, x_s) < e^* \leq \frac{1}{4}\varepsilon(\frac{1}{2}\varepsilon(\delta))$. Thus $d(p, x_s) < \frac{1}{2}\varepsilon(\frac{1}{2}\varepsilon(\delta))$; as $\varepsilon(\delta) < \delta$ and $\varepsilon$ is monotone, $d(p, x_s) < \frac{1}{2}\varepsilon(\delta)$ and $p \in B_{\varepsilon(\delta)}(x_s)$.

Set $B = A(\delta, x_s)$. Let $w_1^m, \ldots, w_{k_m}^m$ be a maximal set of points of $f^m(W^u(p, B)) \cap B$ such that $w_i^m \notin W^u(w_j^m, B)$ for $i \neq j$. We may index the $w_j^m$'s so that $j \leq j_m$ if and only if $W^u(w_j^m, B) \subseteq f^m(W^u(p, B))$; this defines $j_m$. Using 3.4f one easily sees that $k_m$ and $j_m$ are independent of which $w_j^m$'s are chosen.

(4.3) LEMMA. $j_m \leq N_m$.

PROOF. Let $z_j, j \leqslant j_m$, be the periodic point given by 3.5 for point $y_j = f^{-m}(w_j^m)$ and the canonical coordinate neighborhood $B = A(\delta, x)$. For $i \neq j$ we claim that $d(f^k(y_i), f^k(y_j)) > \gamma$ for some $0 \leqslant k \leqslant m$. Since $y_i \in W^u(p, B) = W^u(y_j, B) \subseteq W^u(y_j, \gamma)$, we have

$$d(f^{-t}(w_i^m), f^{-t}(w_j^m)) = d(f^{m-t}(y_i), f^{m-t}(y_j)) \leqslant \gamma$$

for $t \geqslant m$. If our claim were not true, we would have $d(f^{-t}(w_i^m), f^{-t}(w_j^m)) \leqslant \gamma$ for all $t0$ and so $w_i^m \in W^u(w_j^m, \gamma) \cap B = W^u(w_j^m, B)$—a contradiction. Proposition 3.5 gives us $d(f^k(y_i), f^k(z_i)) \leqslant e'$ and $d(f^k(y_j), f^k(z_j)) \leqslant e'$. Recall that $\delta$ was chosen so that $e' \leqslant \frac{1}{3}\gamma$. By the triangle inequality

$$d(f^k(z_i), f^k(z_j)) \geqslant d(f^k(y_i), f^k(y_j)) - d(f^k(y_i), f^k(z_i)) - d(f^k(y_j), f^k(z_j)) \geqslant$$
$$\geqslant \gamma - \tfrac{1}{3}\gamma - \tfrac{1}{3}\gamma > 0.$$

Hence $z_i \neq z_j$ for $i \neq j$.

(4.4) LEMMA. *For some $L$ independent of $m$, $k_m \leqslant j_{m+L}$ for all $m > 0$.*

PROOF. As $p \in B_{\varepsilon(\delta)}(x_s)$ we apply 3.4g to obtain an $L = It$ so that $f^L(W^u(p, B)) \supseteq W^u(p, 4b(\delta))$. Now

$$
\begin{aligned}
y \in f^m(W^u(p, B)) &\Rightarrow f^{-m}(y) \in W^u(p, B) \subseteq W^u(p, 2b(\delta)) \quad \text{by 3.4e} \\
&\Rightarrow W^u(f^{-m}(y), 2b(\delta)) \subseteq W^u(p, 4b(\delta)) \quad \text{by 3.4c} \\
&\Rightarrow f^{m+L}W^u(p, B) \supseteq f^m W^u(p, 4b(\delta)) \supseteq f^m W^u(f^{-m}(y), 2b(\delta)) \\
&\qquad \supseteq W^u(y, 2b(\delta)) \supseteq W^u(y, B) \quad \text{by 3.4e}
\end{aligned}
$$

Applying this to $y = w_j^m$ we get that $k_m \leqslant j_{m+L}$.

(4.5) LEMMA. *We may choose the $C_i^j(m, V)$ as in 4.2 with $D(m, V) = k_m$.*

PROOF. For each $j = 1, ..., k_m$ define

$$E(w_j^m, m) = \bigcup \{W^s(f^{-m}(z), B) : z \in W^u(w_j^m, B) \cap f^m(B)\}.$$

Let $0 \leqslant t \leqslant m$ and $y \in E(w_j^m, m)$. Then $y \in W^s(f^{-m}(z), A) \subseteq W^u(f^{-m}(z), 2b(\delta))$ by 3.4e; so $d(f^t(y), f^{t-m}(z)) \leqslant 2b(\delta) < \frac{1}{2}e'$. As $z \in W^u(w_j^m, B)$, $d(f^{t-m}(z), f^{t-m}(w_j^m)) \leqslant 2b(\delta) < \frac{1}{2}e'$. Thus $d(f^t(y), f^{t-m}(w_j^m)) < e'$ and $f^tE(w_j^m, m)$ lies in the $e'$-ball about $f^{t-m}(w_j^m)$ which in turn lies in some $C_i^j(m, V)$ in $\mathscr{A}$.

We have left to show that every $y \in V \cap f^{-m}(V)$ is in some $E(w_j^m, m)$. As $d(y, x_s) < e^* \leqslant \frac{1}{4}\varepsilon(\frac{1}{2}\varepsilon(\delta))$ and $d(x_s, p) \leqslant \frac{1}{2}\varepsilon(\frac{1}{2}\varepsilon(\delta))$, $d(y, p) < \varepsilon(\frac{1}{2}\varepsilon(\delta))$. Thus let

$$w \in W^s(y, \tfrac{1}{2}\varepsilon(\delta)) \cap W^u(p, \tfrac{1}{2}\varepsilon(\delta)) = W^s(y, B) \cap W^u(p, B).$$

Then $f^m(w) \in W^s(f^m(y), \frac{1}{2}\varepsilon(\delta))$. As $f^m(y) \in V$, $d(f^m(y), x_s) < e^* \leqslant \frac{1}{2}\varepsilon(\delta)$; so $d(f^m(w), x_s) < \varepsilon(\delta)$ and $f^m(w) \in B$ by 3.4a. Thus $f^m(w)$ is in $f^m(W^u(p, B)) \cap B$ and hence in $W^u(w_j^m, B)$ for some $j$ by the very definition of the $w_j^m$'s. Then $y \in E(w_j^m, m)$.

Using 4.3, 4.4, and 4.5 we have

$$\overline{\lim_m} \frac{1}{m} \log D(m, V) = \overline{\lim_m} \frac{1}{m} \log k_m \leqslant \overline{\lim_m} \frac{1}{m} \log j_{m+L}$$
$$\leqslant \overline{\lim_m} \frac{1}{m} \log N_{m+L} = T.$$

(4.6) THEOREM. *Let $f$ be as before. Then the periodic points of $f$ are dense in $\Omega(f)$. Either $\Omega(f)$ is finite or ent $f > 0$.*

PROOF. In that part of the proof of Theorem 4.1 following Lemma 4.2 we constructed a periodic point $p$ when given a small ball $V$ such that $f^n(V) \cap V \neq \varnothing$ for arbitrarily large $n$. One notices that $d(p, V) \to 0$ as $\text{diam}(V) \to 0$. If $y \in \Omega(f)$ and $V_m = B_{1/m}(y)$, we get $p_m \to y$. Thus periodic points are dense in $\Omega(f)$.

Suppose $\Omega(f)$ is not finite. Then there are infinitely many periodic points. Take any cover $\mathcal{A}$ and consider the sets $V \in \mathcal{B}$ as in the last theorem's proof. At least one such $V$ must intersect itself under infinitely many iterates and contain a periodic point $q$ other than the constructed one $p$. As $d(f^k(p), f^k(q))$ takes on finitely many positive values, $q \notin W^u(p, B)$ (we use the notation of the proof of the previous theorem).

Let $M$ be the period of $q$, $I$ that of $p$. Apply 3.4h to $\{p, q\} \subseteq B_{\varepsilon(\delta)}(x_s)$ to obtain a $d_0$ such that for $d \geqslant d_0$ we have $f^{dMI}W^s(q, B) \subseteq W^s(q, B)$ and $f^{dMI} W^s(p, B) \subseteq W^s(p, B)$. Set $t = W^s(q, B) \cap W^u(p, B)$. Then $f^{nMI}(t) \to q$ as $n \to \infty$. Pick $d \geqslant d_0$ such that $d(q, f^{dMI}(t)) < d(q, W^u(p, B))$. Then $t' = f^{dMI}(t) \in f^{dMI}W^u(p, B) \cap B$ but $t' \notin W^u(p, B)$. Thus $k_{dMI} \geqslant 2$.

(4.7) LEMMA. $k_{m+dMI} \geqslant 2j_m$ *for any $m > 0$.*

PROOF. We have $W^u(w_j^m, B) \subseteq f^m W^u(p, B) \cap B$ for $j = 1, \ldots, j_m$ and $w_i^m / W^u(w_j^m, B)$ for $i \neq j$. Set

$$z_1, j = W^s(p, B) \cap W^u(w_j^m, B),$$
$$z_2, j = W^s(q, B) \cap W^u(w_j^m, B) = W^s(t, B) \cap W^u(w_j^m, B).$$

Let $x_{i,j} = f^{dMI}(z_{i,j})$. As $z_{i,j} \in W^s(p, B) \cup W^s(q, B)$, the conditions placed on $d$ guarantee that $x_{i,j} \in W^s(p, B) \cup W^s(q, B) \subseteq B$. Thus $x_{i,j} \in B \cap f^{m+dMI}W^u(p, B)$. We will be done if we show that $x_{i,j} \notin W^u(x_{k,n}, B)$ unless $i = k$ and $j = n$.

Assume $x_{i,j} \in W^u(x_{k,n}, B)$. Then $x_{i,j} \in W^u(x_{k,n}, 2b(\delta))$ by 3.4e; so $z_{i,j} = f^{-dMI}(x_{i,j}) \in W^u(f^{-dMI}(x_{k,n}), 2b(\delta)) \cap B = W^u(z_{k,n}, B)$. Hence $W^u(w_j^m, B) = W^u(z_{i,j}, B) = W^u(z_{k,n}, B) = W^u(w_n^m, B)$ by 3.4f and so $j = n$. Suppose $i \neq k$; we may assume $x_{i,j} = x_{1,j}$ and $x_{k,n} = x_{2,j}$. Using 3.4e $d(f^J(z_{1,j}), f^J(p)) \leqslant 2b(\delta)$ and $d(f^J(z_{2,j}), f^J(t)) \leqslant 2b(\delta)$ for $J \geqslant 0$. As $x_{1,j} \in W^u(x_{2,j}, 2b(\delta))$, we have

$$d(f^S(z_{1,j}), f^S(z_{2,j})) = d(f^{S-dMI}(x_{1,j}), f^{S-dMI}(x_{2,j})) \leqslant 2b(\delta)$$

for $S \leqslant dMI$. Using the triangle inequality we get

$$d(f^S(t), f^S(p)) \leqslant 6b(\delta) \leqslant \gamma \quad \text{for } 0 \leqslant S \leqslant dMI.$$

As $t \in W^u(p, B)$, we have $d(f^S(t), f^S(p)) \leqslant \gamma$ for $S \leqslant 0$. Thus

$$d(f^k(p), f^k(t')) = d(f^{k+dMI}(p), f^{k+dMI}(t)) \leqslant \gamma$$

for $k \leqslant 0$; hence $t' \in W^u(p, \gamma) \cap B = W^u(p, B)$—a contradiction. Hence $i = k$.

Combining Lemmas 4.7 and 4.4 we get

$$j_{m+dMI+L} \geqslant k_{m+dMI} \geqslant 2j_m.$$

As $k_{dMI} \geq 2, j_{dMI+L} \geq 2$. Inductively we now obtain $j_{n(dMI+L)} \geq 2^n$. By Lemma 4.3, $N_{n(dMI+L)} \geq 2^n$. Thus

$$\text{ent} f = \overline{\lim_m} \frac{1}{m} \log N_m \geq \overline{\lim_n} \frac{1}{n(dMI+L)} \log 2^n = \frac{\log 2}{dMI+L} > 0.$$

(4.8) REMARK. From the first part of the previous theorem it follows that if $f$ is an Anosov diffeomorphism, then $\Omega(f)$ has periodic points dense. This is well known and follows immediately from Pugh's closing lemma and structural stability for such $f$. The above theorems apply to $f|\Omega$ or $f|\Omega_i$ when $f$ is a diffeomorphism satisfying Axiom A. This essentially is the only known case to which the theorems apply. Theorem 4.1 implies that the topological and metric entropies are equal for an ergodic automorphism of a torus.

From Theorem 2.4 we obtain immediately:

(4.9) THEOREM. *Let $f$ be a diffeomorphism satisfying Smale's Axiom A. Then* $\text{ent} f = \overline{\lim}_n (1/n) \log N_n(f)$, *and* $\text{ent} f > 0$ *unless $\Omega(f)$ is finite.*

In Smale's [13] study of diffeomorphisms the zeta function $\zeta_f(z) = \exp \sum_{n=1}^{\infty} (1/n) N_n(f) z^n$ is viewed as having some importance.

(4.10) PROPOSITION. *Suppose $f \in \text{Diff}(M)$ satisfies Axiom A. Then $\text{ent} f = -\log R$, where $R$ is the common radius of convergence of $\zeta_f$ and $\zeta_f'/\zeta_f = \sum_{n=1}^{\infty} N_n z^n$. $\zeta_f$ has positive radius of convergence which is less than 1 unless $\Omega(f)$ is finite. If $\zeta_f$ is rational, then $\text{ent} f$ is the logarithm of an algebraic number.*

PROOF. By 4.9, for the first statement we need only show that the two functions do indeed have the same radius of convergence. But from elementary analysis we know that if $a_n \geq 0$, then $\sum_{n=1}^{\infty} a_n z^n$ and $\exp(\sum_{n=1}^{\infty} a_n z^n)$ have the same radius of convergence. That $R > 0$ is just $\text{ent} f < \infty$ or Proposition 2.8. As $R \leq 1$, $R < 1$ unless $R = 1$, i.e. $\text{ent} f = 0$. By 4.6 this happens only if $\Omega(f)$ is finite. The converse is 2.6. The last statement follows from the first.

(4.11) COROLLARY. *If $f$ is an Anosov diffeomorphism, then $\text{ent} f$ is the logarithm of an algebraic number.*

PROOF. For Williams [15] has shown that $\zeta_f$ is rational if $f$ is Anosov. This corollary was proved earlier by Sinai [12].

5. **Inverse limits.** Let $f: X \to X$ be a continuous surjective map on a compact Hausdorff space. Let $\Sigma_f$ be the set of sequences $(x_0, x_1, \ldots)$ of elements of $X$ such that $f(x_{n+1}) = x_n$ for all $n \geq 0$. Define the inverse limit $\sigma_f: \Sigma_f \to \Sigma_f$ of $f$ by $(\sigma_f(x))_n = f(x_n)$. Give $\Sigma_f$ the topology it has as a subset of $\prod_{Z^+} M$. As $M$ is Hausdorff, $f(x_{n+1}) \neq x_n$ specifies an open set; so $\Sigma_f$ is closed in $\prod_{Z^+} M$, hence compact.

(5.1) PROPOSITION [14]. $N_m(\sigma_f) = N_m(f)$ for all $m > 0$.

PROOF. The proof given by Williams: the map given by

$$p \to (p, f^{m-1}(p), f^{m-2}(p), \ldots, f(p), p, f^{m-1}(p), \ldots)$$

is a bijection between the fixed points of $f^m$ and those of $\sigma_f^m$.

(5.2) PROPOSITION. ent $\sigma_f = $ ent $f$.

PROOF. The analogous statement for metric entropy was proven by Rochlin in [10] (there the inverse limit is called a "natural extension"). Let $U(n, B) = \{x \in \Sigma_f : x_n \in B\}$ where $n \geq 0$ and $B$ is open in $X$. Then $\{U(n, B)\}_{n,B}$ is a base for the topology of $X$. To see this one notices that $x_i \in B_i$ for $0 \leq i \leq n$ is equivalent to $x_n \in \bigcap_{i=0}^{n} f^{-i} B_{n-i}$.

Let $\mathscr{A}$ be a finite open cover of $\Sigma_f$. Then there are $U_1, \ldots, U_k$ (some $k$), $U_i = U(n_i, B_i)$ which cover $\Sigma_f$ such that each $U_i$ is contained in some member of $\mathscr{A}$. Let $\mathscr{C} = \{U_1, \ldots, U_k\}$. Then $h(\sigma_f, \mathscr{C}) \geq h(\sigma_f, \mathscr{A})$. Furthermore, taking $n = \max n_i$, we may assume each $n_i = n$ by replacing $B_i$ with $f^{n_i - n}(B_i)$ and $n_i$ with $n$. Then $\mathscr{B} = \{B_1, \ldots, B_k\}$ covers $X$ and $\mathscr{B}_n^* = \{U_1 = U(n, B_1), \ldots, U_k\} = \mathscr{C}$. We have ent $\sigma_f = \sup_{n,f} h(\sigma_f, \mathscr{B}_n^*)$. Now $x = (x_i) \in U_{i_0} \cap f^{-1} U_{i_1} \cap \ldots \cap f^{-m} U_{i_m}$ if and only if $\sigma_f^k(x) \in U_{i_k}$ for $0 \leq k \leq m$. This in turn happens if and only if $f^k(x_n) \in B_{i_k}$ for $0 \leq k \leq m$, which happens if and only if $x_n \in B_{i_0} \cap f^{-1} B_{i_1} \cap \ldots \cap f^{-m} B_{i_m}$. From this we get a correspondence of subcovers of $\mathscr{B}_n^* \vee \sigma_f^{-1} \mathscr{B}_n^* \vee \ldots \vee \sigma_f^{-m} \mathscr{B}_n^*$ with those of $\mathscr{B} \vee f^{-1} \mathscr{B} \vee \ldots \vee f^{-m} \mathscr{B}$ and $M_m(\sigma_f, \mathscr{B}_n^*) = M_m(f, \mathscr{B})$. It follows that $h(\sigma_f, \mathscr{B}_n^*) = h(f, \mathscr{B})$. Thus

$$\text{ent } \sigma_f = \sup_{\mathscr{B}} h(f, \mathscr{B}) = \text{ent } f.$$

(5.3) THEOREM. Suppose $f: M^m \to M^m$ is a differentiable map of a compact manifold satisfying Axiom A, i.e. $\Omega(f)$ has periodic points dense and $T_{\Omega(f)} M$ has a hyperbolic splitting. Then

$$\text{ent } f = \varlimsup_n \frac{1}{n} \log N_n(f).$$

PROOF. First, ent $f = $ ent $f|\Omega(f)$ by 2.4. It is known ([17], less general versions are in [13] and [14]) that there exists a diffeomorphism $g: S^{2m+2} \to S^{2m+2}$ with a hyperbolic set $\Lambda \subseteq \Omega(g)$ having periodic points dense and separated from the rest of $\Omega(g)$ such that $g|\Lambda$ is topologically conjugate to $\sigma_{f|\Omega(f)}$. $g|\Lambda$ has hyperbolic canonical coordinates (3.1). Thus by 4.1 we have

$$\text{ent } g|\Lambda = \varlimsup_n \frac{1}{n} \log N_n(g|\Lambda).$$

As $g|\Lambda$ is conjugate to $\sigma_{f|\Omega(f)}$,

$$\text{ent } \sigma_{f|\Omega(f)} = \varlimsup_n \frac{1}{n} \log N_n(\sigma_{f|\Omega(f)}).$$

Using Propositions 5.1 and 5.2 we have

$$\text{ent } f|\Omega(f) = \varlimsup_n \frac{1}{n} \log N_n(f|\Omega(f)).$$

As period points of $f$ are in $\Omega(f)$, $N_n(f|\Omega(f)) = N_n(f)$ and so

$$\text{ent } f = \overline{\lim_n} \frac{1}{n} \log N_n(f).$$

6. **The zero-dimensional case.** We recall the definition of the Bernoulli shifts. Let $S$ be a finite set with the discrete topology, $X_S$ be the set of functions from $Z$ to $S$ provided with the product topology, and $\alpha : X_S \to X_S$ be the shift map defined by $(\alpha(x))_n = x_{n+1}$ where $x_n$ is the value of $x$ at $n$. Let $d(a, b) = 0$ or $1$ according to whether $a = b$ or not, for $a$, $b$, $S$. We define our metric on $X_S$ by

$$d(x, y) = \sum_{k \in Z} 2^{-|k|} d(x_k, y_k).$$

(6.1) PROPOSITION. $\alpha : X_S \to X_S$ has hyperbolic canonical coordinates.

Suppose $\delta > 0$. Choose $J$ large enough so that

$$\varepsilon = \sum_{|k| > J} 2^{-|k|} \leq \delta.$$

Suppose that $d(x, y) \leq \varepsilon$. Then $x_k = y_k$ for $|k| \leq J$. Define $z \in X_S$ by $z_k = x_k$ for $k \leq J$ and $z_k = y_k$ for $k \geq -J$. Then $\alpha^n(z)$ and $\alpha^n(y)$ agree in places $-J$ through $J$ for all $n \geq 0$. Hence $d(\alpha^n(z), \alpha^n(y)) \leq \delta$ and $z \in W^s(y, \delta)$. Similarly $z \in W^u(x, \delta)$. It follows that $\alpha$ has canonical coordinates.

Let $\delta^* = \frac{1}{2}$ and $\lambda = \frac{1}{2}$. Suppose $y \in W^s(x, \delta^*)$. If $d(\alpha^n(x), \alpha^n(y)) \leq \frac{1}{2}$, then $\alpha^n(x)$ and $\alpha^n(y)$ agree in the zeroth place, i.e. $x_n = y_n$. Hence $x_n = y_n$ for $n \geq 0$. From this it follows that for $m \geq 0$

$$d(\alpha^m(x), \alpha^m(y)) = \sum_{k < 0} 2^{k-m} d(x_k, y_k) = (\tfrac{1}{2})^m d(x, y).$$

We get similarly an equality for $y \in W^u(x, \delta^*)$ and $m \leq 0$. We have hyperbolicity with $c = 1$.

In [13] Smale shows that $X_S$ can be realized as an $\Omega_i$ for a suitable diffeomorphism satisfying Axiom A. The above proposition then also follows from 2.1.

If $X \subseteq X_S$ is a closed $\alpha$-invariant set, then $\alpha|X$ (or $X$) is called a *subshift*. It is of *finite type* if there is an $N$ such that $z \in X$ if for each $n$ there is a $y^n \in X$ such that $z_k = y_k^n$ for $n \leq k < n + N$, i.e. $X$ is determined by specifying the "blocks" of length $N$ its elements contain. These were studies by W. Parry [7] as "intrinsic Markov chains." In [4] Lanford and the author show that their zeta functions are rational.

(6.2) PROPOSITION. *A subshift $\alpha|X$ has canonical coordinates if and only if it is of finite type. The canonical coordinates are then hyperbolic.*

PROOF. Suppose $\alpha|X$ has finite type with $N$ above. The proof for canonical coordinates goes exactly as in 6.1 once one requires that $J$ satisfy $2J + 1 \geq N$.

Suppose that $\alpha|X$ has canonical coordinates. Choose $J$ large enough so that if $x$ and $y$ agree in places $-J$ through $J$ then $d(x, y) \leq \varepsilon(\frac{1}{2})$. Set $N = 2J + 2$. As $X$ is closed and $\alpha$-invariant, it is enough to show that any $m$-block of elements of

$S$ occurs in some element of $X$ provided that all of its $N$-subblocks occur in elements of $X$.

This is obviously true for $m \leqslant N$. We proceed by induction and suppose that $(a_1, ..., a_m, a_{m+1})$ has all $N$-subblocks occurring in elements of $X$. Applying the inductive hypothesis we obtain a point $y$ in $X$ containing the $m$-block $(a_1, ..., a_m)$ and a point $z$ in $X$ containing $(a_2, ..., a_{m+1})$. By the shift invariance of $X$ we may assume that $a_k = y_{k-m+J}$ for $1 \leqslant k \leqslant m$ and $a_k = z_{k-m+J}$ for $2 \leqslant k \leqslant m+1$. As $m - 1 \geqslant 2J + 1$, we see that $y$ and $z$ agree in places $-J$ through $J$; hence $d(y, z) \leqslant \varepsilon(\frac{1}{2})$. Let $x \in W^u(y, \frac{1}{2})$. Then $x_n = z_n$ for all $n \geqslant 0$ and $x_n = y_n$ for all $n \leqslant 0$. Hence $a_k = x_{k-m+J}$ for $1 \leqslant k \leqslant m$ and $x$ contains $(a_1, ..., a_m, a_{m+1})$. We are done by induction.

Hyperbolicity was proved in 6.1.

(6.3) COROLLARY. *Two conjugate subshifts are of finite type or not together.*

(6.4) COROLLARY. *One-sided and two-sided subshifts of finite type satisfy* entropy $= \overline{\lim}_n (1/n) \log N_n$.

PROOF. One-sided shifts are defined as were the Bernoulli shifts (here referred to a two-sided) above except that one considers functions $Z^+ \to S$. For two-sided subshifts of finite type the result follows from 4.1. The inverse limit of a one-sided subshift of finite type is a two-sided one of finite type. The result then holds for one-sided ones by 5.

We next come to a result of Gottschalk which is proved by H. Keynes and J. Robertson in [6].

(6.5) PROPOSITION. *Any expansive homeomorphism of a zero-dimensional compact metric space is topologically conjugate to some subshift.*

From 6.2 we then get the following characterization of our zero-dimensional case (recall that having canonical coordinates is preserved under topological conjugacy).

(6.6) THEOREM. *An expansive homeomorphism of a zero-dimensional compact metric space having canonical coordinates is conjugate to a subshift of finite type.*

(6.7) THEOREM. *A subshift $X$ of finite type with periodic points dense is conjugate to $\Omega(f)$ for some embedding $f: U \to R^2$ with $U$ open in $R^2$ and $f$ satisfying Axiom A.*

PROOF. Suppose $X \subseteq X_S$ is determined by its $N$-blocks. Let $T$ be the set of all $N$-blocks of elements of $X$. Then the map $g: X \to X_T$ given by $(g(x))_n = (x_n, x_{n+1}, ..., x_{n+N-1})$ is a conjugacy of $X$ with a subshift determined by 2-blocks. Thus we may assume $X$ is determined by its 2-blocks.

In [13] Smale shows how to realize the full $X_S$'s through "horseshoe" examples and gives an example of a realization of a proper subshift. We review his "horseshoe" for $\overline{\overline{S}} = K$. His $f$ maps an open square $D$ into a neighborhood of itself such that $f(P_i) = Q_i$ where the $P_i$ and $Q_i$ are certain rectangles:

Smale shows that here $\Omega$ is compact, contained in the interior of $D$, hyperbolic and isomorphic to $X_K$ via the map $h : \Omega \to X_K$ given by $(h(x))_k = j$ where $f^k(x) \in P_j$.

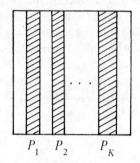 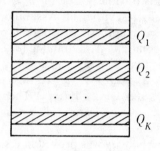

The subshift $X \subseteq X_K$ consists of all elements of $X_K$ such that $(x_i, x_{i+1}) \notin G$ for all $i \in Z$ where $G$ is the subset of $[1, K] \times [1, K]$ of inadmissable 2-blocks. Let $U$ be the square $D$ minus small closed squares about $Q_i \cap P_j$ for $(i, j)$ in $G$ (small means "intersects no other $Q_k \cap P_n$"). Then $C = \Omega \cap \bigcap_{n \in Z} f^n(U)$ satisfies $h(C) = X$.

Let $\Lambda$ be the nonwandering set of $f|U$. Suppose $x \in U\backslash C$. If $x \notin \Omega$, then $x$ wanders under $f$— so it does under $f|U$ also. If $x \in \Omega\backslash C$, then $f^n(x) \in Q_i \cap P_j$ for some $n$, $(i, j) \in G$. As $Q_i \cap P_j$ is bounded away from $U$, taking $n$ minimal, it follows that $x$ is wandering for $f|U$. Thus $\Lambda \subseteq C$. Periodic points of $C$ belong to $\Lambda$; as $C = h^{-1}(X)$ and periodic points are dense in $X$,

$$C \subseteq \text{closure of periodic points of } f|C \subseteq \overline{\Lambda} = \Lambda.$$

Thus $C = \Lambda$. $f|U$ is the desired embedding.

7. **A one-dimensional theorem.** R. L. Adler and M. H. McAndrew show in [3] that the $k$th Tchebychev polynomial considered as a map $T_k$ of $[-1, +1]$ into itself has topological entropy $\log k$. One can check that $\log k = \overline{\lim}_n (1/n) \log N_n(T_k)$. We give now a generalization:

(7.1) THEOREM. *Let $K$ be a finite 1-complex and $f : |K| \to |K|$ a continuous map taking vertices into vertices. Suppose there is a metric $d$ on $K$ and a constant $\tau > 1$ such that $d(f(x), f(y)) \geqslant \tau d(x, y)$ whenever $x$ and $y$ lie in the same 1-simplex. Then* ent $f = \overline{\lim}_n (1/n) \log N_n(f) < \infty$.

PROOF. Obviously $f$ is one-to-one on simplices. From this and the fact that certices are mapped onto vertices, it follows that $f$ maps a 1-simplex onto a union of 1-simplices. We obtain a subdivision $K_1$ of $K$ by taking as vertices those of $K$ and also those points mapped under $f$ onto vertices of $K$; this triangulation also satisfies the conditions of the theorem. If diam $K$ denotes the maximum diameter of a simplex of $K$, we have diam $K_1 \leqslant (1/\tau)$ diam $K$. Iterating this construction we obtain triangulations $K_m$ satisfying the hypotheses of the theorem with diam $K_m \leqslant (1/\tau^m)$ diam $K$. Let $\mathscr{A}$ be an open cover of $K$ consisting of small neighborhoods of the 1-simplices together with open sets which are the isolated vertices.

We will show that $h(f, \mathscr{A}) \leqslant \overline{\lim}_n (1/n) \log N_n(f)$. As $K_m$ fulfills the hypotheses of the theorem, letting $\mathscr{A}_m$ be a covering derived from $K_m$, we will then have $h(f, \mathscr{A}_m) \leqslant \overline{\lim}_n (1/n) \log N_n(f)$. Letting $m \to \infty$, $\operatorname{diam}(\mathscr{A}_m) \to 0$ and we get $\operatorname{ent} f \leqslant \overline{\lim}_n (1/n) \log N_n(f)$.

We proceed with the proof of $h(f, \mathscr{A}) \leqslant \overline{\lim}_n (1/n) \log N_n(f)$. Let $I_1, ..., I_q$ be the open 1-simplices of $K$; $\mathscr{A} = \{A_1, ..., A_q, P_1, ..., P_w\}$ where the $P_i$ are the isolated points and $A_j \supseteq \bar{I}_j$ is open. Let $B(I_{i_0}, ..., I_{i_n}) = I_{i_0} \cap f^{-1} I_{i_1} \cap ... \cap f^{-n} I_{i_n} = \{x : f^k(x) \in I_{i_k} \text{ for } 0 \leqslant k \leqslant n\}$ and $B_n = \{(I_{i_0}, ..., I_{i_n}) : B(I_{i_0}, ..., I_{i_n}) \neq \varnothing\}$. The $B(I_{i_0}, ..., I_{i_n})$'s cover all of $K$ except the finite number of points $x$ with $f^r(x)$ a vertex of $K$ for some $r \leqslant n$. Hence, unless $x$ is an isolated vertex, $x$ is in some $\overline{B(I_{i_0}, ..., I_{i_n})}$; then

$$x \in \overline{B(I_{i_0}, ..., I_{i_n})} \subseteq \bigcap_{k=0}^{n} f^{-k} \overline{I}_{i_k} \subseteq \bigcap_{k=0}^{n} f^{-k} A_{i_k}.$$

We thus get a subcover of $\mathscr{A} \vee f^{-1}\mathscr{A} \vee ... \vee f^{-n}\mathscr{A}$ by taking all $\bigcap_{k=0}^{n} f^{-k} A_{i_k}$ with $(I_{i_0}, ..., I_{i_n}) \in B_n$ together with an open set about each of the $w$ isolated points. Hence $M_n(f, \mathscr{A}) \leqslant \overline{\overline{B}}_n + w$ and $h(f, \mathscr{A}) \leqslant \overline{\lim}_n (1/n) \log \overline{\overline{B}}_n$.

Let $D_{r+1} = \{(I_{i_0}, ..., I_{i_r}) \in B_r : I_{i_0} = I_{i_r}\}$. Set $K = q$ and apply the construction of our fundamental Lemma 2.1. Then $B_n \subseteq F_{n+1}$; for any string of $I_j$'s can be decomposed into at most $q$ substrings, each having equal initial and final elements (see the proof of Lemma 4.2), and any $r$-substring of a string in $B_n$ is in $B_{r-1}$. By Lemma 2.1, to show $\operatorname{ent} f \leqslant \overline{\lim}_n (1/n) \log N_n(f)$, we are left only to show show $\overline{\lim}_r (1/r) \log D_r \leqslant \overline{\lim}_n (1/n) \log N_n(f)$.

(7.2) LEMMA. *If $E$ is an open connected subset of some 1-simplex of $K$, then for any $m$ $C = f^{-1}E \cap I_m$ is also. If $C \neq \varnothing$, then $f(C) = E$.*

PROOF. Say $E \subseteq I_j$. Then $f(I_m) \supseteq I_j \supseteq E$ or $f(I_m) \cap E \subseteq f(I_m) \cap I_j = \varnothing$. There is nothing to show in the second case; in the first case the result follows from the fact that $f|I_m : I_m \to f(I_m)$ is a homeomorphism.

Now suppose $(I_{i_0}, ..., I_{i_r}) \in D_{r+1}$. For $0 \leqslant j \leqslant r$ set $E_j = I_{i_{r-j}} \cap f^{-1} I_{i_{r-j+1}} \cap ... \cap f^{-j} I_{i_r}$. As $(I_{i_0}, ..., I_{i_r}) \in B_r$ we see that $E_j \neq \varnothing$ for each $0 \leqslant j \leqslant r$. As $E_{j+1} = I_{i_{r-j-1}} \cap f^{-1} E_j$, Lemma 7.1 gives us inductively that $E_j$ is an open connected subset of a 1-simplex and $f(E_{j+1}) = E_j$. In particular, $B(I_{i_0}, ..., I_{i_r}) = E_r$ is a segment in $I_{i_0}$ and

$$f^r(E_r) = f^{r-1}(E_{r-1}) = ... = E_0 = I_{i_r} = I_{i_0}.$$

It follows that $\overline{B(I_{i_0}, ..., I_{i_r})}$ contains a point $p$ of period $r$. If $p \notin B(I_{i_0}, ..., I_{i_r})$, then $f^k(p)$ is a vertex of $K$ for some $k$; hence $p$ is a vertex of $K$ as $p$ is periodic and $f$ takes vertices into vertices. If $c$ is the maximum valence of a vertex of $K$ and $K$ has $w$ vertices, then at most $cw$ $B(I_{i_0}, ..., I_{i_r})$'s can have a vertex on their boundaries. Thus at least $D_{r+1} - cw$ $B(I_{i_0}, ..., I_{i_r})$'s contain a point of period $r$. As these sets are disjoint, $D_{r+1} \leqslant N_r(f) + cw$ and $\overline{\lim}_r (1/r) \log D_r \leqslant \overline{\lim}_r (1/r) \log N_r(f)$, which is what we were trying to show.

We now turn toward the inequality in the other direction: $\operatorname{ent} f \leqslant \overline{\lim}_n (1/n) \log N_n(f)$. Fix a positive integer $N$. For $v$ a vertex of $K$, let $A_v$ be $v$ together with the open 1-simplices of $K_N$ adjoining $v$. For any $j$, $A_v \cap I_j$ is either empty of an open 1-simplex of $\supseteq_N$. We claim that $f^k(A_v \cap I_j)$ is contaiened in an open simplex of $K$ for any $0 \leqslant k \leqslant N$. For $f^k(A_v \cap I_j)$ is connected as $A_v \cap I_j$ is; if it intersected two open simplices of $K$, it would contain a vertex of $K$ and hence $A_v \cap I_j$ would contain a vertex of $K_N$—a contradiction.

Let $\mathscr{C}_N = \{I_1, \ldots, I_q, A_v : v \text{ is a vertex of } K\}$. Consider $B = C_0 \cap f^{-1}C_1 \ldots \cap f^{-n}C_n$ where $C_i \in \mathscr{C}_N$. Let $P_n(B)$ be the set of points of period $n$ in $B$ which are not vertices of $K$. If $x \in P_n(B)$, $f^k(x)$ lies in some $I_j$ for any $k \geqslant 0$. Define

$$g : P_n(B) \times [0, n] \to [1, q] \quad \text{by } f^k(x) \in I_{g(x,k)}.$$

(7.3) LEMMA. *If* $g(x, k) \neq g(y, k)$ *and* $g(z, k) = g(u, k)$, *then* $g(z, k + i) = g(u, k + i)$ *for* $0 \leqslant i \leqslant N$.

PROOF. If $C_k$ is some $I_m$, then $g(s, k) = m$ for all $s \in P_n(B)$, contrary to hypothesis. Thus $C_k$ is some $A_v$. Let $j = g(z, k) = g(u, k)$. Then $f^k(z), f^k(u) \in A_v \cap I_j$. We proved above that for $0 \leqslant i \leqslant N$, $f^i(A_v \cap I_j) \in I_{d_i}$ for some $d_i$. Then $g(z, k + i) = g(u, k + i) = d_i$.

(7.4) LEMMA. *If* $x \neq y$ *are two elements of* $P_n(B)$, *then* $g(x, k) \neq g(y, k)$ *for some* $0 \leqslant k \leqslant n$.

PROOF. Otherwise, as $x, y$ have period $n$, $f^k(x)$ and $f^k(y)$ would lie in the same 1-simplex for all $k$. Hence for all positive $k$ we would have, inductively,

$$d(f^k(x), f^k(y)) \geqslant \tau^k d(x, y).$$

As $\tau > 1$, $d(f^k(x), f^k(y)) \to \infty$ as $k \to \mathscr{C}$—contradicting $K$ bounded as compact.

(7.5) LEMMA. $P_n(B)$ *has at most* $q^{(n/N)+1}$ *elements.*

PROOF. We show that this follows from 7.3 and 7.4. Let $G_j$ be the partition induced on $P_n(B)$ by $g(x, k)$ for $0 \leqslant k \leqslant j$, i.e. under the equivalence relation

$$x \underset{j}{\sim} y \quad \text{if and only if} \quad g(x, k) = g(y, k) \quad \text{for } 0 \leqslant k \leqslant j.$$

Now $G_j$ either refines $G_{j-1}$ or $G_j = G_{j-1}$. Lemma 7.3 shows that if $G_j$ refines $G_{j-1}$, then $G_{j+i} = G_j$ for $0 \leqslant i \leqslant N$. Thus, $G_j$ can refine $G_{j-1}$ at most $(n/N) + 1$ times as $j$ goes from 0 to $n$ $(G_{-1} = \{P_n(B)\})$. Also, we clearly have $\overline{\overline{G}}_j \leqslant q \overline{\overline{G}}_{j-1}$. Hence $\overline{\overline{G}}_n \leqslant q^{(n/N)+1}$. By Lemma 7.4 we have $G_n = \{\{x\} : x \in P_n(B)\}$.

We have shown that an arbitrary member of $\mathscr{C}_N \vee f^{-1}\mathscr{C}_N \vee \ldots \vee f^{-n}\mathscr{C}_N$ can contain at most $q^{(n/N)+1}$ points of period $n$ which are not vertices of $K$. Hence there are at most $M_n(f, \mathscr{C}_N)q^{(n/N)+1}$ such points and

$$N_n(f) \leqslant w + M_n(f, \mathscr{C}_N)q^{(n/N)+1}.$$

From this it follows that

$$\overline{\lim}_n \frac{1}{n} \log N_n(f) \leqslant h(f, \mathscr{C}_N) + \frac{1}{N} \log q.$$

The right-hand side is finite as $h(f, \mathscr{A}) \leqslant \bar{\bar{\mathscr{A}}}$ for any open cover $\mathscr{A}$. Let $N \to \infty$; we obtain

$$\overline{\lim_{n}} \frac{1}{n} \log N_n(f) \leqslant \text{ent} f.$$

## REFERENCES

**1.** R. Abraham and S. Smale, *Nongenericity of Ω-stability*, these Proceedings, vol. 14.

**2.** R. L. Adler, A. G. Konheim and M. H. McAndrew, *Topological entropy*, Trans. Amer. Math. Soc. **114** (1965), 309–319.

**3.** R. L. Adler and M. H. McAndrew, *The entropy of Chebychev polynomials*, Trans. Amer. Math. Soc. **121** (1966), 236–241.

**4.** R. Bowen and O. Lanford, *Zeta functions of restrictions of the shift transformation*, these Proceedings, vol. 14.

**5.** M. W. Hirsch and C. C. Pugh, *Stable manifolds and hyperbolic sets*, these Proceedings, vol. 14.

**6.** H. B. Keynes and J. B. Robertson, *Generators for topological entropy and expansiveness* (to appear).

**7.** W. Parry, *Intrinsic Markov chains*, Trans. Amer. Math. Soc. **112** (1964), 55–66.

**8.** C. C. Pugh, *An improved closing lemma and a general density theorem*, Amer. J. Math. **89** (1967), 1010–1021.

**9.** V. A. Rokhlin, *New progress in the theory of transformations with invariant measure*, Russian Math. Surveys **15** (1960), 1–22.

**10.** ———, *Exact endomorphisms of a Lebesgue space*, Amer. Math. Soc. Transl. (2), **39** (1969), 1–36.

**11.** M. Shub, Thesis, University of California, Berkeley, 1967.

**12.** Ja. G. Sinai, *Markov partitions and U-diffeomorphisms*, Functional Anal. Appl. **2** (1968), 64–89.

**13.** S. Smale, *Differentiable dynamical systems*, Bull. Amer. Math. Soc. **73** (1967), 747–817.

**14.** R. Williams, *One-dimensional non-wandering sets*, Topology **6** (1967), 473–487.

**15.** ———, *The zeta function of an attractor*, Proceedings of the Michigan State Topology Conference (Spring, 1967).

**16.** H. Furstenburg, *Disjointness in ergodic theory*, Math. Systems Theory **1** (1967), 1–50.

**17.** J. Guckenheimer, *Endomorphisms of the Riemann sphere*, these Proceedings, vol. 14.

**18.** J. P. Conze, *Points périodiques et entropie topologique*, C. R. Acad. Sci. Paris Sér. A–B **267** (1968), 149–152.

**19.** S. Smale, *The Ω-stability theorem*, these Proceedings, vol. 14.

UNIVERSITY OF CALIFORNIA, BERKELEY

# ZETA FUNCTIONS OF RESTRICTIONS OF THE SHIFT TRANSFORMATION[1]

R. BOWEN AND O. E. LANFORD III

1. **Introduction.** Let $S$ be a finite set with $N \geq 2$ elements, and consider $X = S^Z$. We may regard $X$ as the set of all two-sided sequences $(..., a_{-1}, a_0, a_1, a_2, ...)$ of elements of $S$. Give $S$ the discrete topology and $X$ the product topology; this makes $X$ into a compact space. Let $\alpha$ denote the shift transformation on $X$, i.e., the mapping defined by $(\alpha a)_n = a_{n+1}$. Evidently, $\alpha$ is a homeomorphism of $X$ onto itself. An element $a$ at $X$ is a fixed point of $\alpha^m$ if and only if the sequence $a$ is periodic with period $m$. The number of fixed points of $\alpha^m$ is therefore $N^m$, the number of finite sequences of length $m$.

Now let $Y$ be a closed subset of $X$ which is mapped *onto* itself by $\alpha$. We will denote the number of fixed points of $\alpha^m$ in $Y$ by $N_m(\alpha|Y)$. The *zeta function* of $\alpha|Y$ is by definition

$$\zeta_{\alpha|Y}(z) = \exp\left(\sum_m \frac{N_m(\alpha|Y)}{m} z^m\right).$$

Since $N_m(\alpha|Y) \leq N^m$, $\zeta_{\alpha|Y}$ is holomorphic in the disc $|z| < 1/N$. We will investigate under what circumstances the function $\zeta_{\alpha|Y}$ is rational.

Let $\sigma = (\sigma_0, \sigma_1, ..., \sigma_n)$ be a finite sequence in $S$. We define

$$Y_\sigma = \{a \in X : \text{There is no } i \text{ such that } a_i = \sigma_0, a_{i+1} = \sigma_1, ..., a_{i+n} = \sigma_n\}.$$

In other words, $Y_\sigma$ is the set of all two-sided sequences which nowhere contain a segment of $n + 1$ elements of the form $(\sigma_0, \sigma_1, ..., \sigma_n)$. Any $Y_\sigma$ is a closed subset of $X$ mapped onto itself by $\alpha$, and the same is true of any intersection, finite or infinite, at $Y_\sigma$'s. (It is an amusing fact, which we will not have occasion to use, that every closed $\alpha$-invariant subset of $X$ is the intersection of the $Y_\sigma$'s which contain it.) We say that a closed $\alpha$-invariant subset of $X$ is of *finite type*[2] if it is the intersection of finitely many $Y_\sigma$'s. Our main result, to be proved in §2, is that the zeta function of the restriction of $\alpha$ to any closed invariant subset of finite type is rational.

Not every closed invariant subset of $X$ has a rational zeta function however. In §3, we show that there are only countably many distinct rational zeta functions, but that there is an uncountable family of closed $\alpha$-invariant subsets of $X$ with pairwise distinct zeta functions. Finally, in §4, we give an example of a closed invariant subset of $X$ whose zeta function may be explicitly shown to be non-rational.

[1] Supported in part by NSF contract GP-7176.
[2] See W. Parry, *Intrinsic Markov chains*, Trans. Amer. Math. Soc. **112** (1964), 55–66.

**2. The zeta function of an invariant subset of finite type is rational.** Let $\{\sigma_1, ..., \sigma_m\}$ be a finite collection at finite sequences in $S$. For the purposes of this section, we will say that a sequence, finite or infinite, is *acceptable* if it contains none of the $\sigma_i$'s as a segment. To prove the rationality of the zeta function of $\alpha$ restricted to $Y = Y_{\sigma_1}, ..., Y_{\sigma_m}$, we need a technique for computing the number of periodic acceptable sequences with various periods. We proceed in the following way: To describe a periodic sequence $a$ of period $n$, it is evidently enough to specify $(a_1, ..., a_n)$. However, to determine whether the corresponding infinite periodic sequence is acceptable, it is not enough to look for forbidden segments in $(a_1, ..., a_n)$; one must look at some more terms in the sequence. Let $n_0 + 1$ be the length of the longest sequence in $\{\sigma_1, ..., \sigma_m\}$. It is easy to see that, if $a$ is an infinite sequence which is periodic with period $n$, then $a$ is acceptable if and only if $(a_1, ..., a_{n+n_0})$ is. The sequence $(a_1, ..., a_{n+n_0})$ must be restricted appropriately if it is to come from a periodic sequence of period $n$; for this it is necessary and sufficient that $(a_1, ..., a_{n_0}) = (a_{n+1}, ..., a_{n+n_0})$, i.e., that the initial and terminal segments of length $n_0$ are the same. Thus, we have the following lemma:

LEMMA 1. *The periodic sequences in $Y$ of period $n$ are in a one-one correspondence with the acceptable sequences $(a_1, ..., a_{n_0+n})$ such that $a_1 = a_{n+1}, ..., a_{n_0} = a_{n+n_0}$.*

What we have to do, then, is to compute the number of such finite sequences. To do this, it is actually convenient to compute something more. Let $\tau, \tau'$ be any two acceptable sequences of length $n_0$. We will let $N(n, \tau, \tau')$ denote the number of acceptable sequences of length $n_0 + n$ with $\tau$ as their initial sequence and $\tau'$ as their terminal sequence. It is easy to write a recurrence relation for $N(n, \tau, \tau')$; the idea is simply that an acceptable sequence of length $n_0 + n + 1$ is obtained by adjoining a new first element to an acceptable sequence of length $n + n_0$, being careful not to produce in this way one of the forbidden sequences $\{\sigma_1, ..., \sigma_m\}$ as an initial segment. This adjunction does not change the terminal segment, and whether or not a given element of $S$ can be adjoined to a given acceptable sequence depends only on the initial segment of length $n_0$ of that sequence.

With these remarks in mind, we define a square matrix $T_{\tau\tau'}$ whose rows and columns are labelled by the acceptable sequences of length $n_0$:

$T_{\tau\tau'} = 1$ if the initial segment of length $n_0 - 1$ in $\tau'$ is the same as the terminal segment of length $n_0 - 1$ of $\tau$, and if the sequence obtained by adjoining the first element of $\tau$ to $\tau'$ is acceptable.

$= 0$ otherwise.

Examination of this definition shows that we could alternatively have defined $T_{\tau\tau'} = N(1, \tau, \tau')$.

EXAMPLE. $S = \{0, 1\}$, $m = 1$, $\sigma_1 = (0, 0)$, $n_0 = 1$, $T_{(0),(0)} = 0$, $T_{(0),(1)} = T_{(1),(0)} = T_{(1),(1)} = 1$.

LEMMA 2. *For $n \geq 1$,*

$$N(n, \tau, \tau') = (T^n)_{\tau\tau'},$$

*where $T^n$ means the matrix $T$ raised to the nth power.*

PROOF. We argue by induction on $n$. We have already remarked that the lemma is true for $n = 1$. Suppose it is true for $n$; we will prove it for $n + 1$. Every acceptable sequence of length $n_0 + n + 1$ is obtained by adjoining a first element to an acceptable sequence of length $n_0 + n$. To get an acceptable sequence with initial segment $\tau$ by this process it is necessary and sufficient to start with an acceptable sequence with an initial segment $\tau''$ such that $T_{\tau, \tau''} = 1$ (and, of course, to adjoin the right first element). Hence,

$$N(n + 1, \tau, \tau') = \sum_{\tau''} T_{\tau, \tau''} N(n, \tau'', \tau')$$

$$= \sum_{\tau''} T_{\tau, \tau''} (T^n)_{\tau'', \tau'} \text{ by the induction hypothesis}$$

$$= (T^{n+1})_{\tau, \tau'},$$

so the induction step is proved.

THEOREM 1. *Let* $\{\alpha_1, \ldots, \alpha_m\}$ *be a finite collection of finite sequences; let* $Y = Y_{\sigma_1} \cap \ldots \cap Y_{\sigma_m}$. *Then* $\zeta_{\alpha|Y}$ *is a rational function; in fact, if* $T$ *is the matrix introduced above and if* $\lambda_1, \ldots, \lambda_J$ *are the nonzero eigenvalues of* $T$, *then*

$$\zeta_{\alpha|Y}(z) = \prod_{j=1}^{J} \frac{1}{(1 - \lambda_j z)}.$$

PROOF. By Lemma 1, the number of periodic sequences in $Y$ of period $n$ is just the number of acceptable sequences of length $n_0 + n$ with the same initial and terminal segments of length $n_0$. By the definition of $N(n, \tau, \tau')$ and Lemma 2, we have therefore

$$N_n(\alpha|Y) = \sum_{\tau} N(n, \tau, \tau) = \sum_{\tau} (T^n)_{\tau\tau} = \mathrm{Tr}(T^n) = \sum_{j=1}^{J} \lambda_j^n.$$

The theorem now follows by an elementary calculation from the definition of the zeta function.

In the example given above, where $T$ is the $2 \times 2$ matrix

$$\begin{pmatrix} 0 & 1 \\ 1 & 1 \end{pmatrix},$$

the eigenvalues of $T$ are $\frac{1}{2} \pm \frac{1}{2}\sqrt{5}$, so

$$\zeta_{\alpha|Y}(z) = \frac{1}{1 - z + z^2}.$$

REMARK. Our procedure for calculating $\zeta_{\alpha|Y}$ is frequently very inefficient. For example, if we consider $Y_\sigma$ for a single $\sigma$ of length $n_0 + 1$, $T$ is a $N^{n_0} \times N^{n_0}$ matrix, but it can easily be seen that $T^{n_0}$ has at most $n_0 + 1$ distinct columns and that therefore $T$ has at most $n_0 + 1$ nonzero eigenvalues (counting multiplicity). It would be interesting to have a computation procedure which removes such redundancy.

## 3.  Many invariant subsets have nonrational zeta-functions.

THEOREM 2. (a) *The set of all rational functions given in a neighborhood of zero by a convergent expansion of the form* $\exp\left(\sum_m (N_m/m)z^m\right)$, *where the $N_m$ are integers, is countable.*

(b) *There exists a noncountable collection $\{Y^\gamma\}$ of closed, shift-invariant subsets of X such that $\zeta_{\alpha|Y^\gamma} \neq \zeta_{\alpha|Y^{\gamma'}}$ if $\gamma \neq \gamma'$.*

PROOF. (a) follows immediately from the fact that any rational function of the sort described has a representation as a finite product of factors of the form $(1 - \lambda_i z)^{\pm 1}$, where the $\lambda_i$ are algebraic integers. (For the fact that the $\lambda_i$ must be algebraic integers, see Appendix.)

To prove (b), consider sequences $\gamma = (\gamma_1, \gamma_2, \gamma_3, \dots)$ of integers such that $1 < \gamma_1 < \gamma_2 < \dots$. The set of all such sequences is noncountable. Let $s \in S$, and for any $\gamma$ let $Y^\gamma$ be the subset of $X$, consisting of all sequences having no block of precisely $\gamma_1$ successive $s$'s, no block of precisely $\gamma_2$ successive $s$'s, etc. Then $Y^\gamma$ is a closed subset of $X$ which is mapped onto itself by $\alpha$.

We claim that, if $\gamma \neq \gamma'$, then $\zeta_{\alpha|Y^\gamma} \neq \zeta_{\alpha|Y^{\gamma'}}$. Thus, suppose $\gamma \neq \gamma'$ and choose the first $n$ such that $\gamma_n \neq \gamma'_n$. We can suppose $\gamma_n > \gamma'_n$. We will show that the number of periodic sequences in $Y^\gamma$ of period $\gamma'_n + 1$ is strictly larger than the number of such sequences in $Y^{\gamma'}$. Since $\gamma_1 = \gamma'_1, \dots, \gamma_{n-1} = \gamma'_{n-1}$ every periodic sequence in $Y^{\gamma'}$ of period $\gamma'_n + 1$ is also in $Y^\gamma$. However, if $t \in S$ is different from $s$, then any periodic sequence whose minimal period contains $\gamma'_n$ $s$'s and one $t$ is in $Y^\gamma$ but not in $Y^{\gamma'}$. Thus, $\zeta_{\alpha|Y^{\gamma'}} \neq \zeta_{\alpha|Y^{\gamma'}}$, and the proof of the theorem is complete.

## 4. An example.

Theorem 2 shows that many closed shift-invariant subsets of $X$ have irrational zeta functions, but it does not exhibit a single subset for which the zeta function can be shown to be irrational. We remedy this situation in this section by giving an example in which the zeta function may be computed fairly explicitly and shown to be irrational. The example is the following: Let $S = \{0, 1\}$ and let $Y$ be the subset of $X$ consisting of all periodic sequences (arbitrary period) with at most one "1" in a minimal period, together with all sequences with only one "1". It is clear that $Y$ is shift-invariant; we will prove that it is closed. Thus, let $a$ be in the closure of $Y$. If $a$ contains at most one "1", then it is in $Y$. Otherwise, choose $m > n$ such that $a_n = 1, a_{n+1} = \dots = a_{m-1} = 0, a_m = 1$, and consider the neighborhood $\mathcal{U}$ of $a$ consisting of all sequences $b$ with $b_n = 1, b_{n+1} = \dots = b_{m-1} = 0, b_m = 1$. Then $\mathcal{U}$ contains only one point of $Y$; since $a$ is in the closure of $Y$, it follows that $a$ is actually in $Y$.

It is easy to see that the number of periodic sequences of period $m$ in $Y$ is the sum of the divisors of $m$, plus one. Let $\sigma(m)$ denote the sum of the divisors of $m$; then

$$\zeta_{\alpha|Y}(z) = \exp\left(\sum_m \left(\frac{1 + \sigma(m)}{m}\right) z^m\right)$$

$$= \frac{1}{1 - z} \exp\left(\sum_m \frac{\sigma(m)}{m} z^m\right).$$

It is known [1] that

$$\sum_{m=1}^{\infty} \frac{\sigma(m)}{m} z^m = -\log s(z),$$

where $s(z) = 1 - z - z^2 + z^5 + z^7 - \dots$ (the exponents which appear are those of the form $\frac{1}{2}(3k^2 \pm k)$). Thus we have $s(z)$ exhibited as a power series with arbitrarily long sequences of coefficients equal to zero; such a function cannot be rational unless it is a polynomial. Since

$$\zeta_{\alpha|Y}(z) = \frac{1}{(1-z)s(z)},$$

$\zeta_{\alpha|Y}$ also cannot be rational.

*Appendix.* In the proof of Theorem 2 we used the fact that poles and zeroes of a rational zeta function can occur only at reciprocals of algebraic integers. This fact is well known, but we have been unable to find a satisfactory reference. For the convenience of the reader, we will supply a proof. We start by making a simple reduction. If $\exp\left(\sum_m (N_m/m)z^m\right)$ is rational, so is its logarithmic derivative $\sum_m N_m z^{m-1}$, and the logarithmic derivative has a pole wherever the original function has a pole or a zero. Hence, the statement about zeta functions follows from:

PROPOSITION 1. *Let $f$ be a rational function regular at zero and let the power series expansion of $f$ at zero have the form*

$$f(z) = \sum_m N_m z^m,$$

*where the $N_m$ are integers. Then the poles of $f$ occur at reciprocals of algebraic integers.*

This proposition follows at once from two lemmas:

LEMMA 3. *Let $f$ be a rational function regular at zero; suppose that all derivatives of $f$ at zero are rational numbers. Then $f$ may be written in the form $P/Q$, where $P$ and $Q$ are polynomials with integral coefficients.*

Note that this lemma implies that the poles and zeroes of a rational zeta function must occur at algebraic numbers and that this fact would have been sufficient for the purposes of Theorem 2.

LEMMA 4. *Let $f$ be a rational function which can be written as the quotient of two polynomials with integral coefficients. Suppose that $f$ is regular at zero and that the power series expansion of $f$ at zero has integral coefficients. Then the poles of $f$ occur at reciprocals of algebraic integers.*

PROOF OF LEMMA 3. It evidently suffices to show that $f$ can be written as $P/Q$, where $P$ and $Q$ have rational coefficients. Since $f$ is a rational function regular at zero we can certainly write

$$f(z) = \frac{P(z)}{Q(z)} = \frac{P_0 + P_1 z + \dots + P_n z^n}{1 + Q_1 z + \dots + Q_m z^m}$$

where $P$ and $Q$ have no common factor. We also have $f(z) = \sum_n f_n z^n$, where the $f_n$ are rational numbers, so if we can prove that $Q_1, ..., Q_m$ are rational it will follow that $P_0, ..., P_n$ are rational.

Since $Q \cdot f = P$, we have

$$\sum_{j=1}^{m} Q_j f_{k-j} = -f_k, \quad k = n + 1, n + 2, ....$$

If we consider these equations for $k = n + 1, ..., n + m$, we obtain a system of $m$ linear equations with rational coefficients satisfied by $Q_1, ..., Q_m$. If these equations have a unique solution, then $Q_1, ..., Q_m$ are rational, and we are through. Hence, let $Q'_1, ..., Q'_m$ be such that

$$\sum_{j=1}^{m} Q'_j f_{k-j} = -f_k, \quad k = n + 1, ..., n + m,$$

and let $Q'(z) = 1 + Q'_1 z + ... + Q'_m z^m$. Then

$$Q'P/Q = Q'f = P' + f',$$

where $P'$ is a polynomial of degree at most $n$ and $f'$ has a zero of order at least $n + m + 1$ at zero. Hence, $Qf' = Q'P - QP'$. The left-hand side of this equation has a zero of order at least $n + m + 1$; the right-hand side is a polynomial of degree at most $n + m$, so both sides must be identically zero and we get $f' = 0$. From this and the fact that $Q$ and $P$ have no common factor it follows that $Q' = Q$ and the lemma is proved.

PROOF OF LEMMA 4. This lemma is due to Fatou [2]. we reproduce his proof. We can write $f = P/Q$, where $P$ and $Q$ are polynomials with integral coefficients and no common factor. Since $P$, $Q$ have no common factor we can find other polynomials $A$, $B$ with integral coefficients such that $AP + BQ = N$, $N$ an integer. Then the power series expansion at zero of $N/Q = Af + B$ has integral coefficients. Let us suppose, without changing our notation, that we have made all possible cancellations, i.e., that we have

$$\frac{N}{Q_0 + Q_1 z + ... + Q_m z^m} = C_0 + C_1 z + C_2 z^2 + ...,$$

where $N, Q_0, ..., Q_m, C_0, ...$ are all integers but where

$$1 = \text{G.C.D.} (N, Q_0, ..., Q_m) = \text{G.C.D.} (N, C_0, C_1, ...).$$

It will suffice to show that $Q_0 = 1$. Suppose that this is not the case, i.e., suppose that some prime number $p$ divides $Q_0$. Then, since $N = C_0 Q_0$, $p$ also divides $N$. Hence, if we cross-multiply and reduce mod $p$ we get

$$0 \equiv (Q_0 + ... + Q_m z^m)(C_0 + C_1 z + ...) \quad \mod (p)$$

so either all the $Q_1$'s are divisible by $p$ or else all the $C_i$'s are. In either case we have a contradiction to the assumption that all possible cancellations have been made, so the proof is complete.

*Acknowledgements.* We are grateful to Professor S. Smale for suggesting the problem and for helpful discussions. We also benefited from some discussions with M. Schlessinger concerning the appendix.

### REFERENCES

**1.** G. Pólya and G. Szegö, *Aufgaben und Lehrsätze aus der Analysis,* 3rd ed., Springer, Berlin, 1964, vol. 2, Section VIII, Exercise 75, p. 130.

**2.** P. Fatou, *Séries trigonométriques et séries de Taylor*, Acta Math. **30** (1906), 335–400.

UNIVERSITY OF CALIFORNIA, BERKELEY

# ON THE GENERIC NATURE OF PROPERTY H1 FOR HAMILTONIAN VECTOR FIELDS

MICHAEL A. BUCHNER

In this paper we prove that "almost every" Hamiltonian vectorfield has property H1. (This is stated as a conjecture in Abraham and Marsden [1, §29].)

DEFINITION. Let $M$ be a $c^r$ symplectic manifold ($r \geq 2$) which is finite-dimensional second countable and Hausdorff. Let $H \in \mathscr{F}^r(M)$ be a $c^r$ real-valued function on $M$ and $X_H$ the symplectic gradient of $H$. Then a critical point $m$ of $X_H$ is called $H$-elementary if the following three conditions hold:

(i) Zero is not a characteristic exponent.

(ii) Any pure imaginary characteristic exponent has multiplicity one.

(iii) If $\lambda$ and $\mu$ are characteristic exponents and are pure imaginary with imaginary parts positive, then $\lambda$ and $\mu$ are independent over the integers: i.e., if $n_1\lambda + n_2\mu = 0$ for $n_1, n_2 \in Z$ then $n_1 = n_2 = 0$. A Hamiltonian or its symplectic gradient $X_H$ has *property* H1 if every critical point of $X_H$ is $H$-elementary.

To topologize $\mathscr{F}^r(M)$, the $c^r$ real-valued functions on $M$, choose a countable, locally finite atlas consisting of pseudocompact charts $\{(V_i, \phi_i)\}$, the existence of which can be shown using the fact that second countable, locally compact Hausdorff spaces are paracompact. Denote by $\tilde{\phi}_i$ the extension of $\phi_i$, guaranteed by pseudocompactness and $\bar{V}_i$ the closure of $V_i$ and let $c^r(\tilde{\phi}_i(\bar{V}_i), R)$ denote the $c^r$ real-valued functions on $\tilde{\phi}_i(\bar{V}_i)$. $c^r(\tilde{\phi}_i(\bar{V}_i), R)$ is a Banach space. Consider $X_i c^r(\tilde{\phi}_i(\bar{V}_i), R)$ to have the box topology and define $\Pi: \mathscr{F}^r(M) \to X_i c^r(\tilde{\phi}_i(\bar{V}_i), R)$ to be the map which displays the local representatives via $\tilde{\phi}_i(\bar{V}_i)$.

DEFINITION. The $c^r$ *Whitney Topology* is the smallest topology on $\mathscr{F}^r(M)$ which makes $\Pi$ continuous.

REMARK. This topology is independent of the particular countable, locally finite, pseudocompact atlas.

PROPOSITION. $\mathscr{F}^r(M)$ *is a Baire space.*

PROOF. Adapt the proof of the Baire category theory, observing that essentially the topology on $\mathscr{F}^r(M)$ is given by a countable number of complete pseudo norms.

The idea of "almost every" is made precise by the notion of $c^r$ generic.

DEFINITION. Let $P(f)$ be a property of $c^r$ real-valued functions on $M$. Then $P(f)$ is a $c^r$ *generic property* if $\{f \in \mathscr{F}^r(M) | P(f)\}$ contains a residual set.

REMARK. By Baireness $\{f \in \mathscr{F}^r(M) | P(f)\}$ is dense.

The main result of this paper is the proof of the following.

THEOREM. *Property* H1 *is $c^r$ generic in $\mathscr{F}^r(M)$ for all $r \geq 2$.*

IDEA OF THE PROOF. The condition for a point to be *H*-elementary is translated into a statement about a subset of $J^2(M, R)$ (the two-jet bundle of real-valued functions on *M*). That is, we find a subset $W \subset J^2(M, R)$ such that *m* is an *H*-elementary critical point iff (i) *m* is a critical point and (ii) $j^2 H(m) \notin W$ where $j^2 H$ is the two-jet extension of *H*.

Next we show *W* is the countable union of submanifolds of $J^2(M, R)$ of codimension $\geq 2n + 1$, where dim $M = 2n$. The statement "for all *m*, $j^2 H(m) \notin W$" is shown to be equivalent to a transversality statement $j^2 H \pitchfork W_i$; for each of the submanifolds $W_i$. An appropriate representation machinery is set up and the Transversal Density Theorem (Abraham and Robbin [2, §19]) is used to conclude residuality.

PROOF OF THE THEOREM. Consider the local case first, for $M = U$ an open subset of a Banach space *E*, having dimension $= 2n$. Then $J^2(U, R) = U \times R \times L(E, R) \times L^2_s(E, R)$, $j^2 H(m) = (m, H(m), DH(m), D^2 H(m))$ and *m* is a critical point of *H* iff $j^2 H(m) \in U \times R \times (0) \times L^2_s(E, R)$. We seek a subset $W_1 \subset L^2_s(E, R)$ with the property that if $j^2 H(m) \in U \times R \times (0) \times W_1$ then at least one of three conditions defining *H*-elementary fails. Now in the global case the conditions refer to $X'_H$, the derivative of the symplectic gradient of *H*, and we have $X'_H(m) \in \mathrm{sp}\,(T_m M, \omega(m))$, the infinitesimally symplectic maps with respect to the form $\omega(m)$ (Abraham and Marsden [1, §28.1]). Thus in our local case we must find an appropriate subset $\tilde{W}_1$ of $\mathrm{sp}(E, \omega)$ and then pull back to $L^2_s(E, R)$ via the symplectic form.

CONSTRUCTION OF $\tilde{W}_1$. Let *C* denote the field of complex numbers. Using the symmetric polynomials define the map $v: C^{2n} \to C^{2n}$ such that if $a_i = i$th component of $v(c_1, ..., c_n)$ then $z^{2n} + a_1 z^{2n-1} + ... + a_{2n}$ is the unique monic polynomial whose roots are $c_1, ..., c_{2n}$. Fix some basis *B* for *E*. Let $A_B$ be the matrix for the linear map *A*. Let $\det(A_B - Ix) = x^{2n} + a_1 x^{2n-1} + ... + a_{2n}$ and define $K: L(E, E) \to R^{2n} \subset C^{2n}$ by $K(A) = (a_1, ..., a_{2n})$. Concerning the maps *K* and *v* note that if $(\lambda_1, ..., \lambda_{2n}) \in C^{2n}$, then $K^{-1} \circ v(\lambda_1, ..., \lambda_{2n})$ consists of all those linear maps in $L(E, E)$ which have eigenvalues $(\lambda_1, ..., \lambda_{2n})$. Identify $C^{2n}$ with $R^{4n}$ in the obvious way. It is easy to see that for each pair of integers $(n_1, n_2)$ we can define an algebraic set $V_i$ (i.e., defined by a finite number of polynomial maps from $R^{4n}$ to *R*) such that $K^{-1} \circ v(V_i)$ is the set of linear maps in $L(E, E)$ which have eigenvalues such that at least one of the conditions (i) or (ii) or (iii) for $(n_1, n_2)$ fails. Define $\tilde{W}_1$ to be $V \cap \mathrm{sp}(E, \omega)$, where $V = \bigcup_i K^{-1} \circ v(V_i)$.

PROPOSITION. $\tilde{W}_1$ *is the countable union of $c^\infty$ submanifolds of* $\mathrm{sp}(E, \omega)$.

PROOF. *K* and *v* are algebraic maps and hence by the Tarski Seidenberg Theorem [2, §30.7] $K^{-1}(v(V_i))$ is a semialgebraic set in $L(E, E)$ (i.e., can be defined by a finite number of polynomial equations and polynomial inequalities). Thus $\tilde{W}_1$ is the countable union of semialgebraic sets in $\mathrm{sp}(E, \omega)$ and hence by the Whitney Theorem [2, §30.8] the countable union of $c^\infty$ submanifolds of $\mathrm{sp}(E, \omega)$.

PROPOSITION. *The set $\tilde{W}_1$ as a subset of* $\mathrm{sp}(E, \omega)$ *has no interior.*

PROOF. Construct a complex algebraic set $S_i \subset C^{2n}$ such that $V_i \subset S_i$ by specifying $(\lambda_1, \ldots, \lambda_{2n}) \in S_i$ iff (i) for some $i$, $\lambda_i = 0$. (ii) there exists $i$ and $j$, with $i \neq j$, such that $\lambda_i = \lambda_j$. (iii) there exists $i$ and $j$, with $i \neq j$ and $n_1 \lambda_i + n_2 \lambda_j = 0$, where $n_1$ and $n_2$ are the integers used to define $V_i$. By [1, §30.5] the Newton Map $v$ is closed and so by Narasimhan [3, p. 130] $v(S_i)$ is an analytic subset of $C^{2n}$. Making the usual identification of $C^{2n}$ with $R^{4n}$, $v(S_i)$ becomes a real analytic subset of $R^{4n}$. Hence $[K|\mathrm{sp}(E, \omega)]^{-1}(v(S_i))$ is a real analytic subset of $\mathrm{sp}(E, \omega)$. We can easily find an infinitesimally symplectic matrix which does not obey any of the three conditions defining $S_i$. This means $[K|\mathrm{sp}(E, \omega)]^{-1}(v(S_i))$ can have no interior and so $\tilde{W}_1$ can have no interior.

This shows that the $c^\infty$ submanifolds whose union is $\tilde{W}_1$ have codimension $\geq 1$ in $\mathrm{sp}(E, \omega)$.

CONSTRUCTION OF $W_1$. Identifying $L(E, E)$ with $L(E, E^*)$ using the symplectic form $\omega$, and identifying $L(E, E^*)$ with $L^2(E, R)$ in the usual way, we get an isomorphism between $\mathrm{sp}(E, \omega)$ and $L^2_s(E, R)$. Let $W_1$ be the image of $\tilde{W}_1$ via this isomorphism. Then $W_1$ is the countable union of $c^\infty$ submanifolds of codimension $\geq 1$ in $L_s(E, R)$.

GLOBALIZATION. Let $W \subset J^2(M, R)$ be the set of points $\{\omega \in J^2(M, R)|$ there exists a jet bundle chart $(U, \alpha_0, \alpha)$ around $\omega$ such that $\alpha(\omega) \in \alpha_0(U) \times R \times (0) \times W_1\}$.

REMARK. It is easy to verify that $W$ is independent of the choice of jet bundle chart.

Since $J^2(M, R)$ can be covered by a countable number of charts, $W$ is the countable union of $c^\infty$ submanifolds $W_i$ of $J^2(M, R)$. These submanifolds are of class $c^{r-2}$ and of codimension $\geq 2n + 1$. From the construction it is clear that $m$ is an $H$-elementary critical point of $X_H$ iff (i) $m$ is critical and (ii) $j^2 H(m) \notin W$.

Of course it is sufficient to show for each $i$ $\{H \in \mathcal{F}^r(M)|$ for all critical points $m$ of $X_H$, $j^2 H(m) \notin W_i\}$ is residual in $\mathcal{F}^r(M)$. So from now on we will use $W$ to refer to one of the $W_i$.

*Restatement in terms of transversality.* Using the fact that $W$ has codimension $\geq 2n + 1$ one checks the

PROPOSITION. $j^2 H(m) \notin W$ *for all critical points* $m$ *of* $X_H$ *iff* $j^2 H \pitchfork W$ ($j^2 H$ *is transversal to* $W$).

Let $\{(V_i, \phi_i)\}$ be a countable, locally finite atlas consisting of pseudocompact charts. Let $A_i = \{f \in \mathcal{F}^r(\overline{V}_i)|j^2 f \pitchfork W\}$ and let $\pi_i : \mathcal{F}^r(M) \to \mathcal{F}^r(\overline{V}_i)$ be defined by $\pi_i(f) = f|\overline{V}_i$. Then $\bigcap_i \pi_i^{-1}(A_i)$ is the set of all $c^r$ real valued functions on $M$ having property H1.

The machinery for the application of the transversal density theorem will now be set up in order to prove that $A_i$ is residual in $\mathcal{F}^r(\overline{V}_i)$.

Define $\rho_i : \mathcal{F}^r(\overline{V}_i) \to c^{r-2}(\overline{V}_i, J^2(M, R))$ by $\rho_i : H \to j^2 H$, which makes sense since we may consider $J^2(\bar{V}_i, R) \subset J^2(M, R)$. Now $ev_{\rho_i}$ (the evaluation map) can be factored into two maps $ev_{\rho_i} : (H, m) \to (j^2 H, m) \to j^2 H(m)$. The first of these is $c^\infty$ by 12.2 of [1]. The second of these is $c^{r-2}$ by the evaluation Theorem 12.3 of [1]. Also dim $M$-codim $W \geq -1$. By 12.4 of [1] $ev_{\rho_i} : \mathscr{F}^r(\bar{V}_i) \times \bar{V}_i \to J^2(M, R)$ is a submersion and thus is trivially transversal to $W$.

The hypotheses of the transversal density theorem are satisfied so we may conclude $A_i$ is residual in $\mathscr{F}^r(\bar{V}_i)$.

PROPOSITION. $\pi_i : \mathscr{F}^r(M) \to \mathscr{F}^r(\bar{V}_i)$ is an open mapping.

PROOF. As the details are somewhat involved we omit the complete proof. However, we mention that the basic idea is to use a consequence of the Whitney extension theorem (see Malgrange [4, p. 8]) which shows how to extend a function so that all derivatives up to order $r$ have sup norms which are kept within a constant times the sup norm of the original function.

Using the openness and continuity of $\pi_i$ we get $\pi_i^{-1}(A_i)$ is residual in $\mathscr{F}^r(M)$ so that the set of $c^r$ real valued functions on $M$ having property H1 is residual in $\mathscr{F}^r(M)$.

*Acknowledgment.* To Ralph Abraham, Bob Walton and Jerry Marsden for valuable hints in the course of the proof.

## REFERENCES

1. R. Abraham and J. Marsden, *Foundations of mechanics,* Benjamin, New York, 1967.
2. R. Abraham and J. Robbin, *Transversal mappings and flows,* Benjamin, New York, 1967.
3. R. Narasimhan, *Introduction to the theory of analytic spaces,* Springer-Verlag, Berlin, 1966.
4. B. Malgrange, *Ideals of differentiable functions,* Oxford University Press, 1966.
5. J. Kelley, *General topology,* Van Nostrand, Princeton, New Jersey, 1955.

PRINCETON UNIVERSITY

# LOCATING INVARIANT SETS

R. W. EASTON[1]

The results discussed here represent joint work with Charles Conley. This work concerns firstly one possible method of locating compact invariant sets of "high" dimension for a flow on a manifold and describing the local orbit structure near these sets, and secondly those features of the local orbit structure which are preserved under small $C^0$ perturbation of the vector field. We establish the existence of invariant sets inside submanifolds from properties of the vector field on their boundaries. Our theorems require that these submanifolds be "convex" to the flow in the sense that orbits tangent to these submanifolds "bounce off". The invariant sets which we locate are *isolated* (see Definition 2) and we show that every compact isolated invariant set can be located by our methods.

Let $M$ be a compact, orientable, smooth Riemannian $n$-manifold and let $\mathfrak{X}(M)$ be the set of $C^1$ vector fields on $M$. We topologize this set with the $C^0$ topology (for $\dot{\phi}, \dot{\psi} \in \mathfrak{X}(M)$, $\rho(\dot{\phi}, \dot{\psi}) = \sup_{x \in M}\{|\dot{\phi}(x) - \dot{\psi}(x)|\}$). Each vector field $\dot{\phi} \in \mathfrak{X}(M)$ generates a unique flow $\phi$ on $M$, and for such flows $\phi$ and $\psi$ we measure the distance between them as before, $\rho(\phi, \psi) \equiv \rho(\dot{\phi}, \dot{\psi})$. In what follows we will only consider flows generated by vector fields in $\mathfrak{X}(M)$.

DEFINITION 1. Let $U$ be an open subset of $M$ and let $\phi$ be a flow on $M$. Define $I(U, \phi) = \{x \in U : O(x) \subset U\}$ where $O(x)$ denotes the orbit of $x$ under the flow $\phi$. $I(U, \phi)$ is the maximal invariant set of $\phi$ contained in $U$.

We wish to consider the following problems:

(a) Find open sets $U$ such that $I(U, \phi)$ is nonempty thus "locating" $I(U, \phi)$, and describe the orbit structure of $\phi$ in the set $U$.

(b) Describe the set $I(U, \tilde{\phi})$ where $\tilde{\phi}$ is a flow $C^0$ close to $\phi$ and in particular find topological properties shared by the sets $I(U, \phi)$ and $I(U, \tilde{\phi})$.

We first need the following definitions.

DEFINITION 2. A closed invariant set $J$ of a flow $\phi$ is said to be *isolated* if there exists a neighborhood $U$ of $J$ in $M$ such that $J = I(U, \phi)$.

DEFINITION 3. $N \subset M$ is an *isolating submanifold* for $\phi$ if
(1) $N$ is an $n$-dimensional submanifold of $M$ with boundary.
(2) $\dot{\phi}$ is tangent to $\partial N$ along a submanifold $\tau$ of $\partial N$.
(3) For each $x \in \tau$ there exists $\varepsilon > 0$ such that for $0 < |t| < \varepsilon$, $\phi(x, t) \notin N$.

THEOREM 1. *Let* $U = \operatorname{int} N$ *where* $N$ *is an isolating submanifold for* $\phi$. *Then* $I(U, \phi)$ *is a closed invariant set and hence is isolated.*

[1] Research supported by U.S.A.F. under grant AFOSR-67-0693A and NASA under grant NGR-40-002-015.

THEOREM 2. *Let $J$ be an isolated invariant set of $\phi$. Then there exists an isolating submanifold $N \subset M$ such that $J = I(\text{int } N, \phi)$.*

THEOREM 3. *Given $\varepsilon > 0$ and an isolating submanifold $N$ for $\phi$, there exists $\delta > 0$ such that whenever $\rho(\phi, \psi) < \delta$, there is an imbedding $h : N \to M$ within $\varepsilon$ of the inclusion map such that $N' = h(N)$ is an isolating submanifold for $\psi$, and $\psi$ is tangent to $N'$ along $\tau' = h(\tau)$.*

In what follows $N$ will be an isolating submanifold for $\phi$.

DEFINITION 4. Functions $\sigma^+ : N \to [-\infty, \infty]$ and $\sigma^- : N \to [-\infty, \infty]$ are defined by

$$\sigma^+(x) = \sup \{t \geq 0 : \phi(x, [0, t]) \subset N,$$
$$= +\infty \text{ if the supremum does not exist};$$

$$\sigma^-(x) = \inf \{t \leq 0 : \phi(x, [t, 0]) \subset N\},$$
$$= -\infty \text{ if the infimum does not exist.}$$

THEOREM 4. *$\sigma^+$ and $\sigma^-$ are continuous.*

DEFINITION 5. We define the following sets:

$$A^+ = (\sigma^+)^{-1}(+\infty), \quad A^- = (\sigma^-)^{-1}(-\infty),$$
$$a^+ = A^+ \cap \partial N, \quad a^- = A^- \cap \partial N,$$
$$I = A^+ \cap A^-,$$
$$n = \partial N,$$
$$n^+ = (\sigma^-)^{-1}(0), \quad n^- = (\sigma^+)^{-1}(0).$$

REMARK. $n = n^+ \cup n^-$, $\tau = n^+ \cap n^-$ and $I = I(\text{int } N, \phi)$.

DEFINITION 6. Functions $\pi^+ : N - A^+ \to n^-$ and $\pi^- : N - A^- \to n^+$ are defined by $\pi^+(x) = \phi(x, \sigma^+(x))$ and $\pi^-(x) = \phi(x, \sigma^-(x))$. $\pi^+$ and $\pi^-$ are continuous.

REMARKS. (1) By defining $H_t : N - A^+ \to N - A^+$ for $t \in [0, 1]$ by $H_t(x) = \phi(x, t \cdot \sigma^+(x))$ one sees that $n^-$ is a deformation retract of $N - A^+$. Similarly $n^+$ is a deformation retract of $N - A^-$.

(2) $\pi^+ : n^+ - a^+ \to n^- - a^-$ is a homeomorphism which is the identity on $\tau$.
We will discuss the following questions.

(a) What topological properties of $a^+$, $a^-$ and $I$ are determined a priori by $N$?

(b) What topological properties are shared by the sets $a^+$, $\tilde{a}^+$, $I$, $\tilde{I}$ etc. whenever $\tilde{\phi}$ is a sufficiently small $C^0$ perturbation of $\phi$?

The sets $\tilde{a}^+$, $\tilde{a}^-$, $\tilde{I}$ are defined as follows. By Theorem 2, for $\tilde{\phi}$ sufficiently close to $\phi$ there exists an imbedding $h : N \to M$ such that $N' = h(N)$ is an isolating submanifold for $\tilde{\phi}$. Functions $\tilde{\sigma}^+$ and $\tilde{\sigma}^-$ can be defined as in Definition 4 by replacing $\phi$ and $N$ by $\tilde{\phi}$ and $N'$ in the definition. Then the sets $\tilde{a}^+$, $\tilde{a}^-$, $\tilde{I}$, etc. are defined as in Definition 5.

Regarding question (a), some information is readily available. If $n^+$ and $n^-$ are not homeomorphic, then $a^+ \cup a^-$ and $I$ are nonempty. If $n^-$ is not a deformation retract of $N$, then $a^+$ and $I$ are nonempty.

The main tool used in obtaining more detailed information is the following

DUALITY THEOREM. *Let $X$ be a compact orientable $n$-manifold with or without boundary, and let $C$ be a closed subset of $X$, $C \cap \partial X = \emptyset$. Then $H_K(X, X - C)$ is isomorphic to $\check{H}^{n-K}(C)$ where $H_K(X, X - C)$ denotes the kth singular homology group of the pair $(X, X - C)$ with integer coefficients and $\check{H}^{n-K}(C)$ denotes the $(n - K)$th Čech cohomology group of $C$.*

Since $N$ is given we may assume that the homology groups of the sets $\tau$, $n^+$, $n^-$, $n$, $N$ are known together with the homomorphisms of these groups induced by the diagram of inclusion maps

$$\begin{array}{ccc} & n^+ & \\ \nearrow & & \searrow \\ \tau & & n \to N. \\ \searrow & & \nearrow \\ & n^- & \end{array}$$

A variety of theorems concerning $\check{H}^*(a^+)$ can be proven using this information. For example,

THEOREM 5. *If the inclusion $i_* : H_K(n^-) \to H_K(n)$ has nonzero kernel, then $H_{K+1}(n, n - a^+$ is nonzero and this implies that $\check{H}^{n-K-2}(a^+)$ is nonzero).*

PROOF. Consider the following diagram where the top line is a portion of the exact sequence of the pair $(n, n - a^+)$ and the maps $i_*, j_*, k_*$ are induced by inclusion

$$H_{K+1}(n, n - a^+) \xrightarrow{\partial} H_K(n - a^+) \xrightarrow{k_*} H_K(n)$$
$$\uparrow{\scriptstyle j_*} \qquad \qquad \nearrow{\scriptstyle i_*}$$
$$H_K(n^-)$$

$j_*$ is a monomorphism since $n^-$ is a retract of $n - a^+$. Hence $i_*$ has nonzero kernel implies that $\ker k_* \neq 0$. Hence by exactness $H_{K+1}(n, n - a^+) \neq 0$. By the duality theorem, $H_{K+1}(n, n - a^+)$ is isomorphic to $\check{H}^{n-K-2}(a^+)$. This completes the proof.

An example to which this theorem applies is the following:

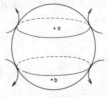

FIGURE 1

Let $(N, \tau)$ be as in Figure 1 and let $z \in H_0(n^-)$ be generated by the zero chain $a - b$. Then $z \in \ker i_* : H_0(n^-) \to H_0(n)$ and hence $\check{H}^1(a^+) \neq 0$.

Information concerning $\check{H}^*(a^+)$ can also be obtained by analyzing the map $\pi^+ : n - a^+ \to n^-$. However, in order to do this some knowledge of the flow is required. An example of this type of theorem is the following.

THEOREM 6. *Let* $z \in H_K(n - a^+)$ *and suppose that* $z = 0$ *in* $H_K(n)$ *but* $\pi_*^+(z) \neq 0$ *in* $H_K(n^-)$. *Then* $\check{H}^{n-K-2}(a^+) \neq 0$.

PROOF. Consider the diagram

$$H_{K+1}(n, n - a^+) \xrightarrow{\ \partial\ } H_K(n - a^+) \xrightarrow{\ k_*\ } H_K(n)$$
$$\downarrow \pi_*^+$$
$$H_K(n^-)$$

The hypothesis implies that $z$ is a nonzero element of kernel $k_*$ and this implies as in the previous theorem that $\check{H}^{n-K-2}(a^+) \neq 0$.

In computing $\check{H}^*(I)$ the following theorem is useful.

THEOREM 7. *There exists an exact sequence*

$$\to H_K(N - A^+, n - a^+) \xrightarrow{\ \mathring{i}\ } H_K(N, n) \xrightarrow{\ \mathring{j}\ } H_K(N, N - I) \xrightarrow{\ \alpha\ }$$

*where* $\mathring{i}$ *and* $\mathring{j}$ *are induced by inclusion and* $\alpha$ *is of degree* $-1$.

$H_*(N, N - I)$ is determined once the groups $H_*(N - A^+, n - a^+)$ and the maps $\mathring{i}$ are known. A theorem proven by analyzing the above exact sequence is the following.

THEOREM 8. *Let* $z \in H_K(n - a^+)$ *be such that*
(1) $z - z' \neq 0$ *in* $H_K(n - a^+)$ *where* $z' = \pi_*^+(z)$
(2) $\widehat{zz}' = 0$ *in* $H_{K+1}(N, n)$ *where* $\widehat{zz}'$ *is the cycle "swept out" by* $z$ *as it is carried across* $N$ *by the flow (this can be made precise). Then* $\check{H}^{n-K-2}(I) \neq 0$.

In dealing with features of the sets $a^+$, $a^-$, $I$ which are preserved under perturbation of the vector field the notion of stable cycle is useful.

DEFINITION 7. $z \in H_K(N, N - I)$ is *stable* if for any sufficiently small perturbation $\hat{\phi}$ of $\phi$ $h_*(z) \neq 0$ in $H_K(N', N' - \tilde{I})$ (where $h : N \to M$ is an imbedding and $N' = h(N)$ is an isolating submanifold for $\hat{\phi}$). $v \in \check{H}^{n-K}(I)$ is *stable* if the corresponding dual cycle $z \in H_{n-K}(N, N - I)$ is stable. Stable cycles in $\check{H}_K(n, n - a^+)$ and cocycles in $H^{n-K}(a^+)$ can be similarly defined.

*Problem.* Find sufficient conditions to insure that a cycle is stable. A result of this sort is the following

THEOREM 9. *If* $z \in H_K(n - a^+)$ *satisfies the hypothesis of Theorem 6, and* $\omega \in H_{K+1}(n, n - a^+)$ *is such that* $\partial \omega = z$, *then* $\omega$ *is a stable cycle.*

This is essentially due to the fact that for $\hat{\phi}$ sufficiently close to $\phi$ the mappings $\pi^+ : n - a^+ \to n^-$ and $h^{-1} \circ \tilde{\pi}^+ : n - h^{-1}(\tilde{a}^+) \to n^-$ can be made arbitrarily close to each other on their common domain.

The previous results can be applied to study the nonlinear Hamiltonian system of equations $\dot{x} = H_y$, $\dot{y} = -H_x$, where

$$H(x, y) = \tfrac{1}{2}(y, y) + v(x)$$

$$v(x) = x_1^3 - \tfrac{3}{2}x_1 x_2^2$$

$$x = (x_1, x_2), \quad y = (y_1, y_2).$$

These equations describe the motion of a point sliding without friction in a monkey saddle. They give rise to a flow on $E^4$ which preserves the constant "energy" surfaces $H = h$. We wish to study the flow restricted to the 3-dimensional surface $M = \{(x, y): H(x, y) = h > 0\}$.

Let $p: E^4 \to E^2$ be the projection $p(x_1, x_2, y_1, y_2) = (x_1, x_2)$. Then $p(M) = V^{-1}((-\infty, h])$. One can choose three straight line segments as in Figure 2 which together with the curve $V = h$ enclose a region $R$ such that $p^{-1}(R)$ is an isolating submanifold for the flow on the energy surface $H = h$.

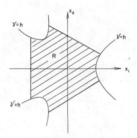

FIGURE 2

$p^{-1}(R)$ is homeomorphic to a 3-dimensional ball with two open 3-balls removed and the flow looks roughly as drawn in Figure 3.

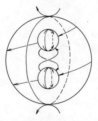

FIGURE 3

In particular the flow is tangent to $N = p^{-1}(R)$ along the three circles forming the equators of the three 2-spheres bounding $N$, and it is fairly easy to prove that orbits exist which cross $N$ as drawn. Theorems 6 and 8 can be applied to show that $\check{H}^1(a^+) \neq 0$ and $\check{H}^1(I) \neq 0$.

BROWN UNIVERSITY

# ANOSOV DIFFEOMORPHISMS

JOHN FRANKS

**Introduction.** There has been considerable interest in recent years in studying the orbit structure of diffeomorphisms. The strongest useful equivalence for this study seems to be topological conjugacy—we say two diffeomorphisms $f: M \to M$ and $g: N \to N$ are *topologically conjugate* if there exists a homeomorphism $h: N \to M$ such that $f \circ h = h \circ g$. It is clear that in this case any orbit of $g$ is mapped by $h$ to an isomorphic orbit of $f$.

An interesting example which arose in this study is the following: Consider a $2 \times 2$ matrix $\bar{f}$ with integer entries, determinant $\pm 1$, and with no eigenvalues of absolute value one.

$$\begin{pmatrix} 2 & 1 \\ 1 & 1 \end{pmatrix}$$

is such a matrix. Then $\bar{f}$ can be thought of as a linear transformation of the plane $R^2$ which preserves the lattice $L$ of points with integer coordinates. $\bar{f}$ induces an automorphism $f$ of the quotient group $R^2/L = T^2$, the two-dimensional torus. The example is of interest in the study of orbit structure because it can be shown that the periodic points of $f$ are dense in $T^2$ and that any $C^1$ perturbation of $f$ also has this property. In fact this diffeomorphism satisfies the much stronger property of being *structurally stable*. A diffeomorphism $f$ is said to be structurally stable if there is a neighborhood $N(f)$ of $f$ in the $C^1$ topology on Diff $(M)$ such that $g \in N(f)$ implies $f$ and $g$ are topologically conjugate.

D. Anosov generalized this example to a much larger class of diffeomorphisms (see [2]).

DEFINITION. An Anosov diffeomorphism $f$ is a $C^\infty$ diffeomorphism from a $C^\infty$ manifold $M$ to itself which satisfies the following:

(a) There is a continuous splitting of the tangent bundle $TM = E^s \oplus E^u$ which is preserved by the derivative $Df$.

(b) There exist constants $C$, $C'$ and $\lambda \in (0, 1)$ and a Riemannian metric $|| \; ||$ on $TM$ such that

$$||Df^n(v)|| \leq C\lambda^n ||v|| \quad \text{for } v \in E^s$$

and

$$||Df^n(v)|| \geq C'\lambda^{-n} ||v|| \quad \text{for } v \in E^u.$$

Anosov diffeomorphisms are considered in some detail in S. Smale's paper [19]. In particular, an appendix of this paper gives an exposition by J. Mather of J. Moser's proof that Anosov diffeomorphisms on compact manifolds are structurally stable. (This theorem was originally proved by D. V. Anosov [2].)

If the manifold $M$ is compact condition (b) is independent of the choice of the Riemannian metric.

The automorphism of the torus mentioned above is an Anosov diffeomorphism because the linear map $\bar{f}$ is hyperbolic (has no eigenvalues of absolute value one).

To a limited extent we will deal here with Anosov coverings ($C^\infty$ self-coverings of a manifold which satisfy conditions (a) and (b) above) as well as diffeomorphisms. All Anosov diffeomorphisms and coverings considered here will be $C^\infty$ maps on compact $C^\infty$ manifolds without boundary.

Our investigation of Anosov diffeomorphisms will be within the context of the conjugacy problem, which can be stated in its most general form as follows: Given a diffeomorphism $f: M \to M$ for what diffeomorphisms $g: N \to N$ does there exist a nontrivial map $h: N \to M$ which makes the diagram

$$\begin{array}{ccc} M & \overset{h}{\leftarrow} & N \\ \downarrow f & & \downarrow g \\ M & \overset{h}{\leftarrow} & N \end{array}$$

commute?

For some diffeomorphisms this question reduces entirely to a homotopy problem. To be precise we make the following definition.

DEFINITION. A $\Pi_1$ *diffeomorphism* ($\Pi_1$ *covering*) $f: M \to M$ is a diffeomorphism ($C^\infty$ covering) with the following properties: Given any homeomorphism (continuous covering) $g: K \to K$ of a compact $CW$ complex and any map $h: K \to M$ such that

$$\begin{array}{ccc} \Pi_1(M) & \overset{h_*}{\leftarrow} & \Pi_1(K) \\ \downarrow f_* & & \downarrow g_* \\ \Pi_1(M) & \overset{h_*}{\leftarrow} & \Pi_1(K) \end{array}$$

commutes, there exists a unique base point preserving map $h': K \to M$, homotopic to $h$, such that $f \circ h' = h' \circ g$.

We prove in Theorem (2.1) that the hyperbolic automorphism of the torus mentioned above is a $\Pi_1$ diffeomorphism. This fact is subsequently used to show in Theorem (7.3) that any Anosov diffeomorphism of a two manifold is topologically conjugate to a hyperbolic toral automorphism like the example above. These, however, are only special cases of more general results.

The main purpose of this paper is to give a partial solution to the following two problems.

(1) Classify all Anosov diffeomorphisms up to topological conjugacy.

(2) Classify all $\Pi_1$ diffeomorphisms up to topological conjugacy.

It seems appropriate to consider the two together because, to the best of my knowledge, the set of all known examples of Anosov diffeomorphisms coincides (up to topological conjugacy) with the set of all known examples of $\Pi_1$ diffeomorphisms. Moreover, this fact is very useful in considering the first problem.

Before going further we must list some more examples of Anosov diffeomorphisms and Anosov coverings.

(1) HYPERBOLIC TORAL ENDOMORPHISMS. This is an immediate generalization of the example given above. Let $\bar{f}$ be a hyperbolic isomorphism of $R^n$ which preserves the integer lattice $L$ (hyperbolic means $\bar{f}$ has no eigenvalues of absolute value one). The induced endomorphism $f$ on the $n$ dimensional torus $T^n = R^n/L$ is an Anosov covering. If $\det(\bar{f}) = \pm 1$ it is a diffeomorphism.

(2) HYPERBOLIC NILMANIFOLD ENDOMORPHISMS. A generalization of the example above is obtained by considering an automorphism $\bar{f}$ of a simply connected nilpotent Lie group $N$ with the property that the derivative $d\bar{f}$ at the identity of $N$ is hyperbolic. If $\bar{f}$ preserves a uniform discrete subgroup $\Gamma$, then the induced map on the nilmanifold $N/\Gamma$ is an Anosov diffeomorphism. A specific example of this type is worked out in Smale's paper [19].

(3) HYPERBOLIC INFRA-NILMANIFOLD ENDOMORPHISMS. Suppose $N$ is a simply connected nilpotent Lie group, $A$ is a finite group of automorphisms of $N$, and $G$ is a torsion free uniform discrete subgroup of the semidirect product $A \cdot N$. Then by a result of L. Auslander in [3], $N \cap G$ is a uniform discrete subgroup of $N$ and $N \cap G$ has finite index in $G$. An element of $A \cdot N$ is a pair $(x, a)$ with $x \in N$, $a \in A$ and it acts on $N$ by first applying $a$ and then left translating by $x$. $G$ acts freely on $N$ because if $g \in G$ and $x \in N$, $g(x) = x$ implies $g^n(x) = x$ for all $n$, but for some $n$, $g^n$ is left translation by an element of $N$. Hence $g^n(x) = x$ would imply $g^n$ is the identity of $G$, but $G$ is torsion free. Thus the space $N/G$ (the quotient space of $N$ under the action of $G$) is a compact manifold.

If $\bar{f}: A \cdot N \to A \cdot N$ is an automorphism for which $\bar{f}(G) = G$ and $\bar{f}(N) = N$ then it induces a covering $f: N/G \to N/G$. We will use terminology due to M. Hirsch and call $N/G$ an infra-nilmanifold and $f$ an infra-nilmanifold endomorphism. If the derivative of $\bar{f}$ is hyperbolic then $f$ will be an Anosov covering. A specific example of this type is worked out in [18].

To the best of my knowledge all known examples of Anosov diffeomorphisms of compact manifolds are topologically conjugate to a diffeomorphism of one of the types above.

In §2 we prove the following

(2.2) THEOREM. *Diffeomorphisms which are hyperbolic infra-nilmanifold automorphisms are $\Pi_1$ diffeomorphisms.*

This immediately suggests the following unanswered question.

*Problem.* Are all Anosov diffeomorphisms $\Pi_1$ diffeomorphisms?

DEFINITION. Two diffeomorphisms (coverings) $f: M \to M$ and $g: N \to N$ will be called $\Pi_1$-conjugate if there is an isomorphism $\phi: \Pi_1(M) \to \Pi_1(N)$ such that

$$\begin{array}{ccc} \Pi_1(M) & \overset{\phi}{\to} & \Pi_1(N) \\ \downarrow f_* & & \downarrow g_* \\ \Pi_1(M) & \overset{\phi}{\to} & \Pi_1(N) \end{array}$$

commutes.

It is not difficult to show in (3.2) that two $\Pi_1$ coverings are topologically conjugate if and only if they are $\Pi_1$-conjugate.

In §3 we consider the problem of classifying $\Pi_1$ diffeomorphisms and obtain the following partial converse to (2.2).

(3.6) THEOREM. *If* $f: M \to M$ *is a* $\Pi_1$ *diffeomorphism and* $\Pi_1(M)$ *is torsion free, then*

(a) *If* $\Pi_1(M)$ *has a nilpotent subgroup of finite index, then* $f$ *is topologically conjugate to a hyperbolic infra-nilmanifold automorphism.*

(b) *If* $\Pi_1(M)$ *is nilpotent, then* $f$ *is topologically conjugate to a hyperbolic nil-manifold automorphism.*

(c) *If* $\Pi_1(M)$ *is abelian, then* $f$ *is topologically conjugate to a hyperbolic toral automorphism.*

If all Anosov diffeomorphisms could be shown to be $\Pi_1$ diffeomorphisms, this result would give a good start on their classification. Our results, however, are somewhat more modest.

DEFINITION. A *codimension one* Anosov diffeomorphism is an Anosov diffeomorphism for which either dim $E^u = 1$ or dim $E^s = 1$.

We consider a class of Anosov diffeomorphisms which we call metrically splitting (see §1 for a definition) and show in §4 that they are $\Pi_1$ diffeomorphisms. In §§5 and 6 we prove that any codimension one Anosov diffeomorphism $f$, whose nonwandering set $NW(f)$ is the entire manifold, is metrically splitting. From this we prove

(6.3) THEOREM. *Let* $f: M \to M$ *be a codimension one Anosov diffeomorphism with* $NW(f) = M$, *then* $f$ *is topologically conjugate to a hyperbolic toral automorphism. If* $g$ *is another such diffeomorphism then* $f$ *and* $g$ *are topologically conjugate iff they are* $\Pi_1$-*conjugate.*

The condition that $NW(f) = M$ will always be satisfied if there is a Lebesque measure on $M$ invariant under $f$. If the dimension of $M$ is $\leq 3$ then any Anosov diffeomorphism will be codimension one, so we have the following result.

(6.4) COROLLARY. *If* $f: M \to M$ *is an Anosov diffeomorphism with* $NW(f) = M$ *and* dim $M \leq 3$, *then* $f$ *is topologically conjugate to a hyperbolic toral automorphism.*

In §7 we show that if dim $M = 2$, then it is always the case that $NW(f) = M$.

(7.3) THEOREM. *If* $f: M^2 \to M^2$ *is an Anosov diffeomorphism then* $f$ *is topologically conjugate to a hyperbolic toral automorphism. Any two such diffeomorphisms are topologically conjugate iff they* $\Pi_1$-*conjugate.*

In the final section we consider expanding maps of compact manifolds. An expanding map is an Anosov covering with $TM = E^u$. Expanding maps have been investigated in some detail by M. Shub [18]. Using similar techniques to that of (3.6) we prove,

(8.2) THEOREM *If $f: M \to M$ is an expanding map and $\Pi_1(M)$ has a nilpotent subgroup of finite index, then $f$ is topologically conjugate to an infra-nilmanifold endomorphism.*

In (8.3) we show that if $f: M \to M$ is an expanding map then $\Pi_1(M)$ has polynomial growth and using this fact to obtain the conclusion of (8.2) when $\Pi_1(M)$ has a solvable subgroup of finite index.

Conversations with a number of people have been very helpful in the preparation of this paper. A few specific acknowledgements are made in the text, but I wish also to thank M. Hirsch, C. Pugh, M. Shub, and especially my adviser S. Smale.

**1.** In this section we develop some geometric properties of Anosov coverings which will be needed for the proofs of later theorems. If $f: M \to M$ is an Anosov covering then $P: \overline{M} \to M$ will denote the simply connected covering of $M$ and $\overline{f}: \overline{M} \to \overline{M}$ will be a lift of $f$. Note that $\overline{f}$ is a diffeomorphism. Let $d$ denote a complete metric on $\overline{M}$ obtained by lifting a Riemannian metric on $M$.

(1.1) PROPOSITION. *$\overline{M}$ has two foliations $U$ and $S$ with the following properties.*

(a) *$\overline{f}$ preserves both $U$ and $S$. If $u(x)$ $(s(x))$ denotes the leaf of $U$ $(S)$ passing through $x$, then $u(x) = u(y)$ iff $\lim_{n \to \infty} d(\overline{f}^{-n}(x), \overline{f}^{-n}(y)) = 0$ and $s(x) = s(y)$ iff $\lim_{n \to \infty} d(\overline{f}^n(x), \overline{f}^n(y)) = 0$.*

(b) *Each leaf of $U(S)$ is a $C^\infty$ injectively immersed copy of $R^k$ $(R^{n-k})$, where $k = \dim E^u$, and the tangent space to $u(x)$ $(s(x))$ at $x$ is $P^*E^u_x$ $(P^*E^s_x)$.*

(c) *Both foliations are oriented.*

PROOF. The existence of $U$ and $S$ and (a) and (b) are essentially the stable manifold theorem for $\overline{f}$ (see [7]). The foliations are oriented because $\overline{M}$ is simply connected and hence $P^*E^u$ and $P^*E^s$ are orientable. q.e.d.

Henceforth $U$ will denote the space of leaves of $U$ and similarly $S$ will be the space of leaves of $S$. $U$ and $S$ have the quotient topology obtained from the maps $u: \overline{M} \to U$ and $s: \overline{M} \to S$ where $u(x)$ is the leaf of $U$ through $x$ and $s(x)$ is the leaf of $S$ through $x$. Since $u(\overline{f}(x)) = \overline{f}(u(x))$, $\overline{f}$ induces a map (also denoted $\overline{f}$) from $U$ to itself. The map $\overline{f}: S \to S$ is defined similarly.

Since $u(x)$ is smoothly immersed in $\overline{M}$, we can obtain a Riemannian metric on $u(x)$ by pulling back the metric on $\overline{M}$. This in turn gives a complete metric on $u(x)$ whose distance function will be denoted $d(u(x); x_1, x_2)$. Alternatively $d(u(x); x_1, x_2)$ is the infimum of the lengths of piecewise smooth paths lying in $u(x)$ and joining $x_1$ and $x_2$. $d(s(x); x, y)$ is defined similarly.

(1.2) LEMMA. *If $u \in U$ and $x, y \in u$, then*

$$d(\overline{f}^n(u); \overline{f}^n(x), \overline{f}^n(y)) \geq C\lambda^{-n} d(u; x, y).$$

*If $s \in S$ and $x, y \in s$ then*

$$d(\overline{f}^n(s); \overline{f}^n(x), \overline{f}^n(y)) \leq C'\lambda^n d(s; x, y).$$

$C$, $C'$ and $\lambda$ are the constants given in the definition of Anosov coverings and $n$ is any integer $\geq 0$.

PROOF. Suppose $x, y \in u$ and $h$ is a piecewise smooth path in $u$ joining $x$ and $y$. Let $l(h) = \int_0^1 h'(t)dt$ be the length of $h$. Then

$$d(u; x, y) = \inf_h l(h), \text{ and } d(\overline{f}^n(u); \overline{f}^n(x), \overline{f}^n(y)) = \inf_h l(f^n h)$$

where both infimums are taken over all piecewise smooth paths in $u$ joining $x$ and $y$. But $l(\overline{f}^n h) = \int_0^1 d\overline{f}^n(h'(t))dt > C\lambda^{-n} \int_0^1 h'(t)dt = C\lambda^{-n}l(h)$. The first inequality is now immediate. The second follows similarly. q.e.d.

We will generally identify the fundamental group $\Pi_1(M)$ with the group of covering transformations for the covering $P: \overline{M} \to M$.

DEFINITION. A compact fundamental domain is a set $K \subset \overline{M}$ which is compact and the closure of its interior and for which $P(K) = M$.

(1.3) COROLLARY. *If $\alpha \in \Pi_1(M)$ is a covering transformation then $\alpha$ maps leaves of $U$ to leaves of $U$ and leaves of $S$ to leaves of $S$.*

PROOF. $\alpha$ is an isometry of the Riemannian metric on $\overline{M}$ and its derivative preserves $P*E^u$ and $P*E^s$. If $x$ and $y$ are in $u$ and $h$ is a path joining them then $\alpha \circ h$ is a path joining $\alpha x$ and $\alpha y$. But because of the properties of $\alpha$ it follows that $\lim_{n \to \infty} l(\overline{f}^{-n} \circ \alpha \circ h) = 0$ so by (1.1) (a), $u(\alpha x) = u(\alpha y)$. The result for $S$ is obtained similarly. q.e.d.

It follows from (1.3) that the foliations $U$ and $S$ are independent of the choice of the lift $\overline{f}$. If $\alpha \in \Pi_1(M)$ then by (1.3) $\alpha$ induces a map (which will also be denoted $\alpha$) of $U$ to itself. We define $\alpha: S \to S$ similarly.

Since $\Pi_1(M)$ preserves $U$ and $S$, we can obtain two foliations on $M$ simply by applying $P$ to the foliations of $\overline{M}$. If $x \in M$ and $x = P(\overline{x})$ then $W^u(x)$ and $W^s(x)$ will denote $P(u(\overline{x}))$ and $P(s(\overline{x}))$ respectively.

(1.4) PROPOSITION. *If $s \in S$ and $P: \overline{M} \to M$ is the covering map then $P$ restricted to $s$ is injective.*

PROOF. $P$ is a local homeomorphism, in fact, there exists $\varepsilon > 0$ such that $0 < d(x, y) < \varepsilon$ implies $P(x) \neq P(y)$. Suppose now that $x, y \in s$ and $P(x) = P(y)$ then for all $i$, $P(\overline{f}^i(x)) = P(\overline{f}^i(y))$. But by (1.1) (a) there is an $n$ such that

$$d(\overline{f}^n(x), \overline{f}^n(x)) < \varepsilon$$

which gives a contradiction. q.e.d.

This shows that $W^s(x)$ is an injectively immersed plane in $M$. We remark that our definition of $W^s(x)$, $x \in \text{Per}(f)$, does not coincide with the usual definition of the stable manifold of a periodic point of an endomorphism (see Shub [17]). In fact $W^s(x)$ is the component of $x$ in the stable manifold of $x$. $W^u(x)$ is precisely the unstable manifold of $x$, and if $f$ is a diffeomorphism then $W^s(x)$ is the stable manifold.

If $p \in M$ let $W^u(p; \varepsilon)$ be the connected component of $p$ in $\{x \mid x \in W^u(p)$ and $d(p, x) \leq \varepsilon\}$ and define $W^s(p; \varepsilon)$ similarly.

(1.5) PROPOSITION (LOCAL PRODUCT STRUCTURE). *There is an $\varepsilon > 0$ and an embedding $h: W^u(p; \varepsilon) \times W^s(p; \varepsilon) \to M$ which is given by $h(x_1, x_2) =$ the unique point of intersection of $W^u(x_1 : \varepsilon)$ and $W^s(x_2; \varepsilon)$. $\varepsilon$ is independent of $p$.*

PROOF. This follows from (1.1), the transversality of $E_p^s$ and $E_p^u$, and the fact that $E_x^s$ and $E_x^u$ vary continuously with $x$. Compare (7.4) of S. Smale's article [18]. $\varepsilon$ is independent of $p$ because of the compactness of $M$. q.e.d.

If $N$ is the image of $h$ then $N$ will be called a *product neighborhood* of $p$. If $N$ is sufficiently small that $P^{-1}$ restricted to $N$ is a homeomorphism then $P^{-1}(N)$ will be called a product neighborhood in $\overline{M}$. Let $u(x; \varepsilon) = P^{-1}(W^u(x; \varepsilon))$ and define $s(x; \varepsilon)$ similarly.

If the local product structure is in fact global we say that $f$ is splitting. More precisely,

DEFINITION. An Anosov covering is said to be *splitting* if $\overline{M}$ is homeomorphic to $R^k \times R^{n-k}$ by a homeomorphism which takes each leaf $u \in U$ to a $k$-plane and each leaf $s \in S$ to an $(n - k)$ plane. Note that in this case $U$ is homeomorphic to $R^{n-k}$ and $S$ is homeomorphic to $R^k$.

DEFINITION. A map $T$ of a complete metric space $X$ to itself will be called *contracting* if there exist constants $C > 0$ and $\lambda \in (0, 1)$ such that $d(T_x^n, T_y^n) \leq C\lambda^n d(x, y)$ for any $n \geq 0$ and any $x, y \in X$.

The following well-known result will be frequently used in later sections.

(1.6) LEMMA. *If $T: X \to X$ is a contracting map, then there is a unique $x \in X$ such that $T(x) = x$.*

We are interested in Anosov coverings which satisfy a somewhat stronger condition than being splitting.

DEFINITION. An Anosov covering will be called *metrically splitting* if it is splitting and $S$ possesses a complete metric such that
(a) $\overline{f}^{-1}: S \to S$ is contracting and
(b) $\Pi_1(M)$ acts as isometries on $S$.

If $x \in M$ then $x$ is said to be wandering if there is a neighborhood $N$ of $x$ such that the sets $f^n(N)$, $n = 0, 1, 2, \ldots$ are all disjoint. A point which is not wandering is called nonwandering. The union of all nonwandering points $NW(f)$ is easily seen to be a closed set invariant under $f$. Clearly $NW(f) \supset \text{Per}(f)$, the periodic points of $f$.

*Question.* If $f: M \to M$ is an Anosov covering, is it always the case that $NW(f) = M$? It is easy to see that for all the examples given in the introduction $NW(f) = M$.

The following result was announced by Anosov for the diffeomorphism case some time ago. The proof given here was shown to me by M. Hirsch.

(1.7) PROPOSITION. *If $f: M \to M$ is an Anosov covering then $\text{Per}(f)$ is dense in $NW(f)$.*

PROOF. Suppose $x \in NW(f)$ and let $V$ be a neighborhood of $x$ in $M$. We must find a periodic point in $V$. Let $N \approx W^u(x; \varepsilon) \times W^s(x; \varepsilon)$ be a product neighborhood of $x$ contained in $V$. Let $N'$ be the smaller neighborhood homeomorphic to $W^u(x; \varepsilon/4) \times W^s(x; \varepsilon/4)$. Since $x$ is nonwandering there exists $x' \in N'$ such that $W^u(x'; \varepsilon/8) \subset N'$ and $f^n(x') \in N'$ for some $n$ which is sufficiently large that $C\lambda^{-n} > 8$ and $C'\lambda^n < 1/8$. Then $D = f^n(W^u(x'; \varepsilon/8)) \cap N$ will contain $W^u(f^n(x'); \varepsilon/2)$.

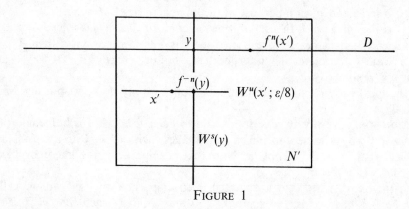

FIGURE 1

Let $D_0$ be the component of $D$ containing $f^n(x')$. We define a map $h: D_0 \to D_0$ by first mapping $D_0$ to $W^u(x'; \varepsilon/8)$ by $f^{-n}$ and then projecting back to $D_0$ along the $W^s$ leaves of $N'$. Since $D_0$ is a cell $h$ has a fixed point $y$. Clearly $f^n(W^s(y)) = W^s(y)$.

Let $D' = W^s(y; 3\varepsilon/2)$ then $f^n(D') \subset W^s(f^n(y); 3\varepsilon/16) \subset D'$ so this map has a fixed point $x_0$ in $N$. $f^n(x_0) = x_0$ so $x_0$ is the desired point. q.e.d.

REMARK. If $\bar{f}: \overline{M} \to \overline{M}$ is a lift to the simply connected covering of an Anosov covering $f: M \to M$ then essentially the same proof shows $\mathrm{Per}(\bar{f})$ is dense in $\mathrm{NW}(\bar{f})$.

(1.8) LEMMA. If $f: M \to M$ is an Anosov covering with $\mathrm{NW}(f) = M$ then for any $x \in \mathrm{Per}(f)$, $W^u(x)$ is dense in $M$.

PROOF. Let $K =$ closure of $W^u(x)$. We show that $K$ is open. Suppose $y \in K$ and let $N$ be a product neighborhood of $y$; say $N \approx W^s(y; \varepsilon) \times W^u(y; \varepsilon)$. By (1.7) it suffices to show the periodic points in $N$ are in $K$. Let $y'$ be a periodic point of $f$ in $N$. If $n$ is the product of the periods of $x$ and $y'$, then $g = f^n$ is an Anosov covering on $M$ which has the same foliations as $f$. $g(y') = y'$ so if we make $y'$ the base point of $M$ and $\bar{y} \in P^{-1}(y)$ the base point of $\overline{M}$, $g$ has a unique lift $\bar{g}: \overline{M} \to \overline{M}$ such that $\bar{g}(\bar{y}) = \bar{y}$.

By construction there exists $\bar{x} \in s(\bar{y}; \varepsilon)$ such that $P(\bar{x}) \in W^u(x)$. Clearly $\lim_{i \to \infty} \bar{g}^i(\bar{x}) = \bar{y}$ so $\lim_{i \to \infty} g^i(P(\bar{x})) = y'$. But $g^i(P(\bar{x})) \in W^u(x)$ for all $i \geq 0$, so $y' \in K$. q.e.d.

(1.9) PROPOSITION. If $f: M \to M$ is an Anosov covering with $\mathrm{NW}(f) = M$, then for any $x$, $W^u(x)$ is dense in $M$.

PROOF. Let $R(x) = \bigcup_{n \geq 0} f^{-n}(W^s(x))$; we show first that for any $x$, $R(x)$ is dense in $M$. By the Lebesque covering lemma there is a $\delta > 0$ such that $d(x_1, x_2) < \delta$ implies there is a product neighborhood $N$ in $M$ containing $x_1$ and $x_2$. Suppose $x_0 \in M$ and $V$ is an open set in $M$; we wish to show $R(x_0) \cap V \neq \varnothing$. Let $y$ be a periodic point of $f$ contained in $V$ (which exists by (1.7)). Choose $\varepsilon$ sufficiently small that $B = W^u(y; \varepsilon)$ is contained in $V$.

Let $m$ be the period of $y$ and let $U_n = \{x | d(x, f^{mn}(B)) < \delta\}$. Since $W^u(y)$ is dense in $M$ and any point of $W^u(y)$ is contained in $f^{mn}(B)$ for some $n$, we have $M = \bigcup_{n=1}^{\infty} U_n$. Since $M$ is compact, there exists an integer $q > 0$ such that $M = U_q$. Hence any $x \in M$ has a product neighborhood $N(x)$ containing both $x$ and a point of $f^{mq}(B)$. From this it follows that (perhaps by enlarging $q$) we can guarantee that $W^s(x) \cap f^{mq}(B) \neq \varnothing$ for all $x \in M$.

In particular $W^s(x_0) \cap f^{mq}(B) \neq \varnothing$ so there exists $y_0 \in B$ such that $f^{mq}(y_0) \in W^s(x_0)$. Clearly $y_0 \in R(x_0)$ and $y_0 \in V$. Thus for any $x \in M$, $R(x)$ is dense in $M$.

Now suppose $x' \in M$ and $V'$ is an open set in $M$. We show $W^u(x') \cap V' \neq \varnothing$. Let $y'$ be a periodic point of $f$ in $V'$ and let $B'$ be a neighborhood of $y'$ in $W^s(y')$ which is sufficiently small that $B' \subset V'$. It is easily checked that $R(y') = \bigcup_{n \geq 0} f^{-n}(B')$. Let $U'_n = \{x | d(x, f^{-n}(B')) < \delta\}$ then since $R(y')$ is dense, $M = \bigcup_{n=0}^{\infty} U'_n$. By the compactness of $M$, there exists an integer $r > 0$ such that $M = U'_r$. Again it follows that (perhaps by enlarging $r$) we can guarantee that $W^u(x) \cap f^{-r}(B') \neq \varnothing$ for all $x \in M$.

In particular let $x'' \in f^{-r}(x')$, then $W^u(x'') \cap f^{-r}(B') \neq \varnothing$. If $y'_0 \in W^u(x'') \cap f^{-r}(B')$ then $f^r(y'_0) \in W^u(f^r(x'')) \cap B' = W^u(x') \cap B'$. Hence $W^u(x') \cap V' \neq \varnothing$. q.e.d.

REMARK. If $f$ is a diffeomorphism then the stable manifolds for $f$ are the unstable manifolds for $f^{-1}$ so it follows that for any $x \in M$, $W^s(x)$ is also dense in $M$. Even when $f$ is a covering, if $x \in \mathrm{Per}(f)$ then $R(x)$ is the stable manifold of $x$ (in the sense of Shub [18]) and the proof shows that it is dense in $M$.

(1.10) COROLLARY If $f: M \to M$ is an Anosov covering with $\mathrm{NW}(f) = M$, then for any $u \in U$, $\bigcup_{\alpha \in \Pi_1(M)} \alpha u$ is dense in $\overline{M}$.

PROOF. Let $u \in U$ and $V$ an open set in $\overline{M}$. Say $W^u(x) = P(u)$, then by (1.9) there exists $y \in P(V) \cap W^u(x)$. Hence there is a $\bar{y}_1 \in V$ such that $P(\bar{y}_1) = y$ and a $\bar{y}_2 \in u$ such that $P(\bar{y}_2) = y$. Consequently there must be an $\alpha \in \Pi_1(M)$ such that $\alpha y_2 = y_1$. q.e.d.

REMARK. If $u(x)$ is one-dimensional and $\hat{u}$ is one component of $u(x) - \{x\}$, then the proofs of (1.9) and (1.10) work for $P(\hat{u})$ and $\hat{u}$. That is, in this case, $\bigcup_{\alpha \in \Pi_1(M)} \alpha \hat{u}$ is also dense in $\overline{M}$.

2. In this section we prove that there exist $\Pi_1$ diffeomorphisms. In fact all the examples of Anosov diffeomorphisms given in the introduction are $\Pi_1$ diffeomorphisms. Throughout the section all spaces will be considered to have base points; in particular the identity element of groups will be chosen as base point. All maps (except covering transformations) will be assumed to be base point preserving.

(2.1) PROPOSITION. *If* $f: T^n \to T^n$ *is a hyperbolic automorphism of a torus then it is a* $\Pi_1$ *diffeomorphism.*

PROOF. Suppose $g: K \to K$ is a homeomorphism of a compact $CW$ complex to itself such that there exists $h_0: K \to T^n$ for which

$$\begin{array}{ccc} \Pi_1(K) & \xrightarrow{\phi} & \Pi_1(T^n) \\ \downarrow g_* & & \downarrow f_* \\ \Pi_1(K) & \xrightarrow{\phi} & \Pi_1(T^n) \end{array}$$

commutes, where $\phi$ is the homomorphism induced by $h_0$.

Let $\bar{f}: R^n \to R^n$ be the linear automorphism of $R^n$ which covers $f$. Let $\bar{K}$ be the simply connected covering of $K$ and let $\bar{g}: \bar{K} \to \bar{K}$ be the lift of $g$. Since $\bar{f}$ is hyperbolic there is an invariant splitting $R^n = E^u \oplus E^s$ with $|\bar{f}^n(v)| \le C\lambda^n|v|$ for $v \in E^s$ and $n \ge 0$ and $|\bar{f}^n(v)| \ge C'\lambda^{-n}|v|$ for $v \in E^u$ and $n \ge 0$, where $C, C' > 0$ and $0 < \lambda < 1$.

Let $Q = \{h | h \in C_0(\bar{K}, R^n)$ and $h(\alpha x) = h(x)$ for all $x \in \bar{K}$ and all $\alpha \in \Pi_1(K)\}$, where $C_0(\bar{K}, R^n)$ is the space of base point preserving continuous maps from $\bar{K}$ to $R^n$. Then $Q$ is a Banach space with the sup norm.

Let $Q^s = \{h | h \in Q$ and $h(\bar{K}) \subset E^s\}$ and let $Q^u = \{h | h \in Q$ and $h(\bar{K}) \subset E^u\}$.

Then one checks easily that $Q \cong Q^u \oplus Q^s$. We define an isomorphism $F: Q \to Q$ by $F(h) = \bar{f}^{-1} \circ h \circ \bar{g}$. It follows easily that $F$ is linear and preserves $Q^u$ and $Q^s$. Also we have, using the sup norm, $\|F^n(h)\| \le C\lambda^n \|h\|$ for $n \ge 0$ and $h \in Q^s$, and $\|F^n(h)\| \ge C'\lambda^{-n}\|h\|$ for $n \ge 0$ and $h \in Q^u$. Let $I: Q \to Q$ be the identity, then $(F - I)$ is an isomorphism because on $Q^s$

$$(F - I)^{-1} = -\sum_{n=0}^{\infty} F^n.$$

The right-hand side converges because $\|F^n\| \le C\lambda^n$. Similarly on $Q^u$ we have $(I - F^{-1})^{-1} = \sum_{n=0} F^{-n}$ so $(F - I)^{-1} = F^{-1}(I - F^{-1})^{-1}$ exists. Hence $(F - I)$ is an isomorphism of $Q$.

Let $\bar{h}_0: \bar{K} \to R^n$ be the lift of $h_0$ and let $h' = F(\bar{h}_0) - \bar{h}_0 = \bar{f}^{-1} \circ \bar{h}_0 \circ \bar{g} - \bar{h}_0$. Then $h' \in Q$ because

$$\begin{aligned} h'(\alpha x) &= F(\bar{h}_0)(\alpha x) - \bar{h}_0(\alpha x) \\ &= f_*^{-1} \circ \phi \circ g_*(\alpha)F(\bar{h}_0)(x) - \phi(\alpha)\bar{h}_0(x) \\ &= \phi(\alpha)F(\bar{h}_0)(x) - \phi(\alpha)\bar{h}_0(x) \\ &= F(\bar{h}_0)(x) - \bar{h}_0(x) \end{aligned}$$

because $\phi(\alpha)$ is simply a translation of $R^n$.

Since $h' \in Q$ there is an $\hat{h} \in Q$ such that $(F - I)\hat{h} = h'$. Let $k: \bar{K} \to R^n$ be given by $k(x) = \bar{h}_0(x) - \hat{h}(x)$, then $k(\alpha x) = \phi(\alpha)k(x)$ so $k$ is the lift of a map $k_0: K \to T^n$ and by a standard homotopy result (see [21 p. 427]) $k_0$ is homotopic to $h_0$.

Moreover, $\bar{f}^{-1} \circ k \circ \bar{g} - k = F(\bar{h}_0 - \hat{h}) - (\bar{h}_0 - \hat{h})$
$$= (F - I)\bar{h}_0 - (F - I)\hat{h} = h' - h' = 0.$$

So $\bar{f} \circ k = k \circ \bar{g}$. If $k'$ is any other map with these properties then one checks that $k - k'$ is in $Q$ and $(F - I)(k - k') = 0$, so $k - k' = 0$, that is, $k = k'$. Hence $k_0$ is the unique map homotopic to $h_0$ such that $f \circ k_0 = k_0 \circ g$.　q.e.d.

(2.1) is a special case of the following result.

(2.2) THEOREM. *Diffeomorphisms which are hyperbolic infra-nilmanifold auto-morphisms are* $\Pi_1$ *diffeomorphisms.*

PROOF. Suppose $f: M \to M$ is a hyperbolic infra-nilmanifold automorphism. Say $M$ is the quotient space $N/G$, where $N$ is a simply connected nilpotent Lie group, $A$ is a finite group of automorphisms of $N$ and $G$ is a uniform discrete subgroup of the semidirect product $A \cdot N$. Let $\bar{f}: N \to N$ be the hyperbolic automorphism of $N$ which covers $f$.

Suppose $g: K \to K$ is a homeomorphism of a compact CW complex to itself such that there exists $h_0: K \to M$ for which

$$\begin{array}{ccc} \Pi_1(K) & \overset{\phi}{\to} & \Pi_1(M) = G \\ \downarrow g_* & & \downarrow f_* \\ \Pi_1(K) & \overset{\phi}{\to} & \Pi_1(M) \end{array}$$

commutes, where $\phi$ is the homomorphism induced by $h_0$. Let $\psi: \Pi_1(K) \to A$ be the composition $\Pi_1(K) \overset{\phi}{\to} G \overset{\rho}{\to} A$ where the second homomorphism is the restriction to $G$ of the natural homomorphism $\rho$ mapping $A \cdot N$ onto $A$. Then if $\alpha \in \Pi_1(K)$, $\psi(\alpha)$ is an automorphism of $N$. Let $\bar{K}$ be the simply connected covering of $K$ and let $Q = \{h | h \in C_0(\bar{K}, N)$ and $h(\alpha x) = \psi(\alpha)h(x)$ for all $\alpha \in \Pi_1(K)$ and all $x \in \bar{K}\}$, where $C_0(\bar{K}, N)$ is the space of continuous base point preserving maps from $\bar{K}$ to $N$.

Each element of $G$ has the form $(y, a)$ where $y \in G \cap N$ and $a \in A$. This element acts on $N$ by first applying the automorphism $a$ and then left translating by $y$. We define a multiplication in $Q$ by $h_1 h_2(x) = h_1(x)h_2(x)$. Note that

$$h_1 h_2(\alpha x) = h_1(\alpha x)h_2(\alpha x) = (\psi(\alpha)h_1(x))(\psi(\alpha)h_2(x))$$
$$= \psi(\alpha)(h_1(x)h_2(x)) = \psi(\alpha)h_1 h_2(x).$$

One checks easily that $Q$ is a nilpotent group.

Let $\bar{g}: \bar{K} \to \bar{K}$ be the lift of $g$, and define a homomorphism $F_0: Q \to Q$ by $F_0(h) = \bar{f}^{-1} \circ h \circ \bar{g}$. This is well defined because by Theorem 2 of [2] the automorphism $f_*: G \to G$ lifts to an automorphism $\bar{\theta}: A \cdot N \to A \cdot N$ which induces an automorphism $\theta: A \to A$, and the diagram

$$\begin{array}{ccc} \Pi_1(K) & \overset{\psi}{\to} & A \\ \downarrow g_* & & \downarrow \theta \\ \Pi_1(K) & \overset{\psi}{\to} & A \end{array}$$

commutes. So

$$
\begin{aligned}
F_0(h)(\alpha x) &= \bar{f}^{-1} \circ h \circ \bar{g}(\alpha x) \\
&= \bar{f}^{-1}(\psi(g_*(\alpha))h \circ \bar{g}(x)) \\
&= \bar{f}^{-1}(f_*(\psi(\alpha))h \circ \bar{g}(x)) \\
&= \psi(\alpha)\bar{f}^{-1} \circ h \circ \bar{g}(x) = \psi(\alpha)F_0(h)(x).
\end{aligned}
$$

We define a map $T: Q \to Q$ by $T(h) = F_0(h)h^{-1}$. We are interested in showing that $T$ is a local homeomorphism at the constant map $c$ (the identity of $Q$) which we do by computing the derivative. $Q$ is a banachable space; in fact, if $\Sigma$ is the Lie algebra of $N$ and $\exp: \Sigma \to N$ is the exponential map, then the map $\text{Log}: Q \to \Delta = \{h|h \in C_0(\bar{K}, \Sigma), h(\alpha x) = d\psi(\alpha)h(x) \text{ all } \alpha \in \Pi_1(K)\}$ given by $\text{Log}(h) = \exp^{-1} \circ h$ is a diffeomorphism of $Q$ onto the Banach space $\Delta$. (Since $N$ is simply connected and nilpotent exp is a diffeomorphism; see [8, p. 136].)

Define $F: \Delta \to \Delta$ by $F = \text{Log} \circ F_0 \circ \text{Log}^{-1}$, then since $\exp \circ df = \bar{f} \circ \exp$ where $df$ is the derivative of $\bar{f}$ at the identity of $N$, it follows that $F(h) = df^{-1} \circ h \circ \bar{g}$. Hence $F$ is a linear map of $\Delta$. Define $T': \Delta \to \Delta$ by $T' = \text{Log} \circ T \circ \text{Log}^{-1}$. Clearly $T$ is a local homeomorphism at $c$ if $T'$ is at $\text{Log}^{-1}(c)$.

Now
$$
\begin{aligned}
T'(h) &= \text{Log} \circ T \circ \text{Log}^{-1}(h) \\
&= \text{Log}(F_0(\exp \circ h)(\exp \circ h)^{-1}) \\
&= \text{Log}((\bar{f}^{-1} \circ \exp \circ h \circ \bar{g})(\exp \circ h)^{-1}) \\
&= \text{Log}((\exp \circ F(h))(\exp \circ h)^{-1})
\end{aligned}
$$
or, if $\text{Exp} = \text{Log}^{-1}$

$$
T'(h) = \text{Log}(\text{Exp}(F(h))\text{Exp}(h)^{-1}).
$$

$\text{Log}(c)$ is clearly the zero of $\Delta$. We now compute the derivative of $T'$ at $0$. In fact, the derivative $dT'_0 = F - I$ where $I: \Delta \to \Delta$ is the identity, because, if $h \in \Delta$

$$
\lim_{t \to 0} \frac{1}{t} T'(th) = \lim_{t \to 0} \frac{1}{t} \text{Log}(\text{Exp}(F(th))\text{Exp}(-th))
$$

and by the Campbell-Hausdorf formula (see [8, p. 112] for example).

$$
\text{Log}(\text{Exp}(F(th))\text{Exp}(-th)) = F(th) - th + t^2 \text{ (higher order terms)}.
$$

Hence $\lim(1/t)T'(th) = F(h) - h$.

We now show that $dT'_0$ is an isomorphism. Since $df$ is hyperbolic, there are subspaces $\Sigma^u$, and $\Sigma^s$ of $\Sigma$ such that

$$
\Sigma = \Sigma^u \oplus \Sigma^s, \quad |df^n(v)| \le C\lambda^n|v| \quad \text{for } n > 0, v \in \Sigma^s, C > 0, 0 < \lambda < 1
$$

and $|df^{-n}(v)| \le C'\lambda^n|v|$ for $n > 0$, $v \in \Sigma^u$. Let $\Delta^u = \{h|h \in \Delta \text{ such that } k(\bar{K}) \subset \Sigma^u\}$ and define $\Delta^s$ similarly. Clearly $\Delta^u$ and $\Delta^s$ are invariant under $F$ and it is easily seen that $\Delta = \Delta^u \oplus \Delta^s$ and $\|F^n(h)\| \le C\lambda^n\|h\|$ for $n > 0$ and $h \in \Delta^s$. Moreover $F$ restricted to $\Delta^u$ is invertible and $\|F^{-n}(h)\| \le C'\lambda^n\|h\|$ for $n > 0$ and $h \in \Delta^u$. On $\Delta^s$ we have $(F - I)^{-1} = -\sum_{n=0}^{\infty} F^n$. The right side converges because $\|F^n\| \le C\lambda^n$. Similarly in $\Delta^u$ we have $(I - F^{-1})^{-1} = \sum_{n=0}^{\infty} F^{-n}$, so $(F - I)^{-1} = F^{-1}(I - F^{-1})^{-1}$ exists. Hence $F - I$ is an isomorphism of $\Delta$.

From this it follows by the inverse function theorem that $T'$ is a local homeomorphism at 0 and hence $T$ is a local homeomorphism at $c$.

We use this to show that $T: Q \to Q$ is a surjection. Since a neighborhood of the identity $c$ of $Q$ generates $Q$ (see [15, p. 76]) it suffices to show that if $h_1, h_2 \in$ Image $(T)$ then $h_1 h_2 \in$ Image $(T)$. Let $Q = Q^n \supset Q^{n-1} \supset \ldots \supset Q^0 = c$ be the derived series of $Q$, then each $Q^i$ is connected because the derived series of $N$ consists of contractible groups. Suppose now $T(h_1)$ and $T(h_2) \in Q^1$ the center of $Q$ then

$$\begin{aligned} T(h_1)T(h_2) &= F_0(h_1)h_1^{-1}T(h_2) = F_0(h_1)T(h_2)h_1^{-1} \\ &= F_0(h_1)F_0(h_2)h_2^{-1}h_1^{-1} \\ &= T(h_1 h_2). \end{aligned}$$

Hence $Q^1 \subset \text{Im}(T)$. Assume inductively that $Q^i \subset \text{Im}(T)$ and suppose $T(h_1)$ and $T(h_2) \in Q^{i+1}$. Then

$$\begin{aligned} T(h_1)T(h_2) &= F_0(h_1)h_1^{-1}T(h_2) \\ &= F_0(h_1)h'T(h_2)h_1^{-1} \quad \text{where } h' = [h_1, T(h_2)^{-1}] \in Q^i \\ &= F_0(h_1)F_0(h_2)h''h_2^{-1}h_1^{-1} \quad \text{where } h'' \in Q^i \end{aligned}$$

and hence there is $h_3$ such that $T(h_3) = h''$. Thus

$$\begin{aligned} T(h_1)T(h_2) &= F_0(h_1)F_0(h_2)F_0(h_3)h_3^{-1}h_2^{-1}h_1^{-1} \\ &= T(h_1 h_2 h_3). \end{aligned}$$

Consequently, $Q^{i+1} \subset \text{Image}(T)$ and by induction $Q \subset \text{Image}(T)$.

We also claim $T$ is an injection because $F$ clearly fixes only 0 and hence $F_0$ has only the fixed point $c$. Thus $T(h) = c$ implies $h = c$ and if $T(h_1) = T(h_2)$ then $T(h_1 h_2^{-1}) = c$ so $h_1 = h_2$.

We return now to the map $h_0: K \to M$ and let $\overline{h}_0: \overline{K} \to N$ be the lift of $h_0$. We let $\overline{h} = F_0(\overline{h}_0^{-1})\overline{h}_0$ and show that $\overline{h} \in Q$.

$$\begin{aligned} \overline{h}(\alpha x) &= F_0(\overline{h}_0(\alpha x)^{-1})\overline{h}_0(\alpha x) \\ &= \overline{f}^{-1}([\overline{h}_0(\overline{g}(\alpha x))]^{-1})(\overline{h}_0(\alpha x)) \\ &= \overline{f}^{-1}([\phi(g_*(\alpha))\overline{h}_0(\overline{g}(x))]^{-1})(\phi(\alpha)\overline{h}_0(x)). \end{aligned}$$

But by hypothesis $\phi \circ g_* = f_* \circ \phi$. Hence

$$\begin{aligned} \overline{h}(\alpha x) &= [f_*^{-1} \circ \phi \circ g_*(\alpha)\overline{f}^{-1}(\overline{h}_0(\overline{g}(x)))]^{-1}(\phi(\alpha)\overline{h}_0(x)) \\ &= [\phi(\alpha)\overline{f}^{-1}(\overline{h}_0(\overline{g}(x)))]^{-1}(\phi(\alpha)\overline{h}_0(x)). \end{aligned}$$

Suppose $\phi(\alpha) = (y, a) \in G$, then $a = \psi(\alpha)$ and

$$\begin{aligned} \overline{h}(\alpha x) &= [ya(\overline{f}\overline{h}_0\overline{g}(x))]^{-1}ya(\overline{h}_0(x))) \\ &= a([F_0(\overline{h}_0)(x)]^{-1})y^{-1}y \cdot a(\overline{h}_0)(x)) \\ &= \psi(\alpha)[(F_0(\overline{h}_0)(x))^{-1}\overline{h}_0(x)] \\ &= \psi(\alpha)\overline{h}(x). \quad \text{So } \overline{h} \in Q. \end{aligned}$$

Let $\hat{h} = T^{-1}(\overline{h})$, and consider the map $\overline{k} = \overline{h}_0 \hat{h}$. The map $T$ can be applied to $\overline{k}$ and $\overline{h}_0$ even though they are not in $Q$. We get

$$T(\overline{k}) = F_0(\overline{h}_0\hat{h})\hat{h}^{-1}\overline{h}_0^{-1} = F_0(\overline{h}_0)T(\hat{h})\overline{h}_0^{-1} = F_0(\overline{h}_0)F_0(\overline{h}_0^{-1})\overline{h}_0\overline{h}_0^{-1} = c.$$

So $\bar{f}^{-1} \circ \bar{k} \circ \bar{g} = \bar{k}$ or $\bar{f} \circ \bar{k} = \bar{k} \circ \bar{g}$.

To complete the proof we need only show that $\bar{k}$ is the lift of a map $k: K \to M$ which is homotopic to $h_0$. However, this is the case because $\Pi_i(M) = 0$ for $i > 1$ and

$$\bar{k}(\alpha x) = \bar{h}_0(\alpha x)\hat{h}(\alpha x)$$
$$= (\phi(\alpha)h_0(x))(\psi(\alpha)\hat{h}(x)),$$
$$= y\psi(\alpha)(\bar{h}_0(x))\psi(\alpha)(\hat{h}(x)), \text{ where } \phi(\alpha) = (y, \psi(\alpha)),$$

so $\qquad \bar{k}(\alpha x) = y\psi(\alpha)(\bar{h}_0 h(x)) = \phi(\alpha)\bar{k}(x).$

If $k'$ is another map with the properties of $k$ then its lift $\bar{k}'$ has all the properties of $\bar{k}$. Consider the map $\bar{k}^{-1}\bar{k}'$,

$$\bar{k}^{-1}\bar{k}'(\alpha x) = [\phi(\alpha)\bar{k}(x)]^{-1}(\phi(\alpha)\bar{k}'(x))$$
$$= \psi(\alpha)(\bar{k}(x)^{-1})y^{-1}y(\psi(\alpha)\bar{k}'(x)),$$

where $\phi(\alpha) = (y, \psi(\alpha))$. So $\bar{k}^{-1}\bar{k}'(\alpha x) = \psi(\alpha)\bar{k}^{-1}\bar{k}'(x)$ and hence $\bar{k}^{-1}k' \in Q$. But $T(\bar{k}^{-1}\bar{k}') = F_0(\bar{k}^{-1})F_0(\bar{k}')\bar{k}'^{-1}\bar{k} = c$, so $\bar{k}^{-1}\bar{k}' = c$. Thus $\bar{k} = \bar{k}'$ and hence $k$ is unique. q.e.d.

REMARK. It follows from the proof above that the semiconjugacy $k: K \to M$ depends continuously on $g: K \to K$.

Question. If $f: M \to M$ is a covering map which is a hyperbolic infra-nilmanifold endomorphism, then is $f$ a $\Pi_1$ covering? The proof given above fails in this case because the map $F_0: Q \to Q$ will not be a homeomorphism and hence it does not follow that $T: Q \to Q$ is a surjection. However, in the case of expanding maps the conclusion can be obtained (see §8).

**3.** We now develop the properties of $\Pi_1$ diffeomorphisms and $\Pi_1$ coverings, and prove a partial converse to (2.2). As before all maps considered are base point preserving.

(3.1) LEMMA. *If* $f: M \to M$ *is a* $\Pi_1$ *covering, then* $\Pi_i(M) = 0$ *for all* $i \geq 2$.

PROOF. Suppose $h: S^i \to M$ represents a nontrivial element of $\Pi_i(M)$, then

$$\begin{array}{ccc} \Pi_1(S^i) & \overset{h_*}{\to} & \Pi_1(M) \\ \downarrow id & & \downarrow f_* \\ \Pi_1(S^i) & \overset{h_*}{\to} & \Pi_1(M) \end{array}$$

commutes.

Since $f: M \to M$ is a $\Pi_1$ covering there exists $h': S^i \to M$ homotopic to $h$ such that $h' \circ id = f \circ h'$. Let $p$ be the base point of $S^i$ and $q$ the base point of $M$. Since $h'$ represents a nontrivial element of $\Pi_i(M)$ there exists $x \in S^i$ such that $h'(x) \neq q$. Let $S^1$ be the great circle in $S^i$ passing through $p$ and $x$ and let $k: S^1 \to M$ be given by $k = h'$ restricted to $S^1$. If $c: S^1 \to q$ is the trivial map then clearly $k$ is homotopic to $c$.

Now $k \circ id = f \circ k$ and $c \circ id = f \circ c$. By the definition of $\Pi_1$ covering there is a unique map with this property, but $k(x) \neq p = c(x)$ which is a contradiction. q.e.d.

(3.2) THEOREM. *A necessary and sufficient condition that two $\Pi_1$ coverings be topologically conjugate is that they be $\Pi_1$-conjugate.*

PROOF. Suppose $f: M \to M$ and $g: N \to N$ are $\Pi_1$ conjugate maps which are $\Pi_1$ coverings. If $\phi: \Pi_1(M) \to \Pi_1(N)$ is the conjugating isomorphism, then since $\Pi_i(M)$ and $\Pi_i(N)$ are trivial for $i > 1$, it follows from standard results of homotopy theory (see [21, p. 427]) that there exist maps $h: M \to N$ and $k: N \to M$ such that $h_* = \phi$ and $k_* = \phi^{-1}$. Since $f$ and $g$ are $\Pi_1$ coverings there are unique maps $h'$ and $k'$ homotopic to $h$ and $k$ respectively such that $f \circ h' = h' \circ g$ and $g \circ k' = k' \circ f$. From this it follows that $f \circ h' \circ k' = h' \circ g \circ k' = h' \circ k' \circ f$. From homotopy theory [21, p. 427] we know $h' \circ k': M \to M$ is homotopic to the identity id$: M \to M$. But clearly id $\circ f = f \circ$ id, so since $f$ is a $\Pi_1$ covering, the uniqueness condition implies $h' \circ k' = $ id. That is, $h'$ and $k'$ are homeomorphisms. q.e.d.

(3.3) LEMMA. *Let $T: R^n \to R^n$ be a linear map with an eigenvalue $\lambda$ of absolute value one, then either $\lambda = \pm 1$ or there is a two-dimensional subspace $V$ of $R^n$ such that $T$ restricted to $V$ is topologically conjugate to a rotation.*

PROOF. If $\lambda$ is real then $|\lambda| = 1$ implies $\lambda = \pm 1$. If $\lambda$ is complex then standard results on canonical forms imply there is a subspace $W$ of $R^n$ invariant under $T$ such that $T$ restricted to $W$ can be represented by the matrix

$$
\begin{bmatrix}
\alpha & \beta & & & & & & \\
-\beta & \alpha & & & & 0 & & \\
1 & 0 & & & & & & \\
0 & 1 & & & & & & \\
& & & & & & & \\
& 0 & & & & 1 & 0 & \alpha & \beta \\
& & & & & 0 & 1 & -\beta & \alpha
\end{bmatrix}
$$

where $\lambda = \alpha + i\beta$. If $e_1, \ldots, e_i$ is the basis of $W$ for which $T$ has this matrix, then let $V$ be the space spanned by $e_{i-1}$ and $e_i$. $T$ restricted to $V$ is a rotation through the angle $\theta$ where $\sin \theta = \beta$, $\cos \theta = \alpha$. q.e.d.

Let $M$ and $N$ denote compact manifolds and let $\bar{M}$ and $\bar{N}$ be their respective universal coverings. If $h: M \to N$ is any continuous map let $\bar{h}: \bar{M} \to \bar{N}$ be a covering of $h$.

(3.4) LEMMA. *If $h: M \to N$ is a homotopy equivalence, then $\bar{h}: \bar{M} \to \bar{N}$ is proper.*

PROOF. Let $p \in \bar{N}$ be the base point and let $K$ be any compact fundamental domain in $\bar{N}$ which contains $p$. We show first that if $Q$ is any compact subset of $\bar{N}$ then $\alpha(K) \cap Q \neq \varnothing$ holds for only finitely many $\alpha$ in $\Pi_1(N)$. Since $Q$ is compact there exists $r$ such that $B_r(p) = \{x \mid x \in \bar{N}$ and $d(p, x) \leq r\}$ contains $Q$. Choose $t > r + \text{diam } K$, then $B_t(p)$ has the property that $\alpha(p) \notin B_t(p)$ implies $\alpha(K) \cap Q = \varnothing$.

But $B_t(p)$ is compact and hence $\alpha(p) \in B_t(p)$ for only finitely many $\alpha$. Thus $\alpha(K) \cap Q \neq \varnothing$ for only finitely many $\alpha$.

Now let $D$ be a compact fundamental domain in $\overline{M}$. Since $h$ is a homotopy equivalence, it is surjective and $\overline{h}(D)$ is a compact fundamental domain in $\overline{N}$. Let $K = \overline{h}(D)$, then $\Psi = \{\alpha | \alpha \in \Pi_1(N) \text{ and } \alpha(K) \cap Q \neq \varnothing\}$ is finite. We know $\alpha(K) = \overline{h}[h_*^{-1}(\alpha)(D)]$. Thus $\overline{h}^{-1}(Q) \subset \bigcup_{\alpha \in \Psi} h_*^{-1}(\alpha)(D)$ which is compact. q.e.d.

The following result was suggested by L. Auslander and communicated to me by M. Hirsch.

(3.5) PROPOSITION (EXISTENCE OF A MODEL). *If $f: K \to K$ is a covering and $\Pi_1(K)$ has a nilpotent subgroup of finite index and is torsion free then there exists a covering $g: M \to M$ which is an infra-nilmanifold endomorphism and which is $\Pi_1$-conjugate to $f$. If $\Pi_1(K)$ is nilpotent then $g$ is a nilmanifold endomorphism and if $\Pi_1(K)$ is abelian then $g$ is a toral endomorphism.*

PROOF. We recall that an infra-nilmanifold is a quotient space $M = N/G$ where $N$ is a simply connected nilpotent Lie group, $A$ is a finite group of automorphisms of $N$ and $G$ is a uniform discrete subgroup of the semidirect product $A \cdot N$. An infra-nilmanifold endomorphism $g$ is a map on $M$ induced by an automorphism $\overline{g}: A \cdot N \to A \cdot N$ such that $\overline{g}(G) \subset G$ and $\overline{g}(N) = N$.

Let $G$ be a group isomorphic to $\Pi_1(M)$ and let $\Gamma$ be the nilpotent subgroup of finite index in $G$. We can assume $\Gamma$ is normal (if not the intersection of $\Gamma$ with all its conjugates will be a normal nilpotent subgroup of finite index in $G$). By a result of Malcev (Theorem 6 of [10]) $\Gamma$ can be embedded as a uniform discrete subgroup in a simply connected nilpotent Lie group $N$. Let $A$ be the finite group $G/\Gamma$.

We then have the exact sequence

$$1 \to \Gamma \to G \xrightarrow{\psi} A \to 1.$$

We will construct an extension $\Delta$ of $N$ by $A$,

$$1 \to N \to \Delta \to A \to 1$$

and then show this extension splits.

Let $\psi: G \to A$ be the natural projection. For each $a \in A$ choose $\sigma(a) \in G$ such that $\psi(\sigma(a)) = a$, choosing in particular $\sigma(1) = 1$. Then conjugation by $\sigma(a)$ yields an automorphism of $\Gamma$ which will be denoted $\theta(a)$. By a result of Malcev (Theorem 5 of [10]) $\theta(a)$ extends uniquely to an automorphism $\overline{\theta}(a): N \to N$. The product $\sigma(a)\sigma(b)$ differs from $\sigma(ab)$ by an element in $\Gamma$ which will be denoted $\varepsilon(a, b)$, that is,

$$\sigma(a)\sigma(b) = \varepsilon(a, b)\sigma(ab).$$

Let $\Delta = \{(n, a) | n \in N, a \in A\}$. We define a multiplication in $\Delta$ by $(n, a) \cdot (m, b) = (n \cdot \overline{\theta}(a)(m) \cdot \varepsilon(a, b), ab)$. It follows from Lemma 8.1 on p. 125 of [9] (or by direct check) that $\Delta$ is a group with this multiplication and that the homomorphisms

$i: N \to \Delta$ given by $i(n) = (n, 1)$ and $\bar{\psi}: \Delta \to A$ given by $\bar{\psi}(n, a) = a$ give an exact sequence

$$1 \to N \overset{i}{\to} \Delta \overset{\bar{\psi}}{\to} A \to 1.$$

Moreover $G$ is isomorphic to the subgroup of $\Delta$ consisting of pairs $(n, a)$ with $n \in \Gamma$. We now have the following commutative diagram:

$$\begin{array}{ccccccccc}
1 & \to & N & \overset{i}{\to} & \Delta & \to & A & \to & 1 \\
& & \cup & & \cup & & \uparrow \text{id} & & \\
1 & \to & \Gamma & \to & G & \to & A & \to & 1
\end{array}$$

We wish to show that the top sequence splits. Let $(A, N, \phi)$ be the abstract kernel of the top extension.

$$\phi: A \to \operatorname{Aut}(N)/\operatorname{In}(N),$$

where $\operatorname{Aut}(N)$ is the group of automorphisms of $N$ and $\operatorname{In}(N)$ is the subgroup of inner automorphisms, is the map induced by the extension $\Delta$.

Let $Z$ be the center of $N$, then by Corollary 5.4 on p. 117 of [9], $H^2(A, Z) = 0$. From this and Theorem 8.8 on p. 128 of [9] it follows that there is a unique extension of $N$ with abstract kernel $(A, N, \phi)$.

To show that this extension splits we induct on the nilpotent length of $N$. If $N$ is abelian then there exists a split extension with kernel $(A, N, \phi)$ so $\Delta$ must be this extension. Assume now that the result holds for all simply connected nilpotent groups of nilpotent length $\leq n$ and let $\overset{\wedge}{N}$ have nilpotent length $n + 1$.

Consider the extension

$$1 \to N/Z \to \Delta/Z \overset{\bar{\psi}'}{\to} A \to 1,$$

where $\bar{\psi}'$ is the natural map induced by $\bar{\psi}$. It has kernel $(A, N/Z, \bar{\phi})$ where $\bar{\phi}: A \to \operatorname{Aut}(N/Z)/\operatorname{In}(N/Z)$ is the map induced by $\phi$. Then by the induction hypothesis this extension splits and hence there exists a homomorphism $\hat{\sigma}: A \to \Delta/Z$ such that $\bar{\psi} \circ \hat{\sigma} = \operatorname{id}$. Let $\phi': A \to \operatorname{Aut}(N)$ be the homomorphism which assigns to $a \in A$ the restriction to $N$ of the inner automorphism of $\Delta$ determined by any element of the coset $\hat{\sigma}(a)$.

We have then the semidirect product extension

$$1 \to N \to A \times_{\phi'} N \to A \to 1.$$

Let $\hat{\phi}: A \to \operatorname{Aut}(N)/\operatorname{In}(N)$ be the map induced by $\phi'$. Then $\hat{\phi}(a)$ is determined by the restriction to $N$ of the inner automorphism of $\Delta$ by any element of the coset $\hat{\sigma}(a)$. Now $\phi(a)$ is determined by the restriction to $N$ of the inner automorphism of $\Delta$ by $\sigma(a)$. Since $\bar{\psi}'(\hat{\sigma}(a)) = a$ and $\bar{\psi}(\sigma(a)) = a$ any element of the coset $\hat{\sigma}(a)$ differs from $\sigma(a)$ at most by an element of $N$. Hence they determine the same element of $\operatorname{Aut}(N)/\operatorname{In}(N)$, that is $\hat{\phi} = \phi$.

Thus the semidirect product extension above has kernel $(A, N, \phi)$ and must be isomorphic to $\Delta$. So the extension $1 \to N \to \Delta \to A \to 1$ splits. Henceforth we will write $A \cdot N$ for $\Delta$.

Let $H$ be the subgroup of $A$ which acts trivially on $N$ and let $A' = A/H$. Then the composition $G \hookrightarrow A \cdot N \to A \cdot N/H$ is injective because $G$ is torsion free and $H$ is finite. If $\Delta' = A \cdot N/H$ then one checks easily $\Delta' \simeq A' \cdot N$, so $G$ is isomorphic to a subgroup of $A' \cdot N$ where $A'$ acts effectively on $N$.

Let $\phi: \Pi_1(K) \to G$ be an isomorphism and define a homomorphism $g_*: G \to G$ by $g_* = \phi \circ f_* \circ \phi^{-1}$. Since $f$ is a finite covering, $f_*(\Pi_1(K))$ has finite index in $\Pi_1(K)$ and $f_*$ is injective, hence $g_*(G)$ has finite index in $G$ and $g_*$ is injective. Since $\Gamma$ is a uniform discrete subgroup of $N$, $G$ is a uniform discrete subgroup of $A' \cdot N$. And since $g_*(G)$ has finite index in $G$, it too is a uniform discrete subgroup of $A' \cdot N$.

$g_*$ is an isomorphism of $G$ to $g_*(G)$, so by a result of Auslander (Theorem 2 of [3]) $g_*$ extends to an automorphism $\bar{g}$ of $A' \cdot N$. Topologically $A' \cdot N$ is $A' \times N$ and $N$ is the connected component of the identity of $A' \cdot N$; hence $\bar{g}(N) = N$.

Let $M$ be the infra-nilmanifold $N/G$ and let $g: M \to M$ be the map induced by $\bar{g}$. Then $\Pi_1(M) \simeq G$ and $g_*: \Pi_1(M) \to \Pi_1(M)$ is simply the map $g_*: G \to G$, so the infra-nilmanifold endomorphism $g: M \to M$ is $\Pi_1$ conjugate to $f: K \to K$.

Clearly if $G$ itself is nilpotent then $\Delta = N$ so $M$ is a nilmanifold. If moreover, $G$ is abelian then $N = \Delta$ is abelian so $M$ is a torus. q.e.d.

(3.6) THEOREM. *If $f: M \to M$ is a $\Pi_1$ diffeomorphism and $\Pi_1(M)$ is torsion free, then*

(a) *If $\Pi_1(M)$ has a nilpotent subgroup of finite index then $f$ is topologically conjugate to a hyperbolic infra-nilmanifold automorphism.*

(b) *If $\Pi_1(M)$ is nilpotent then $f$ is topologically conjugate to a hyperbolic nilmanifold automorphism.*

(c) *If $\Pi_1(M)$ is abelian then $f$ is topologically conjugate to a hyperbolic toral automorphism.*

PROOF. Let $g: M' \to M'$ be the infra-nilmanifold endomorphism which is $\Pi_1$-conjugate to $f$ (which exists by (3.5)). Say $M' = N/G$ and $\bar{g}$ is the automorphism of $N$ which covers $g$. We will show $\bar{g}$ is hyperbolic.

If $\phi: \Pi_1(M') \to \Pi_1(M)$ is the $\Pi_1$-conjugacy, then since $\Pi_i(M) = 0$ for $i > 1$, (by (3.1)) and $\Pi_i(M') = 0$ for $i > 1$, by standard results of homotopy theory (see [21, p. 427]) there exists a homotopy equivalence $h: M' \to M$ such that $h_* = \phi$. Since $f$ is a $\Pi_1$ diffeomorphism we can choose $h$ such that $f \circ h = h \circ g$.

Let $\bar{h}: N \to \bar{M}$ be the lift of $h$, then by (3.4) $\bar{h}$ is proper. Let $\Sigma$ be the Lie algebra of $N$ and $\exp: \Sigma \to N$ the exponential map. $d\bar{g}$, the derivative of $\bar{g}$ at the identity of $N$, is an automorphism of $\Sigma$ and $\exp$ is a topological conjugacy between $\bar{g}$ and $d\bar{g}$ (see [8, p. 136]). Let $k = \bar{h} \circ \exp$, then $k: \Sigma \to M$ is a proper map such that $f \circ k = k \circ d\bar{g}$. We want to show that $\bar{g}$ is hyperbolic, that is, that $d\bar{g}$ has no eigenvalues of absolute value one.

Suppose to the contrary that $d\bar{g}$ has an eigenvalue $\lambda$ such that $|\lambda| = 1$. Then by (3.3) either $\lambda = \pm 1$ or there is a two-dimensional invariant subspace $V$ such that $d\bar{g}$ restricted to $V$ is a rotation.

We consider the two cases separately. If $\lambda = \pm 1$ let $x$ be an eigenvector such that $k(x)$ does not cover the base point of $M$ ($x$ can be found because $k$ is proper).

Let $D$ be the interval $\{tx|t \in [-1, 1]\}$ in $\Sigma$, and define $T: D \to D$ by $T(y) = y$ if $\lambda = 1$ or by $T(y) = -y$ if $\lambda = -1$. Let $k': D \to M$ be given by $k'(y) = P \circ k(y)$ where $P: \bar{M} \to M$ is the covering map.

Clearly $k' \circ T = f \circ k'$, but since $f$ is a $\Pi_1$ diffeomorphism and $D$ is simply connected, the only map with this property is the constant map, which maps all of $D$ to the base point of $M$. By construction, however, $k'(x)$ is not the base point of $M$, so we have a contradiction.

If $\lambda$ is complex let $D'$ be a disk in $V$ with center 0, which is invariant under the rotation of $d\bar{g}$, and sufficiently large that it contains a point $x'$ such that $k'(x')$ is not the base point of $M$ ($x'$ can be found because $k$ is proper). If $T': D' \to D'$ is $d\bar{g}$ restricted to $D'$ then as before $k' \circ T' = f \circ k'$ and $D'$ is simply connected, but $k'$ is not the constant map, so we have a contradiction. Thus it must be the case that $d\bar{g}$ has no eigenvalues of absolute value one, that is, $g$ is a hyperbolic infranilmanifold endomorphism. $\bar{g}$ is an automorphism, so $g$ is a covering map. Since $g_*$ is an isomorphism $g$ is in fact a diffeomorphism.

By (2.2) $g$ is a $\Pi_1$ diffeomorphism which is $\Pi_1$-conjugate to $f$, so by (3.2) $f$ and $g$ are topologically conjugate. q.e.d.

**4.** In this section we develop the properties of metrically splitting Anosov diffeomorphisms. An immediate result is the following.

(4.1) PROPOSITION. *A metrically splitting Anosov covering has a fixed point.*

PROOF. $S$ has a metric $\rho$ with respect to which $\bar{f}^{-1}$ (some lift of $f$) is a contraction. By (1.6) there is a $s \in S$ such that $\bar{f}^{-1}(s) = s$. By (1.2) $\bar{f}$ restricted to $s$ is a contraction in the metric $d(s; \ )$ and hence has a fixed point $x$. Clearly $\bar{f}(x) = x$ so $P(x) \in M$ is a fixed point for $f$. q.e.d.

*Question.* Does every Anosov covering on a compact manifold have a fixed point?

Throughout this section, if $f: M \to M$ is a metrically splitting Anosov covering. We will choose the fixed point $x_0$ of $f$ as a base point of $M$, and $\Pi_1(M)$ will mean $\Pi_1(M, x_0)$. All manifolds in this section will be assumed to have base points and all maps (except covering transformations) assumed to be base point preserving.

(4.2) THEOREM. *Metrically splitting Anosov diffeomorphisms are $\Pi_1$ diffeomorphisms.*

PROOF. Suppose that $f: M \to M$ is a metrically splitting Anosov diffeomorphism and $g: N \to N$ is a homeomorphism of a compact CW complex and $k: N \to M$ is a continuous map such that the diagram

$$\begin{array}{ccc} \Pi_1(N) & \xrightarrow{\phi} & \Pi_1(M) \\ \downarrow{g_*} & & \downarrow{f_*} \\ \Pi_1(N) & \xrightarrow{\phi} & \Pi_1(M) \end{array}$$

commutes, where $\phi = k_*$.

Let $\bar{N}$ be the simply connected covering space of $N$. Consider the function spaces

$$H = \{h: \bar{N} \to \bar{M} | h \circ \alpha = \phi(\alpha) \circ h \quad \text{for all } \alpha \in \Pi_1(N)\}$$
$$H^s = \{h: \bar{N} \to S | h \circ \alpha = \phi(\alpha) \circ h \quad \text{for all } \alpha \in \Pi_1(N)\}$$
$$H^u = \{h: \bar{N} \to U | h \circ \alpha = \phi(\alpha) \circ h \quad \text{for all } \alpha \in \Pi_1(N)\}.$$

$H$ is a closed subspace of the space of continuous mappings from $\bar{N}$ to $\bar{M}$ with the topology of uniform convergence. $H$ is homeomorphic to $H^s \times H^u$ with the homeomorphism given by $h \to (s \circ h, u \circ h)$.

Since $f$ is metrically splitting, $S$ has a complete metric $\rho$ such that

$$\rho(\bar{f}^{-n}s, \bar{f}^{-n}s') \le C\lambda^n \rho(s, s').$$

We define a complete metric $D$ on $H^s$ by $D(h_1, h_2) = \sup_{x \in N} \rho(h_1(x), h_2(x))$. Since $h_i(\alpha x) = \phi(\alpha)h_i(x)$ it follows that if $K$ is a compact fundamental domain in $\bar{N}$ then $D(h_1, h_2) = \sup_{x \in K} \rho(h_1(x), h_2(x))$ which is finite. The other properties of $D$ which must be checked to show that it is a complete metric are quite easy.

Consider the map $F: H \to H$ given by $F(h) = \bar{f}^{-1} \circ h \circ \bar{g}$. $F = F_1 \times F_2$: $H^s \times H^u \to H^s \times H^u$ where $F_1(h^s) = \bar{f}^{-1} \circ h^s \circ \bar{g}$ and $F_2(h^u) = \bar{f}^{-1} \circ h^u \circ \bar{g}$. Note that if $h_1, h_2 \in H^s$

$$
\begin{aligned}
D(F_1^n(h_1), F_1^n(h_2)) &= \sup_{x \in \bar{N}} \rho(\bar{f}^{-n} \circ h_1 \circ \bar{g}^n(x), \bar{f}^{-n} \circ h_2 \circ \bar{g}^n(x)) \\
&= \sup_{x \in \bar{N}} \rho(\bar{f}^{-n} \circ h_1(x), \bar{f}^{-n} \circ h_2(x)) \\
&\le C\lambda^n (\sup_{x \in \bar{N}} \rho(h_1(x), h_2(x)))
\end{aligned}
$$

where $C > 0$ and $\lambda \in (0, 1)$ because $\bar{f}^{-1}$ is contracting on $S$. Hence

$$D(F_1^n(h_1), F_1^n(h_2)) \le C\lambda^n D(h_1, h_2).$$

Thus $F_1$ is contracting on $H^s$ and by (1.6) there is a unique $h_0^s$ such that $F_1(h_0^s) = h_0^s$.

Let $Q = h_0^s \times H^u \subset H$, then $Q$ is a closed subspace of $H$ invariant under $F$ because $F(h_0^s, h^u) = (F_1(h_0^s), F_2(h^u)) = (h_0^s, F_2(h_u)) \in Q$.

We define a metric $\bar{D}$ on $Q$ by $\bar{D}(h_1, h_2) = \sup_{x \in \bar{N}} d(s(h_1(x)); h_1(x), h_2(x))$. This is well defined because $h_1, h_2 \in Q$ implies that $s \circ h_1 = s \circ h_2$ To check that $\bar{D}(h_1, h_2)$ is finite, note that for all $\alpha \in \Pi_1(N)$ and $h_1, h_2 \in Q$

$$
\begin{aligned}
d(s(h_1(\alpha x)); h_1(\alpha x), h_2(\alpha x)) &= d(\phi(\alpha)s(h_1(x)); \phi(\alpha)h_1(x), \phi(\alpha)h_2(x)) \\
&= d(s(h_1(x)); h_1(x), h_2(x)).
\end{aligned}
$$

Hence it suffices to take the sup over the compact fundamental domain $K \subset \bar{N}$. The metric properties of $\bar{D}$ follow easily. As before $d$ denotes the complete metric on $\bar{M}$ obtained from a Riemannian metric on $\bar{M}$ which was lifted from a Riemannian metric on $M$. Let $\bar{d}$ be the complete metric on $Q$ given by $\bar{d}(h_1, h_2) = \sup_{x \in \bar{N}} \bar{d}(h_1(x), h_2(x))$. Since $d(y_1, y_2) \le d(s; y_1, y_2)$ for any $y_1, y_2 \in s \in S$, we have $\bar{d}(h_1, h_2) \le \bar{D}(h_1, h_2)$ for any $h_1, h_2 \in Q$.

From (1.2) it follows that

$$
\begin{aligned}
\bar{D}(F^{-n}(h_1), F^{-n}(h_2)) &= \sup_{x \in \bar{N}} d(\bar{f}^n(s(h_1 \circ \bar{g}^{-n}(x))); \bar{f}^n \circ h_1 \circ \bar{g}^{-n}(x), \bar{f}^n \circ h_2 \circ \bar{g}^{-n}(x)) \\
&\le C'\lambda^n \sup_{x \in \bar{N}} d(s(h_1(x)); h_1(x), h_2(x))) \text{ where } C' > 0, \lambda \in (0, 1), \\
&= C'\lambda^n \bar{D}(h_1, h_2).
\end{aligned}
$$

From this one checks easily that for any $h \in Q$, the sequence $h$, $F^{-1}(h)$, $F^{-2}(h)$, ... is Cauchy in the metric $\bar{D}$ and hence in $\bar{d}$ since $\bar{d} \leq \bar{D}$. Since $\bar{d}$ is complete this sequence has a limit $h_0 \in Q$. Clearly $F(h_0) = h_0$, i.e., $\bar{f}^{-1} \circ h_0 \circ \bar{g} = h_0$. Since $F^{-1}$ is contracting in $\bar{D}$, this fixed point is unique in $Q$. Suppose $h_0'$ is any other fixed point of $F$ in $H$, then since $h_0^s$ is the unique fixed point of $F_1$ in $H^s$ we must have $s \circ h_0' = h_0^s$, so $h_0' \in Q$ which implies $h_0' = h_0$.

Since $h_0 \in H$ it covers a map $h \colon N \to M$ such that $h_* = \phi$. Since $\bar{f}^{-1} \circ h_0 \circ \bar{g} = h_0$ it is clear that $f \circ h = h \circ g$. q.e.d.

(4.3) THEOREM. *If* $f \colon M \to M$ *is a metrically splitting Anosov diffeomorphism and* $\Pi_1(M)$ *has nilpotent subgroup of finite index then* $f$ *is topoologically conjugate to a hyperbolic infra-nilmanifold endomorphism. If* $\Pi_1(M)$ *is abelian then* $f$ *is conjugate to a toral automorphism.*

PROOF. Since $f$ is splitting $\bar{M}$ is homeomorphic to $R^n$. By a result of P. A. Smith (see [4, p. 43]) there is no free action of $Z_p$ on $R^n$, hence $\Pi_1(M)$ is torsion free. The theorem now follows from (4.2) and (3.6). q.e.d.

REMARK. (2.1) is a special case of (4.2), that is, a hyperbolic toral automorphism is metrically splitting. However, it seems unlikely that all hyperbolic nilmanifold automorphisms are metrically splitting. If it should happen that the subspaces $\Sigma^s$ and $\Sigma^u$ mentioned in (2.2) commute, i.e., $[v, w] = 0$ for any $v \in \Sigma^s$, $w \in \Sigma^u$, then it is not difficult to show the hyperbolic nilmanifold automorphism will be metrically splitting. One of S. Smale's nilmanifold examples [19] does not satisfy this condition. It is easy to show that all the examples of the introduction are splitting. If it could be shown that a splitting Anosov diffeomorphism is a $\Pi_1$ diffeomorphism this would supercede (2.1), (2.2) and (4.2).

**5.** The purpose of this section is to prove that a codimension one Anosov diffeomorphism $f \colon M \to M$ is always splitting if $\mathrm{NW}(f) = M$. The proofs in this section presuppose some knowledge of foliations. A. Haefliger's paper [5] is a good reference for this.

(5.1) LEMMA. *Let* $f \colon M \to M$ *be a codimension one Anosov covering, then for any* $x, y \in M$, *there is at most one point in* $u(x) \cap s(y)$.

PROOF. The proof is by contradiction, so we assume there exist $x, y \in \bar{M}$ such that $x, y \in u(x) \cap s(y)$ and $x \neq y$. From this we derive a contradiction. We assume without loss of generality that $s(x)$ has dimension one and show first that there exists a closed loop transversal to the foliation $U$.

Let $h \colon [0, 1] \to s(x)$ be a diffeomorphism onto the arc in $s(x)$ joining $x$ and $y$; say $h(0) = x$, and $h(1) = y$. Let $\bar{f} \colon \bar{M} \to \bar{M}$ be any lift of $f$. Since $u(x) = u(y)$, it follows from (1.2) and (1.7) that there is an integer $n > 0$ such that $\bar{f}^{-n}(x)$ and $\bar{f}^{-n}(y)$ lie in the same product neighborhood $V$. Let $h'(t) = \bar{f}^{-n} \circ h(t)$, $0 \leq t \leq 1$. The arc $h'(t)$ like the arc $h(t)$ is everywhere transversal to $U$. Moreover, $h'$ must intersect $u(\bar{f}^{-n}(x))$ with the same orientation at $h'(0)$ and $h'(1)$, since otherwise there would be a point $h'(t_0)$, $0 < t_0 < 1$, where the orientation changed, i.e., a point where $h'$ is tangent to $U$. It is now easy to see how $h'$ can be altered in $V$

(and left fixed outside of $V$) to form a smooth closed loop $k(t)$ transversal to $U$ (see figure). Since $\overline{M}$ is simply connected $k$ is contractible and we can assume

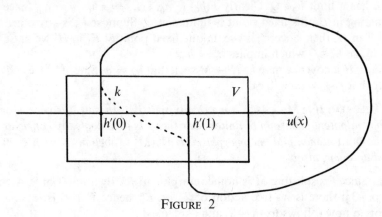

FIGURE 2

dim $M > 1$ (since the lemma is trivial in dim 1) so $k$ bounds a smoothly immersed disk. That is, there is a smooth immersion $\phi: D^2 \to \overline{M}$ such that $\phi|\partial D^2 = k$. Hence for every $t \in \partial D^2$, $\phi$ is transverse at $t$ to the leaf $u(\phi(t))$. It follows that the disk $\phi(D^2)$ is transverse to $U$ on some neighborhood of $\phi(\partial D^2)$. The intersection of $\phi(D^2)$ with leaves of $U$ gives a family of curves (perhaps singular where $\phi(D^2)$ is tangent to the foliation $U$) on $\phi(D^2)$. Since $\phi$ is a local diffeomorphism onto $\phi(D^2)$, these curves can be pulled back to $D^2$ to form, in effect, the integral curves of a vector field (with singularities) on $D^2$. From the construction of $\phi$, this vector field is nonsingular in a neighborhood of $\partial D^2$.

We assign an orientation to each of the curves by orienting $\phi(D^2)$ and assigning to the intersection of a leaf $u$ and $\phi(D^2)$ the orientation determined in the usual way from the orientation of $\overline{M}$, the orientation of $u$, and the orientation of $\phi(D^2)$. Since $\phi(D^2)$ is transverse to $U$ on a neighborhood of $\phi(\partial D^2)$ and $\phi(\partial D^2)$ is connected, the integral curves on $D$ crossing $\partial D$ must either all point inward or all point outward. We assume without loss of generality they point inward.

In the case dim $M = 2$ we immediately derive a contradiction because $\phi$ is a local diffeomorphism and the integral curves on $D$ are $\phi^{-1}$ applied to the leaves of $U$. Hence there can be no singularities on $D$. This and the fact that the integral curves crossing $\partial D$ all come inward contradict the Poincaré-Hopf formula (see [13]).

In the case of higher dimensions we use an argument quite similar to one used by Haefliger in proposition (4.2) of [5].

We wish to alter $\phi$ leaving it fixed near $\partial D$ so that $\phi(D)$ will be transverse to $U$ except at isolated "generic" tangencies. First choose a $C^\infty$ triangulation of $D$ so fine that for any simplex $\sigma$, $\phi|\sigma$ is an embedding and $\phi(\sigma)$ lies in a product neighborhood. We proceed skeleton by skeleton.

Let $D_0$ denote the zero skeleton of $D$ and let $x_0 \in D_0$. Let $Q$ be a compact neighborhood of $\partial D$ such that $\phi(Q)$ is transverse to $U$. We suppose $x_0 \notin Q$ and choose a neighborhood $R$ of $x_0$ which is disjoint from $Q$ and from $D_0 - \{x_0\}$. Let $A = \{\psi | \psi \in C^3(D, \overline{M}), \psi(x_0) = \phi(x_0)$ and $\psi|_{D-R} = \phi|_{D-R}\}$. If ev: $A \times D \to \overline{M}$ is the evaluation map it is clear that ev is transverse to $u(\phi(x_0))$ at $\phi(x_0)$. Hence we can apply a standard transversality theorem (19.1 of [1]) and conclude there is an approximation to $\phi$ which is equal to $\phi$ on $D - R$ and transversal to $U$ at $x_0$. Proceeding similarly for other points of $D_0$ we can construct $\phi_1: D \to \overline{M}$ such that $\phi_1|_Q = \phi|_Q$ and $\phi_1$ is transverse to $U$ at points of $D_0$. Moreover, we can assume that $\phi_1$ is a sufficiently good approximation to $\phi$ that $\phi_1$ is still an embedding when restricted to a simplex and the image of each simplex lies in a product neighborhood in $\overline{M}$.

We now proceed to the one skeleton $D_1$. Let $Q_1$ be a compact neighborhood of $Q \cup D_0$ on which $\phi_1$ is transverse to $U$. Let $L$ be a one simplex of $D$. Then $\phi(L) \subset V$, a product neighborhood in $\overline{M}$. The foliation $U$ is codimension one and hence is a $C^1$ foliation (see [7]). Thus there is a $C^1$ coordinate chart $h: V \to R^n$ which takes leaves of $U$ to $n - 1$ planes in $R^n$. Choose a neighborhood $R$ of $L$ such that $\overline{R}$ is compact. Choose a neighborhood $W$ of the singular points on $L$ such that $\overline{W} \subset R - Q_1$.

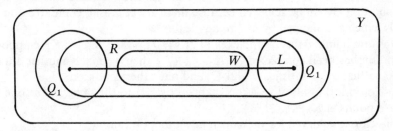

FIGURE 3

Choose $\delta$ such that any $\delta$ $C^1$ approximation to $h \circ \phi_1$ will still be transverse to $n - 1$ planes on $L - W$. Then by a standard approximation theorem (see 4.1 of [14]) if $Y$ is a sufficiently small neighborhood of $\overline{R}$, there is a map $\psi: Y \to R^n$ such that $\psi = h \circ \phi_1$ outside of $R - Q_1$, $\psi$ is a $\delta$ $C^1$ approximation to $h \circ \phi_1$, on $Y$ and $\psi$ is $C^\infty$ on $\overline{W}$.

Let $p: R^n \to R^1$ be projection on the coordinate perpendicular to the $n - 1$ planes. By a standard result of Morse theory (see [11] or [12]) we can find a close approximation $\psi'$ to $\psi$ such that $\psi = \psi'$ outside $W$ and in a neighborhood of $L$ there are only isolated points where $d(p \circ \psi') = 0$. That is, only isolated points where $\psi'$ is tangent to the $n - 1$ planes. We alter $L$ slightly, if necessary, to miss these points. Then let $\phi_1' = \phi_1$ outside $Y$ and $\phi_1'(x) = h^{-1} \circ \psi'(x)$ for $x \in Y$. $\phi_1'$ will be transverse to $U$ on $Q_1 \cup L$. Moreover, we can assume that the approximations were good enough that $\phi_1'$ restricted to any simplex is an embedding and the image of any simplex is still contained in a product neighborhood. Note that $\phi_1 = \phi_1'$ in a neighborhood of the endpoints of $L$.

In precisely the same manner we make an alteration for each of the other one simplexes of $D_1$ and obtain ultimately a map $\phi_2 : D \to \bar{M}$ such that $\phi_2$ is transverse to $U$ on $Q \cup D_1$ and $\phi_2$ restricted to any simplex is an embedding and the image under $\phi_2$ of any simplex is contained in a product neighborhood.

We now proceed to the two skeleton. Let $Q_2$ be a compact neighborhood of $Q$ and $D_1$. If $K$ is a two simplex in $D$ we want to alter $\phi_2$ on its interior so that there will be only isolated "generic" tangencies. Let $W'$ be a neighborhood of the singular points in $K$ such that $W' \subset K - Q_2$. We know there is a product neighborhood $V$ such that $\phi(K) \subset V$. Let $h: V \to R^n$ and $p: R^n \to R^1$ be as before.

Using the approximation theorem as before we get a map $\psi_2 : K \to R^n$ such that $\psi_2$ is $C^1$ close to $h \circ \phi_2$, $\psi_2 = h \circ \phi_2$ on $K \cup Q_2$, $\psi_2$ is $C^\infty$ on $\bar{W}'$ and all singularities of $\psi_2$ are still in the interior of $W'$.

Now using the Morse theory results again we alter $\psi_2$ to $\psi_2'$ which has the following properties:

(1) $\psi_2 = \psi_2'$ outside $W'$ and $\psi_2'$ is an embedding of $K$.

(2) The function $p \circ \psi_2'$ has isolated nondegenerate critical points, all in the interior of $W'$. That is, the critical points are of maximum, minimum or hyperbolic saddle type.

(3) No two critical points of $p \circ \psi_2'$ are mapped to the same leaf in $\bar{M}$ by $h^{-1}$.

Condition (3) can be achieved because there are at most a countable number of levels which are mapped by $h^{-1}$ to any leaf in $\bar{M}$.

We now number the 2-simplexes of $D$ say $K = K_0, K_1, \ldots, K_n$. Suppose the alteration has been completed on $K_0, \ldots, K_{i-1}$ then we make another alteration on $K_i$ so that $\psi_2^{(i)}|_{K_i}$ satisfies (1), (2), (3) and one other condition.

(4) No critical point of $p \circ \psi_2^{(i)}$ is mapped by $h_i^{-1}$ to the same leaf as one of the critical points in $K_0, K_1, \ldots, K_{i-1}$.

This can be achieved for the same reason that (3) can.

Now let $\phi_3 : D \to \bar{M}$ be given by $\phi_3|_{K_i} = h_i^{-1} \circ \psi_2^{(i)}$. Then $\phi_3$ is an immersion and as before induces a family of curves on $D$. From the construction of $\phi_3$ the family of curves has only isolated singularities and these are of maximum, minimum and saddle type. Moreover, because of condition (4) no two singularities are joined by a curve.

We now apply Poincaré-Bendixon theory to this situation (see [6, pp. 150–154]). Let $c: [0, \infty) \to D$ be one of the curves crossing $\partial D$. Let $\omega(c) = \{x | \exists\, t_0, t_2 \ldots$ such that $x = \lim_{n \to \infty} c(t_n)\}$. If $\omega(c) = x_0$, a single point, then $x_0$ is one of the saddle points and it is only possible to have two such curves for each saddle point. Hence we can find a curve $c'$ such that $\omega(c')$ consists of more than one point.

Suppose $\omega(c')$ contains a singular point. Then $\omega(c)$ consists of a finite set of singular points and a collection of curves joining them (see 4.2, p. 154 of [6]). But in our situation no two singular points are connected by a curve. It is also easy to see that $\omega(c')$ is bounded away from $\partial D$ so the only possibility is a figure eight crossing at the singular point. On the other hand if $\omega(c)$ contains no singular points then $\omega(c')$ is a circle which is one of the induced curves on $D$ (see 4.1, p. 151 of [6]).

In either case $\phi_3(\omega(c))$ is a loop in a single leaf of $U$ which represents a nontrivial holomony element of that leaf (see [5] particularly 4.2). But this contradicts the fact that each leaf of $U$ is simply connected and hence has trivial holonomy. This completes the proof of (5.1).

A proof of the following lemma was sketched for me by S. P. Novikov in a personal conversation.

(5.2) LEMMA. *Let $f: M \to M$ be a codimension one Anosov diffeomorphism with* $NW(f) = M$, *then for any* $x, y \in \bar{M}$, $u(x)$ *and* $s(y)$ *intersect in at least one point.*

PROOF. Suppose without loss of generality that dim $u(y) = 1$ (otherwise consider $f^{-1}$). The method of proof is to fix $y = y_0$ and show that $Q = \{x | s(x) \cap u(y_0) \neq \varnothing\}$ is both open and closed in $\bar{M}$ and thus, since $\bar{M}$ is connected, $Q = \bar{M}$.

The fact that $Q$ is open follows immediately from the local product structure of $\bar{M}$ and the well known fact (see [16]) that for any foliation the set of points lying on leaves which intersect a given open set is itself an open set. Hence we need only show that $Q$ is closed. Let $p \in$ closure $(Q)$, then because of the local product structure, $p$ is the limit of a sequence in $Q \cap u(p)$. $u(p)$ is a line and we pick a fixed ordering $<$ for it. Let $x_0 \in Q \cap u(p)$ and suppose $x_0 < p$. Since $Q$ is open, to show $p \in Q$ it suffices to show $p' = \sup\{x|$ if $x \geq y \geq x_0$ then $y \in Q\}$ is in $Q$. Let $\gamma(t)$ be an arc in $s(x_0)$ such that $\gamma(0) = x_0$ and $\gamma(1) = z_1 = s(x_0) \cap u(y_0)$.

We now make a preliminary supposition: For all $t \in [0, 1]$ and for all $x$ in $u(p')$ such that $x_0 \leq x < p'$, $s(x) \cap u(\gamma(t))$ is not empty and moreover if $\{x_i\}$ is any increasing sequence in $[x_0, p') \subset u(x_0)$ then the sequence $y_i'(t) = u(\gamma(t)) \cap s(x_i)$ is

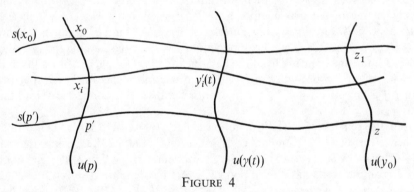

FIGURE 4

convergent in $u(\gamma(t))$. We show that in this case $p' \in Q$. Let $\{x_i'\}$ be an increasing sequence in $[x_0, p')$ converging to $p'$, then clearly for small $t$, $\{y_i'(t)\}$ converges to a point in $s(p')$. Because of the product structure it follows that the set of $t$ for which $\{y_i'(t)\}$ converges to a point in $s(p)$ is both open and closed. Hence in particular if $z = \lim y_i'(1)$ then $z \in u(y_0) \cap s(p')$ so $p'$ is in $Q$.

If, however, the supposition above is false we will derive a contradiction thereby completing the proof. In this case there exists an $r \in [0, 1]$ and an increasing sequence $\{x_i\}$ in $[x_0, p')$ such that the sequence $\{y_i\}$, where $y_i = u(\gamma(r)) \cap s(x_i)$, is divergent in $u(\gamma(r))$.

We remark that from (5.1) it follows that the foliation $S$ is proper and hence Lemma 1 of §3 of [5] shows that for any $x$, $\bar{M} - s(x)$ has two connected components. From this it is easy to see that the sequence $y_i$ is monotonic in $u(y_1)$.

Let $\lim x_i = p_1 \in u(p')$. We choose a product neighborhood $W$ of $p_1$ and assume $\{x_i\} \subset W$. Let $I$ be the line segment $W \cap u(p_1)$.

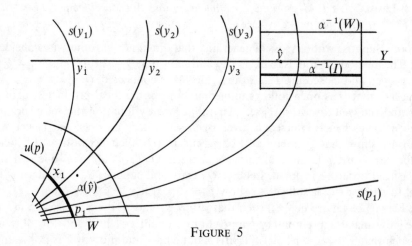

FIGURE 5

If $x \in I$ between $x_1$ and $p_1$ then $s(x)$ intersects $u(y_1)$, because if not, $u(y_1)$ lies entirely in one of the two components of $\bar{M} - s(x)$ but there are $x_i$'s in each component and for all $i$, $s(x_i)$ intersects $s(y_1)$. A similar argument shows $u(p) \cap s(y_1) = \varnothing$. Let $Y = y_1 \cup$ component of $u(y_1) - y_1$ containing the $y_i$'s. Then there is a homeomorphism $j$ from the half open interval $[x_1, p_1) \subset I$ to $Y$ given by $j(x) = s(x) \cap u(y_1)$. Continuity of $j$ and its inverse follow because $j$ is order preserving, $j$ is injective by (5.1), and $j$ is onto because $Y$ is connected. Note that $s(j(x)) = s(x)$.

From the remark after (1.10) it follows that $\{\alpha(Y) | \alpha \in \Pi_1(M)\}$ is dense in $\bar{M}$ and hence there is a $\hat{y} \in Y$ and an $\alpha \in \Pi_1(M)$ such that $\alpha(\hat{y}) \in W$. Consider the map $h: I \to Y$ given by $h(x) = s(\alpha^{-1}x) \cap u(y_1)$. Now define $g: I \to I$ by $g(x) = j^{-1} \circ h(x)$. Since $I$ is an interval $g$ has a fixed point $\hat{x}$. From the definitions $s(g(x)) = s(\alpha^{-1}x)$, so, since $g(\hat{x}) = \hat{x}$, we have $s(\hat{x}) = s(\alpha^{-1}\hat{x})$. But if $P: \bar{M} \to M$ is the covering map $P(\hat{x}) = P(\alpha^{-1}\hat{x})$ which contradicts the assertion of (1.4) that $P$ restricted to $s(\hat{x})$ is injective. q.e.d.

(5.3) THEOREM. *If $f: M \to M$ is a codimension one Anosov diffeomorphism with* $NW(f) = M$, *then $f$ is splitting.*

PROOF. If $u_0 \in U$ and $s_0 \in S$ then the map $h: \bar{M} \to u_0 \times s_0$ given by $h(x) = (s(x) \cap u_0, u(x) \cap s_0)$ is the desired homeomorphism. q.e.d.

**6.** We can now prove that a codimension one Anosov diffeomorphism whose nonwandering set is the entire manifold will be topologically conjugate to a hyperbolic automorphism of a torus. We first show that the fundamental group of a manifold admitting such a diffeomorphism is free abelian and that the diffeomorphism is metrically splitting.

(6.1) PROPOSITION. *If* $f: M \to M$ *is a splitting codimension one Anosov diffeomorphism then* $\Pi_1(M)$ *is free abelian.*

PROOF. Suppose dim $E^u = 1$ then $S$ is homeomorphic to a line. $\Pi_1(M)$ acts as a group of homeomorphisms of $S$. These homeomorphisms are fixed point free because, if $\alpha \neq 0$ and $\alpha(s) = s$ then for any point $x \in s$, $P(x) = P(\alpha x)$ contradicting (1.4).

Moreover since $f$ is a diffeomorphism (1.10) implies every orbit of $\Pi_1(M)$ is dense in $S$. Hence by Theorem 3 of [17] $\Pi_1(M)$ is free abelian. q.e.d.

(6.2) PROPOSITION. *Let* $f: M \to M$ *be an Anosov diffeomorphism with* NW$(f) = M$ *and* dim $E^u = 1$, *then* $f$ *is metrically splitting.*

PROOF. We pick a base point $p$ for $M$, and let $\bar{f}$ be some lift of $f: (M, p) \to (M, f(p))$. By (5.3) $f$ is splitting so $S$ is homeomorphic to $R$. We must construct a complete metric on $S$ with respect to which $\bar{f}^{-1}$ is contracting.

If $s_1, s_2 \in S$ there is a natural map $T(s_1, s_2): s_1 \to s_2$ given by projection along the leaves of $U$. We will define the metric $\rho$ on $S$ by $\rho(s, s') = \sup_{x \in s} d(u(x);$ $x, T(s, s')(x))$. There are several things to check.

For each $\alpha \in \Pi_1(M, p)$ we define a homeomorphism $T(\alpha)$ of $\bar{M}$ to itself by $T(\alpha)(x) = \alpha s(x) \cap u(x)$. By (6.1) $\Pi_1(M, p)$ is abelian so if $\alpha, \beta \in \Pi_1(M, p)$,

$$T(\alpha)(\beta x) = \alpha s(\beta x) \cap u(\beta x) = \beta \alpha s(x) \cap \beta u(x) = \beta T(\alpha)(x).$$

Suppose now that $s' = \alpha s$, then $T(s, s') = T(\alpha)$ restricted to $s$, so

$$\rho(s, s') = \sup_{x \in s} d(u(x); x, T(\alpha)(x))$$
$$\leq \sup_{x \in \bar{M}} d(u(x); x, T(\alpha)(x)).$$

But for any $\beta \in \Pi_1(M, p)$

$$d(u(\beta x); \beta x, T(\alpha)(\beta x)) = d(\beta u(x); \beta x, \beta T(\alpha)(x))$$
$$= d(u(x); x, T(\alpha)(x))$$

because $\beta$ is an isometry of the metric on $\bar{M}$. Thus if $K$ is a compact fundamental domain of $\bar{M}$

$$\sup_{x \in M} d(u(x); x, T(\alpha)(x)) = \sup_{x \in K} d(u(x); x, T(\alpha)(x)) < \infty$$

since $K$ is compact. Hence in this case $\rho(s, s')$ is finite.

Now fix $s$, it follows from (1.10) that $\{\alpha s | \alpha \in \Pi_1(M, p)\}$ is dense in $S$. For an arbitrary $s'$ pick $\alpha \in \Pi_1(M, p)$ so that $s'$ is between $s$ and $\alpha s$, then

$$\rho(s, s') = \sup_{x \in s} d(u(x); x, T(s, s')(x))$$
$$\leq \sup_{x \in s} d(u(x), x, T(\alpha)(x)) = \rho(s, \alpha s) < \infty.$$

Thus $\rho(s, s')$ is always finite. It follows easily from the definition of $\rho$ that it is indeed a metric on $S$. We must check that the topology $\rho$ induces on $S$ is the usual one.

If $u$ is a leaf of $U$, then the usual topology on $S$ is induced by the metric $d_u$ defined by $d_u(s, s') = d(u; y, y')$ where $y = u \cap s$ and $y' = u \cap s'$. Let $B_r(s_0) = \{s | \rho(s, s_0) < r\}$ and $B'_r(s_0) = \{s | d_u(s, s_0) < r\}$. Clearly $B_r(s_0) \subset B'_r(s_0)$. Hence we need only show that given $r$ and $s_0$, there exists $t > 0$ such that $B'_t(s_0) \subset B_r(s_0)$.

If $\alpha \in \Pi_1(M, p)$ then $\bar{f} \circ \alpha \circ \bar{f}^{-1} = f_*(\alpha) \in \Pi_1(M, f(p))$. We define $T(f_*^{-1}(\alpha))$ in the same manner as $T(\alpha)$ and it follows easily that $T(f_*^{-1}(\alpha)) = \bar{f}^{-1} \circ T(\alpha) \circ \bar{f}$.

If $y = \bar{f}^n(x)$ then

$$d(u(x); x, T(f_*^{-n}(\alpha))(x)) = d(u(x); x, \bar{f}^{-n} \circ T(\alpha) \circ \bar{f}^n(x))$$
$$= d(\bar{f}^{-n}u(y); \bar{f}^{-n}(y), \bar{f}^{-n} \circ T(\alpha)(y))$$
$$\leq C\lambda^n d(u(y); y, T(\alpha)(y)) \quad \text{by (1.2)}.$$

So $\sup_{x \in \bar{M}} d(u(x); x, T(f_*^{-n}(\alpha))(x)) \leq C\lambda^n \sup_{y \in \bar{M}} d(u(y); y, T(\alpha)(y))$. Hence we can choose $m > 0$ such that for $f_*^{-m}(\alpha) = \alpha' \in \Pi_1(M, f^{-m}(p))$ we have $\sup_{x \in \bar{M}} d(u(x); x, T(\alpha')(x)) < r$.

$\alpha'$ moves $s_0$ to one side and $\alpha'^{-1}$ moves it to the opposite side. Let $s_1 = \alpha' s_0$ and $s_2 = \alpha'^{-1} s_0$. From the construction of $\alpha'$ we have $\rho(s_0, s_1) < r$ and $\rho(s_0, s_2) < r$. Let $t = \min \{d_u(s_0, s_1), d_u(s_0, s_2)\}$, then $B'_t(s_0) \subset B_r(s_0)$. Thus the topology induced by $\rho$ on $S$ is the usual one. Suppose $\{s_i\}_{i=1}^\infty$ is a sequence in $S$ which is Cauchy in the metric $\rho$, then since $d_u \leq \rho$ the sequence is Cauchy in the complete metric $d_u$ and hence has a limit. Thus $\rho$ is a complete metric.

We now need only check that $\bar{f}^{-1}$ is contracting in $\rho$ and that $\Pi_1(M, p)$ acts isometrically on $S$ in the metric $\rho$. But since $\alpha \in \Pi_1(M, p)$ is an isometry of the Riemannian metric on $\bar{M}$ it follows immediately from the definition of $\rho$ that $\rho(s, s') = \rho(\alpha s, \alpha s')$. Also

$$\rho(\bar{f}^{-n}s, \bar{f}^{-n}s') = \sup_{x \in \bar{f}^{-n}s} d(u(x); x, T(\bar{f}^{-n}s, \bar{f}^{-n}s')(x))$$
$$= \sup_{x \in s} d(u(\bar{f}^{-n}(x)); \bar{f}^{-n}(x), \bar{f}^{-n}T(s, s')(x))$$
$$\leq C\lambda^n \sup_{x \in s} d(u(x); x, T(s, s')(x)), \quad \text{by (1.2)}$$
$$= C\lambda^n \rho(s, s')$$

so $\bar{f}^{-1}$ is contracting. q.e.d.

(6.3) THEOREM. *Let $f: M \to M$ be a codimension one Anosov diffeomorphism with $NW(f) = M$, then $f$ is topologically conjugate to a hyperbolic toral automorphism. If $g$ is another such diffeomorphism then $f$ and $g$ are topologically conjugate iff they are $\Pi_1$-conjugate.*

PROOF. By (6.2) $f$ (or $f^{-1}$) is metrically splitting and hence by (4.3) $f$ is conjugate to an infra-nilmanifold endomorphism. By (6.1) $\Pi_1(M)$ is abelian so in fact, $f$ is topologically conjugate to a hyperbolic toral automorphism. If $g$ is another such diffeomorphism then by (3.2) $f$ and $g$ are topologically conjugate iff they are $\Pi_1$-conjugate. q.e.d.

If dim $M \leq 3$ then any Anosov diffeomorphism must be codimension one. Hence we have the following.

(6.4) COROLLARY. *If $f: M \to M$ is an Anosov diffeomorphism with* $NW(f) = M$ *and* dim $M \leq 3$ *then $f$ is topologically conjugate to a hyperbolic toral automorphism. If $g$ is another such diffeomorphism then $f$ and $g$ are topologically conjugate iff they are* $\Pi_1$-*conjugate.*

**7.** In the case of two-dimensional manifolds the hypothesis that $NW(f) = M$ in (6.4) can be eliminated. S. Smale showed me a direct proof that for an Anosov diffeomorphism of a two manifold the nonwandering set is the entire manifold. We give here essentially his proof.

This section deals with diffeomorphisms satisfying

AXIOM A. (1) $NW(f)$ has a hyperbolic structure.

(2) $Per(f)$ is dense in $NW(f)$

and

AXIOM B. If $\Omega_1$ and $\Omega_2$ are basic sets and $W^s(\Omega_1) \cap W^u(\Omega_2) \neq \varnothing$ then there are periodic points $p \in \Omega_1$, $q \in \Omega_2$ such that $W^s(p)$ and $W^u(q)$ have a point of transversal intersection.

These properties and diffeomorphisms which satisfy them are considered in detail in Smale's paper [19]. We assume a knowledge of the terminology and results of this paper. It follows easily from the definition and (1.7) that all Anosov diffeomorphisms satisfy Axioms A and B.

(7.1) PROPOSITION. *If $f: M \to M$ is a diffeomorphism satisfying Axioms A and B, and $p \in \Omega \subset NW(f)$ where $\Omega$ is a basic set, then $W^s(p)$ and $W^u(p)$ are dense in the component of $p$ in $\Omega$.*

There is a local product structure on $NW(f)$ (see (7.4) of [19]) so the proof of this proposition is exactly the same as the proofs of (1.8) and (1.9).

Let $M^2$ denote a two-dimensional compact manifold without boundary.

(7.2) PROPOSITION. *If $f: M^2 \to M^2$ is an Anosov diffeomorphism then* $NW(f) = M^2$.

PROOF. Assume first that $M^2$ is oriented, then since it admits a nowhere zero line field $M^2$ must be a torus.

Since $f$ satisfies Axioms A and B we can apply the spectral decomposition theorem of [19]. We choose a basic set $\Lambda$ which is a sink. Suppose $\Lambda \neq M^2$, from this we derive a contradiction. From Proposition 5.1 of [20], it follows that there exists a compact neighborhood $V$ of $\Lambda$ such that $f(V) \subset$ int $V$ and if $\Lambda \neq M^2$ then $V \neq M^2$. We can choose $V$ so that $\partial V$ is a finite set of circles embedded in $M^2$.

We wish to show that rank $H_1(V) \geq 2$. Suppose not, then one checks easily that $f_*: H_1(V) \to H_1(V)$ is an isomorphism and by replacing $f$ by $f^2$ if necessary we can assume $f_*: H_1(V) \to H_1(V)$ is the identity. Let $p$ be a periodic point in $\Lambda$, then because (again by (5.1) of [20]) $\bigcap_{n \geq 0} f^n(V) = \Lambda$, it follows that $W^u(p) \subset \Lambda$. Since $W^s(p)$ is dense in $\Lambda$ by (7.1), it must intersect $W^u(p)$ densely.

Since $NW(f) = NW(f^n)$ for all $n$, we can assume that $p$ is a fixed point of $f$.

Let $\gamma(t)$ be a loop in $M^2$ consisting of two arcs, one in $W^s(p)$, the other in $W^u(p)$, such that $\gamma(0) = \gamma(1) = p$. Successive application of $f$ will make the arc in $W^s(p)$ smaller, hence if we replace $\gamma$ by $f^n \circ \gamma$ for some large $n$ then the image of the new $\gamma$ will lie entirely in $V$.

Let $P: \overline{V} \to V$ be the simply connected covering of $V$ and $\overline{f}: \overline{V} \to \overline{V}$ the lift of $\overline{f}$. We choose $p$ as base point of $V$ and $p_1 \in P^{-1}(p)$ as the base point of $\overline{V}$. The loop $\gamma$ lifts to an arc $\overline{\gamma}$ in $\overline{V}$ with $\overline{\gamma}(0) = p_1$; let $p_2 = \overline{\gamma}(1)$. Let $\gamma'(t) = f \circ \gamma(t)$, then since $f_*: H_1(V) \to H_1(V)$ is the identity $\gamma'$ lifts to an arc $\overline{\gamma}'$ in $\overline{V}$ with $\overline{\gamma}'(0) = p_1$ and $\overline{\gamma}'(1) = p_2$.

The foliations on $V$ lift to foliations on $\overline{V}$ (leaves of these foliations will be denoted in the same way as leaves of the foliations of $\overline{M}$). The arc $\overline{\gamma}$ consists of an arc $I$ in $u(p_1)$ and an arc $J$ in $s(p_2)$. If $q$ is the joint endpoint of these two arcs then $q \in u(p_1) \cap s(p_2)$. Also $\overline{f}(q) \in u(p_1) \cap s(p_2)$, so the segment $I' = \overline{f}(I)$ intersects $s(p_2)$ in two places. There is an injection $j: V \to M^2$ and its lift $\overline{j}: \overline{V} \to \overline{M}^2$ will preserve the foliations. Hence $\overline{j}(I')$ is an arc in $u(\overline{j}(p_1))$ crossing $s(\overline{j}(p_2))$ in two places, namely $\overline{j}(q)$ and $\overline{j}(\overline{f}(q))$. However, by (5.1) this is impossible. Hence we have contradicted the assumption that rank $H_1(V) < 2$.

Now let $W$ be the closure of the complement of $V$ in $M^2$. Say $W = \bigcup W_i$ where the $W_i$ are the connected components of $W$. We have $f^{-1}(W) \subset \text{int } W$ and by replacing $f^{-1}$ by $f^{-n}$ for some large $n$ if necessary, we can assure $f^{-1}(W_i) \subset \text{int } W_i$ for all $i$. Thus, precisely the argument used above shows rank $H_1(W_i) \geq 2$ for all $i$.

If $\chi(M^2)$ is the Euler characteristic then we have

$$0 = \chi(M^2) = \chi(V) + \sum \chi(W_i) - \chi(\partial V).$$

But $\chi(V)$ and $\chi(W_i)$ are $< 0$ because rank $H_1(V) \geq 2$ and rank $H_1(W_i) \geq 2$ and $H_2(V) = H_2(W_i) = 0$. Since $\partial V$ is a finite union of circles, $\chi(\partial V) = 0$. Hence $\chi(V) + \sum \chi(W_i) - \chi(\partial V)$ is strictly negative.

Thus we have contradicted the assumption that $\Lambda \neq M^2$. Hence $NW(f) = M^2$.

Now suppose $M^2$ is not oriented, then if $\tilde{f}: \tilde{M}^2 \to \tilde{M}^2$ is the lift of $f$ to the oriented covering then $NW(\tilde{f}) = \tilde{M}^2$. So by (1.7) $Per(\tilde{f})$ is dense in $\tilde{M}^2$ and hence $Per(f)$ is dense in $M^2$. This clearly implies $NW(f) = M^2$. q.e.d.

(7.3) THEOREM. *If $f: M^2 \to M^2$ is an Anosov diffeomorphism then $f$ is topologically conjugate to a hyperbolic toral automorphism. Any two such diffeomorphisms are topologically conjugate iff they $\Pi_1$-conjugate.*

The proof is immediate from (7.2) and (6.4).

**8.** In this final section we apply some previous results to the study of expanding maps, that is, Anosov coverings with $TM = E^u$. Expanding maps have been studied extensively by M. Shub in [**18**] and we will assume some familiarity with his paper.

(8.1) THEOREM. *Expanding maps are $\Pi_1$ coverings.*

This is essentially Theorem 4 of Shub's paper [18].

(8.2) THEOREM. *If $f: M \to M$ is an expanding map and $\Pi_1(M)$ has a nilpotent subgroup of finite index then $f$ is topologically conjugate to an expanding infra-nilmanifold endomorphism. If $\Pi_1(M)$ is nilpotent then $f$ is conjugate to a nilmanifold endomorphism and if $\Pi_1(M)$ is abelian $f$ is conjugate to a toral endomorphism.*

PROOF. Shub shows in [18] that $\Pi_1(M)$ is torsion free. Hence by (3.5) there is an infra-nilmanifold endomorphism $g: M' \to M'$ which is $\Pi_1$ conjugate to $f$.

$\Pi_i(M) = \Pi_i(M') = 0$ for $i > 1$ so by a standard homotopy theorem ([21, p. 427]) there is a homotopy equivalence $h': M' \to M$ such that $h'_*: \Pi_1(M') \to \Pi_1(M)$ is the $\Pi_1$ conjugacy from $g$ to $f$. Since $f$ is a $\Pi_1$ covering, there exists a map $h: M' \to M$ homotopic to $h'$ such that $f \circ h = h \circ g$.

Let $N$ be the simply connected nilpotent Lie group covering $M'$ and let $\bar{f}: \bar{M} \to \bar{M}, \bar{g}: N \to N$, and $\bar{h}: N \to \bar{M}$ be the lifts of $f$, $g$, and $h$ respectively. Then $\bar{f} \circ \bar{h} = \bar{h} \circ \bar{g}$. If $\Sigma$ is the Lie algebra of $N$ and $\exp: \Sigma \to N$ the exponential map then $k = \bar{h} \circ \exp$ will satisfy $\bar{f} \circ k = k \circ d\bar{g}$ where $d\bar{g}: \Sigma \to \Sigma$ is the derivative of $\bar{g}$ at the identity of $N$.

From this it follows that $\bar{f}^n \circ k = k \circ d\bar{g}^n$ for all $n$. We wish to show that all eigenvalues of $d\bar{g}$ are greater than one in absolute value. Suppose $d\bar{g}$ has an eigenvalue $\lambda$ such that $|\lambda| \leq 1$, then if $x \in \Sigma$ is a corresponding eigenvector, $|d\bar{g}^n(x)| \leq |x|$ for all $n \geq 0$. Since $\bar{f}$ is expanding, for all $y \in \bar{M}$ except the base point $p, \bar{f}^n(y)$ tends to infinity as $n \to \infty$ (that is, given any compact set in $\bar{M}$ for all sufficiently large $n, \bar{f}^n(y)$ will be outside that set). Hence since $\bar{f}^n \circ k = k \circ d\bar{g}^n$, $k(x) = p$. The same argument shows that for any real number $r$, $k(rx) = p$. This contradicts the assertion of (3.4) that $\bar{h}$, and hence $k$, is proper (exp is a diffeomorphism, see p. 136 of [8]).

Thus $d\bar{g}$ is expanding and it follows easily that $g$ is an expanding map. By (8.1) $g$ is a $\Pi_1$ covering so by (3.2) $f$ and $g$ are topologically conjugate. q.e.d.

In view of (8.2) it is of interest to find out what kinds of groups can be the fundamental group of a manifold which admits an expanding map.

Given a finitely generated group and a set of generators, the growth function of the group, $\gamma(s)$, is defined to be the number of distinct elements of the group which can be written as words of length $\leq s$ in the generators. A group is said to have polynomial growth if $\gamma(s)$ is majorized by a polynomial.

(8.3) THEOREM. *If $f: M \to M$ is an expanding map, then $\Pi_1(M)$ has polynomial growth.*

PROOF. Let $P: (\bar{M}, p) \to (M, p_0)$ be the simply connected covering. Choose a set of generators $\{\alpha_i\}$ of $\Pi_1(M)$ and let $\mu = \max_i d(p, \alpha_i(p))$. Let the set function $\#(\quad)$ assign to any finite set the number of elements in that set, and let $B(r)$ denote the ball of radius $r$ about $p$ in $\bar{M}$. Let $L = \{\alpha(p) | \alpha \in \Pi_1(M)\}$.

If $\gamma(s)$ is the growth function of $\Pi_1(M)$ corresponding to the generators $\{\alpha_i\}$, then we have $\gamma(s) \leq \#(L \cap B(\mu s))$ for all $s$.

It follows from (1.2) that $d(\bar{f}^n(x), \bar{f}^n(y)) \geq C\lambda^{-n}d(x, y)$ for all $x, y \in \bar{M}$. Hence by replacing $f$ by $f^N$, where $N$ is chosen so $C\lambda^{-N} > 2$, we can assume $d(\bar{f}^n(x), \bar{f}^n(y)) \geq 2^n d(x, y)$.

Choose an integer $t > 1$, such that $B(\mu t)$ is a fundamental domain and let $q = \max_{x \in M} \#(P^{-1}(x) \cap B(\mu t))$.

We wish to get an estimate on $\#(L \cap \bar{f}^n(B(\mu t)))$. Note that $\#(L \cap \bar{f}^n(B(\mu t))) = \#(\bar{f}^{-n}(L) \cap B(\mu t))$. Suppose now that $k = \deg f$, then $k > 1$. The set $f^{-n}(p_0)$ in $M$ contains $k^n$ points. Since $\bar{f}^{-n}(L) = P^{-1}(f^{-n}(p_0))$, we have $\#(\bar{f}^{-n}(L) \cap B(\mu t)) \leq qk^n$. Thus $\#(L \cap \bar{f}^n(B(\mu t))) \leq qk^n$.

One checks easily that $f^n(B(\mu t)) \supset B(2^n \mu t)$ and hence $\#(L \cap B(2^n \mu t)) \leq \#(L \cap \bar{f}^n(B(\mu t)))$. Combining inequalities gives

$$\gamma(2^n t) \leq \#(L \cap B(2^n \mu t)) \leq \#(L \cap \bar{f}^n(B(\mu t))) \leq qk^n.$$

Now choose a positive integer $m$ such that $2^m > k$ and $t^m > q$. Then,

$$\gamma(2^n t) \leq qk^n < 2^{mn} t^m = (2^n t)^m \quad \text{for all } n > 0.$$

Recall that $\gamma(s)$ is monotonic increasing and consider an arbitrary integer $s > 2^m t$. Say $2^i t \leq s < 2^{i+1} t$ where $i \geq m$. Then we have

$$\gamma(s) \leq \gamma(2^{i+1} t) \leq (2^{i+1} t)^m = 2^{mi+m} t^m.$$

Since $i \geq m$, $2^{mi+m} t^m \leq 2^{mi+i} t^{m+1} = (2^i t)^{m+1} \leq s^{m+1}$. Thus for all $s > 2^m t$ we have $\gamma(s) \leq s^{m+1}$. Let $c = \max_{s \leq 2^m t} \gamma(s)$, $\gamma(s) \leq s^{m+1} + c$, so $\gamma(s)$ has polynomial growth. q.e.d.

(8.4) THEOREM. *If $f: M \to M$ is an expanding map and $\Pi_1(M)$ has a solvable subgroup of finite index, then $f$ is topologically conjugate to an expanding infra-nilmanifold endomorphism.*

PROOF. If $G$ is the solvable subgroup of finite index in $\Pi_1(M)$ then since $\Pi_1(M)$ is finitely generated so is $G$. Hence by (8.3) $G$ must have polynomial growth. Results of J. Milnor and J. Wolf (see [22]) show that a solvable group with polynomial growth has a nilpotent subgroup of finite index. Hence $\Pi_1(M)$ has a nilpotent subgroup of finite index. The result now follows from (8.2). q.e.d.

## BIBLIOGRAPHY

**1.** R. Abraham and J. Robbin, *Transversal mappings and flows*, Benjamin, New York, 1967.

**2.** D. V. Anosov, *Roughness of geodesic flows on closed Riemannian manifolds of negative curvature*, Soviet Math. Dokl. 3 (1962), 1068–1070.

**3.** L. Auslander, *Bieberbach's theorems on space groups and discrete uniform subgroups of Lie groups*, Ann. of Math. (2) **71** (1960), 579–590.

**4.** A. Borel et al., *Seminar on transformation groups*, Princeton Univ. Press, Princeton, N.J., 1960.

**5.** A. Haefliger, *Varietes feuilletees*, Ann. Scuola Norm. Sup. Pisa (3) **16** (1962), 367–397.

**6.** P. Hartman, *Ordinary differential equations*, Wiley, New York, 1964.

**7.** M. W. Hirsch and C. Pugh, *Stable manifolds for hyperbolic sets*, to appear.

**8.** G. Hochschild, *The structure of Lie groups*, Holden Day, San Francisco, 1965.

**9.** S. MacLane, *Homology*, Academic Press, New York, 1963.

**10.** A. Malcev, *On a class of homogeneous spaces*, Amer. Math. Soc. Translations No. 39, Amer. Math. Soc., Providence, R.I., 1949.

**11.** J. Milnor, *Lectures on the h-cobordism theorem,* Princeton Univ. Press, Princeton, N.J., 1965.

**12.** ———, *Morse theory,* Princeton Univ. Press, Princeton, N.J., 1963.

**13.** ———, *Topology from the differentiable viewpoint,* Univ. of Virginia Press, Charlottesville, Virginia, 1965.

**14.** J. R. Munkres, *Elementary differential topology,* Princeton Univ. Press, Princeton, N.J., 1961.

**15.** L. Pontrjagin, *Topological groups,* Princeton Univ. Press, Princeton, N.J., 1939.

**16.** G. Reeb, *Sur certaines proprietes topologiques des varietes feuilletees,* Actualités Sci. Indust., Hermann, Paris, 1952.

**17.** R. Sacksteder, *Groups and pseudogroups acting on $S^1$ and $R^1$,* Columbia University.

**18.** M. Shub, *Endomorphisms of compact differentiable manifolds,* Thesis, Univ. of California, Berkeley, 1967.

**19.** S. Smale, *Differentiable dynamical systems,* Bull. Amer. Math. Soc. **73** (1967), 747–817.

**20.** ———, *The $\Omega$-stability theorem,* these Proceedings, vol. 14.

**21.** E. Spanier, *Algebraic topology,* McGraw Hill, New York, 1966.

**22.** J. Wolf, *The growth of finitely generated solvable groups and the curvature of riemannian manifolds,* J. Diff. Geom. **2** (1968), 421–446.

MASSACHUSETTS INSTITUTE OF TECHNOLOGY

# ENDOMORPHISMS OF THE
# RIEMANN SPHERE

JOHN GUCKENHEIMER[1]

**Introduction.** Our primary goal is to give a description of the limit (or basic) sets of rational functions on the Riemann sphere. We characterize these basic sets in an open, dense set of the space of rational functions. For a rational function in this open, dense set, the nonwandering set of the function consists of a finite number of periodic sinks and an indecomposable, perfect set of sources. This perfect set of sources is described as the quotient of a symbol space under a particular kind of quotient map (see Proposition 5.9). An inverse limit construction then gives an example of a diffeomorphism which satisfies Smale's Axioms A and B and whose nonwandering set is not locally the product of a manifold and a Cantor set. This answers in the negative a question of Smale [7]. We give very limited results on decomposing spaces of quadratic functions according to the topological type of the basic sets. Finally, a few comments are made about morphisms of other algebraic manifolds.

The motivation for this problem comes from two directions. Several people, including Fatou, Julia, and Ritt, independently developed an extensive theory concerning the iteration of rational functions in the late 1910's. Their results are elegant, but do not seem to be readily accessible. We attempt to remedy this situation by giving a short, elementary presentation of the results which we need from this theory.

On the other hand, the theory of dynamical systems has undergone remarkable growth in the last decade. The reader is referred to Smale [7] for a survey of many of these developments. The problem of finding generic properties of the space of flows and the group of diffeomorphisms on a compact manifold has been the object of intensive investigation. One of the principle questions in this area is to describe the conjugacy or $\Omega$-conjugacy classes of maps and vector fields. Recent progress has been made by Shub in describing conjugacy classes of expanding maps [6]. Our task is to describe $\Omega$-conjugacy classes of rational functions; we use the techniques of dynamical systems to study the iteration of rational functions.

I wish to thank Stephen Smale without whose encouragement and advice this would not have been written.

**1. Definitions and Statement of Results.** Let $M$ be a smooth Riemannian manifold and $f: M \to M$ be a smooth map.

[1] The author was supported by an NSF Graduate Fellowship while this paper was written.

DEFINITION 1.1. A point $p \in M$ is *nonwandering* if for every neighborhood $U$ of $p$, there exists an $n > 0$ such that $f^n(U) \cap U \neq \emptyset$. $f^n = f \circ f \circ \ldots \circ f$ ($n$-factors) here. The set of nonwandering points of $f$ will be denoted $\Omega(f)$.

REMARK 1.2. $\Omega(f)$ is closed. Thus nonwandering is the weakest possible form of recurrence.

DEFINITION 1.3. Assume $X \subset M$ and $f(x) = X$. $X$ is said to have a *hyperbolic structure* if there is a continuous splitting of the tangent bundle of $M$ restricted to $X$; $T_x(M) = T_x^u(M) + T_x^s(M)$ such that (1) $Df|_x$ preserves this splitting and (2) there exist $0 < \lambda < 1 < \mu$, $c > 0$ such that
(a) $x \in T_x^u(M) \Rightarrow \|Df^n(x)\| \geq c\mu^n\|x\|$,
(b) $x \in T_x^s(M) \Rightarrow \|Df^n(x)\| \leq c\lambda^n\|x\|$.

REMARK 1.4. If $M$ is compact, this definition is independent of the Riemannian metric on $M$.

DEFINITION 1.5. $f$ is *hyperbolic* if $\Omega(f)$ has a hyperbolic structure and the periodic points of $f$ are dense in $\Omega(f)$.

Denote by $\mathscr{S}_d$ the space of holomorphic functions of the Riemann sphere to itself of degree $d$. In terms of a coordinate system on the punctured sphere, the elements of $\mathscr{S}_d$ are given as rational functions with $d$ equal to the maximum of the degrees of numerator and denominator. $\mathscr{S}_d$ has a natural structure as a complex manifold of (complex) dimension $2d - 2$. This structure is defined as follows: the rational functions of degree $d$ form an open, dense set of $P^{2d+1}(C)$ by associating to a rational function the point of projective space with homogeneous coordinates the coefficients of the rational function. The group of fractional transformations acts freely on the space of rational functions by conjugation. The coset space is $\mathscr{S}_d$.

Of primary concern here is the fact that $\mathscr{S}_d$ has a natural topology. Let $\mathscr{H}_d \subset \mathscr{S}_d$ be the set of hyperbolic rational maps of degree $d$.

THEOREM A. $\mathscr{H}_d$ *is dense and open in* $\mathscr{S}_d$. *In particular, hyperbolicity is a generic property of rational functions.*

Now assume we are given two maps of topological spaces $f: M \rightarrow M$, $g: M' \rightarrow M'$.

DEFINITION 1.6. $f$ is *topologically conjugate* to $g$ if there is a homeomorphism $h: M \rightarrow M'$ such that $hf = gh$. $f$ is $\Omega$-*conjugate* to $g$ if there is a homeomorphism $h: \Omega(f) \rightarrow \Omega(g)$ such that $hf|_{\Omega(f)} = g|_{\Omega(g)}h$.

Let $\mathscr{F}$ be a topological space of maps from $M$ into itself.

DEFINITION 1.7. $f \in \mathscr{F}$ is *structurally stable in* $\mathscr{F}$ if $f$ is an interior point of $\{g \in \mathscr{F} | g$ is topologically conjugate to $f\}$. $f \in \mathscr{F}$ is $\Omega$-*stable in* $\mathscr{F}$ if $f$ is an interior point of $\{g \in \mathscr{F} | g$ is $\Omega$-conjugate to $f\}$.

THEOREM B. *The elements of* $\mathscr{H}_d$ *are* $\Omega$-*stable.*

The proof of Theorem B shows more. It exhibits models for the nonwandering sets of elements of $\mathscr{H}_d$.

If $f$ is a diffeomorphism of a smooth, compact manifold, Smale has defined the following axioms [7]:

*Axiom A.* $f$ is hyperbolic.

*Axiom* B. Stable and unstable manifolds of $f$ intersect transversally.

The $\Omega$-stability theorem [8] states that if $f$ satisfies Axioms A and B, then it is $\Omega$-stable.

THEOREM C. *There exists* $g \in \text{Diff}(S^r)$ *satisfying Axioms A and B such that* $\Omega(g)$ *contains a 1-dimensional set which is not locally the product of an interval and a Cantor set.*

DEFINITION 1.8. Let $f: M \to M$ be a map from a topological space to itself and $x \in M$ a periodic point of $f$. Then the *stable manifold of* $x$ is $\{y \in M | f^n(y) \to f^n(x)$ as $n \to \infty\}$. The *unstable manifold of* $x$ is $\{y \in M | \exists$ branches of $f^{-n}$ so that $f^{-n}(g) \to f^{-n}(x)$ as $n \to \infty\}$.

2. **The classical theory.** The results of this section are due to Julia [4], Fatou [2], and Ritt [5], stated in somewhat different language. Ritt ascribes priority to Julia for common results. Most of the material in this section can be found in Julia [4] with the notable exceptions of Propositions 2.8 and 2.13 which are in Fatou [2].

We assume throughout that $f$ is a holomorphic map of the Riemann sphere to itself of degree $d$. If we choose a coordinate system on the complement of a point, $f$ can be written as a rational function of degree $d$ in this coordinate system. At the outset *we shall require that if* $f^n(x) = x$, $|Df^n(x)| \neq 1$. The metric we shall use is the Hermitian metric of the sphere. Note that the set of maps we have eliminated forms a countable collection of (real) hypersurfaces in $\mathscr{S}_d$; hence, it is of first category.

The elements of $\Omega(f)$ are either sources or sinks. Except for critical points where $Df$ is zero, $f$ is a conformal map. Consequently, $\Omega(f)$ can have no saddle points. Where there is little chance of confusion $\Omega(f)$ will be denoted $\Omega$. The set of sources of $\Omega$ will be denoted $\Omega_e$. The unstable manifold of $x \in \Omega$ is $W^u(x)$.

PROPOSITION 2.1. *If* $x \in \Omega_e$ *and* $x$ *is periodic for* $f$, *then* $S^2 - W^u(x)$ *contains at most two points.*

PROOF. We assert that $\{f^n\}_{n \in \mathbf{Z}^+}$ is not a normal family of analytic functions in any neighborhood of $x$. Indeed,

$$|Df^n(x)| \to \infty \text{ as } n \to \infty \text{ and } f^{mn}(x) = x \text{ for some } m.$$

It follows that no subsequence of $\{f^n\}$ converges uniformly in a compact neighborhood of $x$. Therefore, $\{f^n\}$ is not normal in a neighborhood of $x$. The proposition now follows from a theorem of Montel [3] which states that a family of analytic functions defined on a region $\Lambda$ which omits three distinct values is normal.

We can say more about the complement of the unstable manifold of a source. If $C$ is the complement of the unstable manifold of a source, then $f^{-1}(C) = C$. $f^{-1}(C) \subset C$ because unstable manifolds are invariant under $f$. On the other hand, $C$ is finite and $\#f^{-1}(C) \geq \#C$. Equality must hold here. This proves that each point of $C$ is maximally ramified on the Riemann surface of $f^{-1}$.

The significance of Proposition 2.1 is discussed in the final section. Another proof is given which makes no use of Montel's theorem but which assumes $f$ is hyperbolic.

DEFINITION 2.2. If $x \in \Omega$ is a sink, then the *semilocal stable manifold of $f$ at $x$* is the component of $x$ in its stable manifold. We write slsm$(x)$ for the semilocal stable manifold of $f$ at $x$.

If $f$ is of degree 1, then it is a Möbius transformation of the sphere. The structure of $\Omega(f)$ is well known in this case. There are three possibilities:

(1) $f$ is parabolic. $\Omega(f)$ consists of a single point where the derivative of $f$ is 1.

(2) $f$ is hyperbolic. $\Omega(f)$ consists of two points, one a source and the other a sink. Note that our definition of hyperbolic agrees with the usual one in this case.

(3) $f$ is elliptic. $\Omega(f)$ consists of the entire sphere. There are two fixed points about which $f$ is a rotation.

*We assume henceforth that the degree of $f$ is larger than* 1.

PROPOSITION 2.3. *If $x \in \Omega$ is a fixed sink, then* slsm$(x)$ *contains a critical point of $of$.*

PROOF. Assume that $M = $ slsm$(x)$ contains no critical point of $f$. We shall derive a contradiction. The first step is to show that $M$ is simply connected. Since $f(M) = M$ and $M$ contains no critical point of $f$, $f^n|_M$ is a local homeomorphism for all $n > 0$. Let $U$ be a simply connected neighborhood of $x$ such that $f(U) \subset U$. $f^n|_U$ has a well defined inverse $g_n$ which leaves $x$ fixed. $g_n(U) \subset g_{n+1}(U)$. Therefore $\bigcup_{n \geq 0} g^n(U) \subset M$. But $\bigcup_{n \geq 0} g^n(U) = M$ since it is both open and closed in $M$. The boundary of $\bigcup_{n \geq 0} g^n(U)$ is invariant under $f$; therefore, it contains no points of $M$. This shows $\bigcup_{n \geq 0} g^n(U)$ is closed in $M$. Now, $g_n$ is a homeomorphism; so $g_n(U)$ is a disk. Therefore $M$ is an increasing union of disks, hence $M$ is a disk.

The proposition now follows easily from Schwarz's Lemma. $M$ is simply connected and not conformally equivalent to the plane. $f$ has at least two critical points which we have assumed do not belong to $M$. Therefore, there is a conformal homeomorphism $\mu$ mapping $M$ onto the open unit disk with $\mu(x) = 0$. Then $g = \mu \circ f \circ \mu^{-1}$ is a univalent map of the disk onto itself leaving the origin fixed. Therefore, $|g'(0)| = 1$ by Schwarz's lemma. This implies $|f'(x)| = 1$, contradicting the assumption that $f$ is hyperbolic.

COROLLARY 2.4. $f$ *has a finite number of sinks.*

PROOF. There are $2d - 2$ critical points of $f$ where $d = $ degree $f$. Consequently, there can be at most $2d - 2$ periodic orbits which are sinks.

COROLLARY 2.5. $\Omega_e$ *is infinite.*

PROOF. The fundamental theorem of algebra states that $f$ has an infinite number of periodic points. Only a finite number of these are sinks.

PROPOSITION 2.6. $x$ *is not in the closure of the set $E$ of periodic sources if and only if there is a neighborhood $U$ of $x$ such that $\{f^n|_U\}$ is a normal family of analytic functions.*

PROOF. If $x$ is in the closure of $E$, then $\{f^n\}$ is not normal in a neighborhood of $x$. This is Proposition 2.1. In order to obtain the reverse implication, we show $f^{-1}(\bar{E}) \subset \bar{E}$. This proves the proposition as follows: if $x \notin \bar{E}$, then there is a neighborhood $U$ of $x$ such that $U \cap \bar{E} = \varnothing$. But then $\bigcup_{n \geq 0} f^n(U) \cap \bar{E} = \varnothing$. Since $\bar{E}$ contains more than two points, $\{f^n|_U\}$ is a normal family by Montel's theorem.

Let $x \in f^{-1}(E)$, $U$ be a closed disk containing $x$ in its interior and not containing a point which lies outside the unstable manifold of a source (if there is one). Since $f(x) \in \bar{E}$, we require further that $f^m(U)$ covers $f(U)$ univalently for some $m > 1$. We want to apply the Brouwer fixed point theorem to $U$. In order to do so we need to find an $n$ such that $U$ is covered univalently by a disk in $f^n(U)$ for some $n$. It follows from Proposition 2.1 that there is an $n$ such that $f^n(U) \supset U$. There is a neighborhood $U'$ of $x$ and an $n'$ such that $U' \subset U$ and $U'$ is covered univalently by $f^{n'}(U)$. Since $f(U')$ contains a point of $E$, there is an $m$ such that $f^{m+1}(U')$ covers $f(U)$ univalently. It follows that $U'$ is covered univalently by $f^{m+n'}(U')$. The fixed point theorem can now be applied to the branch of $(f^{m+n'}|_{U'})^{-1}$ which has image in $U'$. There is a point $y$ in $U'$ such that $f^{m+n'}(y) = y$.

COROLLARY 2.7. $\bar{E}$ is perfect; i.e., $E$ has no isolated points.

PROOF. Let $U$ be a neighborhood of $x \in E$. Since degree $f > 1$ and $f'(x) \neq 0$, there is a point $y$ not in the period of $x$ such that $f(y) = x$. $y$ lies in the unstable manifold of $x$. Thus there is a $z \in U$ such that $f^n(z) = y$ for some $n$. $z \neq x$ and $z \in \bar{E} \cap U$.

PROPOSITION 2.8. The periodic points of $f$ are dense in $\Omega$. Equivalently $\bar{E} = \Omega_e$.

PROOF. Assume $x$ is a nonwandering point in the complement of $E$. We shall show that $x$ is periodic and therefore a sink. Let $C$ be the component of $x$ in the complement of $\bar{E}$. If $f^m(p) = p$ and $p \in C$, then on a compact neighborhood $U$ of $x$ in $C$, $\{f^{mn}|_U\}_n$ converges uniformly to $p$ by Proposition 2.6. Also $\{f^{mn+k}|_U\}_n$ converges uniformly to $f^k(p)$. Since $x$ is nonwandering, $x = p$.

We assert that there is a periodic point in $C$. Choose sequences $\{z_j\} \subset S$, $\{n_j\} \subset Z^+$ such that $z_j \to x$ and $f^{n_j}(z_j) \to x$. If $\{n_j\}$ is bounded, some $m$ occurs infinitely often in the sequence. Then $f^m(x) = x$ and we are done. Otherwise, choose a compact neighborhood $U$ of $x$ on which $\{f^{n_j}|_U\}$ is normal. There is a subsequence of $\{n_j\}$, again denoted $\{n_j\}$, and an analytic function $g$ defined on $U$ such that $f^{n_j}$ converges uniformly to $g$ on $U$.

$$|g(x) - x| \leq |g(x) - g(x_j)| + |g(x_j) - f^{n_j}(x_j)| + |f^{n_j}(x) - x|.$$

The right-hand side of this inequality tends to 0 as $j \to \infty$. Therefore $g(x) = x$.

Now assume that $C$ does not contain a periodic point. The sequence $\{f^{n_j} - \text{identity}|_U\}$ converges uniformly to $g - \text{identity}$. By assumption, $f^{n_j}(z) - z \neq 0$ if $z \in U$, but $g(x) - x = 0$. Therefore $g(z) = z$ for all $z \in U$. Pick coordinates so that $x = \infty$. Then $f^{n_j}(x) - x = \infty$ and $g(x) - x = 0$, contradicting the fact that $f^{n_j}(x) \to g(x)$.

PROPOSITION 2.9. *If $\Omega$ contains an interior point, then $\Omega = S$.*

PROOF. If $U \subset \Omega$ is open, then $\Omega \supset \overline{\bigcup_{n \geq 0} f^n(U)} = S$.
We digress briefly to state the following two elementary facts.

LEMMA 2.10. *If $f$ is a rational function, $\gamma$ is a simple closed curve on the sphere not passing through any critical point of $f$, and $D_1$, $D_2$ are the components of the complement of $\gamma$, then at most one component of $f^{-1}(D_1 \cup D_2)$ is not simply connected.*

LEMMA 2.11. *If $f$ is a rational function of degree $d$, and $A$ is an open disk in $S$ containing $k$ critical points of $f$, and $f$ has no critical point on the boundary of $A$, then $f^{-1}(A)$ is*

(a) *disconnected with simply connected components if $k < d - 1$,*
(b) *connected and simply connected if $k = d - 1$,*
(c) *connected and multiply connected if $k > d - 1$.*

Proposition 2.3. relates the sinks and critical points of $f$. The location of the critical points in the complement of $\Omega_e$ contains much more information than was obtained there. In order to develop this relationship, we need to interpret $f$ as a mapping from the sphere to a Riemann surface which is given as a branched covering of the sphere.

Specifically, let $X$ and $Y$ be transcendentals over $C$ satisfying $f(Y) = X$. Identify $S$ with the Riemann surface of the field $C(X)$. Denote by $R$ the Riemann surface of $C(X, Y)$. The inclusion map $C(X) \to C(X, Y)$ induces a branched covering $\pi : R \to S$. $\pi$ is ramified at the places $(Y - a)$ for which $Df(a) = 0$. $f$ induces a $1:1$ mapping $\bar{f} : S \to R$ defined by $(X - a) \to (X - f(a), Y - a)$. We can state now:

PROPOSITION 2.12. *Let $x$ be a fixed sink of $f$. Then the stable manifold of $x$ is disconnected, simply connected, or multiply connected as $\pi^{-1}(\text{slsm}(x))$ is disconnected, simply connected, or multiply connected. Moreover, if $\pi^{-1}(\text{slsm}(x))$ is not connected, the stable manifold of $x$ has an infinite number of components and each is simply connected. If the stable manifold of $x$ is multiply connected, it has an infinite order of connection.*

PROOF. $\bar{f}^{-1} : R \to S$ is a homeomorphism, from which the first statement follows immediately. If $\pi^{-1}(\text{slsm}(x))$ is disconnected, $f^{-1}(\text{slsm}(x))$ contains a component $M$ which is disjoint from $\text{slsm}(x)$. Since $f(M) = \text{slsm}(x)$, $f^{-1}(M)$ is disjoint from $M$ and $\text{slsm}(x)$. Similarly, $f^{-1}(M)$ must be disjoint from $f^{-(k-1)}(M), \dots, f^{-1}(M)$, $M$, and $\text{slsm}(x)$. We conclude that the stable manifold of $x$ has an infinite number of components. Lemma 2.11 implies that these components are simply connected.

If the stable manifold $M$ of $x$ is multiply connected, it is connected. Therefore $f^{-1}(M) = M$. $S - M$ contains at least two components $C_1$ and $C_2$. $C_1$ and $C_2$ are simply connected. If $f^{-1}(C_1 \cup C_2) \subset C_1 \cup C_2$, then each of $C_1$ and $C_2$ must contain $d - 1$ critical points of $f$, $d = \text{degree } f$. This certainly cannot happen since $f$ has only $2d - 2$ critical points and some of these belong to $M$. Therefore, there is a component $C_3 \subset S - M$ such that $f(C_3) \subset C_1 \cup C_2$. As above, $\{f^{-k}(C_3)|k \geq 0\}$ has an infinite number of components. $M$ has an infinite order of connection.

REMARK. Lemma 2.11 tells us how to determine the connectedness of the stable manifold of a fixed sink by counting the critical points in its semilocal stable manifold. The location of the critical points in the various stable manifolds is crucial in determining the topological structure of $\Omega$.

PROPOSITION 2.13. *If no critical point of $f$ belongs to $\Omega_e$, then* $\lim_{n \to \infty} |Df^n|_{\Omega_e}| \to \infty$ *uniformly.*

PROOF. $\Omega_e$ has an arbitrarily small compact neighborhood $U$ such that $f^{-1}(U) \subset U$. In particular, we may choose such a $U$ that does not intersect the orbit of any critical point. Cover $U$ by a finite number of disks $U_i$, none of which intersects the orbit of a critical point. On each $U_i$, $f^{-n}$ has $d^n$ well-defined branches $f_i^{-n}$, $0 < i \leq d^n$, $d = $ degree $f$. The family $\mathscr{G} = \{f_i^{-n}|n \geq 0, 0 < i \leq d^n\}$ is normal since its range omits a neighborhood of the critical points of $f$. Any convergent subsequence of $\mathscr{G}$ must converge to a constant because the limit function has range in $\Omega_e$ and $\Omega_e$ contains no interior points. It follows that $|Df_i^{-n}|$ converges uniformly to 0 as $n \to \infty$. Therefore $|Df^n|_{\Omega}| \to \infty$ uniformly.

COROLLARY 2.14. *If no critical point of $f$ belongs to $\Omega_e$, then $f$ is hyperbolic.*

PROOF. We have already verified that the periodic points of $f$ are dense in $\Omega$. By Proposition 2.13, there is an $m > 0$ such that $|Df^m|_{\Omega_e}| \geq v > 1$. Let $c = \min|Df|_{\Omega_e}|$, $c > 0$. Then $|Df^{mn+k}|_{\Omega_e}| \geq c^k v^n$. Setting $C = c^{m-1}/v$, $\mu = (v)^{1/m}$, we obtain $|Df^n|_{\Omega_e}| \geq C\mu^n$. We assumed to begin with that the hyperbolicity condition was satisfied at the sinks.

REMARK 2.15. The converse of Corollary 2.14 is obvious.

3. **The density theorem (Theorem A).** Corollary 2.14 states that if $f \in \mathscr{S}_d$ satisfies
(1) $f^n(x) = x$ implies $|Df^n(x)| \neq 1$, and
(2) $\Omega_e(f)$ contains no critical point of $f$, then $f$ is hyperbolic.
We prove now that $\mathscr{H}_d = \{f \in \mathscr{S}_d|f$ is hyperbolic$\}$ is open and dense in $\mathscr{S}_d$. First we show $\mathscr{H}_d$ is open.

Assume $f \in \mathscr{H}_d$. Then there is an $n > 0$ such that $|Df^n|_{\Omega_e}| > \mu > 1$. Let $U$ be a compact neighborhood of $\Omega_e$ such that $|Df^n|_U| > \mu$. Then $\bigcap_{n \in \mathbf{Z}^+} f^{-n}(U) = \Omega_e(f)$. There is a neighborhood $\mathscr{V}$ of $f$ in $\mathscr{S}_d$ such that if $g \in \mathscr{V}$, $|Dg^n|_U| > \mu$. This implies that $\bigcap_{n \in \mathbf{Z}^+} g^{-n}(U) = \Omega_e(g)$. On the other hand, $S - U$ converges uniformly to the sinks of $f$. There is a neighborhood $\mathscr{V}'$ of $f$ in $\mathscr{S}_d$ such that if $g \in \mathscr{V}'$, $S - U$ converges uniformly to the sinks of $g$, which correspond to the sinks of $f$ in a 1:1 manner. $\mathscr{V} \cap \mathscr{V}'$ is contained in $\mathscr{H}_d$. Therefore $\mathscr{H}_d$ is open in $\mathscr{S}_d$.

It was remarked at the beginning of §2 that the set of $f$ satisfying condition (1) above is of the second category. It is clear that the set of $f$ satisfying condition (2) is open. Thus it remains to show that the set of $f$ satisfying (2) is dense.

Assume $Df(0) = 0$ and $0 \in \Omega_e(f)$. We want to show that $f$ can be perturbed so that the critical point corresponding to 0 becomes periodic. This clearly suffices to show that those $f$ satisfying condition (2) are dense. Consider perturbations of

$f$ of the form $g(x) = f(x) + \varepsilon$. Expand the iterates of $g$ in a Taylor series with respect to $\varepsilon$ at $x$. We compute

$$g^2(x) = f(f(x) + \varepsilon) + \varepsilon$$
$$= f^2(x) + (Df_{f(x)} + 1)\varepsilon + O(\varepsilon^2).$$

Inductively, we obtain the following formula

$$g^n(x) = f^n(x) + Df_{f(x)}^{n-1} \sum_{i=0}^{n-1} \frac{1}{Df_{f(x)}^i} \varepsilon + O(\varepsilon^2).$$

Denote the coefficient of $\varepsilon$ in this expression by $c_n(x)$. Next compute

$$g^n(x + \delta) = f^n(x) + Df^n(x)\delta + c_n\varepsilon + O(\delta, \varepsilon)^2.$$

Observe that $Dg^n(0) = 0$ and that $0$ is the limit of periodic points for $f$. We want to find $\varepsilon$, $\delta$ so that $g^n(0) = 0$.

To terms of $O(\delta, \varepsilon)^2$, $g^n(x + \delta) = x + \delta$ has the solution

$$\delta = (c_n(x)/(1 - Df^n(x))\varepsilon, \quad \text{if } g^n(x) = x.$$

This tells us how far the periodic point corresponding to $x$ moves under perturbation by $\varepsilon$. We need that for $x$ sufficiently small and periodic for $f$, there exists an $\varepsilon$ such that $\delta = x = (c_n(x)/(1 - Df^n(x)))\varepsilon$, $n =$ period of $x$ under $f$. Thus it is necessary to investigate the behavior of $c_n(x)/(1 - Df^n(x))$ as $x$ ranges through a set of periodic points having $0$ as a limit point.

Let $x$ be a periodic point close to the origin of period $n$ for $f$. $|Df^n(x)| > 1$ is the derivative of the period of $x$. $c_n(x) = Df_{f(x)}^{n-1}(\sum_{i=0}^{n-1} 1/Df_{f(x)}^i)$. The sum $\sum_{i=0}^{\infty} 1/Df_{f(x)}^i$ converges absolutely at $x$ because $Df^n(x)$ increases exponentially. Let $m \to \infty$ through multiples of $n$. Then $c_m(x)/(1 - Df^m(x)) \to 1/Df(x) \sum_{i=0}^{\infty} 1/Df_{f(x)}^i$. If $\sum_{i=0}^{\infty} 1/Df_{f(0)}^i \neq 0$, then this limit tends to $\infty$ as $x \to 0$ because $Df(0) = 0$. Therefore, as $x \to 0$ through periodic points, the $\varepsilon$-perturbation necessary to make $g(0) = 0$ tends to $0$. To finish the proof, we need only remark that $\sum_{i=0}^{\infty} 1/Df_{f(0)}^i$ is not constant under perturbation, so that we could have assumed that it was nonzero in the beginning.

4. **Examples.** At this point, it is instructive to consider several examples to illustrate the usefulness of the classical theory developed in §2.

EXAMPLE 1. (Compare with Shub [6].) $f(z) = z^k$. There are two sinks at $0$, $\infty$. Both are critical points of maximal order. The $(k^n - 1)$th roots of unity are all periodic sources. Therefore, the unit circle is contained in $\Omega_e$. On the other hand, if $|z| \neq 1$, then $|z^{k^n}| \to 0$ or $\infty$. Hence, $\Omega_e$ is the unit circle. Moreover, the unstable manifold of $\Omega_e$ omits $0$ and $\infty$. This shows Proposition 2.1 is a sharp result.

EXAMPLE 2. $f(z) = 2z - 1/z$. There is a sink at $\infty$, and the critical points $\pm i$ both belong to its stable manifold, as does the entire imaginary axis. Thus there are no other sinks. We shall prove that $\Omega_e$ is a Cantor set in the next section. In this particular case it is possible to visualize the way the Cantor set arises. It is clear that the stable manifold of $\infty$ contains the complement of the real axis.

On the real axis, $f$ is a covering map of degree 2 with three fixed points at $\pm 1, \infty$. The open arc $-1 \infty 1$ lies in the stable manifold of $\infty$. Consequently, all of the inverse images of this arc lie in the stable manifold of $\infty$. These inverse images form the complement of a Cantor set.

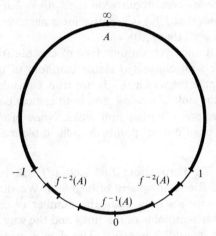

EXAMPLE 3 (JULIA). $f(z) = z^2 - 2$. This example illustrates that it can happen that a critical point lies in $\Omega_e$. It is easy to verify directly that $\Omega_e(f) = [-2, 2]$ on the real line. 0 is a critical point in $\Omega_e$. In the theory developed in the next section, it does not happen that $\Omega_e$ is an interval. This behavior should be thought of as pathological from our viewpoint.

EXAMPLE 4 (M. LATTES). The results of §2 do not rule out the possibility that $\Omega_e = S$ and there are no sinks at all. The following example due to M. Lattes shows that this can indeed happen. Consider the Weierstrass $p$-function with parameters $g_2 = 4, g_3 = 0$. This elliptic function satisfies the doubling formula

$$p(2u) = (p(u)^2 + 1)^2/4p(u)(p(u)^2 - 1).$$

Therefore, if $2^k z \equiv z$ modulo the Gaussian integers, $z$ is periodic for the rational function $f(z) = (z^2 + 1)^2/4z(z^2 - 1)$. Such $z$ are dense in $S$. Therefore $\Omega = S$.

EXAMPLE 5. $f(z) = ((z - 2)/z)^2$. This is another example for which $\Omega = S$. To see this, observe that the critical points of $f$ are 2, $\infty$, $f(z) = 0$, $f(0) = \infty$, $f(\infty) = 1$ and 1 is a fixed source. Therefore, the critical points lie in $\Omega_e$ by Proposition 2.8. Proposition 2.3 now implies that there are no sinks.

EXAMPLE 6 (JULIA). $f(z) = (-z^3 + 3z)/2$. The hyperbolic examples we have considered to this point have had at most two sinks. This example has three sinks $\pm 1, \infty$, all of which are critical points. $f$ is hyperbolic because it satisfies conditions (1) and (2) of §3. Proposition 2.12 implies that $\Omega_e(f)$ divides the sphere into an infinite number of components when $f$ has more than two sinks. In this case, one sink must have a stable manifold which is not connected. We shall discuss the topological structure of $\Omega_e(f)$ in detail in later sections.

EXAMPLE 7. The sinks of the above endomorphisms are fixed. It does happen, however, that periodic sinks occur with periods of arbitrarily high period. Consider the family of endomorphisms $\{f_a(z) = z^2 + a| -z < a < 0\}$. In order that the critical point 0 lie in a period of order $k$, the polynomial equation $f_a^k(0) = 0$ must be satisfied. $f_a^k(0)$ depends continuously on $a$. $f_{-2}^k(0) = 2$ if $k \geq 2$. $f_a^k(0) = 0$ and $df_a^k(0)/da = 1$. It follows that $f_a^k(0)$ is negative for $a$ negative and close to 0. Then there is an $a \in (-2, 0)$ such that $f_a^k(0) = 0$.

EXAMPLE 8. We introduce yet another type of complication. All of the above examples have a sink with connected stable manifold (if there is a sink at all). We now give an example for which this is not true. Consider $f(z) = -\frac{1}{4}(z + 1/z) + 7/16$. $f$ has a periodic sink of order 4, and both critical points of $f$ belong to its stable manifold. Therefore, $f^4(z)$ has four sinks. None of its stable manifolds is connected. The number of critical points in each stable manifold of $f^4(z)$ is 3, 6, 9, 12.

## 5. The structure of $\Omega$.

Throughout this section $f$ will be a fixed, hyperbolic endomorphism of the Riemann sphere of degree $d$. We shall give a topological model for $\Omega_e$ which depends only upon the number of critical points in each component of the stable manifolds of the sinks and the way in which the closures of the semilocal stable manifolds intersect. This data is invariant under perturbation; consequently, hyperbolic endomorphisms are $\Omega$-stable.

It follows from the results of §2 that $\Omega(f)$ has a finite number of periodic sinks and an indecomposable, perfect, nowhere dense set of sources which we denote $\Omega_e(f)$. This section is devoted to an analysis of the topological structure of $\Omega_e(f)$. In particular, if $f$ has a single sink, $\Omega_e(f)$ is a Cantor set (Proposition 5.2). If $f$ has two sinks, each with connected stable manifold, $\Omega_e(f)$ is a simple closed curve. Rarely will this simple closed curve be smooth. In all other cases $\Omega_e(f)$ is not locally the product of a Cantor set and a manifold.

PROPOSITION 5.1. *If $p$ is a fixed sink of $f$ and* slsm$(p)$ *is simply connected, then* $\partial(\text{slsm}(p))$ *is an arc.*

PROOF. Pick an $n$ such that $|Df^n|_{\partial(\text{slsm}(p))}| > 1$. Let $U$ be a neighborhood of $\partial(\text{slsm}(p))$ such that $|Df^n|_U| > 1$. There is a simple closed curve $\gamma \subset U \cap \text{slsm}(p)$.

Then $\{f^{-m}(\gamma) \cap \text{slsm}(p)\}$ converges uniformly to $\partial(\text{slsm}(p))$ as $m \to \infty$. Therefore $\partial(\text{slsm}(p))$ is an arc [10].

It certainly need not be the case that $\partial(\text{slsm}(p))$ is a simple closed curve. Indeed, if we take $f(z) = (-z^3 + 3z)/2$ and $p = \infty$, then $\partial(\text{slsm}(p))$ is all of $\Omega_e$. But $\Omega_e$ divides the plane into an infinite number of regions.

The significance of Proposition 5.1 for our purposes is that it implies that $\partial(\text{slsm}(p))$ is *accessible* from slsm$(p)$. This means that we can join any point of $\partial(\text{slsm}(p))$ to slsm$(p)$ by an arc [10].

Cantor sets with semishift maps play a crucial role in what follows. Our models for $\Omega_e$ will be quotients of Cantor sets of a particular kind. That $\Omega_e$ can be represented as a quotient of a Cantor set is by no means new [12]. What is new is that we can write down the quotient map quite explicitly and then examine the quotient. Moreover, the quotient constructed is invariant under deformation in $\mathcal{H}_d$.

Denote the set of (right) sequences of $d$ symbols $\{0, 1, ..., d - 1\}$ by $C_d$. $C_d$ is endowed with the compact open topology. We write $\{a_i\}_{i \geq 1} \in C_d$. The (left) semi-shift $s: C_d \to C_d$ is defined by $s(\{a_i\}_{i \geq 1}) = \{a_{i+1}\}_{i \geq 1}$. $s$ is a $d:1$ local homeomorphism. For the general theory of "symbolic dynamics", the reader is referred to [14].

PROPOSITION 5.2. *If $f \in \mathcal{H}_d$ has only one sink, then $f|_{\Omega_e(f)}$ is conjugate to $s: C_d \to C_d$.*

PROOF. Let $p$ be the sink of $f$. It is necessarily fixed. The stable manifold of $p$ is connected and contains all of the critical points of $f$. Join $p$ to all of the critical points of $f$ by a simply connected graph $T$ lying in the stable manifold of $p$. Since $T$ contains all of the critical points of $f$, $f|_{S-T}$ is a local homeomorphism. $S - T$ is simply connected since $T$ is connected. Therefore, $f^{-1}|_{S-T}$ has $d$ well defined branches $f_i^{-1}$, $0 \leq i \leq d - 1$. The images of $f_i^{-1}$ partition $f^{-1}(S - T)$ into $d$ disjoint open sets containing $\Omega_e(f)$.

Now define $g: C_d \to \Omega_e$ by $g(\{a_i\}) = \bigcap_\alpha g_i(\Omega_e)$ where $g_i = g_{i-1} \circ f_{a_i}^{-1}$, $i > 1$ and $g_1 = f_{a_1}^{-1}$. If $g$ is a well defined homeomorphism, then it is clear that $g$ is a conjugacy from $s$ to $f|_{\Omega_e}$ since $f(_\alpha g_i(\Omega_e)) = {}_{s(\alpha)} g_{i-1}(\Omega_e)$. It remains to show $g$ is a homeomorphism.

$g_i(\Omega_e) \subset g_{i-1}(\Omega_e)$ since $f_j^{-1}(\Omega_e) \subset \Omega_e$. We have shown that the branches of $f^{-n}$, $n > 0$ form a normal family of analytic functions in a neighborhood of $\Omega_e$ and that a convergent subsequence of these functions converges to a constant. But $g_i$ is a well-defined branch of $f^{-i}$. Therefore $\{g_i\}$ has a subsequence which converges uniformly on $\Omega_e$ to a constant function. This constant is obviously $\bigcap g_i(\Omega_e)$. Therefore $g$ is well defined. If $\alpha = \{a_i\}$, $\beta = \{b_i\}$ and $\alpha \neq \beta$, there is an $i_0$ such that $a_{i_0} \neq b_{i_0}$. Then ${}_\alpha g_{i_0}(\Omega_e) \bigcap {}_\beta g_{i_0}(\Omega_e) = \varnothing$. $g$ is $1:1$. To show $g$ is bicontinuous it suffices to show $g$ is open because $C_d$ and $\Omega_e$ are compact Haussdorff spaces. A base for the topology of $C_d$ is the collection of sequences of the form $\{\alpha | i\text{th term of } \alpha \text{ is } a_i \text{ for } i \subseteq i_0\}$ for some $i_0 > 0$ and $a_1, ..., a_{i_0} \in \{0, 1, ..., d - 1\}$. This set maps onto ${}_\alpha g_{i_0}(\Omega_e)$ which is open in $\Omega_e$.

PROPOSITION 5.3. *If $p_1$ and $p_2$ are fixed sinks for $f$ and $\partial(\text{slsm } p_1) \cap \partial(\text{slsm } p_2) \neq \varnothing$, then $\partial(\text{slsm } p_1) \cap \partial(\text{slsm } p_2)$ contains a fixed point of $f$.*

PROOF. Since $f(\partial(\text{slsm } p_i)) \subset \partial(\text{slsm } p_i)$, $i = 1, 2$, there is some branch of $f^{-1}$ which maps $\partial(\text{slsm } p_1) \cap \partial(\text{slsm } p_2)$ into itself. Let $e$ be this branch of $f^{-1}$. $e|\partial(\text{slsm}(p_1)) \cap \partial(\text{slsm } p_2)$ is a contraction. But $\partial(\text{slsm } p_1) \cap \partial(\text{slsm } p_2)$ is a closed set in $S$; therefore, $e$ has a fixed point in $\partial(\text{slsm } p_1) \cap \partial(\text{slsm } p_2)$. A fixed point of $e$ is also a fixed point of $f$.

PROPOSITION.5.4. *If $f$ has only fixed sinks and a sink $p$ with connected stable manifold $M$, then there is a surjective map $g: C_d \to \Omega_e(f)$ such that $g$ is injective on the complement of the set of sequences which are eventually constant.*

REMARK. The hypothesis that some sink have a connected stable manifold is not necessary, and we shall later eliminate it. The proof is easier to understand in this case, however. Note that this case includes all polynomials since $\infty$ is a fixed sink of maximal ramification for a polynomial.

PROOF. Set $Q = \{f^{-n}(q): n > 0, q \in \Omega_e$ is a fixed point of $f\}$. Let $p$ be the sink of $f$ with connected stable manifold $M$. Denote $S - (\Omega_e - Q)$ by $R$. $R$ is arcwise connected because $\partial M = \Omega_e$. If $C$ is any component of $S - \Omega_e$, $\partial C \cap Q \neq \emptyset$. Recall that the boundary of each complementary region to $\Omega_e$ is accessible from its interior. Thus $M \cup C \cup Q$ is arcwise connected. Now connect all of the critical points of $f$ by a simply connected graph $T$ in $R$.

Mimic the construction of the preceding Proposition 5.2. $f^{-1}|_{S-T}$ has $d$ well defined branches $f_i^{-1}$, $d = $ degree $f$. Define $g: C_d \to \Omega_e$ as before: if $\alpha = \{a_i\} \in C_d$,

$$g(\alpha) = \lim_{i \to \infty} \overline{{}_\alpha g_i(\Omega_e - T)}$$

where ${}_\alpha g_1 = f_{a_1}^{-1}$ and ${}_\alpha g_i = {}_\alpha g_{i-1} \circ f_{a_i}^{-1}$.

To show $g$ is well defined and continuous, consider a sequence $\alpha_j = \{\{a_{ij}\}_i\}_j \in C_d$ converging to a sequence $\gamma = \{c_i\} \in C_d$. Then, given $N \in Z^+$, there exists $N' \in Z^+$ such that $a_{i,j} = c_i$ if $i \leq N$, $j \geq N'$. Therefore if $j \geq N'$, $g_N(\alpha_j) \subset \overline{{}_{\alpha_N} g_N(\Omega_f - T)}$. Since the diameter of $\overline{g_i(\Omega_f - T)} \to 0$ as $i \to \infty$, $g(\alpha_j)$ converges. Moreover, the limit is independent of the sequence $\{\alpha_j\} \to \gamma$. Hence $g$ is well defined.

As before, it is evident from the definition of $g$ that the diagram

$$
\begin{array}{ccc}
C_d & \xrightarrow{s} & C_d \\
g \downarrow & & \downarrow g \\
\Omega_e & \xrightarrow{f} & \Omega_e
\end{array}
$$

commutes. $g$ is $1:1$ on the complement of $Q$ because $x \in \Omega_e - Q$ lies in one and only one of the $\overline{{}_\alpha g_i(\Omega_e - T)}$ for each $i > 0$.

To fix the ideas of the above construction, let us consider two examples in some detail before eliminating the unwanted hypothesis of Proposition 5.4.

EXAMPLE 5.5. $f(z) = z^2$. We already know that $\Omega_e(f)$ is the unit circle $S^1$. We wish to determine the map $g$. We cut out the positive real axis $[0, \infty] = T$ from $S$. On $S - T$, there are two well-defined branches $f_0^{-1}, f_1^{-1}$ of $(z)^{1/2} = f^{-1}(z)$. These are given by $\operatorname{Im}(f_0^{-1}(z)) > 0$ and $\operatorname{Im}(f_1^{-1}(z)) < 0$. Let $\alpha = \{a_i\} \in C_2$. If $a_1 = 0$, then $g_1(S - T)$ is the open upper half plane. If $a_1 = 1$, $g_1(S - T)$ is the lower half plane. Indeed, ${}_\alpha g_j(S^1 - T)$ is the set $\{e^{i\theta}|a < (2^{j-1}/\pi)\theta < a + 1\}$ where $a$ is the integer whose binary representation is $a_1 \ldots a_j$. Therefore, $\alpha \mapsto \exp^{(2\pi i a)}$ where $a \in [0, 1]$ is the number whose binary representation is $a_1 a_2 \ldots$. The shift map on $C_2$ induces multiplication by 2 on $R/Z$ by the map which sends an element in $C_2$ into the point with that binary representation. Multiplication by 2 becomes squaring under the exponential map. Thus we have a very good picture of the map $g$ in this case.

EXAMPLE 5.6. The previous example is amusing, but nothing surprising happens. Now let us consider in some detail $f(z) = (z^3 + 3z)/2$. Recall that $f$ has three sinks at $-1, 1, \infty$. Each sink is a critical point. The stable manifold of $\infty$ is connected while the stable manifolds of $-1$ and $1$ are not connected. 0 is the only fixed source. So we cut $S$ on $T = [-1, 1] \cup \{ia|0 \le a \le \infty\}$. It is easily verified that $T \cap \Omega_e(f) = \{0\}$. Indeed, the graphs of $f|_R$ and $f|_{\text{imaginary axis}}$ look like

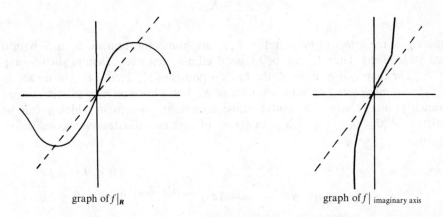

graph of $f|_R$            graph of $f|_{\text{imaginary axis}}$

Note that $\Omega_e$ must have a branch point at 0 because $\Omega_e$ separates the punctured imaginary axis from the punctured real axis locally. Recall that $f|_{\Omega_e}$ is a local homeomorphism. Therefore every point of $Q = \{f^{-n}(0), n \in Z^+\}$ must be a branch point of $\Omega_e$. We know that $Q$ is dense in $\Omega_e$. We are thus dealing with an unfamiliar object. We can, however, use the construction of $g$ in Proposition 5.4 to give a topological description of $\Omega_e$: $\Omega_e$ is to be a quotient space of $C_3$ such that the quotient map is 1:1 on the complement of the set of eventually constant sequences in $C_3$.

First we map $C_3 \overset{u}{\to} S^1 = R/Z$ as in Example 5.5; namely, $u(\{a_i\}) = x \in [0, 1]$ where the ternary expansion of $x$ is $a_1 a_2 \ldots$. Next we shall construct a topological space $K \subset R^2$ such that $K$ is a quotient of $S^1$ and $K$ is homeomorphic to $\Omega_e$.

$K \subset R^2$ is described as follows: Choose two circles of unit diameter which are tangent. At the opposite ends of the diameters through the point of tangency of the two circles, attach tangentially two new circles of diameter $1/2$. After the $n$th stage of the construction, proceed to the $n + 1$st stage by attaching tangentially circles of diameter $1/2^n$. These circles are to be attached at the midpoint of each arc whose endpoints lie on two circles in the $n$th stage and none of whose interior points lie on two circles. $K$ is the closure of all the circles obtained.

We assert that $K$ is homeomorphic to $S/\sim$ where $\sim$ is the equivalence relation defined by $a/2 \cdot 3^n \sim b/2 \cdot 3^n$ if $n, a, b \in Z$, $3{\not|}ab$, and $|a - b| = 1$. $\sim$ is the trivial relation on the complement of the set of points with eventually constant ternary expansions because the only points of $[0, 1]$ with eventually constant expansions are those which are of the form $a/2 \cdot 3^n$, $a, n \in Z$.

Now we construct a homeomorphism $h: S^1/\sim \to K$. To do this, we need more notation. Denote $K_n$ the $n$th state of the construction of $K$. $L_n$ is defined to be $S^1/\sim_n$ where $\sim_n$ is defined by $a/2\cdot 3^m \sim_n b/2\cdot 3^m$ if $m, a, b \in \mathbf{Z}$, $3\nmid ab$, $|a - b| = 1$, and $m \le n$. We construct commutative diagrams

$$\begin{array}{ccc} L_n & \overset{\longrightarrow}{j_n} & L_{n+1} \\ h_n\downarrow & & \downarrow h_{n+1} \\ K_n & \overset{\longrightarrow}{i_n} & K_{n+1} \end{array} \qquad (*)$$

where $j_n$ is the quotient map and $h_n$, $h_{n+1}$ are homeomorphisms. $L_0$ is $S'$ with $0$ and $\frac{1}{2}$ identified. Thus $L_0$ can be mapped affinely onto $K_0$. Denote such a map by $h_0$. The map $i_n$ is to have all the branch points of $K_n$ fixed. Let $\gamma$ be an arc in $K_n$ whose endpoints are branch points of $K_n$ but whose interior points are not branch points. There is a unique affine coordinate system in which $\gamma$ can be written $\{e^{i\theta}|0 \le \theta \le a, a \le 2\pi\}$. In terms of this coordinate system, we define $i_n|_\gamma$ by

$$i_n(e^{i\theta}) = \begin{cases} = \exp^{(i3\theta/2)} & \text{if } 0 \le \theta \le a/3 \\ (1 + a/4\pi)\exp^{(ia/2)} - (a/4\pi)\exp^{(i(a/2 + 6\pi\theta/a))} & \text{if } a/3 \le \theta \le 2a/3 \\ = \exp^{(i(3\theta/2 - a/2))} & \text{if } 2a/3 \le \theta \le a. \end{cases}$$

A picture of $i_n|_\gamma$ is as follows:

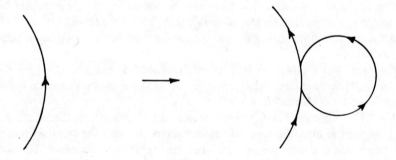

$j_n$, $i_n$ and $h_0$ have been defined thus far. $h_{n+1}$ is defined inductively as $i_n h_n j_n^{-1}$. $j_n$ is 1:1 except at $A = \{a/2\cdot 3^n| 3\nmid a\}$. On $A$, $j_n$ is 2:1, identifying $a/2\cdot 3^n$ with $b/2\cdot 3^n$ if $|a - b| = 1$. One notes that $i_n$ is also 1:1 except at the points which divide an arc between two branch points into three consecutive pieces. Since $h_n$ multiplies arc length by a constant factor, $i_n h_n$ identifies precisely the same points as $j_n$. Therefore $h_{n+1} = i_n h_n j_n^{-1}$ is well defined. Moreover, $h_{n+1}$ multiplies arc length by a constant factor.

We have constructed the commutative diagram (*). By patching these diagrams together, we have

$$
\begin{array}{ccccccc}
L_0 & \xrightarrow{j_0} & L_1 & \xrightarrow{j_1} & L_2 & \xrightarrow{j_2} & \cdots \\
\downarrow h_0 & & \downarrow h_1 & & \downarrow h_2 & & \\
K_0 & \xrightarrow{l_0} & K_1 & \xrightarrow{l_1} & K_2 & \xrightarrow{l_2} & \cdots
\end{array}
$$

We conclude that $S^1/\sim \; = \lim_{\overrightarrow{j_k}} L_k$ is homeomorphic to $K = \lim_{\overrightarrow{i_e}} K_e$.

It remains to show that $K$ is homeomorphic $\Omega_e(f)$. A computation accomplishes this. Recall that we cut $S$ on $T = [-1, 1] \bigcup$ upper half of the imaginary axis. The three inverse images of $S - T$ are separated by a graph which looks like:

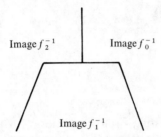

The nine inverse images of $S - T$ under $f^{-2}$ look like:

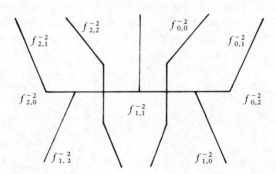

We have arbitrarily chosen the numbering of the branches of $f^{-1}$. Having done so, the branches of $f^{-2}$ are determined by $f_{a_1,a_2}^{-2} = f_{a_1}^{-1} \circ f_{a_2}^{-2}$. It must happen that $g: C_3 \to \Omega_e$ takes the three constant sequences to 0 since $\Omega_e(f)$ has a single fixed point. From the above diagram we can read off that

$$g(0111\ldots) = g(0222\ldots) = g(1000\ldots);$$
$$g(2111\ldots) = g(2000\ldots) = g(1222\ldots)$$

since 0 has three inverse images, two of which are not zero. These inverse images do not belong to $T$, nor do any of their inverse images. Therefore

$$g(a_1...a_n0111...) = g(a_1...a_n0222...) = g(a_1...a_n1000...),$$
$$g(a_1...a_n2111...) = g(a_1...a_n2000...) = g(a_1...a_n1222...),$$

and these are the only identifications which $g$ makes. The identifications we have made on $C_3$ are just those defining $S^1/\sim$. Finally we draw a diagram of $K$ and $S^1/\sim$.

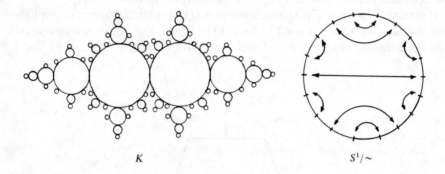

$K$                                                        $S^1/\sim$

The difficulty in eliminating from Proposition 5.4 the hypothesis that some sink have a connected stable manifold is that it is no longer clear that $R$ is connected. Recall that $R$ was defined to be $S - (\Omega_e - Q)$, $Q = \{x | f^n(x)$ is fixed for some $n\}$. Example 4.8 illustrates the problem nicely. $f$ is the fourth iterate of $-\frac{1}{4}(z + 1/z) + 7/16$. The inverse images of the semilocal stable manifolds of the sinks $p_i$, $1 \le i \le 4$, of $f$ look like the following diagram. The numbers in a disk indicate the number of critical points in that disk. The stable manifolds of $p_2$ and $p_4$ are shaded while those of $p_1$ and $p_3$ are not.

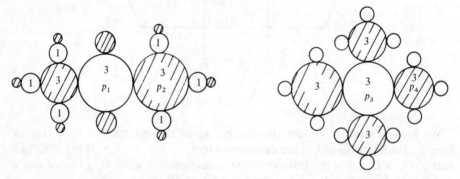

Set $M_i = \bigcup_{n \ge 0} f^{-n}(\mathrm{slsm}(p_i))$. Then $M_1 \cup M_2$; $M_3 \cup M_4$ are each simply connected and invariant under $f^{-1}$. $M_1 \cup M_2$ and $M_3 \cup M_4$ are disjoint. Each contains half the critical points of $f$. While in Proposition 5.4 we had only to pass through a finite number of points of $Q$ to connect the critical points by a graph, here we must go through an infinite chain of components of the $M_i$ to get from $M_1 \cup M_2$ to $M_3 \cup M_4$. This motivates the following definition.

DEFINITION 5.7. Let $f \in \mathcal{H}_d$. An *aggregate* of $f$ is defined inductively as follows:

(1) The closure of a semilocal stable manifold is an aggregate.

(2) If $M_1$ and $M_2$ are aggregates and $M_1 \cap M_2 \neq \varnothing$, then the closure of the component of $M_1 \cup M_2$ in $\bigcup_{n \geq 0} f^{-n}(M_1 \cup M_2)$ is an aggregate.

The way in which we eliminate the unnecessary hypothesis of Proposition 5.4 is to treat aggregates in essentially the same way we have treated semilocal stable manifolds.

LEMMA 5.8. *If* $f \in \mathcal{H}_d$, $R = S - (\Omega_e - Q)$ *and* $Q = \{x \mid f^n(x) \text{ is fixed for some } n\}$, *and all the sinks of* $f$ *are fixed, then* $R$ *is arcwise connected.*

PROOF. If $\Omega_e(f)$ is not connected, then some sink of $f$ has a connected stable manifold $M$. If $C$ is any component of the complement of $\bar{M}$, then $\bar{M} \cup \bar{C}$ contains a point of $Q$ by Proposition 5.3. Moreover $\partial M \cap \partial C$ is accessible from both $M$ and $C$. Therefore $\bar{M} \cup \bar{C}$ is arcwise connected. It follows that $R$ is arcwise connected.

Now assume that $\Omega_e$ is connected. Let $A$ be a proper aggregate; that is, $A \neq S$. We show that there is some aggregate $B$ such that $A \cap B \neq \varnothing$ but int $A \cap$ int $B$ $= \varnothing$. If this is not true then there are aggregates $A = A_1, A_2, \ldots, A_k (k > 1)$ such that each sink belongs to one $A_i$ and the $A_i$ are pairwise disjoint. This implies that the elements of $\{f^{-n}(A_i)\}_{n \geq 0, 1 \leq i \leq k}$ are all pairwise disjoint. In particular, $\Omega_e \cap A$ is a component of $\Omega_e$ since it is separated from every other member of $\{f^{-n}(A_i)\}_{n \geq 0, 1 \leq i \leq k}$. This contradicts the assumption that $\Omega_e$ is connected.

Observe that we can replace "slsm $p_1$" in Proposition 5.3 by "aggregate $A_i$". The proof remains unchanged. If $A_1$ and $A_2$ are aggregates such that $A_1 \cap A_2 \neq \varnothing$, then $A_1 \cup A_2$ is contained in an aggregate $A_3$. If $A_i - A_i \cap (\Omega_e - Q)$, is arcwise connected for $i = 1, 2$, then $A_3 - A_3 \cap (\Omega_e - Q)$ is arcwise connected by the strengthened version of Proposition 5.3. Since semilocal stable manifolds are connected and the only maximal aggregate is $S$, we are done.

We now state

PROPOSITION 5.9. *If* $f \in \mathcal{H}_d$ *has only fixed sinks, there is a surjective map* $g : C_d \to \Omega(f)$ *such that* $g$ *is injective on the complement of the set of sequences which are eventually constant.*

PROOF. Add Lemma 5.8 to the proof of Proposition 5.4.

If $f$ is of degree $d$ and hyperbolic, and $f^n$ has only fixed sinks, then we have constructed a map $g : C_d^n \to \Omega_e(f)$ such that

$$\begin{array}{ccc} C_{d^n} & \xrightarrow{s} & C_{d^n} \\ g \downarrow & & \downarrow g \\ \Omega_e & \xrightarrow{f^n} & \Omega_e \end{array}$$

commutes. $g$ is injective on the complement of the set of eventually constant sequences. If we allow the set on which $g$ is not $1:1$ to be $\{x \mid f^m(x)$ is periodic of period $n$ for some $m\}$, then we obtain $g': C_d \to \Omega_e(f)$ such that

$$
\begin{array}{ccc}
C_d & \overset{s}{\to} & C_d \\
\downarrow {\scriptstyle g'} & & \downarrow {\scriptstyle g'} \\
\Omega_e & \overset{f}{\to} & \Omega_e
\end{array}
$$

commutes, by the same construction.

THEOREM B. *If $f \in \mathcal{H}_d$, then $f$ is $\Omega$-stable.*

PROOF. We seek to establish that the quotient map $g: C_d \to \Omega_e(f)$ can be chosen so that the points $g$ identifies are invariant under perturbation of $f$. How are these points determined? They are given by the action of $f^{-1}$ on the graph $T$ connecting the critical points of $f$. The intersection of this graph with $\Omega_e$ is contained in $Q$, the set of inverse images of certain periodic points. What is more, the number of points of $T$ which lie in the intersection of two aggregates with disjoint interiors is finite. Denote this set by $I$. If $f$ is perturbed to $\bar{f}$, it is clear that there are sinks of $\bar{f}$ corresponding to the sinks of $f$, and that corresponding to the points of $I$ (which satisfy equations of the sort $f^n(x) =$ fixed point) are points of intersection of the aggregates of $\bar{f}$. Therefore, there is a graph $\bar{T}$ close to $T$ which connects the critical points of $\bar{f}$. In particular $\bar{T}$ can be chosen so that $\bar{T}$ is homeomorphic to $T$ and such that the images of $(f|_{S-T})^{-1}$ are homeomorphic to the images of $(\bar{f}|_{S-T})^{-1}$ by homeomorphisms close to the identity. The topology of this data determines which points of $C_d$ are identified under $g$. Since the data for $f$ and $\bar{f}$ are the same, the quotient maps of $C_d$ are the same. We conclude that $f$ is $\Omega$-conjugate to $\bar{f}$.

6. **A counterexample** In this section we prove Theorem C by constructing $g \in \text{Diff}(S^5)$ such that $g$ satisfies Smale's Axioms A and B, but the nonwandering set of $g$ contains a component which is not locally the product of a Cantor set and an interval. The basis for this example is a further analysis of Example 5.5 and an inverse limit construction of Williams [11].

PROPOSITION 6.1. *Suppose $M^n$ is a manifold and $f: M \to M$ is a map such that $f|_{\Omega(f)}$ is a local homeomorphism. Then there is a $g \in \text{Diff}(S^{2n+2})$ such that $\Omega(g)$ contains components homeomorphic to the components of $\varprojlim_{f|_{\Omega(f)}} \Omega(f)$.*

PROOF. Embed $M^n$ in $S^{2n+2}$ and approximate $f$ by an embedding $\bar{g}$. Let $N$ be a tubular neighborhood of $M$ containing the image of $\bar{g}$ (which exists if $\bar{g}$ is sufficiently close to $f$). Then $\bar{g}$ can be extended to a diffeomorphism $g$ of $S^{2n+2}$ such that

(1) $g: N \to N$.

(2) $g$ preserves fibers of $N$.

(3) $g$ is a contraction along each fiber.

$\bigcap_{n \in \mathbf{Z}^+} g^n(N)$ is nonempty since $g(N) \subset N$. The following diagrams from Williams [11] show that $\bigcap_{n \in \mathbf{Z}} g^n(N)$ is homeomorphic to

$$\varprojlim_f M: \quad g^2(N) \xleftarrow{g} g(N) \xleftarrow{g} N \leftarrow \ldots$$

with the diagram structure

$$g(N) \xleftarrow{g} N$$
$$N \qquad \pi$$
$$M \xleftarrow{f} M \xleftarrow{f} M \xleftarrow{f} \ldots$$

In this diagram, $\pi$ is the projection from $N$ to $M$ and the unlabeled vertical arrows are inclusions. Each vertical sequence has inverse limit $\bigcap_{n \in \mathbf{Z}^+} g^n(N)$, giving the diagram

(*)
$$\bigcap g^n(N) \xleftarrow{g} \bigcap g^n(N) \xleftarrow{g} \bigcap g^n(N) \leftarrow \ldots$$
$$\downarrow R \qquad \downarrow R \qquad \downarrow R$$
$$M \xleftarrow{f} M \xleftarrow{f} M \leftarrow \qquad \ldots$$

$g$ is a diffeomorphism, so the inverse limit of the top row is just $\bigcap_{n \in \mathbf{Z}^+} g^n(N)$. Thus there is a commutative diagram

(**)
$$\bigcap g^n(N) \xrightarrow{g} \bigcap g^n(N)$$
$$\downarrow R \qquad \qquad \downarrow R$$
$$\varprojlim_f M \xrightarrow{\tilde{f}} \varprojlim_f M$$

In (*) and (**) $R$ is the map induced by the projection $\pi$. $\tilde{f}$ is the map induced by $f$. Since $g$ contracts each fiber of $N$, it follows that $R$ is a homeomorphism.

Now $\Omega(g|_N) \subset \bigcap_{n \in \mathbf{Z}^+} g^n(N)$. In order to determine the topological type of $\Omega(g|_N)$, (**) tells us that we need only look at $\Omega(\tilde{f})$ where $\tilde{f}$ is the map induced by $f$ on $\varprojlim_f M$.

The points of $\varprojlim_f M$ are sequences $\zeta = (x_0, x_1, \ldots, x_n, \ldots)$ such that $x_i \in M$ and $f(x_i) = x_{i-1}$. A basis for the topology of $\varprojlim M$ is given by the collection of all sets of the form $A = \{(x_i) | x_{i_1} \in U_{i_1}, \ldots, x_{i_n} \in U_{i_n}; U_{i_1}, \ldots, U_{i_n}$ a finite collection of open sets contained in $M\}$. If $m \geq i_n$ for each $i_n$ used in the definition of $A$, then $A \supset \{(x_i) | x_{ij} \in \bigcap_{j=1}^n f^{i_j - m}(U_{i_j})\}$. $\bigcap_{j=1}^n f^{i_j - m}(U_{i_j})$ is open, so we take as a basis for the topology of $\varprojlim M$ the collection of sets of the form $\{(x_i) | x_n \in U; n \in \mathbf{Z}^+$ and $U \subset M$ is open$\}$.

Now it is easy to describe $\Omega(\tilde{f})$: a point $(x_i) \in \varprojlim_f M$ is nonwandering if and only if $x_i \in \Omega(f)$ for all $i \geq 0$. Therefore $\Omega(\tilde{f}) = \varprojlim_{f|_{\Omega(f)}} \Omega(f)$.

We remark that if it is only required that $g$ be defined on a tubular neighborhood $N$ of $M$, and not on the entire sphere, then the above construction can be carried out on $S^{2n+1}$.

Let us return to Example 5.5. $f: S \to S$ is given by $f(z) = -(z^3 + 3z)/2$. $\Omega_e(f)$ is homeomorphic to $S^1/\sim$ where $\sim$ identifies $a/3^i$ with $b/3^i$ if $3|ab$ and $|a - b| = 1$. Recall that $S^1$ is taken to be $R/Z$. Proposition 6.1 tells us how to obtain a diffeomorphism from the map $f$. Moreover, the basic sets we obtain are homeomorphic to $\varprojlim_{f|_\Omega} \Omega(f)$.

What does $\varprojlim \Omega(f)$ look like? Each sink of $f$ gives a sink of $\varprojlim \Omega(f)$. But $\varprojlim \Omega_e(f)$ is a one dimensional set which is not locally homeomorphic to the product of a Cantor set and an interval. This we now prove.

PROPOSITION 6.2. If $f(z) = -(z^3 + 3z)/2$, $\varprojlim \Omega_e(f)$ is not locally the product of a Cantor set and an interval.

PROOF. The arc components of a Cantor set x an interval are intervals. The idea of our proof is to show that the arc components of $\varprojlim \Omega_e$ cannot be intervals. This is done as follows: if $I$ is an open interval and $x \in I$, then $I - x$ has exactly two components. The local arc components of $\varprojlim \Omega_e$ do not have this property.

Let $U$ be the set $\Omega_e(f) - \{0\}$. In §5 we saw that $(f|_U)^{-1}$ is a 3-valued function with well defined branches $f_i^{-1}, i = 0, 1, 2$. Denote by $V$ the set $\{(x_i) \in \varprojlim \Omega_e(f)|x_1 \in U\}$. $V$ is open in $\Omega_e(f)$. Now $f^{-1}|_U$ defines a homeomorphism of $U \times \{0, 1, 2\}$ with $f^{-1}(U)$. Inductively, $f^{-n}|_U$ defines homeomorphism of $U \times \{0, 1, ..., 3^n - 1\}$ with $f^{-n}(U)$. It follows that $V$ is homeomorphic to $U \times C$, where $C$ is a Cantor set. Therefore, the arc components of $V$ are homeomorphic to the components of $U$.

It is clear that $U$ is not a manifold. By looking at the quotient map $C_3 \to \Omega_e$, one sees that removing the point corresponding to $\{\frac{1}{6}, \frac{1}{3}\}$ locally separates $U$ into four components. We conclude that $\varprojlim \Omega_e$ is not locally the product of a Cantor set and a manifold.

PROPOSITION 6.3. If $f \in \mathcal{H}_d$, there is a $h \in \text{Diff}(S^5)$ satisfying Axioms A and B and with a basic set homeomorphic to $\varprojlim \Omega_e$.

PROOF. $S^5 = N_1 \cup N_2$ where $N_1$ and $N_2$ are diffeomorphic to $S^2 \times D^3$ and intersect only along their common boundary which we denote $T$. $T$ is diffeomorphic to $S^2 \times S^2$. On $N_1$ we define a map $g_1: N_1 \to N_1$ which is a diffeomorphism onto its image such that

(1) $g_1$ approximates $f: S^2 \times 0 \to S^2 \times 0$ in $N_1$.
(2) $g_1$ preserves the $D^3$ fibers of $N_1$.
(3) $g_1$ is a contraction along each $D_3$ fiber.

Now look at the map $g_1^{-1}: g_1 T \to T$ in the complement $C$ of $g_1 N_1$. The homology of $C$ is trivial except in dimension 2. $g_1^{-1}$ takes the generator of $H_2(C)$ into the generator of $H_2(N_2)$. Therefore $g_1^{-1}$ extends to a map $g_2: C \to N_2$. As before, this can be done so that $g_2$ preserves and contracts the $D_3$ fibers of $C$ and $g_2$ approximates $f: S^2 \times 0 \to S^2 \times 0$ in $C$.

Define $\bar{h}: S^5 \to S^5$ to be the map defined by

$$\bar{h}(x) = \begin{cases} g_1(x) & \text{if } x \in N_1 \\ g_2^{-1}(x) & \text{if } x \in N_2 \end{cases}$$

$h$ is defined to be a smooth approximation of $\bar{h}$ which equals $\bar{h}$ outside a small neighborhood of $T$. It is evident that the nonwandering points of $h$ are given by $\bigcap_{n>0} g_1(N_1)$, $\bigcap_{n>0} g_2(N_2)$. By Proposition 6.1, each of these sets is homeomorphic to $\lim_{f|\Omega} \Omega(f)$.

Since $f \in \mathcal{H}_d$, $h|_{\Omega(h)}$ has a hyperbolic structure. In the normal direction to $\Omega(h)$, $h$ contracts toward $\Omega(h) \cap N_1$ and expands from $\Omega(h) \cap N_2$. Therefore $h$ satisfies Axiom A. The stable manifolds of points of $\Omega(h) \cap N_1$ contain fibers of the tubular neighbor $N_1$, and the unstable manifolds of points of $\Omega(h) \cap N_2$ contain fibers of the tubular neighborhood $C$. $C \cap N_1 = N_1 - g(N_1)$. By a small change in the bundle structure of $N_1$ (if necessary), we can assure that the fibers of these two bundles intersect transversally. This transversal intersection is Axiom B.

## 7. Real endomorphisms of degree 2.

LEMMA 7.1. *Suppose that $f(z)$ is a rational function on $S$ of degree 2 with real coefficients. Then $f(z)$ can be written in one of the following three forms in a suitable coordinate system:*

(1) $f(z) = z^2 + a$.
(2) $f(z) = a(z - 1/z) + b$.
(3) $f(z) = a(z + 1/z) + b$,

*$a$ and $b$ are real.*

PROOF. The nonreal fixed points of $f$ occur in conjugate pairs. Since $f$ has 3 fixed points (counting multiplicity), $f$ has a real fixed point. By a suitable coordinate change by a real Möbius transformation, we may assume that $\infty$ is a fixed point of $f$. In our new coordinate system, the denominator of $f$ has degree at most one.

If $f$ is a polynomial, $\infty$ is ramified. In this case, the other critical point of $f$ is also real. By a real translation of coordinates, we may assume that 0 and $\infty$ are the critical points of $f$. After a real dilitation $f(z) = z^2 + a$ for some $a \in \mathbf{R}$.

Similarly, if the denominator of $f$ has degree one in a coordinate system in which $\infty$ is a fixed point, then a translation of coordinates along the real axis makes the denominator $z$. After a dilitation, $f(z)$ has equation (2) or (3) for some $a, b \in \mathbf{R}$. If the critical points of $f$ are real, then $f$ has an equation of form (3). $f$ has form (2) if its critical points are not real.

Lemma 7.1 gives examples of 1 and 2 parameter families of rational functions. We can partially classify these spaces of functions according to the topology of the nonwandering sets of its elements.

CASE 1. $f(z) = z^2 + a$.

$\infty$ is a fixed sink. 0 and $\infty$ are the critical points of $f$.

(i) $a > \frac{1}{4}$. $f(z)$ satisfies the inequality $f(z) > z$. Therefore $\{f^n(0)\} \to \infty$ as $n \to \infty$. Consequently, $f$ can have only one sink. $\Omega_e(f)$ is a Cantor set.

(ii) $a < -\frac{1}{2}$. This case is similar to (i). $f(0) = a$; $f(a) = a^2 + a$ and $a^2 + a > -a$. $f(z)$ satisfies the inequality $f(z) > z$ if $z \geq -a$. Therefore $\{f^n(0)\} \to \infty$ and $f$ has only one sink. $\Omega_e(f)$ is a Cantor set.

(iii) $-\frac{3}{4} < a < +\frac{1}{4}$. $\frac{1}{2}(1 - \sqrt{1 - 4a})$ is a fixed sink. Thus $f$ has two fixed sinks and $\Omega_e$ is a simple closed curve.

(iv) $-5/4 < a < -3/4$. In this case $f(z)$ has a periodic sink of order 2. $f^2(z) = (z^2 + a)^2 + a = z^4 + 2az^2 + (a^2 + a)$. We solve $f^2(z) - z = 0 = (z^2 - z + a)(z^2 + z + (a + 1))$. The period of order 2 solves the equation $z^2 + z + (a + 1) = 0$ since the fixed points of $f$ are $\infty$ and the roots of $z^2 - z + a = 0$. Thus the points of the period of order 2 are $\frac{1}{2}(-1 \pm \sqrt{1 - 4(a + 1)})$. The derivative of the period is $4(a + 1)$.

We could have obtained part of this last result by examining the map $f(z) = z^2 - \frac{3}{4}$. This map has a fixed point at $-\frac{1}{2}$ with $f^1(-\frac{1}{2}) = -1$. Therefore $f^2(z)$ has a fixed point at $-\frac{1}{2}$ with $(f^2)^1(-\frac{1}{2}) = +1$. This means that $f^2(z) - z = 0$ has a multiple root at $-\frac{1}{2}$, but $f(z) - z = 0$ has a simple root at $-\frac{1}{2}$. We conclude that there is an "embedded" period of order 2 at $-\frac{1}{2}$ with derivative of absolute value 1. Now the derivative of a particular period depends analytically on the coefficients of $f$. To the right of $a = -\frac{1}{4}$ we know that $z^2 + a$ does not have a periodic sink of order 2. Therefore, to the left of $a = -\frac{3}{4}$, $z^2 + a$ must have a periodic sink of order 2.

This argument allows us to conclude that $f(z) = z^2 + a$ has a periodic sink of order 4 if $a$ lies in some interval $I_2$ with right-hand endpoint $-5/4$. Since there is more than one period of order 4, it is not possible to explicitly solve $(f^4)^1(z_0) = -1$, $z_0$ periodic of order 4, for $a$.

We can say, however, that there is a sequence of intervals $\{I_n\}$ such that the right-hand endpoint of $I_n$ is the left-hand endpoint of $I_{n-1}$ and the $f$ within $I_n$ have a periodic sink of order $2^n$. The lengths of the $I_n$ decrease, and there is a point $p$ which is the greatest lower bound of $\bigcup I_n$. $f(z) = z^2 + p$ must have a critical point in $\Omega_e(f)$ because we have analyzed the behavior of functions lying near a function with a period having derivative of modulus 1.

Example 4.7 showed that there are values of $a$ in $(-2, 0)$ such that $f(z) = z^2 + a$ has a periodic sink of any given order. If $a \in (-2, p)$ and $f(z) = z^2 + a$ is hyperbolic, $f(z)$ has a periodic sink of order not a power of 2. The points where $f(z)$ has a periodic point of derivative $-1$ separate intervals in which there are periodic sinks having periods $q$ and $2q$. The points where $f(z)$ has a critical point in $\Omega_e(f)$ separate regions in which the periodic sink of $f(z)$ has different odd prime components.

CASE 2. $f(z) = a(z - 1/z) + b$.

This is the simplest of the three cases because the real axis is preserved under $f^{-1}$ as well as $f$. It follows that the upper and lower half planes are preserved or interchanged. In either case, Proposition 2.1 implies that $\Omega_e(f)$ is contained in the real axis. The critical points of $f$ are $\pm i$, and these do not belong to $\Omega_e$.

(i) $b^2 + 4a(a - 1) < 0$. $f(z)$ has 2 nonreal fixed points. These fixed points do not belong to $\Omega_e(f)$; therefore, they are sinks. $\Omega_e(f)$ must be the entire real axis since it separates the stable manifolds of the two sinks.

(ii) $b^2 + 4a^3/(a + 1) < 0$. $f^2(z) = a(a(z - 1/z) + b) - 1/(a(z - 1/z) + b) = b$. As before, we calculate that the period of order 2 of $f$ solves the quadratic equation $(a^2 + a)z^2 + (ab + b)z - a^2 = 0$. If $b^2 + 4a^3/(a + 1) < 0$, the roots of this polynomial are not real. We saw above that nonreal periodic points are sinks.

Therefore $f$ has a periodic sink of order 2. $\Omega_e(f)$ is again the entire real axis.

(iii) $b^2 + 4a(a - 1) > 0$ and $b^2 + 4a^3/(a + 1) > 0$. In this case $f(z)$ does not have a periodic point of order 1 or 2 off the real axis. Since $\Omega_e(f)$ is contained in the real axis, there can be no periodic point of order greater than 2 off the real axis. Therefore, there is a sink $p$ on the real axis. Now $f$ maps conjugate points to conjugate points. Hence both critical points $\pm i$ tend to $p$ under iteration. We conclude that $\Omega_e(f)$ is a Cantor set.

We draw a diagram of the $a, b$ plane.

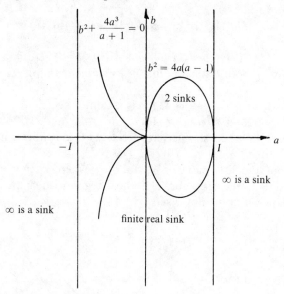

CASE 3. $f(z) = a(z + 1/z) + b$.

Here the picture is much more complicated than in Case 2, and the results are far from complete. We shall go as far as the quadratic formula allows us to in decomposing the $a, b$ plane. We may assume that $b > 0$ in the equation defining $f$ as the change of coordinates $z \to -z$ transforms $a(z + 1/z) + b$ into $a(z + 1/z) - b$.

(i) $a > 1$, $b^2 < 4a(a - 1)$. $\infty$ is a fixed sink if $f(z)$ lies in this region. To show that it is the only sink, it suffices to show that the critical points $\pm 1$ tend to $\infty$ under iteration. Since $b > 0$, $f(z)$ is a monotone increasing function on the ray $[+1, \infty)$ which lies above the diagonal. There $\{f''(1)\} \to \infty$. $f(z)$ is also monotone increasing on $(-\infty, -1]$ and lies below the diagonal. Therefore $\{f''(-1)\} \to \infty$. $\Omega_e(f)$ is a Cantor set.

(ii) $a > 1$, $4a^3/(a + 1) > b^2 > 4a(a - 1)$. The fixed points of $f$ are all real. $\infty$ is a sink. $z = (1/2(a - 1))(-b \pm \sqrt{b^2 - 4a(a - 1)})$ are the other fixed points. We want to evaluate $f'(z) = a(1 - 1/z^2)$ at these fixed points. The inequality $|f'(z)| < 1$ is equivalent to $a/(a + 1) < z^2 < a/(a - 1)$. The inequality $a/(a + 1) < ((1/2(a-1))(-b + \sqrt{b^2 - 4a(a - 1)}))^2 < a/(a - 1)$ is implied by the inequality $4a^3/(a + 1) > b^2 > 4a(a - 1)$. Therefore $(1/2(a - 1))(-b + \sqrt{b^2 - 4a(a - 1)})$ is a sink of $f$. Since $f$ has two fixed sinks, $\Omega_e(f)$ is homeomorphic to a circle.

(iii) $a > 1$, $b > 2a$. As in (i), we verify that $\{f''(1)\} \to \infty$ and $\{f''(-1)\} \to \infty$. $\Omega_e(f)$ is a Cantor set.

(iv) $0 < a < 1$ and $b^2 < 4a^3/(a + 1)$. As in (ii) we verify that both finite real fixed points are sinks. Therefore $\Omega_e(f)$ is a simple closed curve.

(v) $0 < a < 1$ and $b > 2a$. Both $-1$ and $1$ tend to the sink $(1/2(1 - a))(b + \sqrt{b^2 - 4a(a - 1)})$ under iteration (if $b > 0$). Therefore $\Omega_e(f)$ is a Cantor set.

(vi) $a < -1$, $b^2 < 4a(a - 1)$. The finite fixed points and period of order 2 are not real. $\infty$ is a fixed sink and both $-1$ and $+1$ belong to its stable manifold. $\Omega_e(f)$ is a Cantor set.

(vii) $a < -1$, $4a^3/(a + 1) > b^2 > 4a(a - 1)$. As in (ii) $\Omega_e$ is a simple closed curve.

(viii) $0 > a > -1$; $(1 + (2a + 1)^2)(a/(a + 1))^2 > b^2$. There is a periodic sink of order 2 containing both critical points in its stable manifold. $\Omega_e(f)$ is a simple closed curve.

(ix) $0 > a > -1$, $b^2 > 4a(a - 1)$. $\Omega_e(f)$ is a Cantor set.

(x) $a < -1$, $(1 + (2a + 1)^2)(a/(a + 1))^2 > b^2 > 4a^3/(a + 1)$. In this region $\infty$ is a sink, and there is a periodic sink of order 2. $\Omega_e(f)$ is neither a manifold, nor a Cantor set in this case.

The diagram of the $a, b$ plane is as follows, where each region has been numbered according to the cases above. In those regions which are not numbered, $\Omega_e(f)$ is not a manifold or a Cantor set.

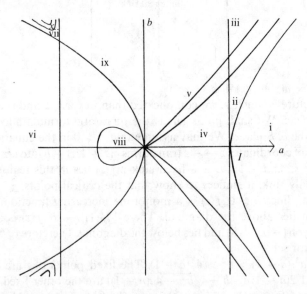

8. **Higher dimensions.** The question naturally arises as to whether the theory that has been developed here can be generalized to spaces of analytic mappings on other algebraic or complex analytic manifolds. There are roughly two directions one can go in seeking generalizations: higher genre and higher dimensions.

There is no extension to higher genre except for genus one. A compact Riemann surface of genus greater than one admits only a finite number of non-constant analytic mappings into itself. These are all automorphisms and all periodic [13]. A periodic map obviously cannot be hyperbolic. The space of analytic self mappings of a torus having a fixed basepoint form a finitely generated group. Other than the automorphisms which are periodic, these mappings are all expanding maps. Expanding maps have been studied by M. Shub [6].

The extension of the theory to higher dimensional projective spaces appears to be an interesting and difficult problem. One immediately encounters the lack of satisfactory generalizations of the theorems of one complex variable to several complex variables. To illustrate this point we consider Proposition 2.1. Recall that Proposition 2.1 says that the unstable manifold as a source omits at most two points.

The proof of Proposition 2.1 used a theorem of Montel on normal families of analytic functions. In order to make the proposition more amenable to generalization, we give two additional proofs.

PROOF 1. Suppose $f(z)$ is a rational function, $f(0) = 0$ and $|f'(0)| > 1$. $f(z)$ has a Taylor series $\sum_{i \geq 1} a_i z^i$ at the origin with $|a_1| > 1$. We seek a local analytic coordinate system in which $f(z)$ has the equation $a_1 z$. Such a coordinate system is given by an analytic function $g$ such that $gf(z) = a_1 g(z)$. To find $g$, it suffices to find its Taylor series at the origin. Suppose $g(z) = \sum b_j z^j$ exists. Then $\sum_j b_j (\sum_i a_i z^i)^j = a_1 \sum b_j z^j$. This equation formally reduces to an infinite system of polynomial equations:

$$a_1 b_1 = b_1 a_1,$$
$$b_2 a_1^2 + b_1 a_2 = a_1 b_2,$$
$$b_3 a_1^3 + b_2(2a_1 a_2) + b_1 a_3 = a_1 b_3,$$
$$\vdots$$
$$b_n a_1^n + P_n(a_1, ..., a_n, b_1, ..., b_{n-1}) = a_1 b_n,$$

where $P_n(a_1, ..., a_n, b_1, ..., b_{n-1})$ is a polynomial in the indicated variables. Since $|a_1| \neq 1$, we can recursively solve these equations for $b_j$ after assigning a nonzero value to $b_1$ (say 1). Thus there is a formal solution to the equation $gf(z) = a_1 g(z)$. In order to show there does exist a solution it suffices to show that the formal solution has a positive radius of convergence. This is implied by $\overline{\lim} \sqrt[n]{P_n} < \infty$. We prove this now.

There is an $r > 0$ such that $|a_i| < r^i$. $P_n$ is linear in $(b_1, ..., b_{n-1})$ and of weighted degree $n$ in the $a_i$. The number of monomials of $\{a_1, ..., a_n\}$ multiplying $b_j$ in $P_n$ is equal to the number of partitions of $n$ into $j$ positive integers. If $s(n)$ is the number of ways of writing $n$ as a sum of positive integers,

$$s(n) = 1 + s(1) + ... + s(n - 1).$$

It follows that $s(n) \leq 2^n$. Therefore $|P_n| \leq (2r)^n \sum_{j < n} |b_j|$ from which it easily follows that $\overline{\lim} \sqrt[n]{P_n} < \infty$.

In some local coordinate system $\alpha$ about the origin, $f(z)$ has the equation $f(z) = a_1 z, |a_1| > 1$. This allows us to construct a meromorphic function $h : C \to S$ whose image is the unstable manifold of 0. Let $\mathscr{U}$ be a disk of radius $r$ about the origin such that $\mathscr{U}$ is contained in the intersection of the domain of the coordinate system $\alpha$ and the unstable manifold of 0. If $\omega \in C$, there is an $n$ such that $|\omega/a_1^n| < r$. We define $h(\omega) = f^n g^{-1}(\omega/a_1^n)$. This is well defined because

$$f^{n+1}g^{-1}(\omega/a_1^{n+1}) = f^n(fg^{-1}(\omega/a_1^{n+1}))$$
$$= f^n g^{-1}(a_1\omega/a_1^{n+1})$$
$$= f^n g^{-1}(\omega/a_1^n)$$

if $\omega/a_1^n \in \mathscr{U}$. The range of $h$ is clearly the unstable manifold of 0. We now invoke Picard's theorem which states that the image of a meromorphic function omits at most two points of the sphere.

PROOF 2. In this proof we make the *a priori* assumption that $f$ is hyperbolic. Suppose $f(z) = z$ and $|f'(z)| > 1$. Then there is a well-defined branch $g$ of $f^{-1}$ on some neighborhood $\mathscr{U}$ of $z$. Moreover, we may choose $U$ to be a disk centered at $x$ such that $|g'|_U| < 1$. Then $g(U) \subset U$ and consequently $\{g^n\}$ is normal on $\mathscr{U}$. In fact, if $|g'(U)| < \lambda < 1$, then $|g^n(y) - x| \le \lambda^n|y - x|$. Therefore $\{g^n\}$ converges uniformly to the constant $x$ on $\mathscr{U}$. Now suppose $\{f^n\}$ is normal in a neighborhood of $z$. Then there would be a subsequence $\{n_k\}$ such that $\{f^{n_k}\} \to h$ uniformly on some compact neighborhood of $x$. $h$ is analytic. But, if $y \ne x$ and $y$ close to $x$, $g^{n_k} \circ f^{n_k}(y) = y \to \lim g^{n_k} \circ h(y) = x$. This is absurd.

Therefore, $\{f^n\}$ is not normal on any neighborhood of $x$. It is an elementary fact that a uniformly bounded family of analytic functions is normal. Therefore the unstable manifold of $x$ is dense in $S$ with totally disconnected complement $C$. It is clear that $f^{-1}(C) = C$.

Moreover, there are arbitrarily small neighborhoods $T$ of $C$ such that $f(T) \subset T$. It follows that $C$ contains an indecomposable nonwandering set $\Lambda$ which must be a sink. The periodic points are dense in $\Lambda$, but periodic sinks are isolated, indecomposable nonwandering sets themselves. Hence $\Lambda$ is finite.

We assert that every point of $C$ is nonwandering. Suppose $z \in C$ is wandering and choose a sequence $\{z_k\}$ such that $z_0 = z$ and $f(z_i) = z_{i-1}$. This sequence has a limit point $w$ in $C$ since $C$ is closed. $w$ is nonwandering, hence a sink. If $z \ne w$, we choose a point $z_k$ in a neighborhood $W$ of $w$ such that $z \notin W$ and $f(W) \subset W$. Then $z \in W$ since $z = f^k(z_k) \in f^k(W) \subset W$. This contradicts the assumption $z \notin W$. It follows then that all points of $C$ are isolated periodic sinks. Hence $C$ is finite. Since $f^{-1}(C) \subset C$, each element of $C$ is totally ramified; i.e., each element of $C$ has only 1 inverse image. Since $f$ has $2(\deg f) - 2$ critical points, $C$ has at most two elements.

Proof 1 demonstrates the relationship of Proposition 2.1 to the Picard theorem. The lack of an adequate generalization of the Picard theorem to higher dimensions prevents us from going much farther. However, there are recent results in this direction, and we shall see how they apply to morphisms of projective spaces.

The significance of Proof 2 is that it is completely elementary. The deepest fact which is used is that a uniformly bounded family of analytic functions is normal. This gives hope that a satisfactory generalization of Proposition 2.1 might be made under the additional hypothesis that the morphism be hyperbolic. The fundamental importance of Proposition 2.1 cannot be stressed too much. It demonstrates the global nature of the behavior of a rational function near a source.

The following is an integrated version of a theorem of Poincaré:

THEOREM (POINCARÉ). *Suppose* $f: C^n \to C^n$ *is defined in some neighborhood of* 0 *and* $f(0) = 0$. *Assume that* $Df(0)$ *is diagonalizable and that the eigenvalues* $\lambda_i$, $1 \le i \le n$, *of* $Df(0)$ *satisfy*

(1) $|\lambda_i| > 1$

(2) $\lambda_j \ne \prod \lambda_i^{m_i}$ *for any nonnegative integers* $m_i$ *such that* $\sum m_i > 1$.

*Then there is an analytic change of coordinates so that* $f$ *becomes linear.*

PROOF. See Sternberg [9].

Now suppose $f(z)$ is a rational endomorphism of $P_n$ with a fixed source at $z_0$ such that $Df(z_0)$ satisfies the conditions of the above theorem. Then there is an analytic map $h: C^n \to P_n$ such that the image of $h$ is the unstable manifold of $z_0$. $h$ is defined as in Proof 1 by $h(\omega) = f^n g^{-1}(A^{-n}\omega)$ where $A = Df(z_0)$, $gfg^{-1}(z) = Az$ and $A^{-n}\omega$ is in the domain of $g^{-1}$.

We are able now to apply a theorem of Bott and Chern [1].

PROPOSITION 8.1. *If* $f: P_n \to P_n$ *is an endomorphism and* $z_0$ *is a fixed point such that* $Df(z_0)$ *satisfies the hypotheses of Poincaré's theorem; i.e.,*

(1) $Df(z_0)$ *is diagonalizable with eigenvalues* $\lambda_i$, $1 \le i \le n$,

(2) $|\lambda_i| > 1$,

(3) $\lambda_j \ne \prod \lambda_i^{m_i}$ *for any nonnegative integers* $m_i$ *such that* $\sum m_i > 1$,

*then the unstable manifold of* $z_0$ *is dense in* $P_n$.

PROOF. We state the following

EQUIDISTRIBUTION THEOREM (BOTT AND CHERN [1]).

*Let* $E$ *be a complex vector bundle of fiber dimension* $n$ *over the complex connected manifold* $X$, *and let* $V \subset \Gamma(E)$ *be a finite dimensional space of holomorphic sections of* $E$. *Assume further that*

(1) $X$ *admits a concave exhaustion* $f$.

(2) $V$ *is sufficiently ample in the sense that*

($\alpha$) *The map* $s \to s(x)$ *maps* $V$ *onto* $Ex$ *for each* $x \in X$.

($\beta$) *There is some* $s \in V$ *and some* $x_0 \in X$ *so that* $s: X \to E$ *is transversal to the zero section.*

Under these circumstances nearly every section in $V$ vanishes the same number of times. Precisely, a Hermitian structure on $V$ defines a Hermitian structure on $E$, and hence a deficiency measure $\delta(s)$ on the generic sections of $V$. The assertion is that except for a set of measure 0, $\delta(s) = 0$.

This theorem applies to our situation with $X = C^n$, $E$ is the trivial line bundle on $C^n$, and $V$ is the space of sections of $E$ determined by the homogeneous coordinates of the image of $h$. For more details, refer to the introduction of [1]. Hypothesis $(2\beta)$ above is the only assumption which is not satisfied in the situation arising from any map $C^n \to P_n$. In our case, this hypothesis is satisfied at the origin.

We give two final examples in which we can describe the nonwandering sets of endomorphisms.

EXAMPLE 8.2. Let $(x, y, z)$ be homogeneous coordinates in $P_2$. Suppose $f: P_2 \to P_2$ is an endomorphism with the equation $f(x, y, z) = (P(x, y), Q(x, y), z^k)$. $P$, $Q$ are homogeneous polynomials of degree $k$ in $x$ and $y$. $f$ is maximally ramified along the line $Z$ given by $z = 0$.

On the line $Z$, $f$ induces the rational function $P/Q$. Therefore $\Omega(f) \cap Z$ is homeomorphic to $\Omega(P/Q)$. In the normal direction to $Z$, the derivative of $f$ is $0$. Therefore, the sinks of $P/Q$ are sinks of $f$ and the sources of $P/Q$ are saddle points of $f$.

The point $\mathcal{O} = (0, 0, 1)$ is a fixed point of $f$ at which $Df$ has rank $0$. Therefore $(0, 0, 1) = \mathcal{O}$ is a sink.

Now consider a line $H$ given by constant $= x/y$. If $(x, y, z) \in H$ and if $f^n(x, y, z) \to Z$, then $f^n(kx, ky, z) \to Z$ for all $k$ such that $|k| \geq 1$. Similarly, if $f^n(x, y, z) \to \mathcal{O}$, $f^n(x, y, z) \to \mathcal{O}$ for all $k$ such that $|k| \leq 1$. It follows that the stable manifolds of $\Omega(f) \cap Z$ and $\mathcal{O}$ are connected disks, separated by $T$, a fibre bundle over $Z = S^2$ with fibre $S^1$. $T$ must be totally invariant under $f$; i.e., invariant under both $f$ and $f^{-1}$.

The nonwandering set of $f|_T$ will lie above $\Omega(P/Q)$ in the fibering. All the points above $\Omega(P/Q)$ are nonwandering because the fibers over points of $Z$ are preserved. A dense set of the points of $T$ lying over periodic points of $Z$ are periodic. As $\Omega$ is closed, it follows that the fibers over $\Omega(P/Q)$ are nonwandering.

Thus $\Omega(f)$ breaks up into three pieces:

(1) a sink at $\mathcal{O}$

(2) a copy of $\Omega(P/Q)$ which attracts in the direction normal to $Z$

(3) a bundle over $\Omega(P/Q)$ with fibre $S^1$ which expands along the lines $x/y = $ constant.

EXAMPLE 8.3. Let $f: P_n \to P_n$ be the map which raises coordinates to the $k$th power, $k > 1$. That is $f(x_0, \ldots, x_n) = (x_0^k, \ldots, x_n^k)$. $f$ is maximally ramified along each of the coordinate hyperplanes defined by $x_i = 0$.

$f$ has one source diffeomorphic to $S^1 \times S^1 \times \ldots \times S^1$ ($n$ factors) defined by the equations $|x_0| = |x_1| = \ldots = |x_n|$. On each coordinate hyperplane we have a copy of the map on $P_{n-1}$ which raises coordinates to the $k$th power. Thus one sees inductively that $\Omega(f)$ splits into $C_{n+1, g+1}(n + 1, j + 1)$ copies of $S^1 \times \ldots \times S^1$ ($j$ factors), $j = 0, \ldots, n$. The unstable manifold of a $j$-torus is a $2j$ (real) dimensional manifold. The stable manifold of a $j$-torus is a $2(n - j)$ dimensional manifold.

ADDED IN PROOF. The author has obtained recent results of M. B. Jacobson (Moscow) which overlap the results of this paper, after its preparation.

## References

**1.** R. Bott and S. S. Chern, *Hermitian vector bundles and the equidistribution of the zeroes of their holomorphic sections,* Acta Math. **114** (1965), 71–112.

**2.** P. Fatou, *Sur les équationes fonctionelles,* Bull. Soc. Math. France **47** (1919), 161–271 and **48** (1920), 33–94 and 208–314.

**3.** E. Hille, *Analytic function theory,* vol. II, Ginn, New York, 1962.

**4.** C. Julia, *Memoire sur l'iteration des fonctions rationelles,* J. Math. Pures Appl. **4** (1918), 47–245.

**5.** J. F. Ritt, *On the iteration of rational functions,* Trans. Amer. Math. Soc. **21** (1920), 348–356.

**6.** M. Shub, Thesis, Univ. of California, Berkeley, 1967.

**7.** S. Smale, *Differentiable dynamical systems,* Bull. Amer. Math. Soc. **73** (1967), 747–817.

**8.** ———, *The Ω-stability theorem,* these Proceedings, vol. 14.

**9.** S. Sternberg, *Local contractions and a theorem of Poincaré,* Amer. J. Math. **79** (1957), 809–824.

**10.** R. Wilder, *Topology of manifolds,* Amer. Math. Soc. Colloq. Publ, vol. 32, Amer. Math. Soc., Providence, R.I., 1949.

**11.** R. F. Williams, *One dimensional non-wandering sets,* Topology **6** (1967), 473–487.

**12.** R. D. Anderson, *On raising flows and mappings,* Bull. Amer. Math. Soc. **69** (1963), 259–264.

**13.** H. Weyl, *The concept of a Riemann surface,* 3rd ed., Addison-Wesley, Reading, Mass., 1955.

**14.** W. Gottschalk and G. Hedlund, *Topological dynamics,* Amer. Math. Soc. Colloq. Publ, vol. 36, Amer. Math. Soc., Providence, R.I., 1955.

UNIVERSITY OF CALIFORNIA, BERKELEY

# EXPANDING MAPS AND TRANSFORMATION GROUPS

MORRIS W. HIRSCH

**Notation.** We denote by $M$ a compact $C^\infty$ Riemannian manifold without boundary. A map $f: M \to M$ is *expanding* if $f$ is $C^\infty$, and there exist constants $Q > 0$ and $\lambda > 1$ such that $|Df^n(v)| \geq Q\lambda^n |v|$ for all tangent vectors $v \in TM$ and positive integers $n \in \mathbf{Z}_+$. We shall assume that $Q = 1$. (John Mather has shown that $M$ has a metric for which $Q = 1$. See 3.1 of [**9**].)

We assume such an $f$ given, and let $g: \overline{M} \to \overline{M}$ be a fixed lifting of $f$ to the universal covering space $p: \overline{M} \to M$ of $M$. M. Shub has shown that $\overline{M}$ is diffeomorphic to a Euclidean space, and $g$ has a unique fixed point $a \in \overline{M}$.

**Infrahomogeneous spaces.** Let $N$ be a 1-connected Lie group, $K$ a finite group of automorphisms of $N$, and $H = N \cdot K$ the semidirect product. Let $\Gamma \subset H$ be a uniform discrete subgroup ($H/\Gamma$ compact).

$\Gamma$ acts on the space $H/K$ of left cosets $hK$ of $K$ in $H$ by left translation. The space $H/K$ is naturally diffeomorphic to $N$. If this action of $\Gamma$ on $N$ is totally discontinuous (i.e., if the map $N \to N/\Gamma$ is a covering space) then the double coset space $\Gamma \backslash H/K \approx N/\Gamma$ is a smooth manifold $M_0$. Such an $M_0$ is called an *infrahomogeneous space*. It has as a covering space the homogeneous space $N/N \cap \Gamma$. An *endomorphism* of $M_0$ is a map $f_0: M_0 \to M_0$ induced by an automorphism of $H$ taking $K$ onto $K$ and $\Gamma$ into $\Gamma$.

If $N$ is nilpotent, we call $M_0$ an *infranil-manifold*. In this case $N \cap \Gamma$ is uniform in $N$ and of finite index in $\Gamma$ (see Auslander) and $M_0$ is covered by the nilmanifold $N/N \cap \Gamma$.

Flat Riemannian manifolds are infranil-manifolds with $N = R^n$ and $\Gamma = Z^n$.

**The main conjecture.** Two maps $f: X \to X$ and $g: Y \to Y$ are *topologically conjugate* if there exists a homeomorphism $h: X \to Y$ such that $g = hfh^{-1}$.

We can now state the main conjecture on expanding maps:

MAIN CONJECTURE. *Every expanding map of a compact manifold is topologically conjugate to an expanding endomorphism of an infranil-manifold.*

It is easy to construct expanding maps which are not *differentiably* conjugate to infranil-endomorphisms. For example let $f: S^1 \to S^1$ be a slight perturbation of the map $z \mapsto z^2$, so that the derivative $f'(1)$ is no longer an integer.

The current state of knowledge concerning the main conjecture is summarized in the following theorem.

THEOREM. *The main conjecture is true in each of the following cases:*

(a) *The fundamental group of $M$ has a solvable subgroup of finite index* (J. Franks).

(b) *There are constants $c_1 > c$, $c_2 > c$ and $\lambda > 1$ such that*

$$c_1 \lambda^n |v| \leq |Df^n(v)| \leq c_2 \lambda^n |v|$$

*for all $v \in TM$ and $n \geq 0$. In this case $f$ is differentiably conjugate to a dilation of a flat Riemannian manifold.*

(c) *$f$ is an endomorphism of an infrahomogeneous space $M_0$. In this case $M_0$ is necessarily infranil.*

Parts (b) and (c) will be discussed below. We shall introduce a transformation group $G(f)$ acting on $\overline{M}$; a necessary and sufficient condition for the main conjecture to be true for $f$ is that $G(f)$ acts *semiproperly* (defined below). Unfortunately we cannot verify this condition on $G(f)$ except under strong hypotheses; we cannot even reprove J. Franks' theorem. We shall, however, indicate a proof of the following weaker result, in order to illustrate the possibilities of this method.

THEOREM 1. *Let the fundamental group of M contain an abelian subgroup of finite index. Then f is topologically conjugate to an affine expanding map of a flat Riemannian manifold.*

We shall also sketch proofs of (b) and (c).

In the final section we shall show how the construction of $G(f)$ naturally leads to an invariant ergodic measure for $f$.

**The group** $G(f)$. Let $\mathcal{H}(X)$ denote the topological group of homeomorphisms of a locally compact space $X$ in the compact open topology. A subgroup $G \subset \mathcal{H}(X)$ is called *semiproper* if for every $x \in X$, the evaluation map $ev_x \colon G \to X$ is a proper map. (Here $ev_x(\gamma) = \gamma(x)$; *proper* means the inverse image of a compact set is compact.) This is a strong condition on $G$; it implies that $G$ is locally compact and that all isotropy subgroups are compact. To give some perspective, we quote the theorem of Van der Waerden and Van Dantzig: *The isometry group of a locally compact metric space is semiproper.*

Consider now an expanding map $f \colon M \to M$. (More generally, $f$ could be merely a self-covering space.) Let $\xi_0$ be the universal covering space $p \colon \overline{M} \to M$; denote by $\xi_n$ the iterated covering space

$$f^n \circ p \colon \overline{M} \to M.$$

Let $\Gamma_n$ denote the group of automorphisms (=deck transformations) of $\xi_n$. Then

$$\pi_1(M) \approx \Gamma_\circ \subset \Gamma_1 \subset \ldots \subset \Gamma_n \subset \Gamma_{n+1} \subset \ldots \subset \text{Diff } \overline{M} \subset \mathcal{H}(\overline{M}).$$

Let $\Gamma_* = \bigcup \Gamma_n$. Put $\tilde{G} = \tilde{G}(f) = $ closure of $\Gamma_*$ in $\text{Diff } \overline{M}$; $G = G(f) = $ closure of $\Gamma_*$ in $\mathcal{H}(\overline{M})$. Let $\psi \colon \mathcal{H}(\overline{M}) \to \mathcal{H}(\overline{M})$ denote conjugation by $g \colon \overline{M} \to \overline{M}$:

$$\psi(\gamma) = g\gamma g^{-1}.$$

Then $\psi(\Gamma_n) \subset \Gamma_n$ and $\psi^{-1}(\Gamma_n) = \Gamma_{n+1}$. Therefore $\psi(\tilde{G}) = \tilde{G}$ and $\psi(G) = G$. Thus $\psi|G$ is a continuous automorphism $\phi \colon G \to G$.

THEOREM 2. *If $G(f)$ is semiproper, then*

(a) *$G$ is a Lie group,*

(b) *$G$ acts transitively and effectively on $\overline{M}$,*

(c) *the isotropy subgroup $K \subset G$ of the fixed point of $g$ is finite,*

(d) *$G$ is the semidirect product $N \cdot K$ where $N \subset G$ is a 1-connected nilpotent closed subgroup,*

(e) *Let $M_0$ denote the double coset space $\Gamma_0 \backslash G / K$, and $f_0 : M_0 \to M_0$ the infranil-endomorphism induced by $\phi : G \to G$. Then $f_0$ is expanding,*

(f) *The natural map $h : M_0 \to M$ induced by $ev_a : G \to \overline{M}$ is a topological conjugacy from $f_0$ to $f$,*

(g) *If $\tilde{G}$ is semiproper then $\tilde{G} = G$ and $h : M_0 \to M$ is a differentiable conjugacy,*

(h) *If $f$ is an infranil-endomorphism then $\tilde{G}$ is semiproper.*

OUTLINE OF PROOF. If $U \subset M$ is a nonempty open set, then $\bigcup_{n \in Z_+} f^{-n}(U)$ is dense in $M$ (see Shub [7]). It follows that if $a \in \overline{M}$ is the fixed point of $g$, then the orbit $\Gamma_*(a)$ of $a$ under $\Gamma_*$ is dense in $\overline{M}$. From the semiproperness of $G$ it follows that $G(a) = \overline{M}$. Thus $G$ is transitive.

We wish to apply the Corollary on p. 243 of Montgomery-Zippin; this implies that $G$ is a Lie group if $G/G_0$ is compact, where $G_0$ is the component of the identity in $G$. Using semiproperness and Lemma 2.3.1 on page 54 of [5], the compactness of $G/G_0$ can be proved, making $G$ a Lie group.

Let $K \subset G$ be the isotropy group of $a \in \overline{M}$, where $a$ is the fixed point of $g$. Using the effectiveness of the action of $G$ on $G/K = \overline{M}$, it can be shown that the automorphism $\phi | K : K \to K$ is an isometry in a suitable invariant metric. We want to show that the eigenvalues of $D\phi$ at $e \in G$ which do not come from $\phi | K$ are larger than 1 in norm. This is proved by a volume argument based on the topological conjugacy of $g^{-1} : \overline{M} \to \overline{M}$ and the contracting map $G/K \to G/K$ induced by $\phi^{-1}$.

Let $N \subset G$ be the subgroup whose Lie algebra is the union of the eigenspaces corresponding to eigenvalues of $\phi$ having norm $> 1$. Then $\phi | N : N \to N$ is expanding; this makes $N$ nilpotent. It can be proved that $N$ is 1-connected. Because $\Gamma_0 \cap N$ has finite index in $N$, we can prove $K$ finite.

This concludes the outline of the proof of Theorem 2.

REMARK 1. Notice how we lose control of the differential structure on $M_0 = \Gamma_0 \backslash G / K$. We can identify $N$ with $\overline{M}$ topologically, but the differential structure on $N/\Gamma_0$ (which is a covering space of $M_0$) comes from the Lie structure of $G$; this in turn comes from the solution to Hilbert's fifth problem in [5]. It is curious that $g : \overline{M} \to \overline{M}$ is a diffeomorphism in both the differential structure pulled back from $M$, and in the differential structure on the Lie group $N$. It is an open question whether $M_0$ is diffeomorphic to $M$.

REMARK 2. The astute reader will wonder why any proof is needed, since each $\Gamma_n$ is a group of isometries of $\overline{M}$, and the group of all isometries of $\overline{M}$ is a Lie group. The more astute reader will observe that $\Gamma_n$ preserves the metric pulled back from $M$ by $f^n \circ p : \overline{M} \to M$; in general, however, this metric varies with $n$.

If it could be shown that $\overline{M}$ has a topological metric invariant under $\Gamma_*$, the proof would be simpler.

REMARK 3. We cannot expect that $\overline{M}$ necessarily has a Riemannian metric invariant under $\Gamma_*$; for then by Myers and Steenrod the closure $\tilde{G}$ of $\Gamma_*$ in $\mathrm{Diff}(\overline{M})$ would be a Lie group acting on $M$ by diffeomorphisms, and by Montgomery-Zippin [5, Theorem 3, p. 208] $\tilde{G}$ would act differentiably. The proof of Theorem 1 would then show that $f$ is *differentiably* conjugate to an infranil-endomorphism. But as was shown above, this is not always true.

THEOREM 3. *The infranil-endomorphism in Theorem 1 is unique up to a differentiable conjugacy induced by an isomorphism of Lie groups. More precisely, let $f_0: M_0 \to M_0$ be an expanding endomorphism of an infranil manifold $M_0 = \Gamma_0 \backslash H_0 / K_0$; let $f_1: M_1 \to M_1$ be an endomorphism of an infrahomogeneous space $M_1 = \Gamma_1 \backslash H_1 / K_1$. Assume $H_i$ acts effectively on $\overline{M}_i = H_i / K_i$. If $h: M_0 \to M_1$ is a topological conjugacy from $f_0$ to $f_1$, then $h$ is induced by an isomorphism $H_0 \to H_1$.*

PROOF. $h$ induces an isomorphism from $G(f_0)$ to $G(f_1)$. On the other hand $G(f_i)$ is the isomorphic image in $\mathcal{H}(\overline{M}_i)$ of $H_i$.

PROOF OF THEOREM 1. It is easy to prove that if $\Gamma' \subset \Gamma_0$ is an abelian subgroup of finite index, then every element $\alpha$ of $\Gamma'$ has only a finite number of conjugates in $\Gamma_0$. If $\Gamma'' \subset \Gamma_0$ is the subgroup comprising elements with only a finite number of conjugates in $\Gamma_0$, then $\phi(\Gamma'') \subset \Gamma''$ and $\phi^{-1}(\Gamma'') \subset \Gamma''$. Therefore $\Gamma_0 / \phi(\Gamma_0)$ has a set $A = \{\alpha_1, ..., \alpha_k\}$ of right coset representatives such that $A \subset \Gamma''$. Moreover it is easy to see that an element $\alpha \in \Gamma_0$ has only a finite number of conjugates if and only if there exists $P > 0$ such that

$$d(\alpha, 1) = \sup_{x \in \overline{M}} d(\alpha x, x) \leq P.$$

We have proved $\Gamma_0 / \phi(\Gamma_0)$ has a set $A = \{\alpha_1, ..., \alpha_k\}$ of right coset representatives such that $d(\alpha_i, 1) < \infty$ for $i = 1, ..., k$.

To show that $G$ is semiproper boils down to proving: *if $x \in \overline{M}$ and $\gamma_n \in \Gamma_n$ for $n = 1, 2, ...$ and $\gamma_n(x)$ converges in $\overline{M}$, then the sequence $\{\gamma_n\}$ has a subsequence converging in $\mathcal{H}(\overline{M})$.*

It is clear that $A_n = \phi^{-n}(A) = \{g^{-n}\alpha_i g^n\}$ is a set of right coset representatives for $\Gamma_n / \Gamma_{n-1}$ for $n \in Z_+$, since $\phi^{-n}(\Gamma_0) = \Gamma_n$. Therefore we may write

$$\gamma_0 = \theta_0$$
$$\gamma_1 = \theta_1 \beta_{11}$$

(1)
$$\vdots$$
$$\vdots$$

$$\gamma_n = \theta_n \beta_{n1} \cdots \beta_{nn},$$

with $\theta_i \in \Gamma_0$ and $\beta_{ij} \in A_j$.

We shall show there is a subsequence of $\{\gamma_n\}$ such that each of the corresponding subsequences of $\{\theta_n\}$, $\{\beta_{n1}\}$, $\{\beta_{n2}\}$, ... eventually becomes constant, and that this subsequence converges in $\mathcal{H}(\overline{M})$.

Let $B_0 = \max\{d(\alpha, 1)|\alpha \in A\}$, and $B_n = \max\{d(\alpha, 1)|\alpha \in A_n\}$. Every element of $A_n$ has the form $g^{-n}\alpha g^n$ with $\alpha \in A$;

$$d(g^{-n}\alpha g^n, 1) \leq \lambda^{-n}d(\alpha g^n, g^n) \leq \lambda^{-n}d(\alpha, 1) \leq \lambda^{-n}B_0.$$

Therefore $B_n \leq \lambda^{-n}B_0$. It follows that if $p \geq n$,

$$(2) \qquad d(\beta_{p,n+1} \cdots \beta_{pp}, 1) \leq \sum_{i=n+1}^{p} \lambda^{-1}B_0 \leq \lambda^{-n}(\lambda - 1)^{-1}B_0 = R\lambda^{-n}.$$

If we put

$$\mu_{pn} = \theta_p\beta_{p1} \cdots \beta_{pn} \in \Gamma_n,$$
$$v_{pn} = \beta_{p,n+1} \cdots \beta_{pp},$$

then

$$\gamma_p = \mu_{pn}v_{pn} \quad \text{and} \quad d(v_{pn}, 1) \leq R\lambda^{-n}.$$

Therefore given $\varepsilon > 0$ we can choose $n_0(\varepsilon) = n_0 \geq 0$ so large that $v_{pn}(x)$ stays in a compact set $X_\varepsilon \subset \overline{M}$ of diameter $< \varepsilon$ for all $p \geq n \geq n_0$. Also there exists $p_0(\varepsilon) = p_0 \geq 0$ such that $\gamma_p(x)$ stays in a compact set $Y_\varepsilon \subset \overline{M}$ of diameter $< \varepsilon$ for all $p \geq p_0$. Since $\gamma_p(x) = \mu_{pn}v_{pn}(x)$, it follows that $\mu_{pn}(x_\varepsilon) \cap Y_\varepsilon \neq \varnothing$ for all $p$, $n$ such that $p \geq n$, $p \geq p_0$, $n \geq n_0$. Since $\Gamma_n$ acts totally discontinuously, the set $\{\mu \in \Gamma_n | \mu(X_\varepsilon) \cap \mu(Y_\varepsilon) \neq \varnothing\}$ is finite; if $\varepsilon$ is sufficiently small, depending on $n$, it has at most one element. If $\varepsilon = 1/q$, let $n_0(\varepsilon) = n_q$, $p_0(\varepsilon) = p_q$. Then $\mu_{pnq} = \mu_{p+1,n} = \mu(q)$ for $p \geq p_q$. Thus the subsequence $\gamma_{p1}, \gamma_{p2}, \ldots$ has the form

$$\gamma_{p1} = \sigma_1\rho_1$$
$$\gamma_{p2} = \sigma_1\sigma_2\rho_2$$
$$\vdots$$
$$\gamma_{pj} = \sigma_1\sigma_2 \cdots \sigma_j\rho_j$$

where $\sigma_1 \ldots \sigma_j = \mu_{p_jn_j}$. Here $\sigma_i \in A_{n_i}$, $1 \leq n_1 < n_2 < \ldots$, and $d(\rho_i, 1) \leq R\lambda^{-i}$. Since also $d(\sigma_i, 1) \leq R\lambda^{-i}$, it follows easily that the subsequence $\gamma_{p_i}$ converges in $\mathcal{H}(\overline{M})$ as $i \to \infty$.

Once we know that $G$ is semiproper we may use Theorem 2. It is not hard to show that the group $N$ of Theorem 2 must be abelian. This completes the proof of Theorem 1.

REMARK 4. An analysis of the proof of Theorem 1 shows that $G(f)$ is semiproper provided that $\Gamma_0/\phi(\Gamma_0)$ has a set $A$ of right coset representatives having the following property: there exists $\mu < \lambda$ such that

$$\limsup_{n \to \infty} d(\alpha g^n(x), g^n(x))^{1/n} \leq \mu$$

for all $\alpha \in A$ and $x \in \overline{M}$. It is not clear whether this condition is necessary; I would guess it is not.

**Conformal expanding maps.** The expanding map $f: M \to M$ is called *almost conformal* if there exist constants $c_1 \geq c_2 > 0$ and $\lambda > 1$ such that

$$c_1 \lambda^n |v| \geq |Df^n(v)| \geq c_2 \lambda^n |v|.$$

If $c_1 = c_2$ then $c_1 = 1$, and in this case $f$ is *conformal*.

THEOREM 4. *If $f: M \to M$ is an almost conformal expanding map of a compact Riemannian manifold, then there exists an affine dilation of a flat Riemannian manifold $f_0: M_0 \to M_0$, which is differentiably conjugate to $f$. If $f$ is conformal then $M$ is flat and $f$ is an affine dilation.*

PROOF. First suppose $f$ is conformal. Then if $\alpha \in \Gamma_0$ and $n \geq 0$ we have

$$|D(g^{-n}\alpha g^n)v| = \lambda^{-n}|D(\alpha g^n)(v)| = \lambda^{-n}|Dg^n(v)| = |v|,$$

for all $v \in T\bar{M}$, since $\alpha$ is an isometry. Thus $\Gamma_* \subset \text{Diff } \bar{M}$ is a group of isometries; hence its closure $\tilde{G}$ is a Lie group acting differentiably on $\bar{M}$ by isometries. The derivative of $\phi$ at the identity of $\tilde{G}$ has all eigenvalues of absolute value $\lambda$; this makes the Lie algebra of $\tilde{G}$ abelian. The rest of the proof is easy.

If $f$ is almost conformal, then for any element $\gamma = g^{-n}\alpha g^n \in \Gamma_*$ we have $c_1 \leq \|D\gamma_x\| \leq c_2$ for all $x \in \bar{M}$. (Here $\|D\gamma_x\|$ is the norm of the linear map $D\gamma_x: T_x\bar{M} \to T_{\gamma x}\bar{M}$.) It follows from the uniform equicontinuity of the derivatives of the elements of $\Gamma_*$ at points of $\bar{M}$ that $\tilde{G}$ acts semiproperly on $\bar{M}$. Therefore, as in the proof of Theorem 1, $\tilde{G}$ is a Lie group, acting *differentiably* on $\bar{M}$ (see Remark 3). The rest of the proof is left to the reader.

**Comparison with the unilateral shift.** Let $A$ be a set of $k$ elements and $S_k$ the Cantor set $A^{Z_+}$ of sequences in $A$ with the product (=compact open) topology. The *shift map* $T: S_k \to S_k$ is defined by $(T\sigma)(n) = \sigma(n+1)$ for $n \in Z_+$.

THEOREM 5. *Let $f: M \to M$ be an expanding map of a compact Riemannian manifold. Suppose $f$ has degree $\pm k$, $k \in Z_+$. Then there is a continuous surjective map $\pi: S_k \to M$ such that the following diagram commutes:*

$$\begin{array}{ccc} S_k & \to & S_k \\ \pi \downarrow & & \downarrow \pi \\ M & \overset{f}{\to} & M. \end{array}$$

COROLLARY 6. *$f$ admits an invariant ergodic measure, positive on open sets and vanishing on points.*

PROOF. Let $g: \bar{M} \to \bar{M}$ be a universal covering $\xi_0$ of $M$. Let $\Gamma_0$ be the automorphism group of $\xi_0$. Let $\phi: \Gamma_0 \to \Gamma_0$ be the endomorphism $\phi(\gamma) = g\gamma g^{-1}$. Let $A = \{\alpha_1, ..., \alpha_k\}$ be a set of right coset representatives for $\Gamma_0/\phi(\Gamma_0)$, with $\alpha_1 = $ identity.

Let $a \in \bar{M}$ be the fixed point of $g$. If $\gamma = \{\gamma_n\}$ is a sequence in $A$, define $\pi(\gamma) \in M$ by

$$\pi(\gamma) = p \lim_{n \to \infty} (g^{-1}\gamma_1 \circ ... \circ g^{-1}\gamma_n)(a) = p\tilde{\pi}(\gamma).$$

To see that the limit exists, let $\lambda > 1$ be the expansion constant of $f$, and observe that

$$d(g^{-1}\gamma_1 \circ \ldots \circ g^{-1}\gamma_n(a), g^{-1}\gamma_1 \circ \ldots \circ g^{-1}\gamma_{n+1}a)$$
$$\leq \lambda^{-1}d(g^{-1}\gamma_2 \circ \ldots \circ g^{-1}\gamma_n(a), g^{-1}\gamma_2 \circ \ldots \circ g^{-1}\gamma_{n+1}(a))$$

$$\leq \lambda^{-n}d(a, g^{-1}\gamma_{n+1}(a))$$
$$\leq \lambda^{-n}\max\{d(a, g^{-1}\alpha(a))|\alpha \in A\} \quad \text{(since } \gamma_1 \text{ is an isometry)}$$
$$= B\lambda^{-n}.$$

Therefore the sequence is Cauchy. This also proves that $d(\tilde\pi(\gamma), a) \leq B(\lambda - 1)^{-1}$. The Lipschitz constant of $g^{-1}\gamma_1 \circ \ldots \circ g^{-1}\gamma_n$ is $\leq \lambda^{-n}$. To see that $\tilde\pi: S_k \to \bar M$ is continuous, suppose $\gamma_r = \alpha_r$ for $1 \leq r \leq n$. Then

$$d(\tilde\pi(\gamma), \tilde\pi(\alpha)) \leq \lambda^{-n}d(\tilde\pi T^n(\gamma), \tilde\pi T^n(\alpha)) \leq \lambda^{-n} \cdot 2B(\lambda - 1)^{-1}.$$

(This proves that $\tilde\pi$ is even Lipschitz, if $S_k$ has the metric $d(\gamma, \alpha) = \sum_r \lambda^{-r}\delta(\gamma_r, \alpha_r)$ where $\delta$ is the Kronecker function.)

Clearly $\pi T = f\pi$, for

$$\begin{aligned}
f\pi(\gamma) &= \lim fp \circ g^{-1}\gamma_1 \circ \ldots \circ g^{-1}\gamma_n(a) \\
&= \lim pg \circ g^{-1}\gamma_1 \circ \ldots \circ g^{-1}\gamma_n(a) \\
&= \lim p\gamma_1 \circ g^{-1}\gamma_2 \circ \ldots \circ g^{-1}\gamma_n(a) \\
&= \lim pg^{-1}\gamma_2 \circ \ldots \circ g^{-1}\gamma_n(a) \qquad \text{(because } p\gamma_1 = p) \\
&= p \lim g^{-1}\gamma_2 \circ \ldots \circ g^{-1}\gamma_n(a) \\
&= \pi(T\gamma).
\end{aligned}$$

Finally, $\pi$ is surjective because $g^n(a) = g$ and $p\alpha = p$ for any $\alpha \in \Gamma_0$. Therefore $\pi(S_k)$ contains every point of the form

$$p(\alpha \cdot g^{-1}\gamma_1 g \cdot g^{-2}\gamma_2 g^2 \cdot \ldots \cdot g^{-n}\gamma_n g^n(a)),$$

letting $\gamma_i = $ identity for $i > n$. Therefore $\pi(S_k)$ contains $p(ev_a\Gamma_*)$. Since $a$ has a dense orbit, $\pi(S_k)$ is dense. But $S_k$ is compact, so $\pi(S_k) = M$.

## BIBLIOGRAPHY

**1.** R. Arens, *Topologies for homeomorphism groups*, Amer. J. Math. **68** (1946), 593–610.

**2.** L. Auslander, *Bieberbach theorems on space groups and discrete uniform subgroups of Lie groups.* Ann. of Math. (2) **71** (1960), 579–590.

**3.** A. Avez, C.R. Acad. Sci. Paris, **266** (1968), 610–612.

**4.** J. Franks, Thesis, University of California, Berkeley, 1968.

**5.** D. Montgomery and L. Zippin, *Topological transformation groups*, Interscience, New York, 1955.

**6.** S. B. Myers and N. Steenrod, *The group of isometries of a Riemannian manifold*, Ann. of Math. **40** (1939), 400–416.

**7.** M. Shub, Thesis, University of California, Berkeley, 1967.

**8.** D. Van Dantzig and B. Van der Waerden, *Über metrische homogene Raüme*, Abh. Math. Sem. Hamburg Univ. **6** (1928), 367–376.

**9.** M. W. Hirsch and C. C. Pugh, *Stable manifolds for hyperbolic sets*, these Proceedings, vol. 14.

UNIVERSITY OF CALIFORNIA, BERKELEY

# STABLE MANIFOLDS AND HYPERBOLIC SETS

MORRIS W. HIRSCH AND CHARLES C. PUGH

0. **Introduction.** Let $U$ be an open set in a smooth manifold $M$ and $f: U \to M$ a $C^1$ map. A fixed point $x$ of $f$ is *hyperbolic* if the derivative $T_x f: M_x \to M_x$ is an isomorphism and its spectrum is separated by the unit circle. If $T = T_x f$, this means that $M_x$ has a unique splitting $E_1 \times E_2$ under $T$ such that $T|E_1$ is expanding and $T|E_2$ is contracting. That is, for suitable equivalent norms on $E_1$ and $E_2$,

$$\max\{\|T^{-1}|E_1\|, \|T|E_2\|\} < 1.$$

The classical stable manifold theory says that this convenient behavior of $T_x f$ is reflected in the behavior of $f$ in a neighborhood $V$ of $x$: there is a submanifold $W^s$ of $M$ tangent to $E_2$ at $x$ such that

$$W^s \cap V = \{y \in V | \lim_{n \to \infty} (f|V)^n y = x\},$$

there is also a submanifold $W^u$ tangent to $E_1$ such that

$$W^u \cap V = \{y \in V | \lim_{n \to \infty} (f|V)^{-n} y = x\}.$$

See for example Kelley [1, Appendix], [15] and [14], which contains further references.

We call $W^s$ and $W^u$ *local stable* and *unstable* manifolds of $f$ at $x$, respectively. It turns out that they enjoy the same differentiability as $f$, and if $f$ is $C^k$ they depend continuously on $f$ in the $C^k$ topologies.

For technical reasons we allow $M$ to be an infinite dimensional manifold modelled on a Banach space.

The notion of hyperbolic fixed point can be generalized to that of a *hyperbolic set* $\Lambda \subset U$. This means that $f(\Lambda) = \Lambda$, and $T_\Lambda M$ (the tangent bundle of $M$ over $\Lambda$) has an invariant splitting $E_1 \oplus E_2$ such that $Tf|E_1$ is expanding and $Tf|E_2$ is contracting. (For this theory $M$ is assumed finite dimensional and $\Lambda$ compact, although generalizations are possible.) In Smale's theory of $\Omega$-stability, and related topics [12], [13], "generalized stable manifold theorem" plays a key role: there is a neighborhood $V$ of $\Lambda$, and submanifolds $W^s(x)$, $W^u(x)$ tangent to $E_2(x)$ and $E_1(x)$ respectively for each $x \in \Lambda$, such that

$$W^s(x) = \{y \in V | \lim_{n \to \infty} d((f|V)^n y, (f|V)^n x) = 0\},$$

$$W^u(x) = \{y \in V | \lim_{n \to \infty} d((f|V)^{-n} y, (f|V)^{-n} x) = 0\}.$$

If $f$ is $C^k$, so are $W^s(x)$ and $W^u(x)$, and they depend continuously on $f$ in the $C^k$ topologies. Moreover $W^s(x)$ and $W^u(x)$ and their derivatives along $W^s(x)$ and $W^u(x)$ up to order $k$ depend continuously on $x$.

The proof of the generalized stable manifold theorem proceeds in the following steps:

(A) Let $E = E_1 \times E_2$ be a Banach space; $T: E \to E$ a hyperbolic linear map expanding along $E_1$ and contracting along $E_2$; $E(r) \subset E$ the ball of radius $r$; and $f: E(r) \to E$ a Lipschitz perturbation of $T|E(r)$. The unstable manifold $W$ for $f$ will be the graph of a map $g: E_1(r) \to E_2(r)$ which satisfies $W = f(W) \cap E(r)$. We are led to consider the following transformation $\Gamma_f$ in a suitable function space $G$ of maps $g$:

$$\text{graph } \Gamma_f(g) = E(r) \cap f(\text{graph } g).$$

We call $\Gamma_f(g)$ the *graph transform* of $g$ by $f$. The fixed point $g_0$ of $\Gamma_f$ gives the unstable manifold of $f$; the existence of a fixed point is proved by the contracting map principle if $f$ is sufficiently close to $T$ pointwise, and the Lipschitz constant of $f - T$ is small enough.

(B) If $f$ is $C^k$ ($k \in \mathbf{Z}_+$) so is $g_0$. This is proved by induction on $k$. The successive approximations $\Gamma_f^n(g)$ converge $C^k$ to $g_0$, but not, apparently, exponentially. The fibre contraction theorem (1.2) is needed to get this convergence.

(C) Let $\Lambda \subset U$ be a hyperbolic set. Let $\mathcal{M}$ be the Banach manifold of bounded maps $\Lambda \to M$, and $i \in \mathcal{M}$ the inclusion of $\Lambda$. Let $\mathcal{U} = \{h \in \mathcal{M} | h(\Lambda) \subset U\}$. Define $f_*: \mathcal{U} \to \mathcal{M}$ by

$$f_*(h) = f \circ h \circ f^{-1}.$$

Then $f_*$ has a hyperbolic fixed point at $i$. By (A), $f_*$ has a stable manifold $\mathcal{W}^s \subset \mathcal{M}$. For each $x \in \Lambda$, define $W^s(x) = ev_x(\mathcal{W}^s) = \{y \in M | y = \gamma(x) \text{ for some } \gamma \in \mathcal{W}^s\}$. It turns out that this definition gives a system of stable manifolds for $f$ along $\Lambda$.

In §1 we collect various facts about maps and function spaces, including the Lipschitz inverse function theorem and the fiber contraction theorem. In §2 we carry out steps (A) and (B), and (C) is done in §3.

In §4 technical criteria for hyperbolicity are established. These are applied in §5 to prove that if $V \subset \Lambda$ is an invariant smooth submanifold then the non-wandering set of $f| V$ is a hyperbolic set for $f|V$.

The smoothness of the subbundles $E^s$ and $E^u$, considered as fields of subspaces of $T_\Lambda M$, is studied in §6, which does not depend on the other sections. It is shown that $E^s$ (also $E^u$) is always Hölder; if the stable manifolds have codimension 1, then $E^s$ is Lipschitz, and is $C^1$ on every invariant $C^2$ submanifold of $\Lambda$ if $f$ is $C^2$. If the stable manifolds have codimension 1 and fill up an open set and $f$ is $C^2$, then they provide a $C^1$ foliation of that set. Both the stable and the unstable foliations are $C^1$ if $f$ is a $C^2$ volume preserving Anosov diffeomorphism of a compact 3-manifold.

§7 deals with hyperbolic sets of perturbations of $f$. There is a compact neighborhood $V$ of $\Lambda$ containing a unique maximal hyperbolic set $\Lambda_0$, and $\Lambda_0$ contains every invariant set in $V$. These maximal hyperbolic sets are structurally stable in a certain sense.

While preparing this paper we were fortunate to have had helpful discussions with participants in the 1968 Conference on Global Analysis in Berkeley. It is a pleasure to record our gratitude to them, especially to M. Shub, R. F. Williams and J. Moser. Special thanks are due to S. Smale for being the source of so many of the ideas developed in this paper.

1. **Lipschitz maps.** A map $f: X \to Y$ between metric spaces is *Lipschitz* if there exists a number $k$ such that $d(fx, fy) \leq kd(x, y)$ for all $x, y \in X$. The smallest such $k$ is the *Lipschitz constant* $L(f)$. If $X = Y$ and $L(f) < 1$ then $f$ is a *contraction*. If $f$ is not Lipschitz then $L(f) = \infty$.

1.1. (CONTRACTING MAP THEOREM). *Let $X$ be a complete metric space and $f: X \to X$ a contraction. Then $f$ has a unique fixed point $x_0$, equal to $\lim_{n \to \infty} f^n(x)$ for all $x \in X$. If $L(f) = k < 1$ and $g: X \to X$ has a fixed point $y_0$ such that $d(fy_0, gy_0) \leq \varepsilon$, then $d(x_0, y_0) \leq \varepsilon/(1 - k)$.*

PROOF. The proof of the first assertion is quite standard and is therefore omitted. The last assertion is true because

$$
\begin{aligned}
d(x_0, y_0) &= d(fx_0, gy_0) \\
&\leq d(fx_0, fy_0) + d(fy_0, gy_0) \\
&\leq kd(x_0, y_0) + \varepsilon.
\end{aligned}
$$

If $X$ is a set and $Y$ a metric space, $\mathcal{M}(X, Y)$ denotes the space of all maps $X \to Y$ with the *uniform topology*, generated by the collection of sets

$$
\{\mathcal{N}_\varepsilon(f) \mid f \in \mathcal{M}(X, Y), \varepsilon > 0\}
$$

where

$$
\mathcal{N}_\varepsilon(f) = \{g \in \mathcal{M}(X, Y) \mid d(fx, gx) < \varepsilon \text{ for all } x\}.
$$

We write $d(f, g) = \sup_x\{d(fx, gx)\} \leq \infty$. If $Y$ is complete and $f_0 \in \mathcal{M}(X, Y)$, the set of maps at a finite distance from $f_0$, namely

$$
\{g \in \mathcal{M}(X, Y) \mid d(f_0, g) < \infty\} = \mathcal{M}(X, Y; f_0)
$$

is a complete metric space with metric $d(f, g)$; it is open and closed in $\mathcal{M}(X, Y)$. A *bounded* subspace of $\mathcal{M}(X, Y)$ means a bounded subspace of some $\mathcal{M}(X, Y; f_0)$. If $X$ is a space, the subset $C(X, Y) \subset \mathcal{M}(X, Y)$ of continuous maps is closed. If $X$ is metric and $\lambda \geq 0$, then the subspace

$$
\mathcal{H}_\lambda(X, Y) = \{f \in \mathcal{M}(X, Y) \mid L(f) \leq \lambda\}
$$

is closed in $\mathcal{M}(X, Y)$.

The contracting map theorem has as a corollary: the function assigning to each contraction of a complete metric space its fixed point is continuous.

A fixed point $x_0$ of $f: X \to X$ is called *attractive* if $\lim_{n \to \infty} f^n(x) = x_0$ for all $x \in X$. The following extension of the contracting map principle is useful for proving maps to be $C^1$.

1.2. FIBER CONTRACTION THEOREM. *Let $X$ be a space and $f:X \to X$ a map having an attractive fixed point $p \in X$. Let $Y$ be a metric space and $\{g_x\}_{x \in X}$ a family of maps $g_x: Y \to Y$ such that the formula $F(x, y) = (fx, g_x y)$ defines a continuous map $F: X \times Y \to X \times Y$. Let $q \in Y$ be a fixed point for $g_p$. Then $(p, q) \in X \times Y$ is an attractive fixed point for $F$ provided*

(a) $\lim \sup_{n \to \infty} L(g_{f^n x}) < 1$ *for each $x \in X$.*

PROOF. For each $(x, y) \in X \times Y$ we must prove $\lim_{n \to \infty} F^n(x, y) = (p, q)$. Therefore we could replace $X$ by $\{p\} \bigcup \{f^n(x) | n \geq n_0\}$ for any $n_0$ and an arbitrary $x$. Hence we may assume instead, by (a), that

(1) $$L(g_x) \leq \lambda < 1 \quad \text{for all } x \in X.$$

The theorem is proved if we show that $\pi_2 F^n(x, y) \to q$.

Call $d(g_{f^n x} q, q) = \delta_n$. Since $f^n(x) \to p$ as $n \to \infty$ and $F$ is continuous at $(p, q)$, $\delta_n \to 0$. By definition of $F$, $\pi_2 F^{n+1}(x, \cdot) = g_{f^n x} \circ \ldots \circ g_x$ and so

$$\begin{aligned}
d(\pi_2 F^{n+1}(x, q), q) &\leq d(\pi_2 F^{n+1}(x, q), g_{f^n x}(q)) + d(g_{f^n x}(q), q) \\
&\leq \lambda d(\pi_2 F^n(x, q), q) + \delta_n \\
&\leq \lambda[\lambda d(\pi_2 F^{n-1}(x, q) + \delta_{n-1}] + \delta_n \\
&\leq \ldots \leq \lambda^n d(\pi_2 F(x, q), q) + \lambda^{n-1}\delta_1 + \ldots + \delta_0,
\end{aligned}$$

which is $\sum_{j=0}^n \lambda^{n-j}\delta_j$. This tends to zero because, for any $k$, $0 < k < n$,

$$\begin{aligned}
\sum_{j=0}^n \lambda^{n-j}\delta_j &= \sum_{j=0}^{k-1} \lambda^{n-j}\delta_j + \sum_{j=k}^n \lambda^{n-j}\delta_j \\
&\leq (\lambda^n + \ldots + \lambda^{n-k+1})\max_j(\delta_j) + (1 + \ldots + \lambda^{n-k})\max_{j \geq k}(\delta_j) \\
&\leq \lambda^{n-k}\max_j(\delta_j)/(1 - \lambda) + \max_{j \geq k}(\delta_j)/(1 - \lambda),
\end{aligned}$$

which tends to zero as $k$ and $n - k \to \infty$.

Hence

$$d(\pi_2 F^n(x, y), q) \leq d(\pi_2 F^n(x, y), \pi_2 F^n(x, q))$$

$$+ d(\pi_2 F^n(x, q), q) \leq \lambda^n d(y, q) + \sum_{j=0}^{n-1} \lambda^{n-j}\delta_j \to 0 \quad \text{as } n \to \infty.$$

REMARK 1. It is easy to see that if $A \subset X$ is such that $f(A) \subset A$, $f^n(x) \to p$ uniformly in $A$, and $L(g_x) \leq \kappa < 1$ for all $x \in A$, then the convergence $f^n(x, y)$ to $(p, q)$ is uniform in $A \times Y$.

REMARK 2. Suppose $f$ is a contraction of $X$ and each $g_x$ is a contraction of $Y$ with $L(g_x)$ bounded away from 1. It is not clear whether there exists a metric on $X \times Y$ for which $F$ is a contraction. The product metric will not suffice as the maps $f(x) = \frac{1}{2}x$, $g_x(y) = |x|^{1/2} + \frac{1}{2}(y - |x|^{1/2})$ show. This is the phenomenon of "shear" in the $y$-direction. It seems likely that no such metric exists in general.

Next we collect certain elementary relations between Lipschitz constants. In order to minimize notation, we assume the existence of all sums, compositions, inverses, etc. that are needed to make sense of the notation. Thus "$f + g$" implies that $f$ and $g$ are maps into a Banach space, and $f + g$ is the map $x \mapsto f(x) + g(x)$.

If $X_1$ and $X_2$ are metric spaces, the metric in $X_1 \times X_2$ is $d(x, y) = \sup\{d(x_1, y_1), d(x_2, y_2)\}$.

1.3. PROPOSITION.
(a) $L(f \circ g) \le L(f)L(g)$.
(b) $L(f + g) \le L(f) + L(g)$, and $L(f) - L(g) \le L(f - g)$.
(c) $d(f_1 \circ g_1, f_2 \circ g_2) \le d(f_1, f_2) + L(f_2)d(g_1, g_2)$.
(d) Recall that $\mathcal{H}_\lambda(Y, Z)$ is the set of maps $h: Y \to Z$ with $L(h) \le \lambda$. Composition defines a map $\mathcal{H}_\lambda(Y, Z) \times C(X, Y) \to C(X, Z)$ having Lipschitz constant $\le 1 + \lambda$.
(e) For fixed $h_0 \in \mathcal{H}_\lambda(Y, Z)$, the map $C(X, Y) \to C(X, Z)$ given by $g \mapsto h_0 \circ g$ has Lipschitz constant $\le L(h_0)$.

PROOF. (a), (b) and (e) are obvious; (c) is proved by writing

$$d(f_1 \circ g_1, f_2 \circ g_2) \le d(f_1 \circ g_1, f_2 \circ g_1) + d(f_2 \circ g_1, f_2 \circ g_2).$$

And (c) implies (d).

1.4. PROPOSITION. Consider maps between topological subspaces of Banach spaces.
(a) If $f$ is injective and $L(f - g) < L(f^{-1})^{-1}$, then $g$ is injective, and $L(g^{-1}) \le [L(f^{-1})^{-1} - L(f - g)]^{-1}$.
(b) $|g^{-1} - h^{-1}| \le L(g^{-1}) \cdot |h - g|$.
(c) Let $\mathcal{G}$ be a space of invertible maps $g$ such that $L(g^{-1}) \le \lambda$. Then the Lipschitz constant of the map $g \mapsto g^{-1}$ is $\le \lambda$.

PROOF. (a): Follows from the two inequalities,

$$|gx - gy| \ge |fx - fy| - |(g - f)x - (g - f)y|$$

and

$$|fx - fy| \ge L(f^{-1})^{-1}|x - y|.$$

(b): Follows from

$$|g^{-1} - h^{-1}| = |g^{-1} \circ h \circ h^{-1} - g^{-1} \circ g \circ h^{-1}| \le L(g^{-1}) \cdot |h - g|.$$

This proves (c).

The standard modern proof of the Inverse Function Theorem (see [9]) deals with a $C^1$ small perturbation of an invertible linear map; a $C^1$ inverse is produced. Abstracting this idea leads to the following result.

1.5. LIPSCHITZ INVERSE FUNCTION THEOREM. Let $E, F$ be Banach spaces, $U \subset E$ and $V \subset F$ open sets and $f: U \to V$ a homeomorphism such that $f^{-1}$ is Lipschitz. Let $h: U \to F$ be such that $L(h)L(f^{-1}) < 1$ and put $g = f + h: U \to F$. Then $g$ is a homeomorphism onto an open set, and

$$L(g^{-1}) \le L(f^{-1})/(1 - L(g - f)L(f^{-1})) = [L(f^{-1})^{-1} - L(g - f)]^{-1}.$$

Note that no assumption on the existence of a linear isomorphism $E \to F$ (to which $f$ might be nearly tangent) is made. In fact it is not clear that a lipeomorphism between open subsets of $E$ and $F$ implies the existence of a linear isomorphism between $E$ and $F$.

PROOF. The injectivity of $g$ and the estimate on $L(g^{-1})$ follow from 1.4a. Since $g^{-1}$ is Lipschitz, it is continuous. It remains to prove that if $x_0 \in U$, then $g$ maps some neighborhood of $x_0$ onto a neighborhood of $g(x_0)$. We may assume that $h(x_0) = 0$ and $g(x_0) = f(x_0)$; if not, replace $h$ by the map $x \mapsto h(x) - h(x_0)$.

Let juxtaposition denote composition, and put $hf^{-1} = v: V \to F$. We shall prove that $I + v: V \to F$ sends a neighborhood $N$ of $f(x_0)$ onto an open set $N'$; then $f + h = (f + h)f^{-1}f = (I + v)f$ maps $f^{-1}N$ onto $(I + v)N = N'$.

We may assume $f(x_0) = 0 \in F$, and hence $v(0) = 0$. Put $L(v) = \lambda < 1$. Let $V$ contain $B_r = B_r(0)$, the ball in $F$ of radius $r$ and center $0$. Put $s = r(1 - \lambda)$ and let $\mathscr{X}$ be the complete metric space of maps $w: B_s \to F$ such that $w(0) = 0$ and $L(w) \leq \lambda/(1 - \lambda)$. We seek a map $w_0 \in \mathscr{X}$ such that $I + w_0$ is a right inverse for $I + v$; this is equivalent to $w_0 = -v(I + w_0)$.

If $w \in \mathscr{X}$ then $(I + w)B_s \subset B_r$, since if $|x| \leq s$, we have

$$|(I + w)x| \leq |x| + L(w)|x| \leq s(1 + \lambda/(1 - \lambda)) = r.$$

Therefore the composition $-v(I + w) = \Phi(w)$ is defined; it is easy to compute that $\Phi(w) \in \mathscr{X}$. The map $\Phi: \mathscr{X} \to \mathscr{X}$ has Lipschitz constant $\leq L(v) < 1$, so that $\Phi$ has a unique fixed point $w_0 \in \mathscr{X}$. Since $(I + v)(I + w)x = x$ for all $x \in B_s$, it follows that $I + v$ maps the open set $(I + v)^{-1}$ (int $B_s$) onto int $B_s$. This completes the proof of 1.5.

1.6. SIZE ESTIMATE. *Let* $X, Y$ *be metric spaces and* $f: X \to Y$ *a bijective map such that* $L(f^{-1})^{-1} \geq \lambda$. *Then* $f B_r(x) \supset B_{\lambda r}(fx)$ *for all* $r > 0$ *and* $x \in X$.

PROOF. If $d(y, x) > r$, then $r < d(y, x) \leq \lambda^{-1}d(fy, fx)$, so that $d(fy, fx) > \lambda r$. This means $f(X - B_r(x)) \subset Y - B_{\lambda r}(fx)$. Since $f$ is bijective the result follows.

REMARK 1. The Lipschitz I.F.T. (1.5) asserts that $|L(g^{-1}) - L(f^{-1})| \to 0$ with $L(g - f)$, but says nothing about $L(g^{-1} - f^{-1})$. If $f$ is a diffeomorphism then $L(g^{-1} - f^{-1})$ does approach $0$ with $L(g - f)$. Consider however the nondifferentiable homeomorphism $f: R \to R$ given by

$$f(x) = \quad x \quad \text{if } x \leq 0,$$
$$= 2x \quad \text{if } x \geq 0.$$

Let $g(x) = f(x) + \varepsilon$. Then $L(f - g) = 0$, but if $\varepsilon \neq 0$ then $L(f^{-1} - g^{-1}) \geq \frac{1}{2}$.

REMARK 2. The Size Estimate can be extended to local homeomorphisms in Banach spaces $E, F$ as follows.

1.7. PROPOSITION. *Let* $U \subset E$ *be open and* $f: U \to F$ *a local homeomorphism whose local inverses all have Lipschitz constants* $\leq \lambda^{-1}$. *If* $B_r(x) \subset U$, *then* $B_{\lambda r}(fx) \subset f B_r(x)$ *and there is a unique continuous map* $g: B_{\lambda r}(fx) \to B_r(x)$ *such that* $gf(x) = x$ *and* $f \circ g = I$. *Moreover* $L(g) \leq \lambda^{-1}$.

PROOF. Left to the reader.

2. **The invariant manifolds of a hyperbolic fixed point.** Let $E$ be a Banach space and $T$ an isomorphism of $E$; that is, a linear homeomorphism of $E$ onto $E$. We call

*T*, hyperbolic if its spectrum lies off the unit circle. If *T* is hyperbolic so is $T^{-1}$ because the spectrum of $T^{-1}$ is the set of reciprocals of elements of the spectrum of *T*.

If *T* is hyperbolic there is a unique splitting $E_1 \times E_2 = E$ invariant under *T* such that the spectrum of $T_1 = T|E_1$ lies outside the unit circle, while that of $E_2$ is inside; see [11]. Moreover $E_1$ and $E_2$ can be renormed so that $\|T_2\| < 1$ and $\|T_1^{-1}\| < 1$; see [8], and also Theorem 3.1 below. We shall always assume that $E_1$ and $E_2$ have such norms, and that on *E* the norm is $|(x_1, x_2)| = \max\{|x_1|, |x_2|\}$, for $x_1 \in E_1$ and $x_2 \in E_2$.

The quantity $\tau = \max(\|T_2\|, \|T_1^{-1}\|) < 1$ is quite useful. We shall call it the *skewness* of *T*.

It is convenient to put $m(T_1) = \|T_1^{-1}\|^{-1}$. Then $|T_1 x| \geq m(T_1)|x|$.

The iterative behavior of a hyperbolic *T* is described by the following result.

2.0. PROPOSITION. *Let* $T: E \to E$ *be a hyperbolic linear map and let* $E = E_1 \times E_2$ *canonically. The points of* $E_1$, $E_2$ *are characterized respectively by* $|T^{-n}x| \to 0$, $|T^n x| \to 0$, *or equivalently by these quantities being bounded, as* $n \to \infty$. *Furthermore, if V is a bounded neighborhood of* 0 *in* $E_1$ *such that* $TV \supset V$ *then* $(\bigcup_{n \geq 0} T^{-n}W) \cup E_2$ *is a neighborhood of* $E_2$ *for any uniform neighborhood W of* $TV - V$ *in E*.

PROOF. A uniform neighborhood of a set *A* in a metric space *X* is any set containing $N_\varepsilon(A) = \{x \in X : d(x, a) \leq \varepsilon$ for some $a \in A\}$ for some $\varepsilon > 0$.

The first statement is trivial. The second statement is proved as follows.

Since *V* is bounded there is, for each $x_1 \in E_1$, $x_1 \neq 0$, a largest value $n = n(x_1)$ such that $T^n x_1 \in V$. Thus, $T^{n+1}x_1 \in TV - V$. Clearly, $n \leq \log(\text{radius } V/|x_1|)/\log (m(T_1))$ since $m(T_1)^n|x_1|$ must be no greater than radius *V*.

Hence, for any $(x_1, x_2) \in E$ with $x_1 \neq 0$, $T^n(x_1, x_2) \in N_\varepsilon(TV - V)$ for $n = n(x_1)$ and $\varepsilon = \|T_2\|^n|x_2|$. This proves 2.0. In fact this proves $(\bigcup_{n \geq 0} T^{-n}W) \cup E_2$ to be a neighborhood of $E_2$ which is nonuniform only at infinity.                Q.E.D.

The object of stable manifold theory is to demonstrate similar behavior for suitable perturbations of *T*, and for the more general situation where the fixed point is replaced by a hyperbolic set.

Throughout the rest of §2, *T* will be a hyperbolic isomorphism $E \to E$ of skewness $\tau$, $0 < \tau < 1$.

Let $E(r) \subset E$ be the closed ball of radius *r* about 0.

We shall be interested in proving that for $f: E(r) \to E$, close enough to *T*,

$$W_1 = \bigcap_{n \geq 0} f^n(E(r)), \qquad W_2 = \bigcap_{n \geq 0} f^{-n}(E(r)),$$

are submanifolds close to $E_1(r)$, $E_2(r)$. We shall assume $r < \infty$.

2.1. LEMMA. *Let* $f: E(r) \to E$ *satisfy* $L(f - T) < \varepsilon < (1 - \tau)/(1 + \tau)$. *If* $x, y \in E(r)$ *and* $|x_1 - y_1| \geq |x_2 - y_2|$ *then*

$$|f_1(x) - f_1(y)| \geq (\tau^{-1} - \varepsilon)|x_1 - y_1|$$
$$\geq (\tau + \varepsilon)|x_1 - y_1|$$
$$\geq |f_2(x) - f_2(y)|.$$

PROOF.
$$\begin{aligned}
|f_1(x) - f_1(y)| &= |T_1(x - y) + (f_1 - T_1)(x) - (f_1 - T_1)(y)| \\
&\geq |T_1(x_1 - y_1) - \varepsilon|x - y| \\
&\geq \tau^{-1}|x_1 - y_1| - \varepsilon|x - y| \\
&= (\tau^{-1} - \varepsilon)|x_1 - y_1|,
\end{aligned}$$

since $|x - y| = \max(|x_1 - y_1|, |x_2 - y_2|) = |x_1 - y_1|$. Similarly, $|f_2(x) - f_2(y)|$ $\leq \tau|x_2 - y_2| + \varepsilon|x - y| \leq (\tau + \varepsilon)|x_1 - y_1|$. Since $\varepsilon < (1 - \tau)/(1 + \tau)$ we have $\varepsilon < 1 - \tau$ and $\varepsilon < \tau^{-1} - 1$ which shows that $\tau + \varepsilon < 1 < \tau^{-1} - \varepsilon$, completing the proof.                                                                                    Q.E.D.

2.2. PROPOSITION. *If $f: E(r) \to E$ and $L(f - T) < \varepsilon < (1 - \tau)/(1 + \tau)$, then $W_1 = \bigcap_{n \geq 0} f^n(E(r))$ is the graph of a function $U_1 \to E_2(r)$ while $W_2 = \bigcap_{n \geq 0} f^{-n}(E(r))$ is the graph of a function $U_2 \to E_1(r)$, where $U_1$, $U_2$ are subsets of $E_1(r)$, $E_2(r)$.*

REMARKS. This permits $U_1$ or $U_2$ to be empty. The notation $f^{-n}(E(r))$ means $\{x \in E(r): f(x), \ldots, f^n(x)$ are defined and in $E(r)\}$. One should regard 2.2 as a uniqueness theorem.

PROOF. We deal first with $W_2$. Let $x, y \in W_2$ with $x_2 = y_2$. We must show $x_1 = y_1$. By assumption $f^n(x), f^n(y) \in E(r)$ for all $n \geq 0$. By 2.1,

$$\begin{aligned}
2r \geq |(f^n)_1 x - (f^n)_1 y| &= |f_1(f^{n-1}x) - f_1(f^{n-1}y)| \\
&\geq (\tau^{-1} - \varepsilon)|(f^{n-1})_1 x - (f^{n-1})_1 y| \geq \cdots \geq (\tau^{-1} - \varepsilon)^n|x_1 - y_1|.
\end{aligned}$$

As $n \to \infty$, $(\tau^{-1} - \varepsilon)^n \to \infty$ and so $x_1 = y_1$. Hence $W_2$ is a graph as claimed.

Now suppose $x, y \in W_1$ and $x_1 = y_1$. We must show $x_2 = y_2$. For every $n \geq 0$ there exist $x', y' \in E(r)$ such that $f^n(x') = x, f^n(y') = y$. Suppose, for some $0 \leq j < n$,

$$|(f^j x')_1 - (f^j y')_1| \geq |(f^j x)_2 - (f^j y')_2|.$$

Then by repeated application of 2.1,

$$|(f^n x')_1 - (f^n y')_1| \geq |(f^n x')_2 - (f^n y')_2|,$$

which is the same as $0 \geq |x_2 - y_2|$, so that $x_2 = y_2$.

On the other hand if

$$|(f^j x')_1 - (f^j y')_1| \leq |(f^j x')_2 - (f^j y')_2|$$

for all $0 \leq j \leq n$ then

$$\begin{aligned}
|x_2 - y_2| &= |(f^n x')_2 - (f^n y')_2| \\
&= |f_2(f^{n-1}x') - f_2(f^{n-1}y')| \\
&\leq \tau|(f^{n-1}x')_2 - (f^{n-1}y')_2| + \varepsilon|f^{n-1}(x') - f^{n-1}(y')| \\
&= (\tau + \varepsilon)|(f^{n-1}x')_2 - (f^{n-1}y')_2| \\
&\leq \cdots \leq (\tau + \varepsilon)^n|x_2' - y_2'| \\
&\leq 2r(\tau + \varepsilon)^n.
\end{aligned}$$

Since $|x_2 - y_2|$ is independent of $n$ and $(\tau + \varepsilon)^n \to 0$, we have $x_2 = y_2$. This shows that $W_1$ is a graph as claimed, proving 2.2.

Alternately, we could have proved the second part of 2.2 by looking at $W_2$ for

$f^{-1}$. Unfortunately, $f^{-1}$ is not defined on $E(r)$, so it was easier to proceed directly. We shall come again to this problem in 2.4.

DEFINITION. A *stable manifold* $W^s$ for $f: E(r) \to E$ is the set of $x \in E(r)$, such that $|f^n x|$ is defined and stays bounded as $n \to \infty$. Clearly $f W^s \subset W^s$. An *unstable manifold* $W^u$ for $f: E(r) \to E$ is the set of $x \in E(r)$ such that $|f^{-n} x|$ is defined and stays bounded as $n \to \infty$. Clearly $f^{-1} W^u \subset W^u$.

For suitable $f$, it turns out that $W^u = W_1$, $W^s = W_2$, and $W^s$ is contracted toward the point $W^u \cap W^s$ by $f$ while $W^u$ is contracted toward the point $W^u \cap W^s$ by $f^{-1}$.

Before stating the basic theorem for unstable manifolds, we define the *graph transform*. If $E = E_1 \times E_2$, $r > 0$, and $f: E(r) \to E$, then we write $\Gamma_f(g) = h$ provided $h, g: E_1(r) \to E_2(r)$ and

$$f(\text{graph}(g)) \cap E(r) = \text{graph}(h).$$

$\Gamma_f$ is called the graph transform for $f$. If $T = T_1 \times T_2$ is hyperbolic respecting $E_1 \times E_2$ then for any $g: E_1(r) \to E_2(r)$, $\Gamma_T(g)$ is defined and equals $T_2 \circ g \circ T_1^{-1} | E_1(r)$. Similarly, if $\Gamma_f(g)$ exists then $\Gamma_f(g) \circ f_1 \circ (1, g) = f_2 \circ (1, g)$ where 1 is the identity map of $E_1(r)$. If $f_1 \circ (1, g)$ is invertible this means that

$$\Gamma_f(g) = f_2 \circ (1, g) \circ [f_1 \circ (1, g)]^{-1} | E_1(r).$$

DEFINITION. $C^k(X, Y)$ is the set of functions $f: X \to Y$ of class $C^k$ (first $k$ derivatives exist at all points of $X$ and are continuous) with bounded $k$ norm

$$|f|_k = \sup_{x \in X} \max(|f(x)|, \|(Df)_x\|, \ldots, \|(D^k f)_x\|).$$

DEFINITION. For $r > 0$ fixed, we let $\mathscr{N}_\varepsilon^k(T) = \{f \in C^k(E(r), E): L(f - T) < \varepsilon$ and $|f(0)| < \varepsilon\}$. This could be called a "Lipschitz neighborhood of $T | E(r)$ in $C^k(E(r), E)$."

The $k$-norm makes $C^k(E(r), E)$ complete and defines the so-called *uniform* $C^k$ *topology*.

2.3. UNSTABLE MANIFOLD THEOREM FOR A POINT. *Given* $0 < \tau < 1$ *and* $r > 0$ *there exist* $\varepsilon > 0$, *independent of* $r$, *and* $0 < \delta < \varepsilon$ *with the following properties. If* $T = T_1 \times T_2$ *is a hyperbolic linear operator on* $E_1 \times E_2$ *of skewness* $\tau$ *and* $f: E(r) \to E$ *is a Lipschitz map satisfying* $L(f - T) < \varepsilon$, $|f(0)| \le \delta$, *then there is a unique map* $g_f: E_1(r) \to E_2(r)$ *whose graph is* $W_1 = \bigcap_{n \ge 0} f^n(E(r))$. *Moreover* $L(g_f) < 1$ *and* $g_f$ *is of class* $C^k$ *iff* $f$ *is. The assignment* $f \mapsto g_f$ *is continuous as a map* $\mathscr{N}_\delta^k(T) \to C^k(E_1(r), E_2(r))$. *The map* $(f | W_1)^{-1}: W_1 \to W_1$ *is a contraction of* $W_1$ *into its interior (that is, into* $\{x \in W_1: |x_1| \le s\}$ *for some* $s < r$).

REMARK 1. By 2.2 uniqueness of $g_f$ is assured; it is only a matter of producing a function $E_1(r) \to E_2(r)$ whose graph is contained in its own $f$-image.

REMARK 2. Choice of $\varepsilon$ and $\delta$ are restricted no more than $\varepsilon < (1 - \tau)/(1 + \tau)$, $\delta < \varepsilon^2 r \tau$.

PROOF OF THEOREM 2.3. We shall first prove 2.3 under the assumption $f(0) = 0$. Afterwards the general case is easily handled.

Choose any

$$0 < \varepsilon < (1 - \tau)/(1 + \tau) = (\tau^{-1} - 1)/(\tau^{-1} + 1).$$

We claim this $\varepsilon$ does the job; three forms of the inequality are used:

$$\varepsilon < 1 - \tau, \varepsilon < \tau^{-1}, \text{ and } (\tau + \varepsilon)/(1 - \tau\varepsilon) < 1.$$

Let $\mathcal{G} = \{g \in \mathcal{M}(E_1(r), E_2(r)): g(0) = 0 \text{ and } L(g) \le 1\}$. We show that $\Gamma_f: \mathcal{G} \to \mathcal{G}$ is a well-defined contraction. Clearly $\mathcal{G}$ is complete. Put

$$\psi_f(g) = f_1 \circ (1, g), \qquad \phi_f(g) = f_2 \circ (1, g).$$

Thus $\psi_f(g)$ is a map $E_1(r) \to E_1$ and $\phi_f(g)$ is a map $E_1(r) \to E_2$. Restricting to $E_1(r)$, we have $L(\psi_f(g) - T_1) \le L(f - T) < \varepsilon$. Hence, by the Lipschitz inverse function theorem (1.5), $\psi_f(g)$ is a lipeomorphism, and

$$L([\psi_f(g)]^{-1}) \le [L(T_1^{-1})^{-1} - L(\psi_f g - T_1)]^{-1} \le [\tau^{-1} - \varepsilon]^{-1}$$

and so, since $(\psi_f g)(0) = 0$,

$$(\psi_f g)(E_1(r)) \supset E_1(r(\tau^{-1} - \varepsilon)) \supset E_1(r).$$

Hence $[\psi_f g]^{-1} E_1(r)$ is a well-defined map into $E_1(r)$ with Lipschitz constant $\le [\tau^{-1} - \varepsilon]^{-1} < 1$. Since $L(\phi_f g) \le (\tau + \varepsilon) < 1$ and $(\phi_f g)(0) = 0$, it follows that $\Gamma_f(g) = \phi_f g \circ [\psi_f g]^{-1} E_1(r)$ is a well defined map $\mathcal{G} \to \mathcal{G}$.

To see that $\Gamma_f$ is a contraction, take $g_1, g_2 \in \mathcal{G}$ and estimate

$$|\Gamma_f(g_1) - \Gamma_f(g_2)| \le |(\phi_f g_1) \circ (\psi_f g_1)^{-1} - (\phi_f g_1) \circ (\psi_f g_2)^{-1}|$$
$$+ |(\phi_f g_1) \circ (\psi_f g_2)^{-1} - (\phi_f g_2) \circ (\psi_f g_2)^{-1}|$$
$$\le L(\phi_f g_1)|(\psi_f g_1)^{-1} - (\psi_f g_2)^{-1}| + |\phi_f g_1 - \phi_f g_2|$$

(by 1.4(b))

$$\le (\tau + \varepsilon)L((\psi_f g_1)^{-1}) \cdot |\psi_f g_1 - \psi_f g_2|$$
$$+ (\tau + \varepsilon)|g_1 - g_2|$$
$$\le (\tau + \varepsilon)(\tau^{-1} - \varepsilon)^{-1}\varepsilon|g_1 - g_2| + (\tau + \varepsilon)|g_1 - g_2|$$
$$= ((\tau + \varepsilon)/(\tau^{-1} - \varepsilon))(\varepsilon + (\tau^{-1} - \varepsilon))|g_1 - g_2|$$
$$= ((\tau + \varepsilon)/(1 - \tau\varepsilon))|g_1 - g_2|.$$

Hence $\Gamma_f$ has a unique fixed point $g_f \in \mathcal{G}$.

As remarked before, 2.2 now implies that $W_1 = \bigcap_{n \ge 0} f^n(E(r)) = \text{graph } g_f$. Consider the following commutative diagram

$$\begin{array}{ccc}
W_1 & \xrightarrow{\quad f \quad} & f(W_1) \\
\pi_1 \downarrow & & \downarrow \pi_1 \\
E_1(r) & \xrightarrow[f_1 \circ (1, g_f)]{} & E_1
\end{array}$$

where $\pi_1(x_1, x_2) = x_1$. Since $L(g_f) \le 1$, we have $|x - y| = |x_1 - y_1|$ for $x = (x_1, x_2)$, $y = (y_1, y_2) \in W_1$. This shows that $\pi_1$ is an isometry from $W_1$ onto $E_1(r)$. Since

$f_1 \circ (1, g) = \psi_f(g_f)^{-1}$ is a contraction, the same is true for $f|W_1$. Therefore $f|W_1$ has the unique fixed point 0. Any fixed point of $f$ in $E(r)$ must belong to the unstable manifold $W_1$, and so 0 is the only fixed point of $f$.

Next we investigate the differentiability of $g_f$. Assume $f \in \mathcal{N}^1_\varepsilon(T)$, with $\varepsilon$ as above, and $f(0) = 0$. Define

$$\Delta f : (E \times E)(r) \to E \times E,$$
$$(x, y) \mapsto (fx, Df_x y);$$

similarly for $g \in \mathcal{G}^1 = \mathcal{G} \cap C^1(E_1(r), E_2(r))$,

$$\Delta g : E_1(r) \times E_1(r) \to E_2(r) \times E_2(r),$$
$$(x_1, y_1) \mapsto (gx_1, (Dg)_{x_1} y_1).$$

Even though $\Delta f$ is not Lipschitz close to $T \times T$, we still claim the graph transform for $\Delta f$ is a well defined fiber contraction of $\mathcal{G} \times \mathcal{H}$, where

$\mathcal{H} = \{h \in C^0(E_1(r), L(E_1, E_2)) \mid \|h(x_1)\| \leq 1 \text{ for all } x_1 \in E_1\}$
  $= $ unit ball in the Banach space of continuous bounded maps $E_1(r) \to L(E_1, E_2)$.

Indeed, for $(g, h) \in \mathcal{G} \times \mathcal{H}$ we put

$$\Gamma_{\Delta f}(g, h) = (\Gamma_f g(x_1), (Df_2)_\xi \circ (1, h(\xi_1)) \circ [(Df_1)_\xi \circ (1, h(\xi_1))]^{-1})$$

where $\xi_1 = (\psi_f g)^{-1}(x_1)$, $\xi_2 = g(\xi_1)$, and $\xi = (\xi_1, \xi_2)$. One can see directly that $\Gamma_{\Delta f}$ is well defined and contracts fibers; but it is instructive to be more formal.

Composition on the left by $T$ defines a hyperbolic map of skewness $\tau$, namely $T_\# : \mathscr{E}(1) \to \mathscr{E}$ for $\mathscr{E} = L(E, E)$; the canonical invariant splitting $\mathscr{E} = \mathscr{E}_1 \times \mathscr{E}_2$ is obtained by setting $\mathscr{E}_i = L(E, E_i)$. (Recall that $\mathscr{E}(1)$ is the unit ball in $\mathscr{E}$.)

If $\|T' - T\| \leq \varepsilon$ then composition on the left by $T'$ defines $T'_\# : \mathscr{E}(1) \to \mathscr{E}$ and $L(T'_\# - T_\#) = L(T' - T) \leq \varepsilon$. Hence $\Gamma_{T'_\#}$, the graph transform for $T'_\#$, is a contraction of

$$\mathcal{G}\mathscr{E} = \{H \in C^0(\mathscr{E}_1(1), \mathscr{E}_2(1)) \mid H(0) = 0 \text{ and } L(H) \leq 1\},$$

well defined by the formula

$$\Gamma_{T'_\#}(H)(S_1) = T'_2 \circ (1, H(S_1)) \circ [T'_1 \circ (1, H(S_1))]^{-1}.$$

For any $h \in \mathcal{H}$ and $\xi_1 \in E_1(r)$ we apply the functor $(\cdot)_\#$ to the linear map $h(\xi_1) : E_1 \to E_2$ to get $h(\xi_1)_\# \in \mathcal{G}\mathscr{E}$. The definition of $\Gamma_{\Delta f}$ can be written

$$\Gamma_{\Delta f}(g, h)(x_1) = ((\Gamma_f g)(x_1), \Gamma_{(Df_\xi)_\#}(h(\xi_1)_\#)).$$

It is now clear that $\Gamma_{\Delta f} : \mathcal{G} \times \mathcal{H} \to \mathcal{G} \times \mathcal{H}$ contracts both base and fibers uniformly. Therefore, by the fiber contraction theorem (1.2) there is a unique attractive fixed point $(g_f, h_f)$ of $\Gamma_{\Delta f}$.

Observe that $\Delta(\Gamma_f g) = \Gamma_{\Delta f}(\Delta g)$; this is a consequence of the naturality of $\Delta$, which is just the tangent functor. Take, then, $g_0 \equiv 0$ and consider the convergence $(\Gamma_{\Delta f})^n(\Delta g_0) \to (g_f, h_f)$ as $n \to \infty$. But $(\Gamma_{\Delta f})^n(\Delta g_0) = \Delta(\Gamma_f^n f_0)$, so that $\Delta(\Gamma_f^n g_0)$ converges to $(g_f, h_f)$; thus $D(g_f)$ exists and equals $h_f$.

Now suppose $f$ is of class $C^k$, $k \geq 2$, $|f|_k \leq M$, and 2.3 holds for $k - 1 \geq 1$. Choose $\varepsilon'$, $\varepsilon < \varepsilon' < (1 - \tau)/(1 + \tau)$, and choose $r'$, $0 < r' \leq r$, so that $Mr' < \varepsilon' - \varepsilon$, $r' \leq 1$. Then let $f' = f|E(r')$. We claim that $\Delta f' \in \mathcal{N}_{\varepsilon'}(T \times T)$. We have $L(\Delta f' - T \times T) < \varepsilon'$ since

$$
\begin{aligned}
&|(\Delta f' - T \times T)(x, y) - (\Delta f' - T \times T)(x', y')| \\
&\leq \max(|(f' - T)(x) - (f' - T)(x')|, |((Df')_x - T)y - ((Df')_{x'} - T)y'|) \\
&\leq \max(\varepsilon|x - x'|, |[(Df')_x - (Df')_{x'}]y| + |[(Df')_{x'} - T](y - y')|) \\
&\leq \max(\varepsilon|x - x'|, M|x - x'|r' + \varepsilon|y - y'|) \\
&< \varepsilon' \max(|x - x'|, |y - y'|),
\end{aligned}
$$

for $|x - x'| \geq |y - y'|$ implies

$$
Mr'|x - x'| + \varepsilon|y - y'| \leq (Mr' + \varepsilon)|x - x'| < \varepsilon'|x - x'|,
$$

while $|y - y'| \geq |x - x'|$ implies

$$
Mr'|x - x'| + \varepsilon|y - y'| < (Mr' + \varepsilon)|y - y'|.
$$

Hence $\Delta f' \in \mathcal{N}_{\varepsilon'}(T \times T)$ so by the induction hypotheses there is a unique solution $g_{\Delta f'}$ of $\Gamma_{\Delta f'}(g_{\Delta f'}) = g_{\Delta f'}$ and $g_{\Delta f'}$ is of class $C^{k-1}$. But clearly, $g_{\Delta f}|E_1(r') \times E_1(r')$ solves this equation too. Hence $g_f$ is of class $C^k$ on $E_1(r')$.

We know that $(f|W)^{-1}: W_1 \to W_1$ is a contraction with $L((f|W_1)^{-1}) \leq (\tau^{-1} - \varepsilon)^{-1}$. So choosing $N > \log(r/r')/\log(\tau^{-1} - \varepsilon)$ we have

$$
\pi_1(f|W)^{-N}(E_1(r)) \subset E(r \cdot (\tau^{-1} - \varepsilon)^{-N}) \subset E(r').
$$

Since $\Gamma_f(g_f) = g_f$ we have

(*)        $$g_f = (f^N)_2 \circ (1, g_{f'}) \circ [f^N \circ (1, g_{f'})]^{-1}|E_1(r).$$

Hence $g_f$ is of class $C^k$ on all $E_1(r)$. This formula (*) holds for any $f \in \mathcal{N}_\varepsilon(T)$ with the same $N$. Hence as $\bar{f} \to f$ in $\mathcal{N}_\varepsilon^k(T)$, it is clear that $\bar{f}' \to f'$, $C^k$, which implies $\Delta \bar{f}' \to \Delta f'$, $C^{k-1}$, which implies by induction (since $\Delta f' \in \mathcal{N}_{\varepsilon'}(T \times T)$) $g_{\Delta \bar{f}'} \to g_{\Delta f'}$, $C^{k-1}$, which implies $g_{\bar{f}'} \to g_{f'}$, $C^k$. Hence, using the formula (*) for $g_{\bar{f}}$, we have $g_{\bar{f}} \to g_f$, $C^k$, as $\bar{f} \to f$, $C^k$.

This completes the proof of 2.3 in case $f(0) = 0$.

The case $f(0) \neq 0$ is obtained formally; the idea is to translate the origin of $E$ to the fixed point of $f$, which we must prove exists. We may assume 2.3 true for all $0 < \tau < 1$ and $r > 0$ provided the map involved takes 0 to 0 and lies in $\mathcal{N}_\varepsilon(T)$, $\varepsilon < (1 - \tau)/(1 + \tau)$.

Choose $0 < \delta < \min(\varepsilon, \varepsilon^2 r \tau)$. For $|f(0)| < \delta$ and $L(f - T) < \varepsilon$, $f$ has a unique fixed point $p_f = (p_1, p_2)$ in $E(r)$. To see this observe that the map

$$
\bar{f}: (x_1, x_2) \mapsto (T_1^{-1}(x_1 - T_1 x_1 - f_1(x_1, x_2)), f_2(x_1, x_2))
$$

is a contraction of $E(r)$ and has the same fixed points as $f$. By the contracting map theorem (1.1) $p_f$ depends continuously on $f \in \mathcal{N}_\delta(T)$, and $|p_f| \leq \delta(1 - \tau - \varepsilon)^{-1}$.

We remark that the inequality $\delta < \varepsilon^2 r\tau$ implies

$$\delta < \frac{(\tau^{-1} - \varepsilon - 1)(1 - \tau - \varepsilon)r}{1 + \tau^{-1} - \varepsilon},$$

for

$$\varepsilon^2 r\tau < \varepsilon^2 r\tau(1 + \tau) < \left(\frac{1 - \tau}{1 + \tau}\right)^2 r\tau(1 + \tau)$$

$$= \frac{(1 - \tau)(1 - \tau)r}{\tau^{-1} + 1} = \frac{(\tau^{-1} - 1)(\tau - \tau^2)r}{1 + \tau^{-1}}$$

$$= \frac{\left(\tau^{-1} - \dfrac{1 - \tau}{1 + \tau} - 1\right)\left(1 - \tau - \dfrac{1 - \tau}{1 + \tau}\right)r}{1 + \tau^{-1}}$$

$$< \frac{(\tau^{-1} - \varepsilon - 1)(1 - \tau - \varepsilon)r}{1 + \tau^{-1} - \varepsilon}.$$

Put $\hat{r} = r - \delta(1 - \tau - \varepsilon)^{-1} \geq r - |p_f|$. Define

$$\hat{f}: E(\hat{r}) \to E, \qquad \hat{f}(x) = f(x + p_f) - p_f.$$

Observe that $\hat{f}$ is well defined, $L(\hat{f} - T) = L(f - T) < \varepsilon$, and $\hat{f}(0) = 0$. Hence there exists a unique $\hat{g} = g_{\hat{f}}: E_1(\hat{r}) \to E_2(\hat{r})$ such that $W_1(\hat{f}) = \bigcap_{n \geq 0} \hat{f}^n E(\hat{r})$ $=$ graph($g_{\hat{f}}$). Define $g = g_f: E_1(r) \to E_2(r)$,

$$g(x_1) = p_2 + \phi_{\hat{f}}(g_{\hat{f}}) \circ [\psi_{\hat{f}}(g_{\hat{f}})]^{-1}(x_1 - p_1).$$

To see that $g$ is well defined, observe that $\psi_{\hat{f}}(\hat{g})E_1(\hat{r}) \supset E_1((\tau^{-1} - \varepsilon)\hat{r})$, since $L(\psi_{\hat{f}}(\hat{g})^{-1}) \leq (\tau^{-1} - \varepsilon)^{-1}$, as was shown earlier. Moreover the estimate for $\delta$ and the definition of $\hat{r}$ imply $(\tau^{-1} - \varepsilon)\hat{r} \geq r + \delta(1 - \tau - \varepsilon)^{-1}$. This implies that $[\psi_{\hat{f}}(\hat{g})]^{-1}$ is defined on $x_1 - p_1$ if $|x_1| \leq r$, and also that $g(E_1(r)) \subset E_2(r)$.

Clearly $L(g) = L(\phi_{\hat{f}}(\hat{g}) \circ [\psi_{\hat{f}}(\hat{g})]^{-1}) = L(\hat{g}) \leq 1$; and $g(p_1) = p_2$.

We claim $f(\text{graph } g) \cap E(r) = \text{graph}(g)$. Let $x_1 \in E_1(r)$ and assume $f_1(x_1, g(x_1)) \in E_1(r)$. Then $|x_1 - p_1| < \hat{r}$, for by the Lipschitz I.F.T. (1.5),

$$|f_1(x_1, g(x_1)) - p_1| \geq (\tau^{-1} - \varepsilon)|x_1 - p_1|,$$

which implies

$$|x_1 - p_1| \leq \frac{r + |p_1|}{\tau^{-1} - \varepsilon} \leq r + \frac{\delta(1 - \tau - \varepsilon)^{-1}}{\tau^{-1} - \varepsilon} \leq \hat{r},$$

by our choices of $\delta$ and $\hat{r}$. Therefore $x_1 - p_1 \in E_1(\hat{r})$, and the verification that $f_2(x_1, g_f(x_1)) = g(f_1(x_1, gx_1))$ follows formally from the corresponding property for $\hat{f}$ and $\hat{g}$:

$$g(f_1(x_1, gx_1)) = p_2 + \hat{g}(\hat{f}(x_1 - p_1, g(x_1) - p_2))$$
$$= p_2 + \hat{g}(\hat{f}_1(x_1 - p_1, \hat{g}(x_1 - p_1))$$
$$= p_2 + \hat{f}_2(x_1 - p_1, \hat{g}(x_1 - p_1))$$
$$= f_2(x_1, p_2 + \hat{g}(x_1 - p_1))$$
$$= f_2(x_1, g(x_1)).$$

By 2.2, $W_1$ is the graph of $g = g_f$. The defining formula for $g_f$, and 2.3 for $\hat{f}$, prove 2.3 for $f$.

REMARK 1. The case $k = \infty$ follows from 2.3 because $\varepsilon$ and $\delta$ are independent of $k$.

REMARK 2. The case $r = \infty$, that is, $f$ is defined on all of $E$, presents difficulty in proving the global continuous dependence of the higher order derivatives of $g_f$ on those of $f$. The trouble is that as $|x_1| \to \infty$, so does the number of iterates of $f^{-1}$ needed to bring $(x_1, g_f(x_1))$ inside the small ball $E_1(r')$ where $g_f$ is known to be $C^k$. Another proof, with a different induction hypothesis, can be devised.

REMARK 3. Suppose $f \in \mathcal{N}_\varepsilon^k(T)$, $\varepsilon < (1 - \tau)/(1 + \tau)$ and $f(0) = 0$. Put $\mathcal{G}_*^k = \{g \in \mathcal{G} \cap C^k : L(g) < 1\}$. Then $\Gamma_f : \mathcal{G}_*^k \to \mathcal{G}_*^k$ has $g_f$ as attractive fixed point. Thus $\Gamma_f^n(g) \to g_f$, $C^k$, as $n \to \infty$, for any $g \in \mathcal{G}_*^k$. For $k = 1$ we proved this already: $\Delta\Gamma_f^n(g) = \Gamma_{\Delta f}^n(\Delta g) \to \Delta g_f$, $C^0$, and hence $\Gamma_f^n(g) \to g_f$, $C^1$. Assume the result holds for $k - 1 \geq 1$. Then for $0 < r' \leq r$ (as in the differentiability part of 2.3) we know $f' = f|E(r')$ has $\Delta f' \in \mathcal{N}_{\varepsilon'}^{k-1}(T \times T)$, $\varepsilon < \varepsilon' < (1 - \tau)/(1 + \tau)$ and $g' = g|E_1(r')$ has $\Delta g' \in \mathcal{G}_*^{k-1}(E \times E)$. Thus $\Gamma_{\Delta f'}^n(\Delta g') \to \Delta g_f$, $C^{k-1}$, by induction. Hence $\Gamma_{f'}^n(g') \to g_{f'}$, $C^k$. But for $N \geq \log(r/r')/\log(\tau^{-1} - \varepsilon)$ we have, as in 2.3, graph $\Gamma_f^{n+N}(g) = f^N(\text{graph } \Gamma_{f'}^n(g')) \cap E(r)$ so that $\Gamma_f^{n+N}(g) \underset{n}{\to} f^N(\text{graph}(g_{f'})) \cap E(r) = \text{graph}(g_f)$, $C^k$.

We needed $L(g) < 1$ to insure $\Delta g' \in \mathcal{G}_*^{k-1}(E \times E)$ for some sufficiently small $r'$. If $L(g) = 1$ then $L(\Delta g')$ may be larger than 1 for all $r' > 0$. But of course the case of $L(g) = 1$ could be handled directly.

REMARK 4. If $f(0) \neq 0$ in the above, $\Gamma_f$ does not map $\mathcal{G}$ into itself. If we enlarge $\mathcal{G}$ by letting $g(0) \neq 0$, with $|g(0)|$ appropriately small, we can prove directly that $\Gamma_f$ is a well defined contraction and so has a fixed point, etc. The estimates are clearer, however, for $f(0) = 0$ and that is why we deduced the case $f(0) \neq 0$ formally from the case $f(0) = 0$ instead of proving both at once.

Now we indicate how to deduce the corresponding stable manifold theorem by inverting $f$.

2.4. STABLE MANIFOLD THEOREM FOR A POINT. *Let* $0 < \tau < 1$ *and* $r > 0$ *be given. There exist* $\varepsilon > 0$ *and* $0 < \delta < \varepsilon$ *with the following property. If* $f : E(r) \to E$ *satisfies*

$$L(f - T) < \varepsilon, \qquad |f(0)| < \delta,$$

*for a hyperbolic isomorphism* $T : E \to E$ *of skewness* $\tau$ *then* $W_2 = \bigcap_{n \geq 0} f^{-n}(E(r))$ *is the graph of a unique function* $g_{2f} : E_2(r) \to E_1(r)$. *Moreover,* $L(g_{2f}) \leq 1$ *and* $g_{2f}$ *is of class* $C^k$ *if* $f$ *is. The assignment* $f \mapsto g_{2f}$ *is continuous as a map* $\mathcal{N}_\delta^k(T) \to C^k(E_2(r), E_1(r))$. *The map* $f|W_2 : W_2 \to W_2$ *contracts* $W_2$ *into its interior.*

PROOF. It suffices to show two things: $1^0$. $L(f^{-1} - T^{-1}) \to 0$ and $f^{-1}(0) \to 0$ as

$L(f - T) \to 0$ and $f(0) \to 0$. $2^0$, $W_1(f^{-1}) = W_2(f)$. Since $T^{-1}$ is linear it is justifiable to write

$$f^{-1} - T^{-1} = T^{-1}Tf^{-1} - T^{-1}ff^{-1} = T^{-1} \circ (T - f) \circ f^{-1},$$

so that

$$L(f^{-1} - T^{-1}) \le \|T^{-1}\|L(T-f)L(f^{-1})$$
$$\le \|T^{-1}\|L(T-f)[\|T^{-1}\|^{-1} - L(T-f)]^{-1}$$

by 1.4b. As $L(f - T) \to 0$ this tends to zero.

As for $f^{-1}(0)$ we have

$$|f^{-1}(0)| \le |f^{-1}(0) - f^{-1}(p_f)| + |f^{-1}(p_f)|$$
$$\le L(f^{-1})|p_f| + |p_f|$$
$$\le ([\|T^{-1}\|^{-1} - L(f - T)]^{-1} + 1)\delta(1 - \tau - \varepsilon)^{-1} \to 0,$$

as $\delta \to 0$ and $L(f - T) \to 0$.

Unfortunately, the second statement is not quite true, because $f^{-1}$ is not defined on $E(r)$ but on $f(E(r))$.

Restricting to an $E(r') \subset fE(r)$ and arguing in a fashion similar to the $f(0) \ne 0$ part of 2.3 will show, for large enough $N$,

$$W_2(f) = f^{-N}(W_2(f|E(r'))) = f^{-N}W_1(f^{-1}|E(r')).$$

The second equality is clear. Since $W_1(f^{-1}|E(r'))$ is well described by 2.3 we have proved 2.4.

3. **Stable manifolds for hyperbolic sets.** In this and the succeeding sections we adopt the following conventions.

$M$ is a finite dimensional $C^\infty$ Riemannian manifold; $U \subset M$ is an open set; $f: U \to M$ is a $C^k$ embedding ($k \in \mathbf{Z}_+$). We denote by $\Lambda \subset U$ a compact *invariant* set of $f$; that is, $f(\Lambda) = \Lambda$.

*Notation.* If $g: V \to V'$ is a $C^1$ map between smooth manifolds, we denote the *differential* of $g$ by $Tg: TV \to TV'$. If $x \in V$, then $V_x$ or $T_xV$ denotes the tangent space to $V$ at $x$, and $T_xg: T_xV \to T_{gx}V'$ is the restriction of $Tg$. For any subset $A \subset V$, $T_Ag: T_AV \to T_{gA}V'$ is the restriction of $Tg$ to the tangent bundle of $V$ over $A$.

If $E$ is a Banach space and $g: V \to E$ a $C^1$ map, then $Dg_x: V_x \to E$ is the *derivative* of $g$ at $x \in V$. This is by definition the composition

$$V_x \xrightarrow{\ T_xg\ } E_{gx} \xrightarrow{\ \xi_{fx}\ } E$$

where $\xi_y$ denotes the canonical identification of the tangent space to $E$ at $y \in E$ with $E$. $Dg: TV \to E$ is the map whose restriction to $V_x$ is $Dg_x$.

If $p: X \to Y$ is a bundle, then $X_y = p^{-1}(y)$, the fiber over $y$. If $p: X \to Y$ and $p': X' \to Y'$ are bundles and $F: X \to X'$ takes fibers into fibers, covering $f: Y \to Y'$, then $F_y: X_y \to X'_{fy}$ is the restriction of $F$ to $X_y$. We call such an $F$ a *bundle map*.

If the vector bundles $X$ and $X'$ have Banach space structures on fibers, and $F: X \to X'$ is a linear bundle map, then we put $\|F\| = \sup_{y \in Y}\|F_y\|$.

We put $N_\alpha(x) = \{y|d(y, x) < \alpha\}$ in any metric space.

DEFINITION. The invariant set $\Lambda \subset U$ is *hyperbolic* for the map $f: U \to M$ if $T_\Lambda M$ has a splitting (Whitney sum decomposition) $T_\Lambda M = E^u \oplus E^s$ satisfying:

(1) $E^u$ and $E^s$ are invariant under $Tf$;

(2) there exist constants $c > 0$ and $0 < \tau < 1$ such that for all $n \in \mathbf{Z}_+$,

$$\max\{\|Tf^n|E^s\|,\ \|Tf^{-n}|E^u\|\} < c\tau^n.$$

We say $\Lambda$ has *skewness* $\tau$.

If the Riemann metric on $M$ is such that in (2) we can take $c = 1$, then the metric is called *adapted* to $\Lambda$.

3.1. (MATHER, [10]). *If $\Lambda \subset U$ is a compact hyperbolic set for $M \supset U \xrightarrow{f} M$, then $M$ has a smooth Riemann metric adapted to $\Lambda$.*

PROOF. Let $|v|$ denote the norm of $v \in TM$ in the metric for which $\Lambda$ is hyperbolic. From (2) we have

$$|Tf^n(v)| \le c\tau^n|v| \quad \text{if } v \in E^s,$$

$$|Tf^n v| \ge c^{-1}\tau^{-n}|v| \quad \text{if } v \in E^u.$$

Let $q \in \mathbf{Z}_+$ be such that $c\tau^q < 1$. Define a new metric $\|v\|$ by

$$\|v\|^2 = \sum_{n=0}^{q-1} |Tf^n(v)|^2 \quad \text{if } v \in E^s$$

$$\|v\|^2 = \sum_{n=0}^{q-1} |Tf^{-n}(v)|^2 \quad \text{if } v \in E^u.$$

If $v \in E^s$, then $\|v\|^2 \le qc^2|v|^2$; and

$$\begin{aligned}
\|Tf(v)\|^2 &= \sum_{n=1}^{q} |Tf^n(v)|^2 \\
&= \|v\|^2 - |v|^2 + |Tf^q(v)|^2 \\
&\le \|v\|^2 - [1 - (c\tau^q)^2]|v|^2 \\
&\le \|v\|^2(1 - [1 - c\tau^q]^2/qc^2).
\end{aligned}$$

Thus $\|Tf(v)\| \le \sigma_0\|v\|$, where $\sigma_0^2 = 1 - [1 - c\tau^q]^2/qc^2$. Since $c\tau^q < 1$ we have $\sigma_0 < 1$. Similarly, replacing $f$ by $f^{-1}$, we find that

$$\|Tf(v)\| \ge \sigma_0^{-1}\|v\| \quad \text{for } v \in E^u.$$

Now let $\sigma_0 < \sigma < 1$ and approximate $\|\cdot\|$ by a $C^\infty$ metric $\|\|\cdot\|\|$ such that

$$\|\|Tf(v)\|\| \le \sigma\|\|v\|\| \quad \text{if } v \in E^s,$$

$$\|\|Tf(v)\|\| \ge \sigma^{-1} \|\|v\|\| \quad \text{if } v \in E^u.$$

The proof (due to Mather) of Theorem 3.1 is complete.

If $x \in U$ and $\beta > 0$ we put

$$\Sigma(x, \beta) = \bigcap_{n \ge 0} f^{-n}N_\beta(f^n x).$$

Thus $y \in \Sigma(x, \beta)$ if and only if $f^n(y)$ is defined, and $d(f^n y, f^n x) < \beta$, for all $n \geq 0$.

Let $\beta \geq \alpha > 0$. A submanifold $W \subset U$ is the *stable manifold through* $x \in U$ *of size* $(\beta, \alpha)$ provided $W = \Sigma(x, \beta) \cap N_\alpha(x)$. If $\beta = \alpha$ we say $W$ has size $\beta$.

We shall show that for some $\beta \geq \alpha > 0$ every $x \in \Lambda$ has a stable manifold of size $(\beta, \alpha)$; if the metric is adapted to $\Lambda$, we can take $\beta = \alpha$.

In order to describe the sense in which the stable manifolds vary continuously, we make the following definition. A family $\{W_x\}_{x \in \Lambda}$ of $C^k$ submanifolds of $M$ is *continuous* if for each $x \in \Lambda$ there exists a neighborhood $A$ of $x$ in $\Lambda$ and a (continuous) map $\phi : A \to C^k(D^n, M)$ such that $\phi_x$ maps $D^n$ diffeomorphically onto a neighborhood of $x$ in $W_x$, for each $x \in A$.

3.2. STABLE MANIFOLD THEOREM FOR A HYPERBOLIC SET. *Let* $\Lambda \subset U$ *be a compact hyperbolic set for a* $C^k$ *embedding* $M \supset U \xrightarrow{f} M$, $k \in \mathbf{Z}_+$. *Then*

(a) *There exist numbers* $\beta \geq \alpha > 0$ *such that through each* $x \in \Lambda$ *there is a stable manifold* $W_x$ *of size* $(\beta, \alpha)$.

(b) $\{W_x\}_{x \in \Lambda}$ *is a continuous family of* $C^k$ *submanifolds.*

(c) *There exist numbers* $K > 0$ *and* $\lambda < 1$ *such that if* $y, z \in W_x$, *then* $d(f^n y, f^n z) \leq K \lambda^n d(y, z)$ *for all* $n \geq 0$.

(d) $W_x \cap W_y$ *is an open subset of* $W_x$ *for all* $x, y \in \Lambda$.

(e) $W_x$ *is tangent to* $E_x^s$ *at* $x \in \Lambda$ *(where* $T_\Lambda M = E^u \oplus E^s$ *is the invariant splitting).*

(f) *If the metric on* $M$ *is adapted to* $\Lambda$, *then* $\alpha = \beta$ *in (a) and* $K = 1$ *in (c).*

PROOF. By Mather's theorem (3.1) we may assume the metric in $M$ adapted to $\Lambda$. The basic idea behind the proof is to consider the Banach manifold $\mathcal{M}(\Lambda, M)$ of bounded maps $\Lambda \to M$. Define $\tilde{F} : \mathcal{M}(\Lambda, U) \to \mathcal{M}(\Lambda, M)$ by $\tilde{F}(h) = f \circ h \circ f^{-1}$. The inclusion $i : \Lambda \to M$ is a hyperbolic fixed point of $\tilde{F}$. If $\mathcal{W} \subset \mathcal{M}(\Lambda, U)$ is the stable manifold of $\tilde{F}$, put $\mathcal{W}_x = ev_x(\mathcal{W}) = $ the set of points $h(x)$ for all $h \in \mathcal{W}$.

This definition is conceptually simple, but in order to prove that $\mathcal{W}_x$ is actually a submanifold it is more convenient to work in the exponential coordinate system of $\mathcal{M}$ at $i$.

Let $r > 0$ have the following property. For each $x \in \Lambda$, the exponential map $\exp_x : M_x(r) \to M$ is defined and maps $M_x(r)$ $(= $ the ball of radius $r$ in the tangent space $M_x$ to $M$ at $x)$ diffeomorphically into $U \cap f^{-1}U$. This is possible since $\Lambda$ is compact.

Let $V_r = \{y \in T_\Lambda M \mid |y| \leq r\}$. Define $\tilde{F} : V_r \to T_\Lambda M$ by

$$\tilde{F} | V_r \cap M_x = \exp_{fx}^{-1} \circ f \circ \exp_x.$$

The following diagram commutes:

$$
\begin{array}{ccc}
 & \tilde{F} & \\
V_r & \longrightarrow & T_\Lambda M \\
(p, \exp) \downarrow & & \downarrow (p, \exp) \\
\Lambda \times f^{-1}U & \xrightarrow{f \times f} & \Lambda \times U \\
\downarrow & & \downarrow \\
\Lambda & \xrightarrow{\quad f \quad} & \Lambda
\end{array}
$$

where $p : T_\Lambda M \to \Lambda$ is the bundle projection.

Let $S$ be the Banach space of all bounded, possibly discontinuous sections of $T_\Lambda M$. Let $S(r)$ be the (closed) ball of radius $r$ in $S$ around $0$. Then $F$ induces a map $F : S(r) \to S$ by $F(\sigma) = \tilde{F} \circ \sigma \circ f^{-1}$. More explicitly, if $x \in \Lambda$ and $\sigma \in S(r)$ then $F(\sigma)x = (\exp_x)^{-1} f \exp_{f^{-1}x} \sigma(f^{-1}x)$.

It is left to the reader to verify that $F$ is $C^k$. The derivative of $F$ at $0 \in S(r)$ is the linear map $DF_0 : S \to S$, defined by the formula $DF_0(\sigma) = (Tf) \circ \sigma \circ (f^{-1}|\Lambda)$.

It is easy to see that $F$ is hyperbolic. The splitting of $S$ is $S^u \oplus S^s$ where

$$S^u = \{\sigma \in S \mid \sigma(\Lambda) \subset E^u\}, \text{ and } S^s = \{\sigma \in S \mid \sigma(\Lambda) \subset E^s\}.$$

Therefore by Theorem 2.3, $F$ has a stable manifold function $G : S^s(r) \to S^u(r)$. (It may be necessary to replace $r$ by a smaller number. We assume this done, and remark that the smallness of $r$ depends only on the constants of hyperbolicity, $C$ and $\tau$, and on the first order "Taylor expansions" of $f$ at points $x \in \Lambda$: if $v \in M_x$, put $f(\exp v) = \exp_{fx}((Tf)v + o(v))$. Hence a single $r$ can be chosen for a whole neighborhood of $f$ in $C^1(U, M)$.)

We recall from 2.2 and 2.4 that the stable manifold function $G$ is characterized as follows: Given $\sigma \in S^s(r)$, $G(\sigma)$ is the unique section $\tau \in S^u(r)$ such that $F^n(\tau, \sigma)$ is defined and lies in $S(r)$ for all $n \geq 0$.

LEMMA. *There is a unique map* $H : E^s(r) \to E^u(r)$ *covering* $1_\Lambda$ *such that* $G(\sigma) = H \circ \sigma$ *for all* $\sigma \in S^s(r)$.

PROOF. Given $y \in T_\Lambda M$, define $\sigma_y \in S$ by

$$\sigma_y(x) = 0 \quad \text{if } x \neq p(y),$$
$$= y \quad \text{if } x = p(y).$$

For $y \in E_x^s(r)$ put $H(y) = G(\sigma_y)x$. Now suppose $\sigma(x) = y$. Then

$$\left| \tilde{F}^n(G(\sigma)x, y) \right| \leq r \quad \text{for all } n,$$

and also

$$\left| \tilde{F}^n(H(y), y) \right| \leq r \quad \text{for all } n.$$

From the characterization of $G(\sigma)$ given above, we must have $G(\sigma)x = H(y)$.

LEMMA. $H$ *is continuous, and* $C^k$ *on each fiber* $E_x^s(r)$.

PROOF. Let $\Sigma(y) = \sigma_y$. Then $H$ is the composition of the continuous maps:

$$H : E^s(r) \xrightarrow{(\Sigma, p)} S^s(r) \times \Lambda \xrightarrow{(G, 1)} S^u(r) \times \Lambda \xrightarrow{ev} E^u(r).$$

Since $\Sigma : E^s \to S^s$ and $ev_x : S^u \to E^u$ are linear and $G$ is $C^k$, it follows that $H$ is $C^k$ on fibers.

Now let $\mathcal{W}$ be the graph of $H$, that is,

$$\mathcal{W} = \{(Hy, y) \in E^u(r) \times E^s(r) \mid y \in E^s(r)\}.$$

For each $x \in \Lambda$ put $\mathscr{W}_x = \mathscr{W} \cap M_x(r)$. Then $\mathscr{W}_x$ is a $C^k$ submanifold of $M_x(r)$. Therefore $W_x = \exp_x(\mathscr{W}_x - \partial\mathscr{W}_x)$ is a $C^k$ submanifold of $U$. It follows from the composition formula for $H$ given above that $\{W_x\}_{x \in \Lambda}$ is a continuous family of $C^k$ submanifolds.

To see that $W_x$ is a stable manifold for $x$, choose $\beta > 0$ so small that $N_\beta(z) \subset \exp_z(M_r(z))$ for all $z \in \Lambda$. If $y \in \Sigma(x, \beta)$, that is, $f^n(y)$ is defined and in $N_\beta(f^n(x))$ for all $n \geq 0$, then define a section $\sigma$ of $T_\Lambda M(r)$ by

$$\begin{aligned} \sigma(z) &= \exp_x^{-1}(y) \quad \text{if } z = x \\ &= 0 \qquad\qquad \text{if } z \neq x. \end{aligned}$$

Then $F^n(\sigma)$ is defined and in $S(r)$ for all $n \geq 0$. Therefore $\sigma$ is in the stable manifold of $G$. This means that $\exp_x^{-1}y = H(\sigma)x$ which in turn means that $y \in W_x$.

Part (c) of Theorem 3.2 follows from the analogous fact for stable manifolds of hyperbolic fixed points; (d), (e) and (f) are left to the reader.[z]

4. **Criteria for hyperbolicity.** In this section, which is basically independent of the preceding ones, we establish tests for a linear map or an invariant set to be hyperbolic. These are based on criteria for a linear operator on a Banach space to be hyperbolic. The *universality* of these criteria is important: they are valid in every Banach space, and no special properties of the Banach space are used. The basic estimate is 4.7; the criteria are found in 4.8–4.10.

NOTATION. If $E$ and $F$ are Banach spaces, $L(E, F)$ is the Banach space of linear maps $E \to F$; its unit ball is $L_1(E, F)$. Inv$(E, F)$ is the open subset of invertible linear maps.

The following universal estimate is well known.

4.1. LEMMA. *Let $P \in L(E, E)$ have norm $< 1$. Then $I + P : E \to E$ is invertible and $\|(I + P)^{-1}\| \leq (1 - \|P\|)^{-1}$.*

Next we state an exercise in differential calculus on noncommutative Banach algebras.

4.2. LEMMA.
(a) *Let $B$ be a ball in a Banach space $V$. Let $f : B \to L(E, F)$ and $g : B \to L(F, G)$ be differentiable at $b \in B$. The map $h : B \to L(E, G)$, defined by $h(x) = g(x) \circ f(x)$ is also differentiable at $b$, and $Dh_b \in L(V, L(E, G))$ is the map assigning to $x \in V$ the linear map*

$$g(b) \circ Df_b(x) + Dg_b(x) \circ f(b) : E \to G.$$

(b) *Define $\iota : \text{inv}(E, F) \to \text{inv}(F, E)$ by $\iota(T) = T^{-1}$. Then $\iota$ is differentiable and if $T \in \text{inv}(E, F)$ then $D\iota_T : L(E, F) \to L(E, F)$ is the linear map $S \mapsto -T^{-1}ST^{-1}$.*

PROOF Left to reader. See [4].

4.3. LEMMA. *Let $E_i$ and $F_i$ be Banach spaces, $i = 1, 2$. Let $T : E_1 \times E_2 \to F_1 \times F_2$ be defined by the matrix of linear maps*

$$\begin{bmatrix} A & B \\ C & D \end{bmatrix}$$

so that $T(x_1, x_2) = (Ax_1 + Bx_2, Cx_1 + Dx_2)$. Suppose $A: E_1 \to F_1$ is invertible. Put $u = \|B\| \cdot \|A^{-1}\|$, and suppose $u < 1$. Then the graph transform

$$\Gamma_T : L_1(E_1, E_2) \to L(F_1, F_2),$$

is well defined by the formula

$$\Gamma_T(P) = (C + DP)(A + BP)^{-1}.$$

Moreover the Lipschitz constant $L(\Gamma_T)$ satisfies

$$L(\Gamma_T) \leq (\|CA^{-1}\| + \|D\| \cdot \|A^{-1}\|)u(1 - u)^{-2} + \|D\| \cdot \|A^{-1}\|(1 - u)^{-1}.$$

PROOF. If $P \in L_1(E_1, E_2)$ then $\|BPA^{-1}\| \leq \|B\| \cdot \|A^{-1}\| = u < 1$. Therefore $I + BPA^{-1}$ is invertible and

(1) $$\|(I + BPA^{-1})^{-1}\| \leq (1 - u)^{-1}$$

by Lemma 4.1.

Thus $(A + BP)^{-1} = A^{-1}(I + BPA^{-1})^{-1}$ exists, so that $\Gamma_T$ is well defined. The derivative $(D\Gamma_T)_P : L(E_1, E_2) \to L(F_1, F_2)$ of $\Gamma_T$ at $P$ takes $X \in L(E_1, E_2)$ into

(2) $$- (C + DP)(A + BP)^{-1}BX(A + BP)^{-1} + DX(A + BP)^{-1},$$

by Lemma 4.2. Put $(A + BP)^{-1} = A^{-1}Q$, $Q = (I + BPA^{-1})^{-1}$. Then $\|Q\| \leq (1 - u)^{-1}$ by (1), and from (2) we have

(3) $$(D\Gamma_T)_P(X) = (CA^{-1} + DPA^{-1})QBXA^{-1}Q + DXA^{-1}Q.$$

Since $\|P\| \leq 1$ and $\|Q\| \leq (1 - u)^{-1}$, the result follows.

For convenience, put

$$\|CA^{-1}\| = w, \quad \|D\| \cdot \|A^{-1}\| = v, \quad \|B\| \cdot \|A^{-1}\| = u.$$

Then Lemma 4.1 implies

(4) $$L(\Gamma_T) \leq (w + v)u(1 - u)^{-2} + v(1 - u)^{-1}.$$

4.4. LEMMA. If $u < 1$ and $P \in L_1(E_1, F_1)$ then $\|\Gamma_T(P)\| \leq (w + v)(1 - u)^{-1}$.

PROOF.
$$\|\Gamma_T(P)\| = \|(CA^{-1} + DPA^{-1})(I + BPA^{-1})^{-1}\|$$
$$\leq (\|CA^{-1}\| + \|D\| \cdot \|A^{-1}\|) \cdot \|Q\|$$
$$\leq (w + v)(1 - u)^{-1}.$$

4.5. PROPOSITION. With the above notation, let $F_i = E_i$ $(i = 1, 2)$. Suppose

(a) $$2u < 1 - v,$$

and

(b) $$u + v + w < 1.$$

Then $\Gamma_T : L_1(E_1, E_2) \to L_1(E_1, E_2)$ is a well-defined contraction of Lipschitz constant $\leq (u + v)(1 - u)^{-1} < 1$. Consequently $\Gamma_T$ has a unique fixed point in $L_1(E_1, E_2)$, which depends continuously on $T$.

PROOF. By Lemma 4.3, $\Gamma_T$ is well defined on $L_1(E_1, E_2)$; by (b) and Lemma 4.4, the image of $\Gamma_T$ lies in $L_1(E_1, E_2)$. Combining (a), (b) and 4.4. yields the estimate of $L(\Gamma_T)$.

Now let $W \subset E_1 \times E_2$ be the graph of the fixed point $G \in L_1(E_1, E_2)$ of $\Gamma_T$. We give $W$ the Banach norm it inherits as a closed subspace of $E_1 \times E_2$.

4.6. LEMMA. *Assume* $\|A^{-1}\|(1 - u)^{-1} < 1$. *Then* $T|W$ *is expanding; in fact*

$$|Ty| \geq (1 - u)\|A^{-1}\|^{-1}|y| \quad \text{if } y \in W.$$

PROOF. Put $y = (x, Gx)$ with $x \in E_1$. Then $|y| = \max(|x|, |Gx|) = |x|$ since $\|G\| \leq 1$. Similarly

$$
\begin{aligned}
|Ty| &= |(Ax + BGx, Cx + DGx)| \\
&= |Ax + BGx| \\
&\geq \|(A + BG)^{-1}\|^{-1}|x| \\
&\geq \|A^{-1}\|^{-1} \cdot \|(I + BGA^{-1})^{-1}\|^{-1} \cdot |x|.
\end{aligned}
$$

The lemma follows since

$$\|(I + BGA^{-1})^{-1}\| \leq (1 - u)^{-1}.$$

We summarize these facts:

4.7. PROPOSITION. *Let*

$$T = \begin{bmatrix} A & B \\ C & D \end{bmatrix} : E_1 \times E_2 \to E_1 \times E_2$$

*be as above. Suppose*

(a) $\|A^{-1}\| < 1$.

(b) $u + v + w < 1$.

(c) $2u < 1 - v$,

*where* $w = \|CA^{-1}\|$, $v = \|D\| \cdot \|A^{-1}\|$ *and* $u = \|B\| \cdot \|A^{-1}\|$. *Then the graph transform*

$$\Gamma_T : L_1(E_1 \times E_2) \to L_1(E_1 \times E_2),$$

*is a well-defined contraction. If* $W \subset E_1 \times E_2$ *is the graph of the fixed point* $G_T$ *of* $\Gamma_T$, *then* $T|W$ *is an expansion. Moreover* $G_T$ *depends continuously on* $T$, *and* $\|G_T\| \leq (w + v)/(1 - u)$.

We now derive a perturbation criterion for hyperbolicity.

4.8. THEOREM. *Given* $0 < \tau < 1$, *there exists* $\varepsilon > 0$ *with the following property. Let* $E_1$ *and* $E_2$ *be Banach spaces,*

$$T = \begin{bmatrix} A & B \\ C & D \end{bmatrix} : E_1 \times E_2 \to E_1 \times E_2$$

*a linear map with* $A : E_1 \to E_1$ *and* $D : E_2 \to E_2$ *invertible. Suppose*

$$\max\{\|A^{-1}\|, \|D\|\} < \tau + \varepsilon \text{ and } \max\{\|B\|, \|C\|\} < \varepsilon.$$

*Then T is hyperbolic (for some splitting of $E_1 \times E_2$).*

PROOF. By choosing $\varepsilon$ small enough we may assume $T$ invertible, and apply Proposition 4.7 to both $T$ and $T^{-1}$ to get the expanding and contracting invariant subspaces for $T$.

Another universal criterion is the following.

4.9. PROPOSITION. *Let $0 < \tau < 1$ and $\varepsilon > 0$. There exists $\delta > 0$ with the following property. Let $E_i$ and $F_i$ be Banach spaces $(i = 1, 2)$, and $T_i : E_i \to F_i$ invertible linear maps such that*

(a) $\max\{\|T_1^{-1}\|, \|T_2\|\} \le \tau < 1.$
*Let*

$$H = \begin{bmatrix} P & Q \\ R & S \end{bmatrix} : F_1 \times F_2 \to E_1 \times E_2$$

*be a linear map with $P : F_1 \to E_1$ invertible, satisfying*
(b) $\max\{\|P^{-1}\|^{-1} - 1, \|Q\|, \|R\|, \|S\| - 1\} < \delta.$
*Then the map $HT : E_1 \times E_2 \to E_1 \times E_2$ is hyperbolic for some splitting $E^u \times E^s$ of $E_1 \times E_2$. Moreover if $G^u : E_1 \to E_2$ and $G^s : E_2 \to E_1$ are the unstable and stable manifold functions of $HT$, then*
(c) $\max\{\|G^u\|, \|G^s\|\} < \varepsilon.$
(d) *Analogous statements hold for*

$$TH : F_1 \times F_2 \to F_1 \times F_2.$$

PROOF. Let

$$HT = \begin{bmatrix} A & B \\ C & D \end{bmatrix} = \begin{bmatrix} PT_1 & QT_2 \\ RT_1 & ST_2 \end{bmatrix}.$$

By Proposition 4.6, $HT$ has an unstable manifold provided

$$\|A^{-1}\| < 1,$$

$$\|A^{-1}\| \cdot \|D\| + \|A^{-1}\| \cdot \|B\| + \|CA^{-1}\| < 1,$$

and

$$2\|A^{-1}\| \cdot \|B\| < 1 - \|A^{-1}\| \cdot \|D\|.$$

This will be true provided

(5)                         $\tau\|P\| < 1,$

(6)                  $\|S\|\tau + \|Q\|\tau + \|R\| \cdot \|P^{-1}\| < 1,$

and

(7)                         $2\|Q\|\tau < 1 - \|S\|\tau.$

If $\delta$ is sufficiently small then (b) implies (5), (6) and (7). Moreover

$$\|G^u\| \le (w + v)(1 - u)^{-1} \le (\|R\| \cdot \|P^{-1}\| + \|Q\|\tau)(1 - \|S\|\tau)^{-1},$$

which is $< \varepsilon$ if $\delta$ is sufficiently small.

Applying this result to $H^{-1}T^{-1}$ shows that if $\delta$ is sufficiently small then $TH : F_1 \times F_2 \to F_1 \times F_2$ has a stable manifold, and a stable manifold function of norm $< \varepsilon$. Similar reasoning shows that $TH$ has an unstable manifold and $HT$ a stable manifold, etc.

REMARK. Notice that the estimate of $\|CA^{-1}\|$ was used (from Proposition 4.7) rather than an estimate of $\|C\| \cdot \|A^{-1}\| = \|RT_1\| \cdot \|T_1^{-1}P^{-1}\|$. The latter requires a knowledge of $\|T_1\|$ whereas the former does not.

Next we apply Theorem 4.8 to get a criterion for a hyperbolic set.

4.10. PROPOSITION  *Let* $M \supset V \overset{g}{\to} M$ *be a* $C^1$ *embedding of an open set* $V$. *Let* $X \subset V$ *be invariant under* $g$ *and* $g^{-1}$. *Let* $E_1 \oplus E_2$ *be a splitting of* $T_X M$. *Put*

$$T_X g = \begin{bmatrix} A & B \\ C & D \end{bmatrix} : E_1 \oplus E_2 \to E_1 \oplus E_2,$$

*where* $A$, $B$, $C$, $D$ *are bundle maps covering* $g$. *If there exist* $0 < \tau < 1$ *and* $\varepsilon > 0$ *satisfying Theorem 4.8, and also*

(a) $\max\{\|A^{-1}\|, \|D\|\} < \tau + \varepsilon$,
(b) $\max\{\|B\|, \|D\|\} < \varepsilon$,

*then* $X$ *is a hyperbolic set.*

PROOF. Let $C_0(T_X M) = C_0$ be the Banach space of bounded continuous sections of $T_X M$. Let $F : C_0 \to C_0$ be induced by $g$. That is, $F(\sigma) = T_X g \circ \sigma \circ g^{-1}$. If we write $C_0 = C_0(E_1|X) \times C_0(E_2|X)$, then Proposition 4.8 shows that $F$ is hyperbolic for some splitting of $C_0$. An obvious extension of a theorem of J. Mather [12, Appendix] concerning Anosov diffeomorphisms shows that this suffices for $X$ to be a hyperbolic set. Alternatively, the proof of the existence theorem (3.2) for stable manifolds could be imitated. (This idea is due to S. Smale; in fact it suggested the proof of 3.1 given here.)

5. **Hyperbolicity of submanifolds.** Let $\Lambda \subset U$ be a hyperbolic set for $M \supset U \overset{f}{\to} M$, and $V \subset \Lambda$ a smooth submanifold invariant under $f$ and $f^{-1}$. No examples are known for which $f|V : V \to V$ is not Anosov; on the other hand there is no proof that $f|V$ must be Anosov (which means that $V$ is a hyperbolic set for $f|V$). We prove a partial result in this direction.

5.1. THEOREM.  *Let* $p : E \to V$ *be a finite dimensional vector bundle. Let* $T : E \to E$ *be a linear bundle automorphism covering* $f : V \to V$. *Let* $F \subset E$ *be an invariant subbundle over* $V$. *Let* $\Omega \subset V$ *be the set of nonwandering points. If* $T$ *is hyperbolic, then* $T|F_\Omega$ *is hyperbolic.*

PROOF. Let $E^u \oplus E^s$ be the hyperbolic splitting of $E$. Let the metric on $E$ be adapted to $T$: there exists $0 < \tau < 1$ such that $|Tv| \le \tau|v|$ if $v \in E^s$, $|Tv| \ge \tau^{-1}|v|$ if $v \in E^u$. (See 3.1.)

We shall prove that if $x \in \Omega$, then $F_x = (F_x \cap E_x^u) \times (F_x \cap E_x^s)$; this suffices to prove $T|F_\Omega$ hyperbolic. We do this by proving

(1)                     $\dim E_x^u \cap F_x + \dim E_x^s \cap F_x \geq \dim F_x.$

Let $x \in \Omega$, and let $W \subset V$ be a set over which the bundle pair $(E, F)$ is trivial. Let $\phi:(E_W, F_W) \to (R^n, R^m)$ be a trivialization. For each $y \in W$, $\phi_y = \phi|E_y$ maps $E_y$ isomorphically onto $R^n$, and $F_y$ onto $R^m$.

Suppose $y \in W$ and $z = f^k(y) \in W$. Then

$$T_{k,y} = \phi_y^{-1} \phi_z (F^k)_y : (E_y, F_y) \to (E_y, F_y),$$

is a linear automorphism. In terms of the splitting $E_y = E_y^u \times E_y^s$, $T_{k,y}$ is represented by a matrix

$$\begin{bmatrix} A & B \\ C & D \end{bmatrix}.$$

We want to be in the situation of Proposition 4.9, with $T$ and $H$ of 4.9 represented by $(F^k)_y$ and $\phi_y^{-1}\phi_z$ respectively.

Since $x$ is nonwandering, we can find $y \in U$ and $k \in Z_+$ such that $y$ and $z = f^k(y)$ are as close to $x$ as desired. In particular we can make $\phi_y^{-1}\phi_z$ as close to an isometry as desired, so that (b) of 4.9 will hold; we can simultaneously take $k$ as large as necessary for 4.9(a) to hold. It follows that $T_{k,y} : E_y \to E_y$ will be hyperbolic. Since $F_y \subset E_y$ is invariant under $T_{k,y}$, and $E_y$ is finite dimensional, a simple eigenvalue argument shows that $T_{k,y}|F_y$ is hyperbolic. Moreover the stable and unstable manifolds of $T_{k,y}|F_y$ must be the intersection of $F_y$ with the stable and unstable manifolds $B_y^s$, $B_y^u$ of $T_{k,y}$. And (c) of 4.9 shows that as $y$ and $z$ approach $x$, $B_y^s$ and $B_y^u$ approach $E_x^s$ and $E_x^u$. Since $\dim B_y^u \cap F_y + \dim B_y^s \cap F_y = \dim F_y$, (1) follows and the theorem is proved.

5.2. COROLLARY. *Let $V \subset M$ be a compact invariant $C^1$ submanifold of an Anosov diffeomorphism $f: M \to M$. Then the nonwandering set $\Omega_0$ of $f|V$ is hyperbolic.*

REMARK 1. If $F \subset E$ is a subspace invariant under $T$ and $T^{-1}$, then every eigenvalue of $T$ is clearly an eigenvalue of $T|F$. Therefore $T|F$ is hyperbolic if $T$ is hyperbolic and spectrum $(T)$ consists entirely of eigenvalues as in the finite dimensional case.

REMARK 2. The following infinite dimensional example, due to W. Badé, shows that $T|F$ may fail to be hyperbolic.

Let $C$ be the complex field. Let

$$A = \{z \in C | \tfrac{1}{4} \leq |z| \leq \tfrac{1}{2}, \qquad B = \{z \in C | 2 \leq |z| \leq 3\}.$$

Let $E$ be the Hilbert space of complex functions which are continuous on $A \cup B$ and analytic on $\text{int}(A \cup B)$. Define $T: E \to E$ by $T(f)z = zf(z)$. Then $T$ is hyperbolic; the invariant splitting is found by setting

$$E^u = \{f \in E | f(A) = 0\}, \qquad E^s = \{f \in E | f(B) = 0\}.$$

Let $F \subset E$ be the subspace comprising those functions that extend to a function analytic on $\{z \mid \frac{1}{4} < |z| < 3\}$.

Then $TF = F = T^{-1}F$. However, the constant function 1 belongs to $F$, but is not in the image of $(T - I)|F : F \to F$. Hence the complex number 1 is in spectrum $(T|F)$.

6. **Smoothness of splittings.** In order to study the smoothness of the splitting $T_\Lambda M = E^u \oplus E^s$, or more precisely, the smoothness of the functions assigning to each $x \in \Lambda$ the subspaces $E^u_x$ and $E^s_x$, we must study the smoothness of certain sections of vector bundles. There is no "natural" metric on a vector bundle $p : E \to X$ in which to express Hölder conditions, but there is a natural class of metrics which we now define.

DEFINITION. Let $p : E \to X$ be a vector bundle over a metric space $X$. A metric $d$ on $E$ is *admissible* if there is a complementary bundle $E'$ over $X$, and an isomorphism $h : E \oplus E' \to X \times A$ to a product vector bundle, where $A$ is a Banach space, such that $d$ is induced from the product metric on $X \times A$.

6.1. THEOREM. *Let $p : Y \to X$ be a vector bundle over a metric space $X$ endowed with an admissible metric. Let $D \subset Y$ be the unit ball bundle, and $F : D \to D$ a map covering a homeomorphism $f : X \to X$. Suppose $0 \le \kappa < 1$ and that for each $x \in X$, the restriction $F_x : D_x \to D_{fx}$ has Lipschitz constant $\le \kappa$. Then*

(a) *There is a unique section $g_0 : X \to D_0$ whose image is invariant under $F$.*

(b) *Let $L(f^{-1}) = \lambda < \infty$; let $0 < \alpha \le 1$ be such that $\kappa\lambda^\alpha < 1$. Then $g_0$ satisfies a Hölder condition of exponent $\alpha$.*

(c) *Suppose $X$ is a smooth manifold, $E$ is a smooth vector bundle, and $F, f$ are $C^1$. If $\kappa\lambda < 1$ then $g_0$ is $C^1$.*

PROOF. Let $\mathcal{G}$ be the unit ball in the Banach space of bounded continuous sections of $Y$. Define $\Phi : \mathcal{G} \to \mathcal{G}$ by $\Phi(g) = Fgf^{-1}$. Then $L(\Phi) \le \kappa < 1$; hence $\Phi$ has a unique fixed point $g_0$. This proves (a).

Before proving (b) and (c) we remark that $Y$ may be assumed trivial. For let $Z$ be a bundle over $X$ such that $Y \oplus Z$ is trivial, and define $F'$ to be the composition

$$F' : Y \oplus Z \xrightarrow{\pi} Y \xrightarrow{F} Y \xrightarrow{i} Y \oplus Z$$

where $\pi$ is the projection and $i$ the inclusion. Then $F'$ satisfies the same hypotheses as $F$, and the unique invariant section of $F$ is $ig_0$. Henceforth, we assume $Y = X \times E$ where $E$ is a Banach space. We write $F(x, y) = (fx, f_x y)$.

The proof of (c) is more intuitive than that of (b), so we do it first. Assume $X$ is a Riemannian manifold; let $B \subset E$ be the unit ball. Sections are now maps; $g_0 : X \to B$ is the unique map whose graph is invariant under $F : X \times B \to X \times B$.

Let $\mathcal{H}$ be the Banach space of linear bundle maps $H : TX \to X \times E$ covering $1_X$ and having finite norm $\|H\| = \sup_{x \in X} \|H_x\|$, where $H_x : T_x X \to E$ is defined by setting $H(v) = (x, H_x v)$ for $v \in T_x X$.

For each $g \in \mathcal{G}$ define a linear map $\psi_g : \mathcal{H} \to \mathcal{H}$ as follows. If $v \in T_x X$ put $f^{-1}(x) = y \in X$ and $(Tf^{-1})v = w \in T_y X$. If $H \in \mathcal{H}$ define

$$(\Psi_g H)_x v = D(\pi_2 \circ F)_{(y, gy)}(w, H_y w).$$

Here $\pi_2 : X \times E \to E$ is the projection; $D(\pi_2 \circ F)_{(y, gy)}$ is the linear map from the tangent space at $X \times E$ at $(y, gy)$ to $E$ that is the derivative of $\pi_2 \circ F$; and $(w, H_y w)$ is a tangent vector to $X \times E$ at $(y, gy)$. Observe that if $g \in \mathcal{G}$ is $C^1$, then $\Psi_g(Dg) = D(\Phi g)$.

We claim $L(\Psi_g) \leq \lambda \kappa$. By definition if $H, K \in \mathcal{H}$ then $\|H - K\| = \sup_{x \in X_0} \|H_x - K_x\|$. Observe that

$$(\psi_g H - \psi_g K)_x v = (DF_y)_{gy}(H_y - K_y)(T_x f^{-1})v.$$

Since $\|(DF_y)_{gy}\| \leq \kappa$ if $L(F_y) \leq \kappa$, and $\|T_x f^{-1}\| \leq L(f^{-1}) \leq \lambda$, we have

$$\|\Psi_g H - \Psi_g K\| \leq \kappa \lambda \|H - K\|.$$

Therefore $\Psi_g$ is a contraction if $\kappa \lambda < 1$. By the Fiber Contraction Theorem (1.2) the map $\Psi : \mathcal{G} \times \mathcal{H} \to \mathcal{G} \times \mathcal{H}$, defined by $(g, H) \mapsto (\Phi g, \Psi_g H)$, has an attractive fixed point $(g_0, H_0)$. If $g \in \mathcal{G}$ is $C^1$, then $\Psi^n(g, Dg) = (\Phi^n g, D\Phi^n g)$. Therefore $D(\Phi^n g)$ converges, and so $g_0$ is $C^1$. This proves 6.1(c).

To prove (b) we assume $F$ is defined on all of $X \times E$. To see that there is no loss of generality, let $r : B \to E$ be the radial retraction

$$r(x) = x \quad \text{if } |x| \leq 1$$
$$= x/|x| \text{ if } |x| \geq 1.$$

Then $L(r) \leq 2$, and $L((Fr)_x^n) = L((F^n)_x r) = 2\kappa^n$, which is $< 1$ if $n$ is large enough. If $\kappa \lambda^\alpha < 1$, then $(2\kappa^n)(\lambda^n)^\alpha = 2(\kappa \lambda^\alpha)^n$ is also $< 1$ if $n$ is large enough. Therefore $f$ and $F$ may be replaced by $f^n$ and $F^n r$; $g_0$ and $\alpha$ stay the same, and also $L((F^n r)_x)L(f^{-n})^\alpha < 1$. Therefore we assume $F : X \times E \to X \times B$ given covering $f$, with $L(F_x) \leq \kappa < 1$; and $L(f^{-1}) \leq \lambda$ with $\kappa \lambda^\alpha < 1$.

In order to imitate the proof of (c), we replace $Dg$ by $\Delta g : X \times X \to E$, defined by $\Delta g(x, y) = g(x) - g(y)$. We proceed as follows. Let $\mathcal{G}$ and $\Phi : \mathcal{G} \to \mathcal{G}$ be as before. Let $\mathcal{H}$ be the Banach space of bounded continuous maps $H : X \times X \to E$, such that $H(x, x) = 0$ for all $x$ and the following norm is finite:

$$\|H\| = \sup_{x \neq y} |H(x, y)|/d(x, y)^\alpha + \sup_{(x,y)} |H(x, y)|.$$

The natural map from $\mathcal{H}$ to the Banach space of bounded continuous maps $X \times X \to E$ is continuous and takes closed bounded sets onto closed bounded sets. If $H \in \mathcal{H}$ and $x \in X$ define $H_x : X \to E$ by $H_x(y) = H(x, y)$.

Given $g \in \mathcal{G}$, define $\Psi_g : \mathcal{H} \to \mathcal{H}$ by

$$(\Psi_g H)_x = \Phi(g + H_{f^{-1}x}) - \Phi(g).$$

If $g \in \mathcal{G}$ satisfies an $\alpha$-Hölder condition, define $\Delta g \in \mathcal{G}$ by $\Delta g(x, y) = g(x) - g(y)$. Observe that $\Psi_g(\Delta g) = \Delta(\Phi g)$. Define $\Psi : \mathcal{G} \times \mathcal{H} \to \mathcal{G} \times \mathcal{H}$ by $\Psi(g, H) = (\Phi g, \Psi_g H)$.

We show now that $L(\Psi_g) \leq \kappa \lambda^\alpha$:

$$d(x, y)^\alpha \|\Psi_g(H) - \Psi_g(K)\| = \sup_{x \neq y} |(\Psi_g H)_x y - (\Psi_g K)_x y|$$

$$= \sup_{x \neq y} |F_{f^{-1}x}(g + H_{f^{-1}x})y - F_{f^{-1}x}(g + K_{f^{-1}x})y|$$

$$\leq \kappa \sup_{x \neq y} |H_{f^{-1}x}(f^{-1}y) - K_{f^{-1}x}(f^{-1}y)|$$

$$\leq \kappa \|H - K\| [d(f^{-1}y, f^{-1}x)]^\alpha$$

$$\leq \kappa \|H - K\| \lambda^\alpha d(x, y)^\alpha.$$

Therefore $\|\Psi_g(H) - \Psi_g(K)\| \leq \kappa \lambda^\alpha \|H - K\|$.

Since $\kappa \lambda^\alpha < 1$ it follows from the Fiber Contraction Theorem (1.2) and the completeness of $\mathcal{H}$ that $\Psi : \mathcal{G} \times \mathcal{H} \to \mathcal{G} \times \mathcal{H}$ has a unique attractive fixed point. If $g \in \mathcal{G}$ is $\alpha$-Hölder, then $\Phi^n g \to g_0$, while $\Delta(\Phi^n g)$ converges in $\mathcal{H}$, as $n \to \infty$. But therefore $\Delta(\Phi^n g)$ converges in the uniform topology, so it must converge to $\Delta(g_0)$. Therefore $g_0$ is $\alpha$-Hölder. Q.E.D.

It is useful to generalize 6.1 to the case where $f$ maps a subspace $X_0$ of $X$ homeomorphically onto $X$. For simplicity we deal only with the smooth case.

6.2. THEOREM. *Let $p : E \to X$ be a $C^1$ vector bundle over a $C^1$ manifold $X$. Let $X_0 \subset X$ be an open set and $f : X_0 \to X$ a $C^1$ diffeomorphism. Let $D \subset E$ be the unit ball bundle in an admissible metric and put $D_0 = p^{-1} X_0$. Suppose $F : D_0 \to \text{int } D_0$ is a $C^1$ map covering $f$ such that $L(F_x) \leq \kappa < 1$ for each $x \in X_0$. Then*

(a) *there exists a unique section $g_0$ of $D_0$ that is invariant under $F$ in the sense that $g_0(X_0) \subset F g_0(X_0)$.*

(b) *Let $L(f^{-1}) = \lambda < \infty$. If $\kappa \lambda < 1$ then $g_0$ is $C^1$.*

PROOF. The proof is practically identical to that of 6.1(c) and is left to the reader.

Now let $\Lambda \subset U$ be a compact hyperbolic set for $M \supset U \xrightarrow{f} M$. Let $T_\Lambda M = E^s \oplus E^u$ be the invariant splitting, with $E^s$ contracting and $E^u$ expanding. We define four quantities:

$$a = \|Tf^{-1}|E^u\| < 1 \qquad b = \|Tf|E^s\| < 1$$
$$c = \|Tf|E^u\| > 1 \qquad d = \|Tf^{-1}|E^s\| > 1.$$

If $\Lambda = U = M$ then $f$ is called an *Anosov* diffeomorphism of $M$. In this case the stable and unstable manifolds give two topological foliations of $M$. According to [3] these are not always $C^1$, although they are "absolutely continuous."

6.3. THEOREM. *Let $f : M \to M$ be a $C^2$ Anosov diffeomorphism. Then*

(a) *The stable foliation is $\alpha$-Hölder where $0 < \alpha \leq 1$ and $abc^\alpha < 1$; the unstable foliation is $\beta$-Hölder where $0 < \beta \leq 1$ and $abd^\beta < 1$.*

(b) *If $abc < 1$ then the stable foliation is $C^1$. In particular, the stable foliation is $C^1$ in these two cases: (i) the stable manifolds have codimension one; (ii) $\dim M = 3$ and $f$ preserves the Riemannian measure in $M$; in this case the unstable foliation is also $C^1$.*

PROOF. Give $TM$ a $C^1$ splitting $F^s \oplus F^u$ approximating $E^s \oplus E^u$. For each $x \in M$ put $L_x = L(F^s, F^u)$; then $E^s_x$ is the graph of an element $\lambda_x \in L_x$. Define $\Gamma_x : L_x(1) \to L_y(1)$, $y = f^{-1}x$, to be the graph transform induced by

$$T_x f^{-1} = \begin{bmatrix} A_x & B_x \\ C_x & D_x \end{bmatrix} : F_x^s \times F_x^u \to F_y^s \times F_y^u.$$

Provided the splitting $F^s \oplus F^u$ is sufficiently close to $E^s \oplus E^u$, the map $\Gamma_x$ is well defined by the formula

$$\Gamma_x(\mu_x) = (C_x + D_x \mu_x) \circ (A_x + B_x \mu_x)^{-1};$$

and given $\varepsilon > 0$ we may assume $\|C_x\|$ and $\|B_x\|$ so small that $L(\Gamma_x) \le L(D_x) L(A_x^{-1}) + \varepsilon \le ab + \varepsilon = \kappa$. Choose $\varepsilon$ so that $\kappa < 1$ and $\kappa C^\alpha < 1$.

Let $L(F^s, F^u)$ be the vector bundle over $M$ whose fiber over $x$ is $L_x$; let $D$ be its unit ball bundle. Then $\Gamma : D \to D$ is a bundle map covering $f^{-1}$; and $\Gamma$ is $C^1$ if $f$ is $C^2$. Moreover $L(\Gamma_x) \le \kappa < 1$. Clearly $L(f^{-1}) \le \max\{a, c\} = c$. By 6.1(b) the unique $\Gamma$-invariant section of $D$, which is $\lambda$, is $\alpha$-Hölder; this proves 6.3(a). The proof of 6.3(b) follows similarly from 6.1(c). The proof of 6.3(bii) is left to the reader.

Let $q$ be the fiber dimension of the bundle $E^s$, let $G_q(X)$ be the bundle of $q$-planes in $T_X M$ for any subset $X \subset M$. Let $\theta : \Lambda \to G_q(\Lambda)$ assign $E_x^s$ to $x \in \Lambda$.

6.4. THEOREM. *Let* $\Lambda \subset U$ *be a compact hyperbolic set for the* $C^1$ *embedding* $M \supset U \xrightarrow{f} M$.

(a) *If* $abc^\alpha < 1, 0 < \alpha \le 1$, *then* $\theta : \Lambda \to G_q(\Lambda)$ *is* $\alpha$-*Hölder.*

(b) *Let* $\{V_i\}$ *be a collection of* $C^1$ *submanifolds of* $\Lambda$ *such that* $\bigcup_i V_i$ *is invariant under* $f$. *If* $f$ *is* $C^2$ *and* $abc < 1$, *then each map* $\theta | V_i : V_i \to G_q$ *is* $C^1$.

PROOF. Almost identical to that of 6.3 and left to the reader.

Now let $\{W_x^s\}_{x \in \Lambda}$ be the stable manifold system for $\Lambda$, and put $W^s = \bigcup_{x \in \Lambda} W_x^s$. Then $f(W^s) \subset W^s$ (assuming the metric on $M$ adapted to $\Lambda$; or $f$ could be replaced by $f^n$ for $n$ sufficiently large). Define $\text{glob}(W^s) = \bigcup_{n \ge 0} f^{-n}(W^s)$. Equivalently, $\text{glob}(W^s) = \bigcup_{x \in \Lambda} \{y \in U | \lim_{n \to \infty} d(f^n y, f^n x) \to 0\} = \bigcup_{x \in \Lambda} \text{glob}(W_x^s)$ where

$$\text{glob}(W_x^s) = \bigcup_{n \ge 0} f^n(W_x^s).$$

Each set $\text{glob}(W_x^s)$ is a disjoint union of $q$-dimensional submanifolds of $M$, and so is $\text{glob}(W^s)$. Define $\theta : \text{glob}(W^s) \to G_q$ by $\theta(y) = Tf^{-n} T_{f^n y}(W^s x)$ if $y \in f^{-n} W_x^s$, $x \in \Lambda, n \ge 0$. Then $\theta$ is well defined.

It may happen that $W^s$, and hence $\text{glob}(W^s)$, is open in $M$. This is the case for the 1-dimensional attractors of R. F. Williams [17].

6.5. THEOREM. *Assume* $W^s$ *open in* $M$. *Let* $f$ *be* $C^2$. *Then* $\theta : \text{glob}(W^s) \to G_q(M)$ *is* $C^1$ *provided* $abc < 1$. *In particular this is the case if* $\Lambda$ *is a 1-dimensional attractor.*

PROOF. It suffices to prove that $\theta$ is $C^1$ in some neighborhood $N$ of $\Lambda$ in $W^s$; for if $x$ is any point of $\text{glob}(W^s)$ there exists $n \ge 0$ such that $f^n(x) \in N$, and $\theta = \theta \circ f^n$.

Let $\tilde{E}^u \subset T_N M$ be a subbundle over a neighborhood $N \subset W^s$ of $\Lambda$ extending $E^u$; we do not assume $\tilde{E}^u$ invariant. Let $\tilde{E}^s \subset T_N M$ be the subbundle whose fiber over $y$ is $\theta(y)$. Choose $N$ so that $f(N) \subset N$ (see 8.4 below). Choose $\varepsilon > 0$ so small that $(a + \varepsilon)(b + \varepsilon)(c + \varepsilon) < 1$. We may choose $N$ so small that

$$\|Tf^{-1}|\tilde{E}^u\| < a + \varepsilon, \qquad \|Tf|\tilde{E}^s\| < b + \varepsilon, \qquad \|Tf|E^u\| < c + \varepsilon.$$

(For example given any compact $N_0$ such that $fN_0 \subset N_0$ and $\bigcap_{n \geq 0} f^n(N_0) = \Lambda$, let $N = f^p(N_0)$ for a large value of $p$.) Now the proof of 6.3(b) may be applied, replacing 6.1(c) by 6.2. We leave the details to the reader.

REMARK. In his paper [16], Williams assumed as an axiom that if $\Lambda$ is a one dimensional attractor then the local projections $W^s \to \Lambda$, defined locally by mapping $W^s_x$ to $x$, are Lipschitz. It is easy to see that this is in fact a consequence of Theorem 6.5 if $f$ is $C^2$.

7. **Perturbations of hyperbolic sets.** We continue the standing hypothesis $M \supset U \xrightarrow{f} M$ is a $C^k$ embedding and $\Lambda \subset U$ is a compact invariant hyperbolic set of skewness $\tau < 1$. For simplicity we assume the metric on $M$ adapted to $\Lambda$.

7.1. THEOREM. *Let $\varepsilon > 0$. There exists a neighborhood $V \subset U$ of $\Lambda$ and a neighborhood $\mathcal{N} \subset C^k(U, M)$ such that if $g \in \mathcal{N}$ then any invariant set of $g$ in $V$ is hyperbolic of skewness $\sigma$ with $|\tau - \sigma| < \varepsilon$.*

If $K \subset V$ is compact, then $\bigcap_{n \in \mathbf{Z}} g^n K$ is the unique maximal invariant subset of $K$. Hence

7.2. COROLLARY. *In Theorem 7.1, every compact subset $K \subset V$ contains a unique maximal $g$-hyperbolic subset; this subset contains every $g$-invariant subset of $K$. In particular every compact $g$-invariant subset of $V$ is hyperbolic.*

PROOF OF THEOREM 7.1. Let $W \subset U$ be a neighborhood of $\Lambda$ over which the invariant splitting $E^u \oplus E^s$ of $TM$ can be extended to a splitting $E_1 \oplus E_2$ of $T_W M$. Let $V \subset W$ be a neighborhood of $\Lambda$ so small that if $x \in W$ then $T_x f^n$ is represented by a matrix

$$\begin{bmatrix} A & B \\ C & D \end{bmatrix} : (E_1 \oplus E_2)_x \to (E_1 \oplus E_2)_{fx},$$

satisfying

$$\max\{\|A^{-1}\|, \|D\|\} < \tau + \varepsilon/2, \qquad \max\{\|B\|, \|C\|\} < \varepsilon/2,$$

where $\varepsilon$ is as in Theorem 4.8. Let $\mathcal{N}$ be a neighborhood of $f$ in $C^1(U, M)$ so small that if $g \in \mathcal{N}$ and $x \in V$ then $T_x g$ is represented by a matrix

$$\begin{bmatrix} A & B \\ C & D \end{bmatrix}$$

satisfying (a) and (b) of Proposition 4.10. By that Proposition any $g$-invariant in $V$ is hyperbolic.

The next result shows that maximal hyperbolic sets enjoy a type of structural stability.

7.3. THEOREM. *Let* $\Lambda \subset U$ *be a compact hyperbolic set for the* $C^1$ *embedding* $M \supset U \overset{f}{\hookrightarrow} M$. *Given* $\varepsilon > 0$ *there is a compact neighborhood* $B \subset U$ *of* $\Lambda$ *and a neighborhood* $\mathcal{N}$ *of* $f$ *in* $C^1(U, M)$ *with the following properties: if* $g_i \in \mathcal{N}$ *for* $i = 1, 2$, *then* $g_i$ *has a unique maximal hyperbolic set* $\Lambda_i \subset V$ *containing every invariant set of* $g_i$ *in* $V$; *and there is a unique homeomorphism* $h_1 : \Lambda_1 \to \Lambda_2$ *such that* $h_1 g_1 h_1^{-1} = g_2 | \Lambda_2$; *and* $d(h_1, 1) \leq \varepsilon$. *Moreover* $h_1$ *depends continuously on* $(g_1, g_2)$ $\in \mathcal{N} \times \mathcal{N}$.

PROOF. Choose a compact neighborhood $V \subset U$ of $\Lambda$ and a neighborhood $\mathcal{N}$ of $f$ in $C^1(U, M)$ as in 7.1 and 7.2. Given $g_1$ and $g_2$ in $\mathcal{N}$, let $\Lambda_1$ and $\Lambda_2$ be their respective compact maximal hyperbolic sets in $V$. Let $\mathscr{C}(\Lambda_1, M)$ be the Banach manifold of continuous maps $h : \Lambda_1 \to M$. Define $\Phi : \mathscr{C}(\Lambda_1, U) \to \mathscr{C}(\Lambda_1, M)$ by $\Phi(h) = g_2 \circ h \circ g_1^{-1}$.

Let $i_1 : \Lambda_1 \to M$ be the inclusion of $\Lambda_1$. If $V$ and $\mathcal{N}$ are sufficiently small, depending only on $f$, then Proposition 4.8 shows that the derivative of $\Phi$ at $i_1$ will be hyperbolic. Moreover $\Phi$ has a unique fixed point $h_1$ by Theorem 3.2. Therefore $g_2 h_1 = h_1 g_1$. Clearly $h_1(\Lambda_1)$ is $g_2$-invariant, and so $h_1(\Lambda_1) \subset \Lambda_2$. Similarly there exists $h_2 : \Lambda_2 \to \Lambda_1$ such that $g_1 h_2 = h_2 g_1$.

Observe that $h_1 h_2 g_2 = h_1 g_1 h_2 = g_2 h_1 h_2$. Therefore, by uniqueness, $h_1 h_2$ = identity map of $\Lambda_2$. Similarly $h_2 h_1$ = identity map of $\Lambda_1$; so $h_1$ and $h_2$ are homeomorphisms. The continuity of $h_1$ in $(g_1, g_2)$ depends on the universal estimates for continuity of fixed points of contractions (see 1.1) and is left to the reader.

The next theorem means that the stable manifolds of a hyperbolic set move only slightly under perturbations.

7.4. THEOREM. *Referring to Theorem 7.3, let* $f$ *be* $C^k$. *Let the stable manifold system of* $\Lambda$ *be of size* $\beta$. *Then the neighborhoods* $V \subset U$ *of* $\Lambda$ *and* $\mathcal{N} \subset C^1(U, M)$ *can be chosen so that if* $g_1, g_2 \in \mathcal{N} \cap C^k(U, M)$, *then the following conditions hold. Let* $h : \Lambda_1 \to \Lambda_2$ *be as in* 7.3; *let* $x \in \Lambda_1$ *and put* $h(x) = y \in \Lambda_2$. $\Lambda_i$ *has a stable manifold system* $\{W_x^i\}_{x \in \Lambda_i}$ *for* $g_i$ *of size* $\geq \beta - \varepsilon$. *Moreover if* $x \in \Lambda_1$ *and* $h(x) = y \in \Lambda_2$, *where* $h : \Lambda_1 \to \Lambda_2$ *is as in* 7.3, *there is a* $C^k$ *diffeomorphism* $\theta_x : W_x^1 \to W_y^2$ *which is* $\varepsilon$-*close to the inclusion* $W_x^1 \to M$ *in* $C^k(W_x^1, M)$. *Moreover* $\theta_x$ *depends continuously on* $(g_1, g_2)$.

PROOF. We assume familiarity with the proof of 3.2. It suffices to prove the theorem with $g_1 = f$. Put $g = g_2$. Let $\mathscr{C}(\Lambda, U)$ be the Banach manifold of continuous maps $h : \Lambda \to U$. Define $\Phi_g : \mathscr{C}(\Lambda, U) \to \mathscr{C}(\Lambda, M)$ by $\Phi_g(h) = ghf^{-1}$. Then $\Phi_g$ has the unique hyperbolic fixed point $h_g : \Lambda \to \Lambda_2$. If $\mathscr{W}_g \subset \mathscr{C}(U, M)$ is the stable manifold for $\Phi_g$, it can be shown, imitating the proof of 3.2, that $W_{y,g}$ = $\{h(x) | h \in \mathscr{W}_g\}$ is the stable manifold through $y = h(x) \in \Lambda_2$ for $g$. Now $\Phi_g$ depends continuously on $g$ respecting the $C^k$ topologies. The stable manifold function of $\Phi_g$ is a $C^k$ map $\mathscr{G}_g : \mathscr{C}(E^s) \to \mathscr{C}(E^u)$, where $T_\Lambda \mathcal{N} = E^s \oplus E^u$ and $\mathscr{C}$ denotes the Banach space of bounded continuous sections. Since $\mathscr{G}_g$ depends $C^k$ continuously on $g$, so does its graph, which is $\mathscr{W}_g$. A $C^k$ diffeomorphism $\theta : \mathscr{W}_f \to \mathscr{W}_g$ is defined by $\theta_g(u, \mathscr{G}_f(u)) = (u, \mathscr{G}_g(u))$. (Here we identify $\mathscr{C}(\Lambda, U)$ with $\mathscr{C}(E^s) \times \mathscr{C}(E^u)$

by exponential coordinates.) Define $\theta_x : W_x \to W_{y,g}$ by $\theta_x(v(x)) = \theta(v)(x)$, for $x \in \Lambda$ and $v \in \mathscr{W}_f$. The details are left to the reader.

REMARK. $\{W_x^f\}_{x \in \Lambda}$ and $\{W_y^g\}_{g \in \Lambda_2}$ are "continuous families" of $C^k$ submanifolds. The proper way to state Theorem 7.4 is to define the concept of a *continuous family* of $C^k$ diffeomorphisms $\{\theta_x : W_x \to W_{h(x)}\}_{x \in \Lambda}$. We leave this task to the reader.

## REFERENCES

1. Abraham and Robbin, *Transversal mappings and flows*, Benjamin, New York, 1967.
2. Arnold and Avez, *Problèmes ergodiques de la mecanique classique*, Gauthier-Villars, Paris, 1966.
3. Anosov, *Ergodic properties of geodesic flows . . .*, Soviet Math. 3 (1962), 1068–1070.
4. Dieudonné, *Foundations of modern analysis*, Academic Press, New York, 1960.
5. Artin and Tate, *Class field theory*, Benjamin, New York, 1967.
6. Hartman, *Ordinary differential equations*, Wiley, New York, 1964, Chapter IX.
7. Hirsch and Pugh, *Stable manifolds for hyperbolic sets*, Bull. Amer. Math. Soc. 75 (1969), 149–152.
8. Holmes, *A formula for the spectral radius of an operator*, Amer. Math. Monthly 75 (1968), 163–166.
9. Lang, *Introduction to differentiable manifolds*, Interscience, New York, 1962.
10. Mather, *Characterization of Anosov Diffeomorphisms*, Nederl. Akad. van Wetensch. Proc. Ser. A, Amsterdam 71 = Indag. Math. 30 (1968), no. 5.
11. Riesz and Nagy, *Leçons d'analyse functionelle*, Budapest, 1953.
12. Smale, *Differentiable dynamical systems*, Bull. Amer. Math. Soc. 73 (1967), 747–817.
13. ———, *The Ω-stability theorem*, these Proceedings, vol. 14.
14. ———, *Stable manifolds for differential equations and diffeomorphisms*, Ann. Scuol. Norm. Sup. Pisa 17 (1963), 97–116.
15. Sternberg, *Local contractions and theorem of Poincaré*, Amer. J. Math. 79 (1957), 809–824.
16. Williams, *One dimensional nonwandering sets*, Topology 6 (1967), 473–487.
17. ———, *Classification of one dimensional attractors*, these Proceedings, vol. 14.

UNIVERSITY OF CALIFORNIA, BERKELEY

# COMMUTING DIFFEOMORPHISMS

NANCY KOPELL [1]

**Introduction.** This paper describes some properties of differentiable actions of noncompact abelian Lie groups and all differentiable actions of these groups on the circle $S$. It is shown that, for almost all contracting local diffeomorphisms $f$ of $R^n$ satisfying $f(0) = 0$, the set of local diffeomorphisms which commute with $f$ is a finite dimensional Lie group. Because of this local phenomenon, there are some properties of globally defined diffeomorphisms which are unstable for single diffeomorphisms, but stable for pairs of commuting diffeomorphisms (see Theorems 1 and 2 below). Other properties that result from the rigidity of commuting diffeomorphisms in the neighborhood a common fixed point are shown in Theorems 7 and 8. This paper also discusses the embedding of diffeomorphisms into differentiable flows and shows that, for almost all diffeomorphisms $f$ of $S$, any differentiable commuting homeomorphism is an iterate of $f$.

Before the theorems can be stated, we need some preliminaries:

A diffeomorphism of a manifold $M$ is a $C^\infty$ homeomorphism of $M$ with a $C^\infty$ inverse. $\text{Diff}(M)$ is the group of all such diffeomorphisms. If $M$ is compact, for any $s \in Z_+$, the uniform $C^s$ topology on $\text{Diff}(M)$ is induced by a distance function we shall write as $d_s( \ , \ )$. ($Z_+$ denotes the positive integers.) $f, g \in \text{Diff}(M)$ are close in this topology if they are pointwise close, and all derivatives of order $\leq s$ are close.

If $f \in \text{Diff}(M)$ and $x \in M$, $x$ is a periodic point of $f$ if there is an integer $n$ such that $f^n x = x$. The smallest such positive integer $n$ is called the period of $x$; $x$ is a fixed point if $n = 1$. $\text{Per}(f)$ is the set of all periodic points of $f$, and $\text{Fix}(f)$ is the set of all fixed points of $f$.

The centralizer of $f \in \text{Diff}(M)$ is $\mathscr{C}(f) = \{g \in \text{Diff}(M): gf = fg\}$. If $g \in \mathscr{C}(f), f$ and $g$ induce a homomorphism $\rho: Z + Z \to \text{Diff}(M)$ defined by $\rho(n, m) = f^n g^m$. In general, let $G$ be any Lie group and $M$ a $C^\infty$ manifold. Then a homomorphism $\rho: G \to \text{Diff}(M)$ such that the induced map $G \times M \to M$ is differentiable is called a differentiable action of $G$ on $M$. If $M$ is compact, the set of such actions may be topologized as follows: Let $K$ be any compact subset of $G$ which generates $G$. If $s \in Z_+$ and $\rho, \bar{\rho}: G \to \text{Diff}(M)$ are differentiable actions, let

$$d_s(\rho, \bar{\rho}) = \sup\{d_s(\rho(g), \bar{\rho}(g)): g \in K\}.$$

Finally, suppose that $f$ is an orientation preserving homeomorphism of $S$. There is a real number, $0 \leq \tau(f) < 1$, called the rotation number of $f$, which is

[1] This research was partially supported by NSF and AFOSR contract number F44620-67-C0029.

defined as follows: Let $\tilde{f}: R^1 \to R^1$ be a lifting of $f$ such that $\tilde{f}(x+1) = \tilde{f}(x) + 1$, $\forall x \in R^1$. Let $y \in R^1$. Then $\tau(f) = \lim_{n \to \infty}(1/n)[\tilde{f}^n(y)] \bmod 1$ exists and is independent of $y$ [11]. It is well known that $\tau(f)$ is rational if and only if $f$ has a periodic point. If $f$ is orientation preserving, all periodic points of $f$ have the same period.

The first few theorems describe differentiable actions of $Z^k \times R^l$ on the circle. (Here $Z^k \times R^l$ means $k$ copies of the integers and $l$ copies of the real line.) It follows from the work of M. Peixoto [6] that there is an open dense set $\mathfrak{A} \subset \text{Diff}(S)$ such that $f \in \mathfrak{A}$ if and only if $f$ has a finite, nonzero number of periodic points, all of which are transversal. ($x \in \text{Per}(f)$ is transversal if $f^n x = x$ and $(f^n)'(x) \neq 1$.) It might be conjectured that $\{(f, g) \in \mathfrak{A} \times \mathfrak{A} : fg = gf\}$ is dense in the set $\{(f, g) \in \text{Diff}(S) \times \text{Diff}(S) : fg = gf\}$. The first two theorems show that this is far from true.

THEOREM 1. *There are commuting diffeomorphisms $f$, $g$ of the circle such that if $\bar{f}, \bar{g}$ are sufficiently close to $f$ and $g$, respectively, and $\bar{f}\bar{g} = \bar{g}\bar{f}$, then both $\bar{f}$ and $\bar{g}$ have isolated nontransversal fixed points.*

THEOREM 2. *There are commuting diffeomorphisms $f$, $g$ of the circle such that $f$ is the identity on an open set, and if $\bar{f}$ and $\bar{g}$ are sufficiently close to $f$ and $g$, and satisfy $\bar{f}\bar{g} = \bar{g}\bar{f}$, then $\bar{f}$ must be the identity on an open set.*

To show Theorem 2, we will first prove:

THEOREM 3.[2] *Let $\beta = \{f \in \text{Diff}(S) : \mathscr{C}(f) = \{f^n : n \in Z\}\}$ and let $\text{Diff}(S)$ be given the topology induced by $d_s(\ ,\ )s \geq 2$. Then $\beta$ is open and dense in $\text{Diff}(S)$.*

However, it is possible to obtain a simple description of almost all differentiable actions of abelian groups on the circle. For simplicity we consider only orientation preserving diffeomorphisms.

DEFINITION. Let $G = Z^k \times R^l$ and let $\rho: G \to \text{Diff}(S)$ be a homomorphism. If $x \in S$, let $G_x = \{a \in G : \rho(a)x = x\}$. $\rho$ is an admissible action if $G$ does not act transitively, and if for every $x \in S$, $\rho_x: G/G_x \to S$ is a homeomorphism onto its image, where $\rho_x(g) = \rho(g)x$. (It will be shown below that abelian actions which are not admissible are always topologically equivalent to actions of subgroups of the rotation group.)

Let $\rho: G \to \text{Diff}(S)$ be an admissible action. Let $\mathfrak{G} = \rho(G)$, $\mathfrak{G}_x = \rho(G_x)$, $\mathcal{O}_x$ = orbit of $x$ and $L_x = \text{Cl}\mathcal{O}_x - \mathcal{O}_x$. (Cl denotes closure.) If $\mathcal{O}_x$ is compact, let $W_x = \mathcal{O}_x$. Otherwise, let $W_x = \{z \in S : L_x = L_z\}$. $W_x$ is the invariant manifold of $x$. It is clear that the $\{W_x\}$ are disjoint and invariant under $G$.

THEOREM 4. *Let $\rho: G \to \text{Diff}(S)$ be admissible, $\mathfrak{G} = \rho(G)$ and $x \in S$. Then*
(i) *Every $g \in \mathfrak{G}$ has a rational rotation number.*
(ii) *$W_x$ is compact if and only if $x \in \bigcap_{g \in \mathfrak{G}} \text{Per}(g)$.*
(iii) *$L_x$ is contained in the union of all compact orbits.*
(iv) *$W_x$ is a finite union of points or intervals.*
(v) *If $y \in W_x$ then $G_x = G_y$.*

---

[2] It has come to my attention that some related theorems have been announced by P. F. Lam [13] for diffeomorphisms of the unit interval.

THEOREM 5. *Let* $\rho : G \to \text{Diff}(S)$ *be an admissible action. Given* $\varepsilon > 0$ *and* $s \in Z_+$ *there is an action* $\tilde{\rho} : G \to \text{Diff}(S)$ *such that* $d_s(\rho, \tilde{\rho}) < \varepsilon$ *and such that* $\tilde{\rho}$ *has only finitely many distinct invariant manifolds. Furthermore, for each* $g \in G$, $\text{Per}(\tilde{\rho}(g))$ *is a finite union of points and intervals.*

The rest of the theorems describe actions on higher dimensional manifolds. The first of these is a theorem about the centralizer of a local diffeomorphism $f$ of $R^n$ such that $f(0) = 0$ and $Df(0)$ is a contraction (i.e. all eigenvalues of $Df(0)$ have absolute value less than 1). By a theorem of Sternberg [8], for almost all such local diffeomorphisms there is a differentiable change of coordinates such that, in the new coordinates, $f$ is linear. Hence we restrict ourselves to proving:

THEOREM 6. *Let* $f : R^n \to R^n$ *be a linear diffeomorphism. Let* $\{\lambda_i\}$ *be the eigenvalues of* $f$, $\underline{\lambda} = \min |\lambda_i|$, $\overline{\lambda} = \max |\lambda_i|$, *and assume that* $0 < \underline{\lambda} \le \overline{\lambda} < 1$. *Let* $m$ *be the least integer such that* $\overline{\lambda}^m < \underline{\lambda}$. *If* $g$ *is a homeomorphism of* $R^n$ *such that* $gf = fg$, *and* $g \in C^m$, *then* $g$ *is a polynomial of degree less than* $m$. *Furthermore, if* $f$ *satisfies the inequalities:*

$$\lambda_i \ne \lambda_1^{\alpha_1} \lambda_2^{\alpha_2} \dots \lambda_n^{\alpha_n}, \quad \text{where } \alpha_i \in Z, \, \alpha_i \ge 0, \, \sum \alpha_i > 1$$

*and* $g$ *is as above, then* $g$ *is linear.*

We also show that this theorem is the sharpest possible. In particular, it is not true unless $Df(0)$ is a contraction.

Finally, we derive the following global implications of this theorem.

THEOREM 7. *There is an open set of diffeomorphisms* $f$ *of any compact* $C^\infty$ *manifold* $M$ *with the property that if* $g_1, g_2 \in \text{Diff}(M)$, $g_i f = fg_i$, $i = 1, 2$, *and* $g_1 | V = g_2 | V$ *for any open set* $V \subset M$, *then* $g_1 \equiv g_2$.

THEOREM 8. *On any two-dimensional manifolds there are commuting vector fields* $F$, $G$ *such that* $F$ *and* $G$ *are collinear on an open set, and if* $\overline{F}$ *and* $\overline{G}$ *are sufficiently close to* $F$ *and* $G$, *and satisfy* $[\overline{F}, \overline{G}] = 0$, *then* $\overline{F}$ *and* $\overline{G}$ *are collinear on an open set.*

This paper is essentially my doctoral dissertation which was written at the University of California, Berkeley. I wish to thank Professor Stephen Smale for suggesting and encouraging this work. I would also like to thank the following people for many stimulating conversations: M. Hirsch, B. Kripke, I. Kupka, R. Moore, R. Palais, J. Palis, M. Shub and R. Solovay.

1. This section contains a few lemmas concerning the behavior of commuting diffeomorphisms of $R^1$ in the neighborhood of a common fixed point.

DEFINITION. Let $M$ be a manifold. A local diffeomorphism $f$ is a diffeomorphism of a neighborhood of a point $x \in M$ onto a neighborhood of $fx$. By a local diffeomorphism at $x$ we shall mean a local diffeomorphism $f$ such that $fx = x$. We are concerned with the behavior of such $f$ only in a neighborhood of $x$, and shall sometimes identify two local diffeomorphisms at $x$ if they agree on some neighborhood of $x$. Finally, if $M$ is one-dimensional, it is convenient to consider a neighborhood of $x$ to be one-sided; that is, neighborhoods are sets of the form $(x - \delta, x]$, $[x, x + \delta)$. All local diffeomorphisms will be orientation preserving.

DEFINITION. Let $U$ be a neighborhood of $x \in R^1$, and let $f, g: U \to R^1$ be local diffeomorphisms such that $fx = gx$. $f$ has contact of order $k$ with $g$ at $x$ if $f^{(j)}(x) = g^{(j)}(x)$ $0 \le j \le k$, and $k$ is the largest such integer. Here $f^{(j)}(x)$ denotes the $j$th derivative of $f$. If $f^{(j)}(x) = g^{(j)}(x)$, $\forall j \in Z_+$, then $f$ has infinite contact with $g$ at $x$.

REMARK 1. Suppose that $f$ is a local diffeomorphism at the origin in $R^1$. It follows from Taylor's formula that if there is a sequence of points $x_n \to 0$ such that $fx_n = x_n$, then $f$ has infinite contact with $I$ at $0$. ($I$ denotes the identity.)

Conversations with B. Kripke have been very helpful in proving the following:

LEMMA 1. (a)[3] Let $f$ be a local diffeomorphism at $0 \in R^1$ and suppose that $f$ is a topological contraction; i.e. if $V$ is a neighborhood of $0$ in the domain of $f$, then $\bigcap_{n \ge 0} f^n(V) = \{0\}$. Suppose that $g$ is a local diffeomorphism at $0$ such that $gf = fg$ and there is $y_0 \ne 0$ such that $gy_0 = y_0$. Then $g = I$.

(b) Let $f, g$ be local diffeomorphisms at $0 \in R^1$ such that $gf = fg$. If $f$ has contact of order $k \le \infty$ with $I$ at $0$, and $f \ne I$ on any neighborhood of $0$, then $g = I$ or $g$ has contact of order $k$ with $I$ at $0$. Furthermore, if $k < \infty$, $g$ is completely determined by the relation $gf = fg$ and the $(k+1)$st coefficient in its Taylor expansion.

PROOF. If $f'(0) \ne 1$, the lemma follows immediately from Sternberg [7]. He shows that there exists a local diffeomorphism $\alpha$ such that $\alpha(0) = 0$ and $L = \alpha^{-1} f \alpha$ is linear. But if $h$ is any local diffeomorphism at $0 \in R^1$ such that $h$ commutes with a linear map $L \ne I$, then $h$ is linear. (This is easy to see: Suppose that $L(x) = \lambda x$, $0 < \lambda < 1$; otherwise consider $L^{-1}$. $hL^n = L^n h$, $\forall n \in Z$ implies $h'(\lambda^n x) = h'(x)$, $\forall n$. But $\lambda^n x \to 0$ as $n \to \infty$, so $h'(x)$ is constant.) Hence, if $f'(0) \ne 1$, there is a unique one-parameter group $f^t = \alpha L^t \alpha^{-1}$ such that $f^1 = f$, where $L^t = \lambda^t x$. Furthermore, any local diffeomorphism $g$ which commutes with $f$ belongs to this group and is completely determined by $g'(0)$. Hence we may assume that $f'(0) = 1$.

(a) Suppose that $f$ is a topological contraction and that $gf = fg$. By the above, $g'(0) = 1$. $g'$ is continuous at $0$, so for any $x \in V$, $\lim_{k \to \infty} g'(f^k x) = 1$. Since $gf = fg$,

$$g'(f^k x) = \frac{(f^k)'(gx)}{(f^k)'(x)} \cdot g'(x) = \prod_{j=0}^{k-1} \frac{f'(f^j gx)}{f'(f^j x)} \cdot g'(x).$$

Let $A(x, y) = \prod_{j=1}^{\infty} (f'(f^j y)/f'(f^j x))$. Then $g$ must satisfy the differential equation $g'(x) = 1/A(x, gx)$. Suppose, for definiteness, that the domain of $f$ contains $[0, \delta]$. We will show that $A(x, y)$ is well defined and continuous on $(0, \delta] \times (0, \delta]$, and that if a solution $g$ satisfies the initial condition $gy_0 = y_0 \ne 0$, then $g = I$.

To show that $A$ is well defined and continuous, it suffices to show that $\sum_{j=0}^{\infty} |f'(f^j y)/f'(f^j x) - 1|$ converges uniformly on compact sets $K \subset (0, \delta] \times (0, \delta]$. Let $K$ be such a compact set, and let

$$k_0 = \sup\{|k| : f^k x < y < f^{k-1} x, (x, y) \in K\}.$$

Let

$$M = \sup_{0 \le x \le \delta} 1/f'(x), \qquad \overline{M} = \sup_{0 \le x \le \delta} |f^{(2)}(x)|.$$

---

[3] G. Reeb has asked me to remark that this lemma implies that Example 2 of his paper [14] corresponds to a foliation of class $C^0$ but not class $C^2$.

Given $\varepsilon > 0$, we will find $n_0$ such that

$$\sum_{j=n_0}^{\infty} \left| \frac{f'(f^j y)}{f'(f^j x)} - 1 \right| < \varepsilon, \quad \forall (x, y) \in K.$$

$$\left| \frac{f'(f^j y)}{f'(f^j x)} - 1 \right| \le M |f'(f^j y) - f'(f^j x)| \le M\overline{M} |f^j y - f^j x|.$$

There is a $k \in Z$ such that $|k| \le k_0$ and $f^k x \le y \le f^{k-1} x$. Assume, for definiteness, that $k > 0$. Then $\forall j \in Z, |f^j y - f^j x| \le |f^{j+k} x - f^j x|$ and

$$\sum_{j=n_0}^{\infty} \left| \frac{f'(f^j y)}{f'(f^j x)} - 1 \right| \le M\overline{M} \sum_{j=n_0}^{\infty} |f^{j+k} x - f^j x|$$

$$\le M\overline{M} \sum_{j=n_0}^{\infty} \sum_{i=1}^{k} |f^{j+i} x - f^{j+i-1} x| \le M\overline{M} \sum_{i=1}^{k} f^{n_0+i} x$$

$$\le k_0 M\overline{M} f^{n_0} x.$$

Let $z_0 = \sup\{x : (x, y) \in K\}$, and choose $n_0$ sufficiently large such that $f^{n_0} z_0 \le \varepsilon k_0 M\overline{M}$.

Now suppose that $g$ is a solution of $g'(x) = 1/A(x, gx) = A(gx, x)$, and that $g(y_0) = y_0 \ne 0$. If $g \ne I$, then $\{y : gy \ne y\}$ is open and nonempty. Hence there is $x_0 \ne 0$ and $\delta' > 0$ such that $gx_0 = x_0$ and $gy \ne y$ for any $y \in (x_0, x_0 + \delta']$. For definiteness, we may assume that $gy < y$, $\forall y \in (x_0, x_0 + \delta']$. Now $g$ and all of its iterates $g^n$ satisfy the differential equation $(g^n)'(x) = A(g^n x, x)$, with initial condition $g^n x_0 = x_0$. That is, $\forall n \in Z_+, g^n y = x_0 + \int_{x_0}^{y} A(g^n t, t) dt$. But, as $n \to \infty$, $g^n y \to x_0$ uniformly in $y \forall y \in [x_0, x_0 + \delta']$. Since $A(x, y)$ is uniformly continuous on compact subsets of $(0, \delta] \times (0, \delta]$, and $A(x, y) > 0 \forall (x, y)$, $\int_{x_0}^{y} A(g^n t, t) dt \to \int_{x_0}^{y} A(x_0, t) dt \ne 0$. Hence we have a contradiction.

(b) Suppose that $f$ has contact of order $1 \le k \le \infty$ with $I$ at 0, and $gf = fg$. We will show that for any $j \le k$, $\lim_{x \to 0} |gx - x|/|x|^j = 0$. This implies that $g$ has contact of order $k' \ge k$; by symmetry, that suffices.

Assume first that there is a neighborhood of the origin which contains no other fixed points. Then $f$ or $f^{-1}$ is a topological contraction; for definiteness, assume that $f$ is a topological contraction. Given $\varepsilon > 0$ and $j \le k$, $j < \infty$, we will find a neighborhood $V$, containing the origin, such that $y \in V$ implies $|gy - y|/|y|^j < \varepsilon$. By part (a), if $g \ne I$, $g$ has no fixed points in a neighborhood of 0. Hence we may assume that $g$ is a topological contraction. Let $x_0$ be in the domain of $f$, and let $D = [fx_0, x_0]$. There is an integer $i$ such that $f^i x < gx < x$, $\forall x \in D$. Since $f$ is monotone, $|gf^n x - f^n x| = |f^n gx - f^n x| \le |f^{n+i} x - f^n x|$, $\forall n \in Z_+$, $\forall x \in D$. Now $f^i$ has contact of order $k$ with $I$ at the origin, so there is a neighborhood $\overline{V}(0)$ such that $y \in \overline{V}$ implies $|f^i y - y|/|y|^j < \varepsilon$. Choose a neighborhood $V \subset \overline{V}$ of the form $V = \{0\} \cup \bigcup_{n > N} f^n(D)$. Then $y \in V$ implies that $|gy - y|/|y|^j = |gf^n x - f^n x|/|f^n x|^j$ for some $n > N$, some $x \in D$. But

$$|gf^n x - f^n x|/|f^n x|^j \le |f^{n+i} x - f^n x|/|f^n x|^j$$
$$= |f^i(f^n x) - f^n x|/|f^n x|^j < \varepsilon.$$

Hence, this case is done.

Now suppose that there is a sequence $x_n \to 0$ such that $fx_n = x_n$. Then $f$ has infinite contact with $I$ at the origin. To show that the same is true of $g$, it suffices to show that there is a sequence $y_n \to 0$ such that $gy_n = y_n$. Suppose not. Then, in some neighborhood of the origin, either $g$ or $g^{-1}$ is a topological contraction. By part (a), $fx_n = x_n \ \forall n$ implies that $f$ is the identity on a neighborhood of 0; but this contradicts the hypothesis.

Finally, suppose that $f$ has contact of order $k < \infty$ with $I$ at 0, $g_i f = fg_i$, $i = 1, 2$, and $g_1^{(k+1)}(0) = g_2^{(k+1)}(0)$. Then $g_1 g_2^{-1}$ commutes with $f$ and has contact of order $\geq k + 1$ with $I$. Since $f$ has contact of order $< k + 1$, we must have $g_1 g_2^{-1} = $ identity.

REMARK 2. Suppose that $f$ is a diffeomorphism of an interval $[x, y]$ onto itself, and that $f|[x, y)$ is a topological contraction onto $\{x\}$, i.e. $\bigcap_{n>0} f^n[x, y) = \{x\}$. Let $x_0 \in (x, y)$, and let $g$ be a diffeomorphism of $[x, y]$ which commutes with $f$. Then $g$ is completely determined by its values on $[fx_0, x_0]$ and the formulas $gf^n = f^n g \forall n \in Z$. In particular, if $g$ is the identity on any neighborhood of $x$ or of $y$, then $g$ is the identity on the entire interval $[x, y]$.

The next lemma is algebraic, and concerns the infinitesimal behavior of commuting diffeomorphisms at a common fixed point.

LEMMA 2. *Let $f$ be a formal power series in one variable such that $f(x) = x + ax^{k+1} + $ higher order terms.*

(a) *Let $g_1, g_2$ also be formal power series such that $g_i(x) = x + b_i x^{k+1} + $ higher order terms $(i = 1, 2)$. If $g_i f = fg_i$, and $b_1 = b_2$, then $g_1 = g_2$.*

(b) *There is a vector field $X$, defined in a neighborhood of 0, and singular at 0, such that the Taylor expansion at 0 of $\exp_1(X)$ is equal to $f$. ($X$ gives rise to a one-parameter group $\{g^t\}$ of local diffeomorphisms at 0; $\exp_1(X)$ denotes $g^1$.)*

PROOF. (a) Equating the coefficients of the $x^i$ in the formula $gf = fg$ yields a recursion formula for all of the coefficients if the first $k$ coefficients of $g$ are those of the identity, and the $(k + 1)$st is given. (The first nontrivial equation is obtained by equating the coefficients of $x^{2k+2}$.)

(b) The following proof is due to I. Kupka. We wish to find a formal power series $F(x)$, and an arc of formal power series $g^t(x)$, such that $g^1 = f$, $g^0 = $ identity and $F(g(x, t)) = (\partial/\partial t)g(x, t)$, where $g(x, t) = g^t(x)$. If $\bar{F}$ is any $C^\infty$ function whose Taylor expansion at 0 is $F(x)$, the required vector field is $X = \bar{F}\partial/\partial x$.

$$g(x, t) = \sum_{n=1}^{\infty} a_n(t)x^n \text{ and } F(x) = \sum_{n=1}^{\infty} b_n x^n.$$

We wish to determine the coefficients $a_n(t)$ and $b_n$ inductively. Suppose that the Taylor expansion of $f$ is $\sum_{n=1}^{\infty} c_n x^n$. Comparing coefficients of $x$ in the equation $F(g(x, t)) = (\partial/\partial t)g(x, t)$, we get $a_1'(t) = b_1 a_1(t)$. From the boundary conditions $a_1(0) = 1$, $a_1(1) = c_1$, we see that $a_1(t) = e^{n_1 t}$, $b_1 = \log c_1$.

Suppose, by induction, that $a_j(t)$ and $b_j$ are uniquely determined for $j \leq n - 1$. Comparing coefficients of $x^n$, we get

$$a_n'(t) = b_1 a_n(t) + b_n[a_1(t)]^n + Q_n(b_1, ..., b_{n-1}, a_1(t), ..., a_{n-1}(t))$$

where $Q_n$ is a polynomial, and $[a_1(t)]^n = c_1^{nt}$. The general solution to this equation is

$$a_n(t) = c_1^t [d_n + P_n(t) + b_n(c_1^{(n-1)t} - 1)/b_1(n-1)]$$

where $d_n$ is an arbitrary constant, and $P_n$ is a function of $b_1, ..., b_{n-1}, a_1(t), ..., a_{n-1}(t)$. From the boundary conditions $a_n(0) = 0$, $a_n(1) = c_n$, we get that $a_n(t)$ and $b_n$ are both uniquely determined.

COROLLARY 1. *Let $f$ and $\bar{f}$ be local diffeomorphisms at $0 \in R^1$ with the same Taylor expansions at 0. Suppose that $f, \bar{f}$ have contact of order $k$ with the identity at 0. If $g, \bar{g}$ are local diffeomorphisms at 0 such that $gf = fg$, $\bar{g}\bar{f} = \bar{f}\bar{g}$, and $g^{(k+1)}(0) = \bar{g}^{(k+1)}(0)$ then the Taylor expansions of $g$ and $\bar{g}$ agree.*

PROOF. By Lemma 1, the expansions agree up to order $k$. By Lemma 2, the rest of the expansions agree.

2. In this section we prove Theorems 1–3. Theorem 1 can be proved by the following simple example: Let $x_0 \in (0, 1)$ and $f \in \text{Diff}([0, 1])$ satisfy
   (a) $f(x) > x$ for $0 < x < x_0$; $f(x) < x$ for $x_0 < x < 1$.
   (b) $f'(0) \neq 1 \neq f'(1)$.
   (c) $(f - I)^{(j)}(x_0) = 0$, $\forall j \in Z_+$.
Let $g \in \text{Diff}([0, 1])$ satisfy (a), (b) and (c) and also
   (d) $\dfrac{\log g'(0)}{\log g'(1)} \neq \dfrac{\log f'(0)}{\log f'(1)}$.
(Such a $g$ exists; e.g. let $g|[0, x_0] = f|[0, x_0]$, $g|[x_0, 1) = f^2|[x_0, 1]$.) Let $U(f)$ and $V(g)$ be neighborhoods in $\text{Diff}([0, 1])$ such that $\bar{f} \in U$, $\bar{g} \in V$ implies $\bar{f}, \bar{g}$ satisfy (b) and (d). We will show that if $\bar{f} \in U$, $\bar{g} \in V$, there is a $y \in (0, 1)$ such that $y \in \text{Fix}(\bar{f}) \cap \text{Fix}(\bar{g})$ and

$$(\bar{f} - I)^{(j)}(y) = 0 = (\bar{g} - I)^{(j)}(y) \quad \forall j \in Z_+.$$

First we show that there is a point $z \in (0, 1) \cap \text{Fix}(\bar{f})$ such that $\bar{f}$ has infinite contact with the identity at $z$. Suppose not; then by Remark 1, $\text{Fix}(\bar{f})$ is finite. Suppose that $\text{Fix}(\bar{f}) = \{0, x_1, x_2, ..., x_n = 1\}$. Consider $\bar{g}[0, x_1]$. By Lemma 1 (b), $\bar{g}[0, x_1]$ is completely determined by $\bar{g}'(0)$. By hypothesis, $\bar{f}$ has $k$th order contact with the identity at $x_1$, where $k < \infty$. $\bar{g}[0, x_1]$ determines $\bar{g}^{(k+1)}(x_1)$, which determines $\bar{g}|[x_1, x_2]$. By induction, $\bar{g}$ is completely determined by $\bar{g}'(0)$. It is easy to see that if $\bar{g}'(0) = [\bar{f}'(0)]^t$ for some $t \in R^1$, then $\bar{g}'(x_1) = [\bar{f}'(x_1)]^t$ if $k = 0$, or $\bar{g}^{(k+1)}(x_1) = t\bar{f}^{(k+1)}(x_1)$ if $k \geq 1$. Hence we find, by induction, that $\bar{g}'(1) = [\bar{f}'(1)]^t$. But this violates condition (d).

Now let $y \in$ boundary of $\{z \in \text{Fix}(\bar{f}) : (\bar{f} - I)^{(j)}(z) = 0, \forall j \in Z_+\}$. Then there is a one-sided neighborhood of $y$ on which $\bar{f}$ is not the identity. By Lemma 1 (a) and (b), $y \in \text{Fix}(\bar{g})$ and $(\bar{g} - I)^{(j)}(y) = 0, \forall j \in Z$.

REMARK. 3. The above example shows that the existence of a common nontransversal fixed point is a stable phenomenon; the uniqueness of such a point is not.

We now go to the proof of Theorem 3. Let us first consider the subset $\mathscr{S} \subset \text{Diff}([0, 1])$, defined by $g \in \mathscr{S}$ if and only if $g(x) < x$ for all $0 < x < 1$ and $g'(0) \neq 1 \neq g'(1)$.

LEMMA 3. *If $f \in \mathscr{S}$, $\mathscr{C}(f)$ is either infinite cyclic or isomorphic to $R^1$.*

PROOF. By Sternberg [7], $f$ imbeds locally in a $C^\infty$ one-parameter group defined in a neighborhood of $x = 1$. By commutativity with $f$, this flow extends to a differentiable flow $\{f^t\}$ on $(0, 1]$. Furthermore, if $g \in \text{Diff}([0, 1])$, and $gf = fg$, then $g|(0, 1]$ belongs to this flow. Hence, $\mathscr{C}(f) = \{f^t : f^t$ extends to a differentiable function on $[0, 1]\}$. Let $\mathscr{T} = \{t \in R^1 : f^t \in \mathscr{C}(f)\}$. To prove Lemma 3, it suffices to show that $\mathscr{T}$ is a closed subgroup of $R^1$. Suppose that $t_n \in \mathscr{T}$ and $t_n \to t_0$. We must show that $f^{t_0}$ extends to a differentiable function on $[0, 1]$.

Since $f'(0) \neq 1$, $f$ also embeds a unique one-parameter group $\{\tilde{f}^t\}$ of diffeomorphisms, defined on $[0, 1)$. By uniqueness, $f^t$ extends to a diffeomorphism of $[0, 1]$ if and only if $f^t|(0, 1) = \tilde{f}^t|(0, 1)$, or equivalently, if and only if $f^t|[fx_0, x_0] = \tilde{f}^t|[fx_0, x_0]$ for any $x_0 \in (0, 1)$. By hypothesis, $t_n \in \mathscr{T}$, and so $f^{t_n}|[fx_0, x_0] = \tilde{f}^{t_n}|[fx_0, x_0]$. Since $[fx_0, x_0]$ is compact, given $\varepsilon > 0$, there is $N \in Z_+$ such that $n > N$ implies that $d_s(f^{t_n}|[fx_0, x_0], f^{t_0}|[fx_0, x_0]) < \varepsilon$, and similarly for $\tilde{f}^{t_n}, \tilde{f}^{t_0}$. Hence $f^{t_0}|[fx_0, x_0] = \tilde{f}^{t_0}|[fx_0, x_0]$ and so $f^{t_0} \in \mathscr{C}(f)$. |

Suppose that $f \in \mathscr{S}$ and $\{f^t\}$ is the unique differentiable one-parameter group defined on $(0, 1]$ such that $f^1 = f$. For each $t$, let $T_t(f) = \lim_{x \to 0}[\sup\{(f^t)'(y) : 0 < y \leq x\} - \inf\{(f^t)'(y) : 0 < y \leq x\}]$. Let $B(f) = \inf\{t > 0 : T_t(f) = 0\}$. Let $\text{Diff}([0, 1])$ and $\text{Diff}((0, 1])$ be given the compact-open $C^2$ topology, and regard $B$ as a map $B : \mathscr{S} \to R^1$.

LEMMA 4. *$B$ is an upper semicontinuous function.*

Before we prove Lemma 4, we show that it implies that $\{f \in \mathscr{S} : \mathscr{C}(f) = \{f^n : n \in Z\}\}$ is open in $\mathscr{S}$. Notice that $f^t$ extends to a diffeomorphism of $[0, 1]$ if and only if $\lim_{x \to 0}(f^t)'(x)$ exists, i.e. if and only if $T_t(f) = 0$. Hence, $\{t > 0 : T_t(f) = 0\} = \{t > 0 : \exists h \in \mathscr{C}(f)$ such that $h'(0) = [f'(0)]^t\}$. Thus, $f \in \mathscr{S}$ commutes only with its integral powers if and only if $B(f) = 1$. To show that $\{f \in \mathscr{S} : \mathscr{C}(f) = \{f^n : n \in Z\}\}$ is open in $\mathscr{S}$, we must find a neighborhood $V(f)$ such that if $g \in V$, then $B(g) = 1$. Choose $0 < \varepsilon < \frac{1}{2}$. By Lemma 4 there is a neighborhood $V$ such that if $g \in V$, $B(g) > 1 - \varepsilon > \frac{1}{2}$. Suppose that $g \in V$ and $\mathscr{C}(g) \not\subset \{g^n : n \in Z\}$. By Lemma 3 $\mathscr{C}(g)$ is cyclic or isomorphic to $R^1$. In either case, there is an $h \in \mathscr{C}(g)$ such that $h^k = g$, $k > 1$. But then $B(g) \leq 1/k \leq \frac{1}{2}$, which is a contradiction.

PROOF OF LEMMA 4. Given $\varepsilon > 0$, we must find a neighborhood $V(f)$ such that if $g \in V$, then $B(g) > B(f) - \varepsilon$. If $B(f) - \varepsilon \leq 0$, there is nothing to prove. Otherwise, let $t_1 = B(f) - \varepsilon$, $t_0 = \frac{1}{2}t_1$. Then for each $t \in [t_0, t_1]$, $T_t(f) \neq 0$. Suppose that, for each $t \in [t_0, t_1]$, we can choose a neighborhood $V_t(f)$ such that, if $g \in V_t(f)$, then $T_t(g) \neq 0$. Then, by the compactness of $[t_0, t_1]$, there is a neighborhood $V(f)$ such that, if $g \in V(f)$, $T_t(g) \neq 0$, $\forall t \in [t_0, t_1]$; this is the required $V$. For suppose that $g \in V$, and there is some $t \leq B(f) - \varepsilon$ such that $T_t(g) = 0$. There is an integer $k > 0$ such that $kt \in [t_0, t_1]$. But $\{t > 0 : T_t(g) = 0\}$ is closed under addition, so $T_{kt}(g) = 0$, which is impossible.

We will now find the neighborhood $V_t(f)$ for fixed $t \in [t_0, t_1]$. By Sternberg [7], we may assume that, on some initial subinterval $J = [0, a]$, $f$ is a linear map $f(x) = \lambda x$. Let $x_0 \in (0, a)$, and $D = [fx_0, x_0]$. Suppose that $T_t(f) = \delta \neq 0$.

Since $f|J$ is linear, and $f^t$ commutes with $f$, $(f^t)'(\lambda x) = (f^t)'(x)$, $\forall x \in J$. Therefore $T_t(f) = \sup\{(f^t)'(x): x \in D\} - \inf\{(f^t)'(x): x \in D\}$. Suppose, now, that $g$ is close to $f$ in the $C^2$ topology. Again, by Sternberg [7], there is a diffeomorphism $\alpha:[0, 1] \to [0, 1]$ such that $\alpha'(0) = 1$, $\alpha^{-1}g\alpha$ is linear on $J$, and $\alpha$ is close to the identity in the $C^1$ topology. More specifically, there is a neighborhood $W_t(f)$ such that if $g \in W_t(f)$, there is an $\alpha \in \text{Diff}([0, 1])$ such that $\alpha^{-1}g\alpha$ is linear on $J$ and $d_1(\alpha^{-1}f^t\alpha|D, f^t|D) < \delta/8$. Choose a neighborhood $V_t(f) \subseteq W_t$ such that if $g \in V_t$ and $\alpha$ is chosen as before, then $d_1(\alpha^{-1}g^t\alpha D, \alpha^{-1}f^t\alpha|D) < \delta/8$, where $g^t$ is the unique diffeomorphism of $(0, 1]$ such that $(g^t)'(1) = (g'(1))^t$. Suppose that $T_t(g) = 0$, i.e. that $g^t$ extends to a diffeomorphism of $[0, 1]$. Then $\alpha^{-1}g^t\alpha$ is also a diffeomorphism of $[0, 1]$, so $T_t(\alpha^{-1}g\alpha) = 0$. Now $\sup\{(\alpha^{-1}g^t\alpha)'(x): x \in D\} \geq \sup\{(f^t)'(x): x \in D\} - \delta/4$ and $\inf\{(\alpha^{-1}g^t\alpha)'(x): x \in D\} \leq \inf\{(f^t)'(x): x \in D\} + \delta/4$. But $\alpha^{-1}g\alpha$ is a linear map $x \to \sigma x$, and $\alpha^{-1}g^t\alpha$ commutes with $\alpha^{-1}g\alpha$, so $(\alpha^{-1}g^t\alpha)'(\sigma^n x) = (\alpha^{-1}g^t\alpha)'(x)$, $\forall x \in J$, $n \in Z$. Therefore $T_t(\alpha^{-1}g\alpha) \neq 0$, which is a contradiction.

LEMMA 5. *If $\mathscr{S}$ is given the $C^s$ topology, any $s \geq 1$, then $\{f \in \mathscr{S}:\mathscr{C}(f) = \{f^n: n \in Z\}\}$ is dense in $\mathscr{S}$.*

PROOF. Let $f \in \mathscr{S}$. Given $\varepsilon > 0$ and $s \in Z_+$, we will find a $g \in \mathscr{S}$ such that $d_s(f, g) < \varepsilon$ and $\mathscr{C}(g) = \{g^n: n \in Z\}$.

We may again assume that on some initial subinterval $J = [0, a]$, that $f$ is the linear map $f(x) = \lambda x$, $0 < \lambda < 1$. Let $x_0 \in (0, a)$, $D = [fx_0, x_0]$. Let $J'$ be a final subinterval, $J' = [b, 1]$, such that $J \cap J' = \varnothing$. Let $h \in \text{Diff}([0, 1])$ satisfy $h|J \cup J' = f|J \cup J'$, $h$ embeds in a $C^\infty$ one-parameter flow $\{h^t\}$. [Such an $h$ exists: $f|[0, 1)$ embeds in a unique differentiable flow which defines a vector field $X_0$ on $[0, 1)$. Similarly, there is a vector field $X_1$ on $(0, 1]$ such that $f = \exp_1(X_1)$. Let $X$ be any vector field such that $X|J = X_0|J$, $X|J' = X_1|J'$, and $X(x) \neq 0$ for $x \neq 0, 1$. Let $h = \exp_1(X)$.] Let $\psi:[0, 1] \to [0, 1]$ be defined by $\psi|J' = $ identity, $f = \psi^{-1}h\psi$. It can be verified that $\psi$ is a homeomorphism, and $\psi|(0, 1]$ is a diffeomorphism.

We will now alter the homeomorphism $\psi$ on $J$. For any $\delta > 0$, there is a diffeomorphism $\bar{\beta}: D \to \psi(D)$ such that:
  (i) $\bar{\beta}$ has infinite contact with $\psi$ at $x_0$ and $fx_0$.
  (ii) There is no $0 < t < 1$ such that $\bar{\beta}(\lambda^t x) \equiv \lambda^t \bar{\beta}(x)$, $\forall x \in D$.
  (iii) $d_s(\bar{\beta}|D, \psi|D) < \delta$.
Define $\beta:[0, 1] \to [0, 1]$ such that
  (i) $\beta|[x_0, 1] = \psi|[x_0, 1]$.
  (ii) $\beta|D = \bar{\beta}|D$.
  (iii) $\beta h^n(x) = h^n\beta(x)$, $\forall n \in Z_+$, $\forall x \in D$; $\beta(0) = 0$.
Now let $g = \beta^{-1}h\beta$. For suitable $\delta$, this $g$ has the required properties. First of all, $g \in \text{Diff}([0, 1])$. To prove this, it suffices to show that $\lim_{x \to 0} g'(x)$ exists. But in a neighborhood of $0$, $g = h$. Secondly, $\mathscr{C}(g) = \{g^n: n \in Z\}$. For suppose that $\{g^t\}$, $\{f^t\}$ are the unique differentiable one-parameter groups, defined on $(0, 1]$, such that $g^1 = g$, $f^1 = f$. By uniqueness, $g^t = \beta^{-1}h^t\beta$. Suppose that there is some $0 < t < 1$ such that $g^t$ extends to a diffeomorphism of $[0, 1]$. $h|J$ is linear, since $h|J = f|J$, and $\beta^{-1}h\beta$ is linear on $[0, \min(x_0, \beta^{-1}x_0)]$. Since $\beta^{-1}h^t\beta$ commutes with $\beta^{-1}h\beta$, and is differentiable, $\beta^{-1}h^t\beta|[0, \min(x_0, \beta^{-1}x_0)]$ is linear. Similarly,

$h^t|J$ is linear and, on a suitable neighborhood of 0, $h^t = \beta^{-1}h^t\beta$, where $h^t x = \lambda^t x$ for any $x$ sufficiently small. But $\beta$ was constructed such that, on any domain of the form $[h^k x_0, h^{k-1}x_0]$, there is no $0 < t < 1$ such that $\beta(\lambda^t x) \equiv \lambda^t \beta(x)$, $\forall x \in [h^k x_0, h^{k-1}x_0]$. This is a contradiction. It remains to be shown that $g$ can be constructed such that $d_s(f, g) < \varepsilon$. Notice that, in the above construction, $g$ differs from $f$ only on the interval $[\min(x_0, \psi^{-1}x_0), \max(\psi^{-1}f^{-1}x_0, f^{-1}x_0)]$. By choosing $\delta$ sufficiently small, we can ensure that $d_s(f, g) < \varepsilon$.

We now go back to Theorem 3.

PROOF OF THEOREM 3. Let Diff($S$) be given the $C^s$ topology, $s \geq 1$. Let $\mathfrak{A} = \{f \in \text{Diff}(S): \text{Per}(f) \text{ is finite and nonempty, and each } x \in \text{Per}(f) \text{ is transversal}\}$. It follows from Peixoto [6] that $\mathfrak{A}$ is open and dense in Diff($S$). Hence, it suffices to show that $\mathfrak{A} \cap \beta$ is open and dense in $\mathfrak{A}$.

Suppose that $f \in \mathfrak{A}$ and $\text{Per}(f) = \{p_0, p_1, \ldots, p_m\}$ where $(p_i, p_{i+1}) \cap (p_j, p_{j+1}) = \varnothing$, $i \neq j$. Let $F = f^n$ be the first iterate of $f$ such that $\tau(F) = 0$. By making a small perturbation in $f$, we may assume that $F'(p_i) \neq F'(p_j)$ unless $\exists k \in Z$ such that $f^k p_i = p_j$. By Lemma 5, one can choose $\tilde{F}_0 : [p_0, p_1] \to [p_0, p_1]$ arbitrarily close to $F|[p_0, p_1]$ such that $\tilde{F}_0(p_i) = F'(p_i)$, $i = 0, 1$, and $\mathscr{C}(\tilde{F}_0) = \{\tilde{F}_0^n : n \in Z\}$. Let $\tilde{f}|[p_0, p_1] = f^{1-n}\tilde{F}_0$ and $\tilde{f}|S - [p_0, p_1] = f|S - [p_0, p_1]$. Let $\tilde{F} = \tilde{f}^n$. We will show that $\mathscr{C}(\tilde{f}) = \{\tilde{f}^n : n \in Z\}$, i.e. that $\tilde{f} \in \mathfrak{A} \cap \beta$.

First notice that for any $h \in \mathfrak{A}$, $g \in \mathscr{C}(h)$, $g$ is completely determined by $g(p_0)$ and $g'(p_0)$. For let $H = h^n$ be the first iterate of $h$ such that $\tau(H) = 0$. If $g_1, g_2 \in \mathscr{C}(h)$, $g_1(p_0) = g_2(p_0)$, and $g_1'(p_0) = g_2'(p_0)$, then $g_1 g_2^{-1} \in \mathscr{C}(h)$, $p_0 \in \text{Fix}(g_1 g_2^{-1})$ and $(g_1 g_2^{-1})'(p_0) = 1$. It follows immediately from Lemma 1 (b) that $g_1 g_2^{-1} = I$.

It is easily checked that if $g \in \mathscr{C}(\tilde{f})$ then $gp_0 \in \text{Per}(\tilde{f}) = \text{Per}(f)$, and $\tilde{F}'(p_0) = \tilde{F}'(gp_0)$. Hence $gp_0 = \tilde{f}^k p_0$ for some $k \in Z$. Let $\mathscr{C}_k = \{g \in \mathscr{C}(\tilde{f}) : gp_0 = \tilde{f}^k p_0\}$. It suffices to show that $\mathscr{C}_k = \{\tilde{f}^k \tilde{F}^i : i \in Z\}$, i.e. that $g \in \mathscr{C}_k \Rightarrow g'(p_0) = (\tilde{f}^k)'(p_0)(\tilde{F}^i)'(p_0)$ for some $i \in Z$. But $g\tilde{f}^{-k}|[p_0, p_1] \in \mathscr{C}(\tilde{F}_0) \Rightarrow (g\tilde{f}^{-k})'(p_0) = (\tilde{F}^i)'(p_0)$ for some $i \in Z$. Hence we are done.

The proof that $\mathfrak{A} \cap \beta$ is open in $\mathfrak{A}$ uses similar arguments and Lemma 4.

REMARK 4. The differentiability of the elements of $\mathscr{C}(f)$ is crucial to the above discussion. For any $f \in \mathscr{S}$, there are uncountably many homeomorphisms $g : [0, 1] \to [0, 1]$ such that $gf = fg$. In particular, $f$ always embeds in uncountably many one-parameter groups of homeomorphisms. (cf. Fine and Schweigert [3].)

REMARK 5. Given $f \in \mathscr{S}$, and $\varepsilon > 0$, it is possible to find a $g \in \text{Diff}([0, 1])$ such that $d_0(f, g) < \varepsilon$, and $g$ embeds in a $C^\infty$ flow. However, in order to do this, it is necessary to make a large change in the derivative of $f$ in the vicinity of one of the fixed points.

We are now ready to prove Theorem 2, i.e. to show the existence of a pair of commuting diffeomorphisms $f, g$ such that Fix($f$) contains an interval, and such that if $\tilde{f}$ and $\tilde{g}$ are sufficiently close to $f$ and $g$, and $\tilde{f}\tilde{g} = \tilde{g}\tilde{f}$, then Fix($f$) also contains an interval. First, choose $g \in \mathscr{S} \subset \text{Diff}([0, 1])$ such that $\mathscr{C}(g) = \{g^n : n \in Z\}$ and let $f = $ identity. There is a neighborhood $V(g) \subset \text{Diff}([0, 1])$ such that if $h \in V(g)$, then $\mathscr{C}(h)$ is generated by $h$. We can choose $\varepsilon > 0$ such that $g'(0) + \varepsilon < 1$, and choose $V(g)$ sufficiently small such that, if $h \in V$ then $h'(0) < g'(0) + \varepsilon$. Let $W(f)$

be any neighborhood such that if $\gamma \in W$, then $g'(0) + \varepsilon < \gamma'(0) < (g'(0) + \varepsilon)^{-1}$. If $h \in V(g)$, the only element $\gamma$ of $W(f)$ which satisfies $\gamma h = h\gamma$ is $\gamma =$ identity. This example may easily be extended to a pair of commuting diffeomorphisms of $S$ such that $f \not\equiv$ identity.

REMARK 6. Theorems 1 and 2 follow from the fact that commuting diffeomorphisms $f, g$ of an interval are forced to obey an algebraic relationship $g = f^t$ for some $t$. It follows from this that any perturbations of $f$ and $g$ which preserve commutativity cannot be local. In particular, the notion of a "generic fixed point" is no longer a local one.

3. This section contains the proofs of Theorems 4 and 5 and a discussion of inadmissible actions. First we need a

DEFINITION. Let $G = Z^k \times R^l$ and let $K$ be a subset of $G$ of the form $K = \{a_1, ..., a_k, tb_1, ..., tb_l\}$, where $b_1, ..., b_l$ are $l$ independent vectors in $R^l$, $0 \le t \le 1$, and $a_1, ..., a_k$ generate a complementary discrete subgroup. We shall call this $K$ a special compact generating subset. If $\rho: Z^k \times R^l \to \text{Diff}(M)$, $\mathcal{K} = \rho(K)$ is a set of generators of $\mathfrak{G} = \rho(G)$. We shall refer to $\rho(t_0 b_i)$ as a continuous generator and $\rho(a_i)$ as a discrete generator.

PROOF OF THEOREM 4. (i) Suppose that there is a $b \in G$ such that $\tau(\rho(b))$ is irrational. Let $g = \rho(b)$. For any $x \in S$, $\{g^n x : n \in Z\}$ is dense in $S$ [11]. If $G$ does not act transitively, $\rho_x : G/G_x \to S$ is not a homeomorphism onto its image.

(ii) A compact orbit under the action of $G$ must be a finite set of points. Hence, if $\mathcal{O}_x$ is compact, $x \in \bigcap_{g \in \mathfrak{G}} \text{Per}(g)$. Conversely, let $K$ be a special compact generating subset of $G$. If $x \in \bigcap_{g \in \mathfrak{G}} \text{Per}(g)$, then $x$ is fixed by every $g$ such that $\tau(g) = 0$. In particular, if $g \in \rho(K)$ is a continuous generator, then $\tau(g) = 0$. For suppose that $g$ has a periodic point of period $n > 1$, and let $\{g^t\}$ be a one-parameter subgroup of $\mathfrak{G}$ containing $g$. Since $x$ is not fixed and $g^n x = x$, $\{g^t\}$ must act transitively on $S$, which is not allowed. Hence, it suffices to consider the orbit of $x$ under the subgroup generated by the discrete generators $f_1, ..., f_k$. By hypothesis, $x \in \text{Per}(f_i)$ and $f_i f_j = f_j f_i$, $1 \le i, j \le k$. This implies that $\mathcal{O}_x$ is finite.

Statements (iii), (iv) and (v) follow trivially if $W_x$ is compact; hence we may assume that $W_x$ is not compact. By (ii), there is a $f \in \mathfrak{G}$ such that $\tau(f) = 0$ and $x \notin \text{Fix}(f)$. Let $y_0 = \lim_{n \to \infty} f^n x$, $z_0 = \lim_{n \to \infty} f^{-n} x$. Since $f$ is monotone, and has a fixed point, these limits exist. Assume, for definiteness, that $y_0 < x < z_0$. (By $(y_0, z_0)$ we mean an interval starting at $y_0$ and moving clockwise to $z_0$; by $y_0 < x < z_0$ we mean $x \in (y_0, z_0)$.) We will show that if $h \in \mathfrak{G}$ and $\tau(h) = 0$, then $x_0, y_0 \in \text{Fix}(h)$ and, furthermore, that if $y \in (y_0, z_0)$ and $h(x) \ne x$, then $h(y) \ne y$. This will finish the theorem, for it shows that $y_0, z_0 \in \bigcap_{g \in \mathfrak{G}} \text{Per}(g)$, and that $L_x = \mathcal{O}_{y_0} \cup \mathcal{O}_{z_0}$. Furthermore, if $y \in g(y_0, z_0)$ for some $g \in \mathfrak{G}$, then $\lim_{n \to \infty} f^n y = g y_0$, $\lim_{n \to \infty} f^{-n} y = g z_0$, and so $L_x = L_y$. Hence $W_x = \mathfrak{G}(y_0, z_0) = \bigcup_{g \in \mathfrak{G}} g(y_0, z_0)$. Since $y_0$ and $z_0$ have compact orbits, $W_x$ is a finite union of intervals. Finally, if $y \in g(y_0, z_0)$, then by the above, $G_y = G_{gx}$. But $G_{gx} = G_x$ since $G$ is abelian.

Now suppose that $\tau(h) = 0$. If $h(y_0) \ne y_0$, then $y_1 = \lim_{n \to \infty} h^n y_0 \in \text{Fix}(h)$. Also, since $y_0 \in \text{Fix}(f)$, $h^n y_0 \in \text{Fix}(f)$, $\forall n$, and so $y_1 \in \text{Fix}(f)$. Hence $f$ and $h$ each induces a local diffeomorphism at $y_1$, defined on some half-open interval $V$ containing $y_0$

in its interior, and such that $h|V$ is a topological contraction. Lemma 1 (a) implies that $f|V = I$. But this is not true; therefore, $hy_0 = y_0$ and similarly, $hz_0 = z_0$. Now $f|[y_0, z_0)$ is a topological contraction. Again by Lemma 1 (a), if there is a $y \in (y_0, z_0)$ such that $hy = y$, then $h|[y_0, z_0] =$ identity.∎

Before Theorem 5 can be proved, we need

LEMMA 6. *Let $f, g$ be homeomorphisms of $S$ such that $fg = gf$. Then $\tau(fg) = (\tau(f) + \tau(g))$ mod 1.*

PROOF. First suppose that $\tau(g)$ is rational. Let $\alpha: R^1 \to S$ be a covering map such that $\alpha(x + 1) = \alpha(x)$. Let $\tilde{f}$ and $\tilde{g}$ be liftings of $f$ and $g$ to $R^1$; (i.e. $\tilde{f}$ and $\tilde{g}$ satisfy $\alpha\tilde{f} = f\alpha$, $\alpha\tilde{g} = g\alpha$) and choose $\tilde{f}, \tilde{g}$ such that $0 \leq f(0), g(0) < 1$. It is easy to check that $\tilde{f}\tilde{g}$ and $\tilde{g}\tilde{f}$ are liftings of $fg$, and $\tilde{f}\tilde{g} = \tilde{g}\tilde{f}$. Also, if $(fg)^{\sim}$ is any lifting of $fg$, then $(fg)^{\sim}(x) = \tilde{f}\tilde{g}(x)$ mod. 1. Hence $(\tilde{f}\tilde{g})^n(x) = \tilde{f}^n\tilde{g}^n(x)$ mod 1 $\forall n \in Z$. Now $\tau(fg) = \lim_{n \to \infty} 1/n(\tilde{f}\tilde{g})^n(x)$ mod 1 for any $x \in S$. Suppose that $\tau(g) = k/m$, where $k$ and $m$ are relatively prime, and $x_0 \in \mathrm{Per}(g)$. It is easy to see that the period of $x_0$ is $m$. Hence $g^m x_0 = x_0 + k$, and $g^{mn} x_0 = x_0 + nk$. Then

$$\tau(fg) = \lim_{n \to \infty} 1/mn(f^{mn}(x_0) + nk) \bmod 1 = (\tau(f) + n/k) \bmod 1.$$

Now suppose that $\tau(f)$ and $\tau(g)$ are irrational. There are homeomorphisms $\beta, \bar{\beta}$ such that $\beta f \beta^{-1}$ and $\bar{\beta} g \bar{\beta}^{-1}$ are rotations of the circle through angles of $2\pi\tau(f)$ and $2\pi\tau(g)$ [2, p. 414]. It follows from the fact that $fg = gf$ that one can choose $\beta = \bar{\beta}$. Clearly, $\tau(\beta fg\beta^{-1}) = \tau(\beta f\beta^{-1}) + \tau(\beta g\beta^{-1})$ mod 1. Since conjugation by a homeomorphism does not change rotation number, the formula is proved.

LEMMA 7. *Suppose that $\rho: G \to \mathrm{Diff}(S)$ is admissible. There is $h \in \mathfrak{G}$ with the property that, for any $f \in \mathfrak{G}$, there is an integer $p$ such that $\tau(h^p) = \tau(f)$.*

PROOF. Consider $RN = \{m/n : \exists f \in \mathfrak{G}$ such that $\tau(f) = m/n$, $m, n$ relatively prime$\}$. $RN$ is a finite set. For if $f_1, ..., f_k$ are discrete generators of $\mathfrak{G}$, then by Lemma 6, $RN$ is an additive group (mod 1) generated by $\tau(f_i)$, $1 \leq i \leq k$. Let $N = \max\{n : m/n \in RN\}$, $M = \min\{m > 0 : m/n \in RN\}$. Let $h \in \mathfrak{G}$ be any diffeomorphism such that $\tau(h) = m/n$. It is easy to check that this is the required $h$.∎

REMARK 7. We shall regard $S$ as the unit interval $[0, 1]$ with $0$ and $1$ identified. The uniform $C^s$ topology on $\mathrm{Diff}(S)$ is given by the distance function

$$d_s(f, g) = \sup\{|f^{(j)}(x) - g^{(j)}(x)| : x \in S, 0 \leq j \leq s\}.$$

LEMMA 8. *Let $\rho: G \to \mathrm{Diff}(S)$ be admissible. Let $f_1, f_2 \in \mathfrak{G}$ such that $\tau(f_1) = \tau(f_2)$. Let $\{g^t : 0 \leq t \leq 1\}$ be a compact subset of a one-parameter subgroup of $\mathfrak{G}$. Let $\varepsilon > 0$ and $s \in Z_+$. Then there are at most finitely many invariant manifolds $W_x$ such that*

(i) $\sup\{|f_1^{(j)}(y) - f_2^{(j)}(y)| : y \in W_x, 0 \leq j \leq s\} \geq \varepsilon$

*or*

(ii) $\sup\{|(g^t)^{(j)}(y) - I^{(j)}(y)| : y \in W_x, 0 \leq j \leq s, 0 \leq t \leq 1\} \geq \varepsilon$.

PROOF. Suppose that there are infinitely many such distinct invariant manifolds.

Label these $W_1, W_2, \ldots$. Then there is an integer $0 \leq j_0 \leq s$, a sequence $x_i \in W_i$, and a sequence $t_i \in R^1$ such that

(i) $|f_1^{(j_0)}(x_i) - f_2^{(j_0)}(x_i)| \geq \varepsilon, \forall i$ or

(ii) $|(g^{t_i})^{(j_0)}(x_i) - I^{(j_0)}(x_i)| \geq \varepsilon, \forall i$.

Choose subsequences, again called $\{x_i\}$ and $\{t_i\}$, such that $x_i \to x_0$ and $t_i \to t_0$. We first wish to show that $x_0$ is the limit of a sequence $\{y_i\}$ such that $y_i \in \bigcap_{g \in \mathfrak{G}} \mathrm{Per}(g)$. If an infinite subsequence of the $x_i$ has this property, we are done. Otherwise, we may assume that none of the $x_i$ has a compact orbit. For each $i$, choose $h_i \in \mathfrak{G}$ such that $\tau(h_i) = 0$ and $h_i(x_i) \neq x_i$. Let $y_i = \lim_{n \to \infty} h_i^n(x_i)$. Then $y_i \to x_0$ and, by Theorem 4, $y_i \in \bigcap_{g \in \mathfrak{G}} \mathrm{Per}(g)$.

Now suppose that (i) holds. Since $\tau(f_1 f_2^{-1}) = 0$, $f_1 f_2^{-1}$ fixes each $y_i$. Hence, by Remark 1, $f_1 f_2^{-1}$ has infinite contact with $I$ at $x_0$. Then, $\forall j \in Z_+, f_1^{(j)}(x_0) = f_2^{(j)}(x_0)$. But this contradicts $|f_1^{(j_0)}(x_i) - f_2^{(j_0)}(x_i)| \geq \varepsilon, \forall i$. If (ii) holds, again $g^{t_i}$ has infinite contact with $I$ at $x_0$, $\forall i$. Hence, for $i$ sufficiently large,

$$\left|(g^{t_i})^{(j_0)}(x_i) - I^{(j_0)}(x_i)\right| \leq \left|(g^{t_i})^{(j_0)}(x_i) - (g^{t_0})^{(j_0)}(x_i)\right| + \left|(g^{t_0})^{(j_0)}(x_i) - I^{(j_0)}(x_i)\right| < \varepsilon.$$

Again we have a contradiction. ∎

LEMMA 9. *Let $\rho: G \to \mathrm{Diff}(S)$ be an admissible action, and let $h$, $N$ and $M$ be constructed as above. Let $J \subset S$ be an open interval, $J = (x_0, x_1)$, such that $\mathcal{O}_{x_i}$ is compact, $i = 0, 1$, and such that if $g \in \mathfrak{G}$ and $\tau(g) \neq 0$, then $g(J) \cap J = \varnothing$. ($\varnothing$ denotes the empty set.) For each $g$ such that $\tau(g) = 0$, assume that $g$ has infinite contact with the identity at $x_1$. Let $K$ be a special compact generating subset of $G$, and $\mathcal{K} = \rho(K)$. For each $f \in \mathcal{K}$, let $p(f)$ denote the first nonnegative integer such that $\tau(h^{p(f)}) = \tau(f)$. Let $\varepsilon > 0$ and $s \in Z_+$. Assume, finally, that $\forall f \in \mathcal{K}$,*

$$d_s(f|\mathfrak{G}(J), h^{p(f)}|\mathfrak{G}(J)) < \varepsilon/4,$$
$$d_s(h^{p(f)}|\mathfrak{G}(J), h^{p(f)(1-N)}|\mathfrak{G}(J)) < \varepsilon/4.$$

*Then there is an admissible action $\tilde{\rho}: G \to \mathrm{Diff}(S)$ such that:*

(i) *For any $b \in G$, $\rho(b)|S - \mathfrak{G}(J) = \tilde{\rho}(b)|S - \mathfrak{G}(J)$;*

(ii) *$\mathfrak{G}(J)$ is a single invariant manifold for $\tilde{\rho}$;*

(iii) *For any $b \in K$, $d_s(\rho(b), \tilde{\rho}(b)) < \varepsilon$.*

PROOF. It suffices to define $\tilde{\rho}|K$. For each $b \in K$, we shall define $\tilde{\rho}(b)|\mathfrak{G}(J)$, and condition (i) define $\tilde{\rho}(b)$ on $S - \mathfrak{G}(J)$. We will first define a diffeomorphism $\tilde{h}$ such that $d_s(h, \tilde{h}) < \varepsilon$ and $\mathrm{Per}(\tilde{h}) \cap \mathfrak{G}(J) = \varnothing$.

Let $J_0 = \mathrm{Cl}\, J$, $J_i = h^i(J)$. Then $\bigcup_{i=1}^{N-1} h^i(J) = \mathrm{Cl}\, \mathfrak{G}(J)$. For any $\gamma \in \mathrm{Diff}(S)$, $\gamma_i$ now denotes $\gamma|J_i$, $0 \leq i \leq N - 1$. Let $H = h^N$. Note that $\tau(H) = 0$. For any $\delta > 0$, we can choose a vector field $X$, defined on $J_0$, with the following properties:

(1) $X(x_0) = X(x_i) = 0$, $X(y) \neq 0$ for $y \in J$.

(2) If $\tilde{H}_0$ is defined by $\tilde{H}_0 = \exp_1(X)$, then $H_0^{(j)}(x_i) = \tilde{H}_0^{(j)}(x_i)$, $\forall j \in Z_+$, $i = 0, 1$. (It is possible to find such an $X$ by Lemma 2.)

(3) $d_s(\tilde{H}_0^t, I) < \delta$, $\forall 0 \leq t \leq 1$.

Define $\tilde{h}$ by $\tilde{h}_0 = h^{-N+1}\tilde{H}_0$, $\tilde{h}_i = h_i$, $i > 0$. Let $\tilde{H} = \tilde{h}^N$. By construction $\tilde{H}_0$ imbeds in a unique one-parameter group $\{\tilde{H}_0^t\}$ such that $\tilde{H}_0^1 = \tilde{H}_0$. It is easy to see that $\tilde{H}_i$ also imbeds in a unique one-parameter group, namely $\tilde{H}_i^t = \tilde{h}^i \tilde{H}_0^t \tilde{h}^{-i}$.

Now $\tilde{h}|\mathrm{Cl}\mathfrak{G}(J)$ may be factored through a commuting periodic diffeomorphism. That is, there are diffeomorphisms $r: J_i \to J_{i+1}$, $\gamma: J_i \to J_i$ such that $\tilde{h} = r\gamma = \gamma r$ and $r^n = I$. $\gamma_i$ is just $\tilde{H}_i^{1/N}$ and $r_i = \tilde{h}_i \tilde{H}_i^{-1/N}$. If $\tilde{H}^t$ is the map defined by $\tilde{H}^t|J_i = \tilde{H}_i^t$, it is easy to check that $r$ commutes with $\tilde{H}^t$.

Now let $f$ be a discrete generator, and $\{g^t : 0 \le t \le 1\}$ be an arc of continuous generators. We will construct $\tilde{f}$ and $\{\tilde{g}^t\}$ so that $\tilde{f}$, $\tilde{g}^t$ all commute with $\tilde{h}$ and each other, and $d_s(f, \tilde{f}) < \varepsilon$, $d_s(g^t, \tilde{g}^t) < \varepsilon$. Suppose that $n$ is the least positive integer such that $F = f^n$ has a fixed point. If $\tilde{H}_0$ does not have infinite contact with the identity at $x_0$, let $\tilde{F}_0$ (respectively $\tilde{g}_0^1$) be the unique element of the one-parameter group $\{\tilde{H}_0^t\}$ such that $\tilde{F}_0$ (respectively $\tilde{g}_0^1$) has infinite contact with $F_0$ (respectively $g_0^1$) at $x_0$. (Such $\tilde{F}_0$, $\tilde{g}_0^1$ exist by Corollary 1.) Otherwise, let $\tilde{F}_0 = \tilde{g}_0^1 = \tilde{H}_0$. If $\tilde{g}_0^1 = \tilde{H}_0^u$, let $\tilde{g}_0^t = \tilde{H}_0^{ut}$. Let $\tilde{f}_0 = r^{p(f)} \tilde{F}_0^{1/n}$. Then $\tilde{f}$ and $\tilde{g}^t$ may be defined by $\tilde{f}_i = r^i \tilde{f}_0 r^{-i}$ and similarly for $\tilde{g}^t$.

Let $\tilde{p}: G \to \mathrm{Diff}(S)$ be the action generated by the $\tilde{f}$, $\tilde{g}^t$. Then this action satisfies the conclusions of the theorem. First we must show that $\tilde{f}$, $\tilde{g}^t$ are differentiable at $\mathrm{Cl}\mathfrak{G}(J) - \mathfrak{G}(J)$. $\tilde{h}$ is clearly differentiable because of (2). To show that $\tilde{f}_0$ is differentiable, it suffices to show that $\tilde{f}_0$ has infinite contact with $f_0$ at $x_0$ and $x_1$, or equivalently, that $\tilde{h}^{-p(f)} \tilde{f}_0$, $h^{-p(f)} f_0 : J_0 \to J_0$ have infinite contact at $x_0$ and $x_1$. At $x_1$, $\tilde{H}_0$ and $H_0$ both have infinite contact with the identity, and thus, by Lemma 1 (b), so do $\tilde{h}^{-p(f)} \tilde{f}_0$ and $h^{-p(f)} f_0$. At $x_0$, $(\tilde{h}^{-p(f)} \tilde{f}_0)^{Nn} = \tilde{H}^{-np(f)} \tilde{F}_0^N$ has infinite contact with $(h^{-p(f)} f_0)^{Nn} = H^{-p(f)} F_0^N$ by construction. By Corollary 1, $\tilde{h}^{-p(f)} \tilde{f}_0$ and $h^{-p(f)} f_0$ have infinite contact at $x_0$, and similarly for $g_0^t$ and $\tilde{g}_0^t$. It follows easily that $\tilde{f}$, $\tilde{g}^t$ are differentiable.

By construction, all $\tilde{f}$, $\tilde{g}^t$ commute, and agree with the corresponding generators on $S - \mathfrak{G}(J)$. Also, $\mathfrak{G}(J)$ is a single invariant manifold for $\tilde{p}$ because $\tilde{h}$ has no periodic points on $\mathfrak{G}(J)$. It remains to show that $f$, $\tilde{f}$ are close, and $g^t$, $\tilde{g}^t$ are close for $0 \le t \le 1$.

In the following, unless otherwise noted, $d_s(\alpha, \beta)$ will always mean $d_s|(\alpha \mathfrak{G}(J), \beta|\mathfrak{G}(J))$, where $\alpha, \beta \in \mathrm{Diff}(S)$.

If $f \in \mathcal{K}$, again let $F = f^n$ be the first iterate of $f$ such that $\tau(F) = 0$, and suppose that $\tilde{F}_0$ is constructed as above. Let $t(f) \in R^1$ be defined by $\tilde{F}_0 = \tilde{H}_0^{t(f)}$. (Note that $t(f)$ depends only on the Taylor expansions of $F_0$ and $H_0$ at $x_0$ and hence is independent of the choice of $\tilde{H}_0$.) Then $\sup\{t(f) : f \in \mathcal{K}\} \le \infty$. Choose $\delta > 0$ such that if $d_s(\tilde{H}_0^t, I|J_0) < \delta$, $\forall 0 \le t \le 1$, then

(i) $d_s(h^{p(f)(1-N)}, h^{p(f)(1-N)} \tilde{H}^{p(f)}) < \varepsilon/4$, $\forall f \in \mathcal{K}$.

(ii) $d_s(\tilde{h}^{p(f)}, \tilde{h}^{p(f)} \tilde{H}^{(t(f)/n) - (p(f)/N)}) < \varepsilon/4$, $\forall f \in \mathcal{K}$.

Then, for any $f \in \mathcal{K}$, $d_s(f, \tilde{f}) \le d_s(f, h^{p(f)}) + d_s(h^{p(f)}, h^{p(f)(1-N)}) + d_s(h^{p(f)(1-N)}, \tilde{h}^{p(f)}) + d_s(\tilde{h}^{p(f)}, \tilde{f})$. Now the first two terms are each less than $\varepsilon/4$ by hypothesis. $\tilde{h}^{p(f)} = h^{p(f)(1-N)} \tilde{H}^{p(f)}$ so the third term is less than $\varepsilon/4$ by (i). Finally,

$$\tilde{f} = r^{p(f)} \tilde{F}^{1/n} = r^{p(f)} \tilde{H}^{t(f)/n} = \tilde{h}^{p(f)} \tilde{H}^{(t(f)/n) - (p(f)/N)}$$

so the fourth term is less than $\varepsilon/4$ by (ii). ∎

We are now ready to complete the proof of Theorem 5. By Lemma 8, there are only finitely many distinct noncompact invariant manifolds $W_x$ such that for any $f \in \mathcal{K}$,

$$d_s(f|W_x, h^{p(f)}|W_x) > \varepsilon/4 \text{ or } d_s(h^{p(f)}|W_x, h^{p(f)(1-N)}|W_x) > \varepsilon/4.$$

(If $f$ is a continuous generator, then $p(f) = 0$, so $h^{p(f)} = I$.) Label these $W_1, W_2, \ldots,$ and let $B = S - \bigcup_i W_i$. Then $B = S$ or $B$ is a finite union of intervals. If $B = S$, we can divide $S$ into a finite union of intervals as follows: Choose $x \in S$ such that $\mathcal{O}_x$ is compact. Such an $x$ exists; in fact, if $y \in S$ and $\mathcal{O}_y$ is not compact, then $L_y \neq \varnothing$ and $L_y \subset \cup$ compact orbits by Theorem 4. Then $\mathcal{O}_x$ is a finite set which partitions $S$ into intervals. In either case, $B$ may be written as a finite union of closed intervals $\{J\}$ such that $\mathcal{O}_x$ is compact for $x \in$ boundary $(J)$ and $g(\text{int } J) \cap \text{int } J = \varnothing$ (int denotes interior). For each such $J$, $\mathfrak{G}(J)$ is either already a finite union of invariant manifolds, or else, by Remark 1, there is a $y \in J$ such that $g$ has infinite contact with $I$ at $y$ for each $g \in \mathfrak{G}$ such that $\tau(g) = 0$. By applying Lemma 9, we can perturb $\rho$ on $\mathfrak{G}(\text{int } J)$ such that $\mathfrak{G}(\text{int } J)$ becomes a single invariant manifold. After a finite number of applications, we get the required $\tilde{\rho}$. It follows immediately from Theorem 4, parts (d) and (e), that for each $b \in G$, $\text{Per}(\tilde{\rho}(b))$ is a finite union of points and intervals.|

So far, we have been discussing only those elements of $\text{Diff}(S)$ with periodic points. Now, we shall consider diffeomorphisms with irrational rotation numbers.

If $f$ is a $C^2$ diffeomorphism and $\tau(f)$ is irrational, then the orbit of $x$ under $f$ is dense for every $x \in S$ [11]. For such an $f$, there is a homeomorphism $\beta$ such that $\beta f \beta^{-1}$ is rotation through $2\pi\tau(f)$ [2, p. 414]. Let $SO_2$ act on $S$ by rotations. Then each $r \in SO_2$ commutes with $\beta f \beta^{-1}$. Furthermore, if $g$ is any homeomorphism which commutes with $\beta f \beta^{-1}$, and $x_0 \in S$, then there is some $r \in SO_2$ such that $g(x_0) = r(x_0)$; but then $g = r$ on a dense subset of $S$, so $g \equiv r$. This shows that if $\beta$ is a diffeomorphism, then $\mathscr{C}(f) = \{g \in \text{Diff}(S): gf = fg\} = \{\beta^{-1}r\beta : r \in SO_2\}$. However, if $\tau(f)$ satisfies some small divisor properties (cf. Arnold [1]), there may exist no $C^1$ diffeomorphism $\beta$ such that $\beta f \beta^{-1}$ is a rotation. In this case, the following proposition shows that $\mathscr{C}(f)$ cannot be isomorphic to $SO_2$.

PROPOSITION 1. *Let $\beta$ be a homeomorphism of $S$ such that for every $r \in SO_2$, $f_r = \beta^{-1}r\beta$ is a $C^1$ diffeomorphism. Then $\beta$ is a $C^1$ diffeomorphism.*

PROOF. $\beta$ is a continuous monotonic function, so $\beta'(x)$ exists almost everywhere, in particular at some $x_0 \in S$. Then, since $\beta f_r = r\beta$, $\beta$ is differentiable at $f_r x$. But $SO_2$ act transitively, so $\beta$ is differentiable everywhere, and similarly for $\beta^{-1}$.

Now $\beta'$ is the pointwise limit of continuous functions. Hence $\beta'$ is continuous on a set of second category, in particular at some $y_0 \in S$. By the same argument, $\beta'$ and $(\beta^{-1})'$ must be continuous everywhere.|

REMARK 8. Suppose that $f$ is a homeomorphism of $S$ with an irrational rotation number. Then the closure of $\{f^n : n \in Z\}$, in the set of homeomorphisms of $S$ with the uniform $C^0$ topology, is isomorphic to $SO_2$. However, if $f$ is a diffeomorphism such that $\tau(f)$ is irrational, Proposition 1 shows that the closure of $\{f^n : n \in Z\}$, in the set of diffeomorphisms with the $C^1$ topology, need not be isomorphic to the circle group. I conjecture that if $f$ is conjugate to a rotation, but not conjugate by a diffeomorphism, then $\{f^n : n \in Z\}$ is closed in Diff$(S)$ with the $C^1$ topology.

REMARK 9. The following question remains open: Suppose that $f, g \in$ Diff$(S)$, $fg = gf$, $\tau(f)$ or $\tau(g)$ irrational. Given $\varepsilon > 0$, are there $\bar{f}, \bar{g} \in$ Diff$(S)$ such that $\bar{f}\bar{g} = \bar{g}\bar{f}$, $d_s(f, \bar{f}) < \varepsilon$, $d_s(g, \bar{g}) < \varepsilon$ and $\tau(\bar{f})$, $\tau(\bar{g})$ are rational?

4. This section contains the proofs of the theorems describing actions on higher dimensional manifolds. I wish to thank R. Palais for reformulating the proof of the following theorem in coordinate free language.

PROOF OF THEOREM 6. We may write $g$ as $g = P + R$, where $P$ is a polynomial of degree less than $m$, and $\|R(x)\|/\|x\|^m$ is bounded in a neighborhood of $x = 0$. We will show that $R \equiv 0$. By hypothesis, $g = f^{-k}gf^k = f^{-k}Pf^k + f^{-k}Rf^k$, $\forall k \in Z$. Conjugation by a linear map preserves the homogeneous components of $g$, so $P = f^{-k}Pf^k$ and $R = f^{-k}Rf^k$. Hence, it suffices to prove that, for any $x$ in the domain of $g$, $\lim_{k \to \infty} \|f^{-k}Rf^k(x)\| = 0$. By hypothesis, $\|R(x)\| = \|x\|^m h(x)$, where $h(x)$ is bounded as $x \to 0$. Then

$$\|f^{-k}R(f^k x)\| \le \|f^{-k}\| \cdot \|f^k x\|^m \cdot h(f^k x)$$
$$\le \|f^{-k}\| \cdot \|f^k\|^m \cdot \|x\|^m \cdot h(f^k x).$$

Since $h(f^k x)$ is bounded as $k \to \infty$, it is sufficient to show that $\|f^{-k}\| \, \|f^k\|^m \to 0$ as $k \to \infty$, or equivalently, that $(\|f^{-k}\| \cdot \|f^k\|^m)^{1/k} \to c < 1$. But

$$\lim_{k \to \infty} (\|f^{-k}\| \cdot \|f^k\|^m)^{1/k} = (\lim_{k \to \infty} \|f^{-k}\|^{1/k})(\lim_{k \to \infty} \|f^k\|^{1/k})^m = \bar{\lambda}^m/\underline{\lambda} < 1.$$

Hence $g$ is a polynomial of degree less than $m$. If we consider the polynomials $gf$ and $fg$, and equate the coefficients, we find that if $g$ is of degree larger than one, there are relationships of the form $\lambda_i = \lambda_1^{\alpha_1}\lambda_2^{\alpha_2} \dots \lambda_n^{\alpha_n}$, $\alpha_i \in Z$, $\alpha_i \ge 0$, $\Sigma \alpha_1 > 1$, among the eigenvalues of $f$.

REMARK 10. Sternberg [8] proves that if the eigenvalues $\{\lambda_i\}$ of a local diffeomorphism of $f$ satisfy the conditions:

$$|\lambda_i| < 1, \lambda_i \ne \lambda_1^{\alpha_1}\lambda_2^{\alpha_2} \dots \lambda_n^{\alpha_n} \quad \text{where } \alpha_i \in Z, \alpha_i \ge 0, \sum_i \alpha_i > 1$$

then there is a differentiable change of coordinates $\beta: R^n \to R^n$ such that $\beta^{-1}f\beta$ is linear. The above theorem shows that, under the same conditions, the change of coordinates is unique up to a linear map.

REMARK 11. The conditions on the eigenvalues are necessary in order that only linear maps commute with $f$. For example, if $f: R^2 \to R^2$ is defined by $f(x_1, x_2) = (\lambda x_1, \lambda^2 x_2)$, then all diffeomorphisms of the form $g(x_1, x_2) = (ax_1, bx_2 + cx_1^2)$ commute with $f$. The requirement that we consider only those $g$ such that $g \in C^m$ is necessary in order to get a finite dimensional centralizer. For consider $f: R^2 \to R^2$ given by $f(x_1, x_2) = (\lambda x_1, \lambda^t x_2)$, where $0 < \lambda < 1$, $t > 1$. If $g: R^2 \to R^2$ is any diffeomorphism of the form $g(x_1, x_2) = (ax_1, bx_2 + h(x_1))$, where $h: R^1 \to R^1$

satisfies the relation $h(\lambda y) = \lambda^t h(y)$, it is easy to see that $gf = fg$. We will show that

$$\{h: R^1 \to R^1 : h \in C^j, j < t, h(\lambda y) = \lambda^t h(y) \text{ for fixed } 0 < \lambda < 1\}$$

is not finite dimensional.

For simplicity, we consider $h: [0, \infty) \to [0, \infty)$. We can construct a large family of such $h$ satisfying $h(\lambda y) = \lambda^t h(y)$ as follows: Let $y_0 > 0$, and let $V$ be a closed neighborhood of $y_0$ such that $V \cap \lambda V = \varnothing$. ($\lambda V = \{\lambda x : x \in V\}$.) Let $\overline{h}|V$ be any orientation preserving $C^\infty$ diffeomorphism of $V$ onto $W \subset (0, \infty)$ where $W$ satisfies $W \cap \lambda^t W = \varnothing$ and let $\overline{h}|\lambda V$ be defined by $\overline{h}(\lambda x) = \lambda^t \overline{h}(x)$. Finally, let $h|[\lambda y_0, y_0]$ be any diffeomorphism which agrees with $h$ in a neighborhood of $y_0$ and $\lambda y_0$. Then, since each $x > 0$ may be written as $x = \lambda^n y$ for some $n \in Z$, $y \in [\lambda y_0, y_0]$, $h: [0, \infty) \to [0, \infty)$ may be defined by $h(\lambda x) = \lambda^t h(x)$, $h(0) = 0$. Clearly, $h|(0, \infty) \in C^\infty$, and $h$ satisfies the required functional relationship. We wish to show that if $j < t$, then $h^{(j)}(0) = 0$ and the $j$th derivative is continuous at 0. It suffices to show that, given $\varepsilon > 0$, there is a neighborhood $W$ of the origin (in the positive reals), such that if $y \in W$, then $|h^{(j)}(y)| < \varepsilon$. Let $k$ be the largest integer strictly less than $t$. Let $D = [\lambda y_0, y_0]$, and let $K = \sup_{y \in D; 1 \le j \le k} |h^{(j)}(y)|$. Now $\lambda^t/\lambda^k < 1$, so there is $N_0 \in Z$ such that if $n > N_0$, $(\lambda^t/\lambda^k)^n K < \varepsilon$. Let $W = \bigcup_{n > N_0} \lambda^n D \cup \{0\}$. Note that $W$ is a neighborhood of the origin. Let $x \in W$, $x \ne 0$. Then $x = \lambda^n y$ for some $y \in D$, $n > N_0$. Differentiating the equation $h(\lambda^n y) = \lambda^{nt}(y)$ $j$ times, we get $\lambda^{nj} h^{(j)}(\lambda^n y) = \lambda^{nt} h^{(j)}(y)$ or $h^{(j)}(\lambda^n y) = (\lambda^t/\lambda^j)^n h^{(j)}(y) \le (\lambda^t/\lambda^k)^n K < \varepsilon$.

REMARK 12. If the origin is a saddle point for $f$, i.e., if there are proper invariant subspaces $V$, $W$ of the tangent space such that the eigenvalues of $Df(0)|V$, $Df^{-1}(0)|W$ have absolute value less than one, then the centralizer is always infinite dimensional. Consider, for example, the linear map $f: R^2 \to R^2$ given by $f(x_1, x_2) = (\lambda x_1, \sigma x_2)$, where $0 < \lambda < 1 < \sigma$. There exists a unique $t \in R^1$ such that $\sigma = \lambda^{-t}$. Let $h_1, h_2$ be two $C^\infty$ functions of $R^1$ such that $h_i(0) = 0$, and $h_i$ has infinite contact with the zero function at the origin. Then if $a, b \ne 0$, any $g: R^2 \to R^2$ of the form $g(x_1, x_2) = (x_1(a + h_1(x_1^t x_2)), x_2(b + h_2(x_1^t x_2))$ is a $C^\infty$ diffeomorphism which commutes with $f$.

Among the implications of Theorem 6 for globally defined diffeomorphisms are Theorems 7 and 8 below. To state Theorem 7 more precisely we need the following:

DEFINITION. $f \in \text{Diff}(M)$ is Morse-Smale if it satisfies

(1) the set $\Omega$ of nonwandering points is finite,

(2) each periodic point of $f$ is hyperbolic. (If $f^n(x) = x$, $x$ is hyperbolic if no eigenvalue of $Df^n(x)$ has absolute value equal to 1.)

(3) for each $x, y \in \Omega$, $W^s(x)$, $W^u(y)$ have transversal intersection. ($W^s(x)$ is the stable manifold of $x$; $W^u(y)$ is the unstable manifold of $y$. For definitions of stable and unstable manifolds and nonwandering points, see [10].)

We will call $x$ a contracting periodic point if $x \in \text{Fix}(f^n)$ and $Df^n(x)$ is a contracting linear diffeomorphism; $x$ is an expanding periodic point if it is a contracting point for $f^{-1}$. $f$ is a *special* M.-S. diffeomorphism if $f$ is M.-S. and in a neighborhood of each contracting periodic point with period $k$, a coordinate system may be chosen such that $f^k$ is linear in those coordinates.

J. Palis [4] proved that, if Diff($M$) is given the $C^1$ topology, $\{f \in \text{Diff}(M): f$ is M.-S.$\}$ is an open subset of Diff($M$). By Sternberg [8], the special M.-S. diffeomorphisms are open and dense in the set of M.-S. diffeomorphisms.

Let $f$ be a M.-S. diffeomorphism. If $U$ is any neighborhood of all of the contracting periodic points, then for almost every $y \in M$, there is $n \in Z$ such that $f^n(y) \in U$ [9]. Also, a $C^\infty$ local diffeomorphism commuting with a contracting local diffeomorphism is completely determined by its values on any open set. These are the basic reasons why the following theorem is true. Conversations with M. Shub and J. Palis have been very helpful in simplifying the proof.

THEOREM 7. *Let $M$ be a compact, $C^\infty$ manifold without boundary, and let $f$ be a special M.-S. diffeomorphism of $M$. Let $V \subset M$ be an open subset. If $g_1, g_2 \in \mathscr{C}(f)$ and $g_1|V = g_2|V$, then $g_1 \equiv g_2$.*

PROOF. Suppose $x_1$ and $x_2$ are contracting periodic points. We will say that $x_1$ and $x_2$ are directly connected if there is an expanding periodic point $y$ such that $W^s(x_i) \cap W^u(y) \neq \varnothing$, $i = 1, 2$. If $x$ is another contracting periodic point, $x_1$ is connected to $x$ if there is a sequence $x_1, x_2, ..., x_n = x$ such that $x_i$ and $x_{i+1}$ are directly connected.

It follows from the work of Smale [9] that there is a contracting periodic point $x$ in $\text{Cl} \bigcup_{n \in Z} f^n(V)$. Suppose $x$ has period $k$. $f^k$ may be considered to be a contracting local diffeomorphism at $x$. By hypothesis, there is a neighborhood $U(x) \subset W^s(x)$ such that $f^k|U$ is linear in an appropriate coordinate system. Now $g_1|V = g_2|V$, and by commutativity, $g_1|f^n(V) = g_2|f^n(V)$, $\forall n \in Z$. Since $U(x) \cap \bigcup_n f^n(V)$ is nonempty and open in $U(x)$, it follows from Theorem 3.1 that $g_1|U(x) = g_2|U(x)$. But since any $y \in W^s(x)$ satisfies $y = f^m$ for some $z \in U$, $m \in Z$, $g_1|W^s(x) = g_2|W^s(x)$.

If $x$ is a contracting periodic point which is directly connected to $x$, by a similar argument, $g_1|W^s(x_1) = g_2|W^s(x_1)$. By induction, $g_1|W^s(y) = g_2|W^s(y)$ for any contracting periodic point $y$ which is connected to $x$. Hence, it suffices to show that $\{z \in W^s(y): y$ is a contracting periodic point connected to $x\}$ is dense in $M$. Since $\{z \in W^s(y): y$ is a contracting periodic point$\}$ is dense in $M$ [9], it suffices to show that every contracting periodic point is connected to $x$.

Let $M_1 = \{z \in W^s(y): y$ is a contracting periodic point connected to $x\}$.

$M_2 = \{z \in W^s(y): y$ is a contracting periodic point not connected to $x\}$. Now $\text{Cl} M_1 \cup \text{Cl} M_2 = M$. Since $M$ is connected, there is a $z \in \text{Cl} M_1 \cap \text{Cl} M_2$. If $U'(z)$ is any neighborhood of $z$, there are points $z_i \in M_i \cap U'$ and contracting periodic points $y_i \in M_i$ with period $k_i$ such that $f^{nk_i} z_i \to y_i$ as $n \to \infty$. It follows from Palis [4] that there is an expanding point $y_0$ such that $W^s(y_i) \cap W^u(y_0) \neq \varnothing$, $i = 1, 2$. But this implies that $M_2$ must be empty. $\blacksquare$

We have seen, in §2, that there are commuting diffeomorphisms, $f, g \in \text{Diff}(S)$ such that $f$ is the identity on an open set, and this phenomenon cannot be perturbed away. Similar phenomena occur in higher dimensions. Theorem 8 gives such an example.

PROOF OF THEOREM 8. Let $D$ be a closed embedded disc in a single coordinate patch in $M$, and let $\mathcal{O}$ be the boundary of $D$. Let $F$ be a vector field on $M$ such that $\mathcal{O}$ is a generic closed orbit of $F$. (A closed orbit $\mathcal{O}$ is said to be generic if the Poincaré transformation of a cross-section at $x \in \mathcal{O}$ has a transversal fixed point. For the definition of the Poincaré transformation, see [5].) Assume that there is a unique singularity $p$ of $F$ inside $D$, and choose a coordinate system with $p$ at the origin. Assume that, in a sufficiently small neighborhood of $p$, $\exp_1 F$ is linear, with eigenvalues $\lambda_1, \lambda_2$ satisfying $0 < \lambda_1 < \lambda_2 < 1$, $\lambda_1 \neq \lambda_2^k$ for any $k \in Z_+$. Also assume that $F$ has no closed orbits in int $D$. Finally, assume that $F$ generates a global one-parameter group $\{f^t\}$ such that $f^1 = f$. Such a vector field $F$ may be constructed on any 2-manifold.

Let $G$ be any vector field on $M$ such that $[F, G] = 0$, $G$ has no singularities on $\mathcal{O}$ and $G$ generates a global one-parameter group $\{g^t\}$. Here $[\ ,\ ]$ denotes Lie bracket. Then $G(p) = 0$ and $\mathcal{O}$ is a closed orbit for $G$. We can see this as follows: $g^t f = f g^t$, $\forall t \in R^1$, so $g^t(p)$ is a fixed point of $f$ for each $t$. But $p$ is an isolated fixed point of $f$, so $g^t p = p$, $\forall t$, or $G(p) = 0$. Similarly, if $y \in \mathcal{O}$, there is a $t_0 > 0$ such that $f^{t_0} y = y$. Then $g^t y = g^t f^{t_0} y = f^{t_0}(g^t y)$, $\forall t \in R^1$, so $g^t y$ belongs to a closed orbit of $F$. Since $\mathcal{O}$ is an isolated closed orbit, $g^t y \in \mathcal{O}$. Hence $g^t : \mathcal{O} \to \mathcal{O}$. Since $G$ has no singularities on $\mathcal{O}$, $\mathcal{O}$ is a closed orbit for $G$. We will show that there is a $c \in R^1$ such that $G|D = cF|D$.

First, there is no vector field $H$ defined on $M$ such that $\mathcal{O}$ is a closed orbit for $H$, $p$ is the only fixed point, and $p$ is a hyperbolic saddle point for $\exp_1 H = h$. This is true because the sum of the indices of the singularities inside a closed orbit of a vector field must be $+1$ [12, p. 147], and a saddle point has index $-1$.

Now consider the map $f = \exp_1 F$ and let $V(p)$ be a neighborhood such that $f|V$ is linear. Let $g$ be any $C^\infty$ local diffeomorphism such that $gf = fg$. By Theorem 3.1, $g$ is linear. Hence, in a neighborhood of $p$, and in appropriate coordinates with $p$ at the origin, the one-parameter groups $\{f^t\}$, $\{g^t\}$ must be given by $f^t(x_1, x_2) = (\lambda_1^t x_1, \lambda_2^t x_2)$, $g^t(x_1, x_2) = (\sigma_1^t x_1, \sigma_2^t x_2)$. We will show that $\sigma_1/\lambda_1 = \sigma_2/\lambda_2$. Suppose not. Then for some $s, t \in R^1$, $g^s f^t$ has a hyperbolic saddle point at $p$ and $g^s f^t|\mathcal{O} \neq I$. Let $h = g^s f^t$; let $\{h^r\} = \{g^{rs} f^{rt}\}$ be a one-parameter group containing $h$, and $H$ the associated vector field. It may be checked that $\mathcal{O}$ is a closed orbit for $H$ and $p$ is the only singularity inside $D$. Since this is impossible, $\sigma_1/\lambda_1 = \sigma_2/\lambda_2 = c$. Equivalently, $G|V = cF|V$. Now $f^c = \exp_1 cF$, $g^1 = \exp_1 G$ both commute with $f$. Since $D = \mathrm{Cl} \bigcup_{n \in Z} f^n(V)$ and $f^c|V = g|V$, it follows that $f^c|D = g|D$. Similarly, $f^{ct}|D = g^t|D$, $\forall t \in R^1$. Hence $G|D = cF|D$.

Since $p$ is a hyperbolic fixed point and $\mathcal{O}$ is a generic closed orbit for $F$ and $G$, there are neighborhoods $U(F)$ and $U'(G)$ such that if $\bar{F} \in U$ and $\bar{G} \in U'$, then $\bar{F}$ and $\bar{G}$ each have a unique closed orbit near $\mathcal{O}$ and a unique singularity inside that closed orbit. If $[\bar{F}, \bar{G}] = 0$, then the same proof applies and shows that there is an open set $\bar{V}$ and a real number $\bar{c}$ such that $\bar{G}|\bar{V} = \bar{c}\bar{F}|\bar{V}$. |

## REFERENCES

**1.** V. I. Arnold, *Small denominators*, I. *Mapping the circle onto itself*, Izv. Akad. Nauk. SSSR Ser. Mat. **25** (1961), 21–86.

**2.** E. Coddington and N. Levinson, *Theory of differential equations*, McGraw-Hill, New York, 1955.

**3.** N. Fine and G. E. Schweigert, *On the group of homomorphisms of an arc*, Ann. of Math. **62** (1955), 237–253.

184

**4.** J. Palis, Thesis, University of California, Berkeley, 1967.

**5.** M. Peixoto, *On an approximation theorem of Kupka and Smale*, J. Differential Equations, **3** (1967), 214–227.

**6.** ———, *Structural stability on two-dimensional manifolds*, Topology **1** (1962), 101–120.

**7.** S. Sternberg, *Local $C^n$ transformation of the real line*, Duke Math. J. **24** (1957), 97–102.

**8.** ———, *Local contractions and a theorem of Poincaré*, Amer. J. Math. **79** (1957), 809–824.

**9.** S. Smale, *Morse inequalities for a dynamical system*, Bull. Amer. Math. Soc. **60** (1966), 43–49.

**10.** ———, *Differentiable dynamical systems*, Bull. Amer. Math. Soc. **73** (1967), 747–817.

**11.** E. R. Van Kampen, *The topological transformation of a simple closed curve into itself*, Amer. J. Math. **57** (1936), 142–152.

**12.** P. Hartman, *Ordinary differential equations*, Wiley, New York, 1964.

**13.** J. Auslander and W. Hedlund, *Topological dynamics*, Benjamin, New York, 1968.

**14.** G. Reeb, *Sur les structures feuilletés de codimension un et sur un theorème de M. A. Denjoy*, Ann. Inst. Fourier (Grenoble) **11** (1961), 185–200.

UNIVERSITY OF CALIFORNIA, BERKELEY

MASSACHUSETTS INSTITUTE OF TECHNOLOGY

# A GENERIC PHENOMENON IN CONSERVATIVE HAMILTONIAN SYSTEMS

KENNETH R. MEYER[1] AND JULIAN PALMORE[2]

In a conservative Hamiltonian system two of the characteristic multipliers of any periodic solution must be $+1$. It has been conjectured that generically all periodic solutions in such a system must have the other characteristic multipliers not equal to $+1$. (See [1], p. 182.) We wish to propose an example where this is not the case.

Consider a conservative Hamiltonian system of two degrees of freedom (i.e. a four dimensional system). If a periodic solution has two characteristic multipliers different from $+1$, a classical theorem asserts that this periodic solution lies locally in a smooth cylinder filled with periodic solutions. The energy manifold at an energy $h$ (i.e. the manifold defined by taking the Hamiltonian $H = h$ to be a constant) intersects this cylinder in a circle which is a periodic solution. See Figure 1. Thus there exists a smooth one parameter family of periodic solutions

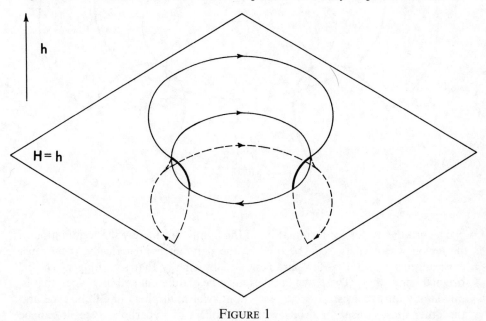

FIGURE 1

[1] This research was supported in part by NONR 3776(00), School of Mathematics, University of Minnesota.

[2] This research was supported in part by NGR 24–005–063, Center for Control Sciences, University of Minnesota.

near any periodic solution with two characteristic multipliers not equal to $+1$ and the parameter may be taken as energy. The period and characteristic multipliers are smooth functions of the parameter. If the period remains bounded the family can be extended until it reaches an equilibrium point or a periodic orbit $\gamma_1$ which has all characteristic multipliers equal to $+1$. If the least periods also converge to the least period of $\gamma_1$, then one has generically the following picture, Figure 2.

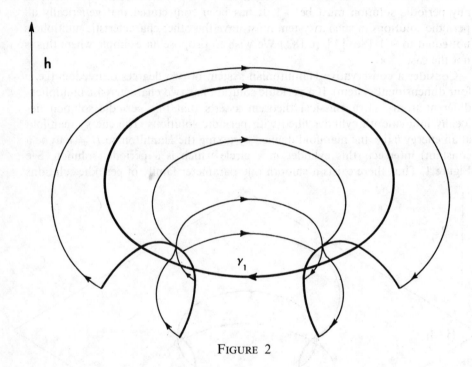

FIGURE 2

The orbit $\gamma_1$ is not the termination of the family but merely the termination of the region where energy can be used as the parameter. Generically, there must exist another smooth family parameterized by energy that smoothly meets the original family at $\gamma_1$. These two families must be of different stability types, that is, one family must consist of orbits with nontrivial multipliers of elliptic type and the other must consist of orbits with multipliers of hyperbolic type. It can be shown that the period can be used as a parameter near $\gamma_1$. Thus the periodic solutions lie in a one parameter family where the parameter can be taken as either energy or period and the family achieves a maximum (or minimum) of energy at an orbit having all characteristic multipliers equal to $+1$.

Once this picture is understood it is natural to suspect that one could construct a one parameter family that does not terminate. See Figure 3 below. This figure shows a torus filled with periodic solutions. An energy level meets the torus in two

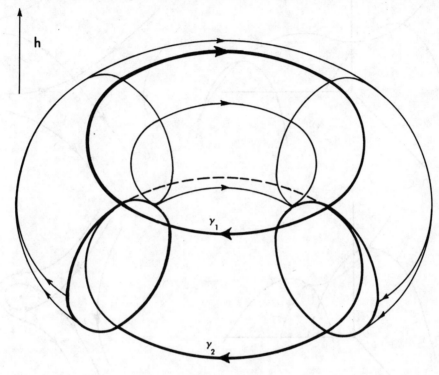

FIGURE 3

periodic orbits, one of elliptic type and one of hyperbolic type, except at the two orbits $\gamma_1$ and $\gamma_2$ where energy takes its maximum and minimum values. The orbits $\gamma_1$ and $\gamma_2$ have all characteristic multipliers $+1$. We propose that such an example exists and cannot be destroyed by a small perturbation.

In order to see how to construct such an example consider a three dimensional local cross section to the flow at some periodic solution. The intersection of an energy level with the cross section is a two dimensional manifold which can be taken as a disk. The flow defines an area preserving diffeomorphism of the disk into itself. Thus the problem of studying the local behavior near a periodic solution in a Hamiltonian system reduces to studying a one parameter family of area preserving diffeomorphisms of the disk. The figures on the right in Figure 4 indicate the changes in the map as the parameter is varied. The cross section indicated by this sequence is taken near the orbit $\gamma_1$. In the first figure there are two fixed points corresponding to the periodic solutions of elliptic and hyperbolic type. The second figure is similar except that the two fixed points are closer together. The bottom figure shows the local behavior of the cross section in the

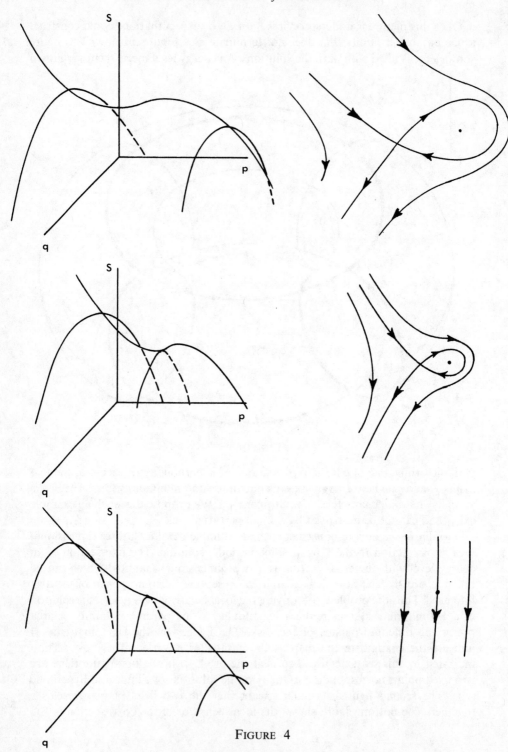

FIGURE  4

energy level of $\gamma_1$. From this point of view we see that the two families come together at $\gamma_1$ and then disappear as the parameter energy is increased.

Poincaré has given a simple method for studying area preserving maps (see [2], Chapter XXX, Vol. III). If $T:(q, p) \rightarrow (Q, P)$ defines an area preserving mapping of the plane into itself with the origin as a fixed point then the form (in $q$ and $p$) given by

$$\Omega = (Q - q)d(P + p) - (P - p)d(Q + q)$$

is exact. Thus there exists a function $S(q, p)$ such that $dS = \Omega$. If $P + p$ and $Q + q$ can be used as local coordinates, that is if $-1$ is not an eigenvalue of the Jacobian matrix of $T$, then one sees that a critical point of $S$ corresponds to a fixed point of $T$ and vice versa. By some elementary algebra one can show that saddle points of $S$ correspond to hyperbolic fixed points of $T$ and maxima and minima of $S$ correspond to elliptic fixed points of $T$. One can construct $T$ from $S$ in the obvious way except in the degenerate case where the Hessian determinant of $S$ is $-1/4$. Thus in order to construct the example one needs to construct a function $S$ with extrema as shown on the left in Figure 4 and then suspend the map in a flow. The function $S(q, p; h) = q^2 + ph - p^3$ is sufficient. The usual methods of jet transversality can be used to show that the above phenomenon persists under perturbation. (See [3] for details.)

The behavior near a maximum in energy of a family of periodic solutions is well known to people working in numerical computations in the restricted three body problem [4]. The example of a torus has been found only recently in the restricted problem near the Lagrangian triangular equilibrium by numerical experimentation. (See [5].)

## REFERENCES

1. R. Abraham, *Foundations of mechanics,* Benjamin, New York, 1967.
2. H. Poincaré, *Les méthodes nouvelles de la mécanique céleste,* Vol. 3, Dover, New York, 1957.
3. H. I. Levine, *Singularities of differential mappings.* I, Mathematisches Institut der Universität Bonn, 1959.
4. A. Deprit and J. Henrard, "A manifold of periodic orbits," in *Advances in astronomy and astrophysics,* Vol. 6, Academic Press, New York, 1968.
5. J. Palmore, *Bridges and natural centers in the restricted three body problem,* University of Minnesota Report, Center for Control Sciences, 1968.

UNIVERSITY OF MINNESOTA

# NONDENSITY OF AXIOM A(a) ON $S^2$

SHELDON E. NEWHOUSE[1]

1. In [6], Smale introduced some theorems which give much information on the structure of certain diffeomorphisms of a compact manifold. A basic question which arose was whether Axiom A [6, §I.6] was dense on any compact manifold. Subsequently, Abraham and Smale showed that Axiom A(a) was not $C^1$ dense on $T^2 \times S^2$ where $T^2$ is the two torus and $S^2$ is the two sphere [1]. We show in this paper that Axiom A(a) is not $C^2$ dense on $S^2$.

We consider the set of all diffeomorphisms of $S^2$ with the uniform $C^r$ topology, $1 \leq r \leq \infty$. Recall from [6] that for $f \in \mathrm{Diff}^r(S^2)$ a point $x \in S^2$ is said to be *non-wandering* if the following is true. For each neighborhood $U$ of $x$ there is a positive integer $n$ such that $f^n(U) \cap U \neq \varnothing$. The set of nonwandering points will be denoted by $\Omega(f)$. Following Smale, we say that $f$ satisfies Axiom A if (a) $\Omega(f)$ has a hyperbolic structure, and (b) the periodic points of $f$ are dense in $\Omega(f)$. One says that a diffeomorphism $g$ of $S^2$ is *topologically conjugate* ($\Omega$-conjugate) to $f$ if there is a homeomorphism $h: S^2 \to S^2$ ($h: \Omega(f) \to \Omega(g)$) satisfying $gh = hf$. $f$ is called $C^r$ *structurally stable* ($C^r \Omega$-stable) if there is a $C^r$ neighborhood $N$ of $f$ such that any $g \in N$ is topologically conjugate ($\Omega$-conjugate) to $f$. The main result we have is the following.

(1.1) THEOREM. *There is an open set $N$ in $\mathrm{Diff}^2(S^2)$ such that if $f \in N$, then $f$ does not satisfy Axiom A(a) and $f$ is not $C^2$ structurally stable.*

The basic idea of the proof of this theorem is to modify Smale's "horseshoe" example [6, §I.5] to produce a diffeomorphism $\bar{L}$ of $S^2$ such that, for some $x \in \Omega(\bar{L})$, the stable manifold $W^s(x, \bar{L})$ is tangent to the unstable manifold $W^u(x, \bar{L})$ at $x$ (see [6] for definitions). One then shows that this phenomenon is preserved under small $C^2$ perturbations of $\bar{L}$. From this, nondensity of Axiom A(a) is immediate, and, with some slight argument, nondensity of structural stability also follows.

We should make several remarks. First, in [1], Abraham and Smale also prove that $\Omega$-stable diffeomorphisms are not $C^1$ dense on $T^2 \times S^2$. Second, Williams has used the "DA" examples of Smale to show that structurally stable diffeomorphisms are not $C^1$ dense on $T^2$ [7]. Third, C. Pugh has shown that Axiom A(b) is $C^1$ dense on all compact manifolds [4]. Finally, the main references for this paper are the papers of Smale, [5] and [6]. In fact, I would suggest that the reader be reasonably familiar with §§I.5 and I.6 of [6] before proceeding.

In §2, we construct the diffeomorphism $\bar{L}$ as a natural extension of a diffeomorphism $L$ of the plane. §3 contains some results about Cantor sets which will be

[1] This work was supported by a National Science Foundation Graduate Fellowship.

needed. §4 is largely motivation for §5 where the main results about $C^2$ perturbations of $L$ are proved. In §6 we prove that all sufficiently small $C^2$ perturbations of $L$ do not satisfy Axiom A(a), and in §7 we sketch a proof that these small perturbations of $L$ are not $C^2$ structurally stable. We conclude §7 with some remarks about the $\Omega$-instability of these small perturbations of $L$.

I wish to express my thanks to the many mathematicians with whom I discussed this paper. Particular thanks go to M. Hirsch, C. Pugh, and S. Smale for much encouragement and many valuable conversations, and to R. Williams, who read a preliminary version of the paper and made many valuable suggestions.

2. To prove Theorem (1.1), we shall define a $C^\infty$ diffeomorphism $\bar{L}$ of $S^2$ such that there is a $C^2$ neighborhood $N$ of $\bar{L}$ such that no $f \in N$ satisfies Axiom A(a). We first construct a diffeomorphism $L$ of the plane $R^2$ so that $L(x) = x$ outside some compact subset $H$ of $R^2$. Then $L$ induces the diffeomorphism $\bar{L}$ of $S^2$ as follows. Let $\phi: R^2 \to U$ be a $C^\infty$ diffeomorphism where $U$ is a coordinate patch on $S^2$. Let $\bar{L}(x) = \phi \circ L \circ \phi^{-1}(x)$ for $x \in U$ and $\bar{L}(x) = x$ for $x \notin U$.

We now construct $L$.

Consider the square $Q = \{(x_1, x_2) \in R^2 : |x_1| \leq 1, |x_2| \leq 1\}$. Let $L: Q \to R^2$ be such that

(2.1) $L(Q) \cap Q$ has two components $A_1 \subset (x_1 < 0)$ and $A_2 \subset (x_1 > 0)$.

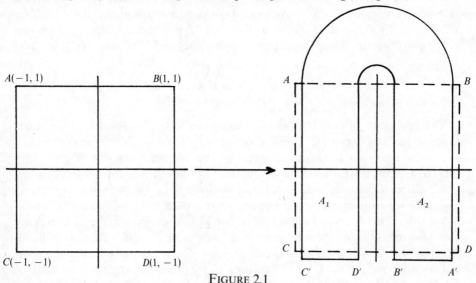

FIGURE 2.1

(2.2) $L(AB) \cup L(CD) \subset (x_2 < 0)$.

$$\begin{aligned} &L_*\big|_{L^{-1}(A_1)} = \begin{pmatrix} \alpha & 0 \\ 0 & 1/\alpha \end{pmatrix}, \quad \alpha < 1/2, \\ (2.3)\quad &L_*\big|_{L^{-1}(A_2 \cup (EFA'B'))} = \begin{pmatrix} -\alpha & 0 \\ 0 & -1/\alpha \end{pmatrix} \text{where } HA' = L(A),\ B' = L(B), \text{ etc., and } EF = A_2 \cap CD. \end{aligned}$$

(2.4) $l\pi_1(A_1) = l\pi_1(A_2) > \text{dist}(A_1, A_2) > 2l\pi_1(L(A_1) \cap A_1)$ where $l$ means horizontal length and $\pi_1: R^2 \to (x_2 = 0)$ is the vertical projection.

(2.5) $\text{dist}(A_2, BD) = \text{dist}(A_1, AC) = \beta$ where $\beta$ is small.

For instance, suppose $4\alpha\beta + \text{dist}(A_1, A_2) < l\pi(A_1) - (\beta + 2\alpha\beta)$. Suppose also that estimates analogous to those in (2.4) and (2.5) are valid for $L^{-1}(A_1)$ and $L^{-1}(A_2)$.

Define $L$ on $EFA'B'$ such that there is a subrectangle $Q_1$ with sides parallel to the coordinate axes such that

(2.6) $L(Q_1)$ is fibered by concentric semicircles which are the images of the vertical line segments in $Q_1$.

(2.7) The images of the top and bottom sides of $Q_1$ are on the same horizontal line.

(2.8) $(L|_{Q_1})_*(E_1)$ is perpendicular to $(L|_{Q_1})_*(E_2)$ where $E_1$ is the horizontal tangent space and $E_2$ is the vertical tangent space. We identify these with $(x_2 = 0)$ and $(x_1 = 0)$, respectively.

(2.9) The image under $L$ of the horizontal line segment through the midpoint of $Q_1$ connecting both vertical sides of $Q_1$ is vertical. Call this image $\rho_0$.

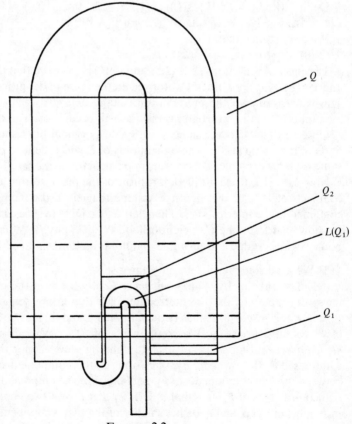

FIGURE 2.2

3

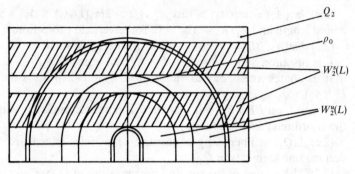

FIGURE 2.3

Let $Q = W^s_{-1}(L) = W^u_{-1}(L)$ and, for $i > -1$, let $W^s_i(L) = L^{-1}(W^s_{i-1}(L) \cap L(Q))$ and $W^u_i(L) = L(W^u_{i-1}(L)) \cap Q$.

Let $Q_2$ be a small rectangle bounded on the left by part of the boundary of $A_1$, bounded on the right by part of the boundary of $A_2$, and such that $L(Q_1) \subset$ interior $Q_2$.

Define $\overline{W}^u_i(L) = L^2(W^u_{i-2}(L)) \cap Q_2$, $i \geq 2$, and $\overline{W}^s_i(L) = W^s_i(L) \cap Q_2$, $i \geq 2$. Let $\overline{W}^s(L) = \bigcap_{i \geq 2} \overline{W}^s_i(L)$ and $\overline{W}^u(L) = \bigcap_{i \geq 2} \overline{W}^u_i(L)$.

We also assume that

(2.10) $\overline{W}^s_2(L) \cap \rho_0 = \overline{W}^u_2(L) \cap \rho_0$.

The effect of conditions (2.4), (2.5), and (2.10) is to insure that the sets $\overline{W}^s(L) \cap \rho_0$ and $\overline{W}^u(L) \cap \rho_0$ are $k$-thick Cantor sets in the sense of definition (3.6), with $k > 1$. This is necessary for the preservation of tangencies between stable and unstable manifolds of small perturbations of $L$, as will become clear in the sequel.

Observe that $L|_Q$ is the same as in Smale's original horseshoe example, so any parts of the stable and unstable manifolds of $L$ which depend only on $L|_Q$ are the same as in his example. This is true, in particular, for the set $\overline{W}^s(L)$ defined above.

Now extend $L$ to be a diffeomorphism of the plane so that outside some large compact set $H$, $L$ is the identity. Assume that all of the picture in Figure 2.2 is contained in the interior of $H$. Then, letting $\overline{L} \in \text{Diff}^\infty(S^2)$ be as above, we proceed to show that for some $C^2$ neighborhood $N$ of $\overline{L}$, any $f \in N$ has a nonwandering point $x$ whose stable and unstable manifolds are tangent at $x$.

3. We need some results about Cantor sets.

(3.1) DEFINITION. By a *two component Cantor set* $F$ in the interval $I \subset R^1$, we mean one obtained as the intersection of a decreasing sequence of closed sets $F_i \subset I$, $i \geq 2$, where (a) $F_2$ has two components, and (b) for $i \geq 2$, each component of $F_i$ contains precisely two components of $F_{i+1}$. We call such a sequence a *defining sequence* for $F$. If $\{F_i\}_{i \geq 2}$ is a defining sequence for the two component Cantor set $F$, the *two-gap*, $g_2$, of $F$ or $F_i$ is the component of $I - F_2$ which is between the two components of $F_2$. An *$i$-gap*, $g_i$, is a component of $I - F_i$ between two components of $F_i$ such that $g_i \subset F_{i-1}$. Let $F$ be a two component Cantor set in $I$, $g_i$ be an $i$-gap, and $c_i$ be the smaller of the two components of $F_i$ adjacent to

$g_i$. Let $lg_i = \sup g_i - \inf g_i$, and $lc_i = \sup c_i - \inf c_i$. Let $k > 0$. Say that a two component Cantor set $F \subset I$ is a $k$-Cantor set in $I$ if there is a defining sequence $\{F_i\}_{i \geq 2}$ for $F$ satisfying

(3.2) $\sup F_2 = \sup I$, $\inf F_2 = \inf I$.

(3.3) For any $i$-gap $g_i$ of $F_i$, $lc_i/lg_i > k$, $i \geq 2$.

(3.4) If $c$ is a component of $F_i$, $i \geq 2$, then $c - F_{i+1}$ is an open interval in $c$.

Thus a $k$-Cantor set is one which has a defining sequence $\{F_i\}_{i \geq 2}$ such that $F_{i+1}$ is obtained by removing open intervals from the components of $F_i$ in such a way that (3.3) is satisfied. For instance, the standard Cantor middle third set is a $k$-Cantor set in the unit interval for any $0 < k < 1$. When we speak of a defining sequence for a $k$-Cantor set, we mean one which satisfies conditions (3.2), (3.3), and (3.4).

(3.5) LEMMA. *If $F$ is a $k$-Cantor set in $I$, $G$ is a $k$-Cantor set in $J$, $k > 1$, $I \cap J \neq \varnothing$, and neither $F$ nor $G$ is completely contained in a gap of the other, then $F \cap G \neq \varnothing$.*

PROOF. Let $\{F_i\}_{i \geq 2}$ and $\{G_i\}_{i \geq 2}$ be defining sequences for $F$ and $G$, respectively. If $F \cap G = \varnothing$, then there is an $m \geq 2$ such that $F_m \cap G_m = \varnothing$. Let $g$ be the smallest gap of $F_m$ or $G_m$ which contains a component of the other. Suppose for definiteness $g$ is a gap of $F_m$, and $g$ contains the component $c$ of $G_m$. In the interval $I \cap J$, either there is a gap $g_r$ of $G_m$ to the right of $c$ such that $lg_r \geq lg$ or max $G_m \in g$. If $g_r$ exists, assume it to be the first such gap; if not, let $g_r = \max G_m$. Similarly, either there is a first gap $g_l$ of $G_m$ to the left of $c$ such that $lg_l \geq lg$ or min $G_m \in g$. Adjust the definition of $g_l$ accordingly. It follows from the hypothesis of the lemma that at least one of $g_l$ and $g_r$ exists. If they both exist, let $\min\{lg_l, lg_r\}$ be the smaller of the two lengths; if one does not exist, let $\min\{lg_l, lg_r\}$ be the length of the existing gap. Let $c_1$ be the interval from the right endpoint of $g_l$ to the left endpoint of $g_r$. Now it follows from the choices of $g$, $g_l$, and $g_r$ that $c_1 \subset g$ so that $lc_1 < lg$. Further, if $i \leq m$ is the first integer such that both gaps $g_l$ and $g_r$ appear in $G_i$, then there are a $c_i$ and $g_i$ for $G$, as defined above, satisfying $lc_i \leq lc_1$ and $\min\{lg_l, lg_r\} \leq lg_i$. In fact, $g_i$ can be chosen as either $g_l$ or $g_r$, and $c_i$ will then be its adjacent component which is contained in $c_1$. Thus

$$lc_i \leq lc_1 < lg \leq \min\{lg_l, lg_r\} \leq lg_i < lc_i/k$$

which is a contradiction.

We will need to generalize the notion of $k$-Cantor set to the situation in which the endpoints of the components of the defining sequence are not in the Cantor set.

(3.6) DEFINITION. Let $F \subset I$ be a two component Cantor set with defining sequence $\{F_i\}$. Let $g_i$, $c_i$ be as above. For $m \geq i$, let $g_{im}$ be the component of $I - F_m$ which contains $g_i$. Let $c_{im}$ be the union of the components of $F_m$ which are contained in $c_i$. Let $lc_{im} = \sup c_{im} - \inf c_{im}$ and $lg_{im} = \sup g_{im} - \inf g_{im}$. Let $k, k_1, k_2 > 0$. We say that $F$ is $(k_1, k_2)$-thick if there is a defining sequence $\{F_i\}$ for $F$ such that for all $i$-gaps, $g_i$, $i \geq 2$, and all $m \geq i$, $k_1 < lc_{im}/lg_{im} < k_2$. We say that $F$ is $k$-thick if there is a defining sequence $\{F_i\}$ for $F$ such that for all $i$-gaps, $i \geq 2$, and all $m \geq i$, $k < lc_{im}/lg_{im}$. Then we have the following lemma, the proof of which is similar to that of Lemma (3.5).

(3.7) LEMMA. *Let $F$ and $G$ be two $k$-thick Cantor sets in $I$. Suppose $k > 1$, neither $F$ nor $G$ is completely contained in a gap of the other, and* max(min $F$, min $G$) < min(max $F$, max $G$). *Then $F \cap G \neq \varnothing$.*

4. To motivate the use of thick Cantor sets in establishing the stability of the tangency condition in §2, and to motivate the proof of Lemma (5.1), we consider a perturbation problem in a one-dimensional setting. The result of this section will not be used in the sequel, so the reader who wishes to skip the section may do so.

Let $V$ be a closed bounded interval contained in the reals $R^1$. We assume that all functions which we consider in this section are defined on $V$ and map $V$ into itself. All closed intervals $I$, $J$, etc., which we consider are assumed to be contained in the interior of $V$.

Let $I \subset V$ be a closed interval, and let $f_1$ and $f_2$ be two contracting (derivative everywhere between 0 and 1) $C^r$ diffeomorphisms which map $I$ into itself such that $f_1(I) \cap f_2(I) = \varnothing$. Let $F_1(f_1, f_2) = I$ and, for $i \geq 2$, $F_i(f_1, f_2) = f_1(F_{i-1}(f_1, f_2)) \cup f_2(F_{i-1}(f_1, f_2))$. Then $F(f_1, f_2) \equiv \bigcap_{i \geq 2} F_i(f_1, f_2)$ is a two component Cantor set in $I$. We say that $F(f_1, f_2)$ is *defined* on $I$. By a $C^r$ perturbation or approximation $F(h_1, h_2)$ of $F(f_1, f_2)$ we mean a Cantor set defined as above on an interval $J \subset V$ where

(4.1) the endpoints of $J$ are close to those of $I$, and

(4.2) $h_1$ and $h_2$ are $C^r$ close on $V$ to $f_1$ and $f_2$, respectively.

Now we ask the following question.

(4.3) Given $F(f_1, f_2)$, can one find perturbations $F(h_1, h_2)$ and $F(\overline{h}_1, \overline{h}_2)$ arbitrarily close to $F(f_1, f_2)$ such that $F(h_1, h_2) \cap F(\overline{h}_1, \overline{h}_2) = \varnothing$?

We do not intend to discuss this question in detail, but rather to consider only those aspects of it which relate to the diffeomorphism $L$ of §2. In this connection we have the following proposition.

(4.4) PROPOSITION. *Let $k > k' > 1$. Assume that $F(f_1, f_2)$ is defined as above, that $F(f_1, f_2)$ is a $k$-Cantor set in $I$, and that $f_1$ and $f_2$ are linear contracting diffeomorphisms of $I$ into itself, i.e. the second derivatives $f_1''(x)$ and $f_2''(x)$ are identically zero on $I$. Then any $F(h_1, h_2)$ which is sufficiently $C^2$ close to $F(f_1, f_2)$ is a $k'$-Cantor set on its interval of definition.*

We observe that combining this proposition and Lemma (3.1), we obtain that, in general, the answer to question (4.3) is no.

PROOF. Let $c = \frac{1}{2} \inf \{\min(f_1'(x), f_2'(x)) : x \in I\}$. Let $g_i(h)$ be an $i$-gap of any approximation $F(h_1, h_2)$ of $F(f_1, f_2)$, and $c_i(h)$ be its adjacent component. Clearly, if $h_1$ and $h_2$ are close enough to $f_1$ and $f_2$, then $lg_2(h)/lc_2(h) < 1/k < 1/k'$.

Fix $i > 2$. Then there is a sequence $n_3, n_4, \dots, n_i$ where each $n_j = 1$ or 2, $3 \leq j \leq i$, such that

$$h_{n_i} \circ h_{n_{i-1}} \circ \dots \circ h_{n_3}(g_2(h)) = g_i(h)$$

and

$$h_{n_i} \circ h_{n_{i-1}} \circ \dots \circ h_{n_3}(c_2(h)) = c_i(h).$$

For $3 \leq j \leq i$, let $g'_j(h) = h_{n_j} \circ \ldots h_{n_3}(g_2(h))$ and $c'_j(h) = h_{n_j} \circ \ldots \circ h_{n_3}(c_2(h))$.

Let $\alpha_j = lg'_j(h)/lg'_{j-1}(h)$ and $\beta_j = lc'_j(h)/lc'_{j-1}(h)$.

Then $lg_i(h) = \alpha_i \cdot \alpha_{i-1} \ldots \alpha_3 lg_2(h)$ and $lc_i(h) = \beta_i \cdot \beta_{i-1} \ldots \beta_3 lc_2(h)$.

Let $m_j = l(g_{j-1}(h) \cup c'_{j-1}(h))$.

Now, $\alpha_j$ is the derivative of $h_{n_j}$ at some point of $g'_{j-1}$, and $\beta_j$ is the derivative of $h_{n_j}$ at some point of $c'_{j-1}$. If $\varepsilon > 0$ is given, and $h_1$ and $h_2$ are $C^2$ close to $f_1$ and $f_2$, then the mean value theorem yields (since $f_1$ and $f_2$ are linear)

$$(\alpha_j - \beta_j)/m_j < \varepsilon.$$

Thus, for $3 \leq j \leq i$,

$$\alpha_j/\beta_j < (\beta_j + m_j\varepsilon)/\beta_j.$$

Thus,

$$\frac{lg_i(h)}{lc_i(h)} < \left( \prod_{j=3}^{\infty} 1 + \frac{m_j\varepsilon}{c} \right) \frac{lg_2(h)}{lc_2(h)}.$$

Now one can easily check that $\sum_{j=3}^{\infty} m_j < \infty$. Hence, choosing $\varepsilon$ appropriately, the proposition follows.

5. In what follows, all of our approximations will be with respect to the $C^2$ metric $d$ on $\mathrm{Diff}^2(H)$. We shall apply the results of §3 to obtain some results about diffeomorphisms $C^2$ near $L$.

Recall $\overline{W}^s(L) = \bigcap_{i \geq 2} \overline{W}^s_i(L)$ and $\overline{W}^u(L) = \bigcap_{i \geq 2} \overline{W}^u_i(L)$. Then, if $\gamma$ is a $C^1$ compact arc in $Q_2$ which is $C^1$ near $\rho_0$, $\overline{W}^s(L) \cap \gamma$ and $\overline{W}^u(L) \cap \gamma$ are $(k_1, k_2)$-thick Cantor sets for some $1 < k_1 < k_2$. The next lemma asserts that this is true for a perturbation of $L$. Note that for $f$ close to $L$ we may define $\overline{W}^s_i(f)$, $\overline{W}^u_i(f)$, $\overline{W}^s(f)$ and $\overline{W}^u(f)$ as we did for $L$. We observe that if $x \in \overline{W}^s(f) \cap \overline{W}^u(f)$, $x \in \Omega(f)$, since it is an accumulation point of homoclinic points of a fixed point of $f$ (see [6] for definitions). For $f$ close to $L$, $\gamma$ $C^1$ near $\rho_0$, let $\overline{W}^s_{i\gamma}(f) = \overline{W}^s_i(f) \cap \gamma$. Let $g^s_{i\gamma}(f)$ be an $i$-gap of $\overline{W}^s_{i\gamma}(f)$, $c^s_{i\gamma}(f)$ be an adjacent component to $g^s_{i\gamma}(f)$. Make similar definitions for $g^s_{im\gamma}(f)$, $c^s_{im\gamma}(f)$, $\overline{W}^u_{i\gamma}(f)$, $\overline{W}^u_\gamma(f)$, etc. For $\gamma$ $C^1$ near $\rho_0$, let $|\gamma - \rho_0|$ denote the $C^1$ distance between $\gamma$ and $\rho_0$.

(5.1) LEMMA. *Let $1 < k_1 < k_2$ be such that $\overline{W}^s_{\rho_0}(L)$ and $\overline{W}^u_{\rho_0}(L)$ are $(k_1, k_2)$-thick Cantor sets. Let $1 < k'_1 < k_1 < k_2 < k'_2$. Then there is an $\alpha > 0$ and a $C^2$ neighborhood $N$ of $L$ such that for any compact arc $\gamma$ in $Q_2$ with $|\gamma - \rho_0| < \alpha$, and any $f \in N$, $\overline{W}^s_\gamma(f)$ and $\overline{W}^u_\gamma(f)$ are $(k'_1, k'_2)$-thick Cantor sets.*

PROOF. We first observe that by $C^1$ dependence of the stable and unstable manifolds on $f$ near $L$ (Smale [5]), it is sufficient to prove that for $f$ $C^2$ near $L$, $\overline{W}^s_{\rho_0}(f)$ and $\overline{W}^u_{\rho_0}(f)$ are $(k'_1, k'_2)$-thick Cantor sets. We prove this for $\overline{W}^s_{\rho_0}(f)$. The proof for $\overline{W}^u_{\rho_0}(f)$ is similar to that for $\overline{W}^s_{\rho_0}(f)$. One does the estimates on $f^{-2}(\overline{W}^u_{\rho_0}(f))$ and then carries them over to $\overline{W}^u_{\rho_0}(f)$.

Let $c = \frac{1}{2} \inf \{|D(L^{-1})_x(v)| : v \text{ is a unit vector in } E_1 \times E_2, x \in H\}$.

We first prove that for $i > 2$, $f$ close to $L$, $lg^s_i(f)/lc^s_i(f) < 1/k'_1$ where $g^s_i(f)$ is an $i$-gap of $\overline{W}^s_{\rho_0}(f)$ and $c^s_i(f)$ is an adjacent component. We have that

$$lg^s_2(L)/lc^s_2(L) < 1/k_1 < 1/k'_1.$$

Fix $i > 2$. Define $g_i^{s,'}(f) = g_i^s(f)$, $c_i^{s,'}(f) = c_i^s(f)$, $g_{j-1}^{s,'}(f) = f(g_j^{s,'}(f))$, and $c_{j-1}^{s,'}(f) = f(g_j^{s,'}(f))$, for $3 \leq j \leq i$.

Let $\varepsilon_1$ be such that if $d(f, L) < \varepsilon_1$, then

$$lg_2^{s,'}(f)/lc_2^{s,'}(f) < 1/k_1 < 1/k_1'.$$

This can be done since, by taking $\varepsilon_1$ small, independent of $i$, if $d(f, L) < \varepsilon_1$, the arcs $g_j^{s,'}(f)$ and $c_j^{s,'}(f)$ are nearly vertical [5], and thus $lg_2^{s,'}(f)/lc_2^{s,'}(f)$ is close to $lg_2^s(f)/lc_2^s(f)$.

Let $\varepsilon_2$ be such that

$$\prod_{i=0}^{\infty} \left(1 + \frac{\varepsilon_2}{2^i c}\right) \frac{1}{k_1} < \frac{1}{k_1'}.$$

Define $\alpha_j = lg_j^{s,'}(f)/lg_{j-1}^{s,'}(f)$ and $\beta_j = lc_j^{s,'}(f)/lc_{j-1}^{s,'}(f)$ for $3 \leq j \leq i$. Let $m_j(f) = l(g_{j-1}^{s,'} \cup c_{j-1}^{s,'})$. Thus,

$$lg_i^{s,'}(f) = \alpha_i \cdot \alpha_{i-1} \ldots \alpha_3 lg_2^{s,'}(f)$$

and

$$lc_i^{s,'}(f) = \beta_i \cdot \beta_{i-1} \ldots \beta_3 lc_2^{s,'}(f).$$

Now $\alpha_j$ may be thought of as the derivative of $f^{-1}|_{c_{j-1}^{s,'}(f)}$ at some point of $c_{j-1}^{s,'}(f)$, and $\beta_j$ may be thought of as the derivative of $f^{-1}|_{c_{j-1}^{s,'}(f)}$ at some point of $c_{j-1}^{s,'}(f)$. Since $D^2 L^{-1} = 0$ on $A_1$, for $\varepsilon_1$ smaller, if necessary, $|D^2 f^{-1}| < \varepsilon_2$ on $A_1$. Since the arcs are nearly vertical, we have, with the proper orientations, if $\alpha_j > \beta_j$, then $(\alpha_j - \beta_j)/m_j(f) < \varepsilon_2$ by the one-dimensional mean value theorem. Thus for $\alpha_j > \beta_j$,

$$\frac{\alpha_j}{\beta_j} < \frac{\beta_j + m_j \varepsilon_2}{\beta_j} = 1 + \frac{m_j \varepsilon_2}{\beta_j}.$$

But again for $\varepsilon_1$ small, $\beta_j > c$ and $m_j < \frac{1}{2} m_{j-1}$ for $3 \leq j \leq i$.
So $\alpha_j/\beta_j < 1 + m_j \varepsilon_2/c$, for $3 \leq j \leq i$ and $m_j < (1/2^{j-2})m_2 < 1/2^{j-3}$. Thus,

$$\frac{lg_i^s(f)}{lc_i^s(f)} < \prod_{j=3}^{i} \left(1 + \frac{m_j \varepsilon_2}{c}\right) \frac{lg_2^{s,'}(f)}{lc_2^{s,'}(f)}$$

$$< \prod_{j=3}^{\infty} \left(1 + \frac{\varepsilon_2}{2^{j-3} c}\right) \frac{1}{k_1} < \frac{1}{k_1'}.$$

To prove that for $m \geq i$, $lg_{im}^s(f)/lc_{im}^s(f) < 1/k_1'$ for $f$ close to $L$, we need only to make estimates of the first derivative of $f$ at certain points. Note that by the above argument, we may prove that $lg_i^s(f)/lc_i^s(f) < \lambda$ where $1/k_1 < \lambda < 1/k_1'$ for $f$ $C^2$ close to $L$. We observe that for $f$ $C^1$ close to $L$, $\sup g_{im}^s(f) - \sup g_i^s(f)$ and $\inf g_i^s(f) - \inf g_{im}^s(f)$ are small enough so that $lg_{im}^s(f)$ is close to $lg_i^s(f)$. Similarly, $lc_{im}^s(f)$ is close to $lc_i^s(f)$. Making these estimates refined enough we may prove $lg_{im}^s(f)/lc_{im}^s(f) < 1/k_1$.

Now notice that we may similarly prove that for $f$ $C^2$ close to $L$, $lc_{im}^s(f)/lg_{im}^s(f) < k_2'$ for all $i$ and all $m \geq i$. This proves Lemma (5.1).

Our goal is to prove that for $f$ $C^2$ near $L$, $\overline{W}^u(f)$ and $\overline{W}^s(f)$ have a point of tangency. Using Lemmas (5.1) and (3.2), we can show that for $\gamma$ $C^1$ near $\rho_0$ and $f$ $C^2$ near $L$, $\overline{W}_\gamma^u(f) \cap W_\gamma^s(f) \neq \varnothing$. Let us orient the arcs of $\overline{W}^u(L)$ and $\overline{W}^s(L)$ so that for $\gamma$ on the left of $\rho_0$, the angles between intersecting arcs of $\overline{W}^u(L)$ and $W^s(L)$ are all less than zero. Then for $\gamma$ on the right of $\rho_0$, the corresponding angles of intersection will all be greater than zero. It is clear that for $f$ $C^1$ close to $L$, an analogous result is true about angles of intersection of $\overline{W}_\gamma^u(f)$ and $\overline{W}_\gamma^s(f)$ for appropriate $\gamma$.

This makes it plausible that the desired points of tangency should exist. However, to prove they actually do exist, we need to know that, for any $i \geq 2$, each component of $f^{-2}(\overline{W}_i^u(f))$ has a nearly constant horizontal width. This is essentially the content of Lemma (5.3). To prove this lemma we will need the following theorem.

(5.2) THEOREM (M. HIRSCH AND C. PUGH [2]). *Let $f$ be a $C^2$ Anosov diffeomorphism of the two torus $T^2$. Let the tangent bundle have the continuous hyperbolic splitting $T(T^2) = E^s \oplus E^u$ where $\|(Df|E^s)\| < 1$ and $\|(Df|E^u)^{-1}\| < 1$. Then $E^s$ and $E^u$ are $C^1$ subbundles of $T(T^2)$, and if $f_1$ is $C^2$ close to $f$, the unit ball bundles of the invariant subbundles $E_1^s$ and $E_1^u$ are $C^1$ close to those of $E^s$ and $E^u$, respectively.*

I should remark that the statement about $C^1$ dependence in the above theorem is not actually written down in [2]. However, both M. Hirsch and C. Pugh informed me that it follows from their methods.

I am indebted to C. Pugh for telling me about the above theorem and for a conversation which was very helpful for the proof of the following lemma.

Notice that the definitions of $W_i^u(f)$, $W_i^s(f)$, etc. make sense for any diffeomorph $\tilde{Q}$ of $Q$ which is $C^1$ close enough to $Q$ and any $f$ $C^1$ close enough to $L$. That is, let $\tilde{W}_i^s(f) = f^{-1}(\tilde{W}_{i-1}^s(f) \cap f(\tilde{Q}))$, etc. We shall call any such $\tilde{Q}$ near $Q$ a "square" near $Q$.

(5.3) LEMMA. *Let $0 < \delta_1 < 1 < \delta_2$. Then there exists a neighborhood $N$ of $L$ and a $C^2$ "square" $\tilde{Q}$ near $Q$ such that the following is true. Let $f \in N$, $i \geq 2$, $\gamma_1$ and $\gamma_2$ be $C^1$ curves nearly horizontal in $Q$, and define $c_{i\gamma_1}^u(f)$, $c_{i\gamma_2}^u(f)$ to be components of $\tilde{W}_i^u(f) \cap \gamma_1$ and $\tilde{W}_i^u(f) \cap \gamma_2$ contained in the same strip of $\tilde{W}_i^u(f)$. Then*

$$\delta_1 < lc_{i\gamma_1}^u(f)/lc_{i\gamma_2}^u(f) < \delta_2.$$

PROOF. We may assume $\gamma_1$ and $\gamma_2$ are horizontal in $Q$. Let $f_0$ be an Anosov diffeomorphism of the two torus $T^2$ (see Smale [6]) such that

(5.4) There is a subset $Q'$ of $T^2$ such that $f_0(Q') \cap Q'$ has two components $A_1'$, and $A_2'$.

(5.5) There is a diffeomorphism $d_1 : Q \to Q'$ such that $d_1(A_1) = A_1'$, $d_1(A_2) = A_2'$, $d_1(L^{-1}(A_1)) = f_0^{-1}(A_1')$, and $d_1(L^{-1}(A_2)) = f_0^{-1}(A_2')$.

Let $e$ be the isometry $A_2 \to A_2$ which is rotation by $\pi$ about the midpoint of $A_2$.

(5.6) There are disjoint neighborhoods $U_1$ of $A_1$, $U_2$ of $A_2$ and a diffeomorphism $d_2 : U_1 \cup U_2 \to T^2$ such that $d_2|_{U_1} = d_1$, and $d_2|_{U_2} = d_1 \circ e$.

(5.7) $d_1(AC)$ and $d_1(BD)$ are segments of unstable manifolds of $f_0$, and $d_1(AB)$ and $d_1(CD)$ are segments of stable manifolds of $f_0$.

(5.8) $f_0 = d_2 \circ L \circ d_1^{-1}$ on a neighborhood of $f_0^{-1}(A_1' \cup A_2')$.

It is easy to see that such an Anosov diffeomorphism exists. Now let $f$ be a slight $C^2$ perturbation of $L$. Then for some neighborhood $U_3$ of $f_0^{-1}(A_1' \cup A_2')$, $d_2 \circ f \circ d_1^{-1}|_{U_3}$ is a slight $C^2$ perturbation of $f_0|_{U_3}$. Let $\bar{f}_0$ be an Anosov diffeomorphism of $T^2$ which is close to $f_0$ such that $\bar{f}_0 = d_2 \circ f \circ d_1^{-1}$ on a neighborhood of $f_0^{-1}(A_1' \cup A_2')$ and $\bar{f}_0 = f_0$ on $T^2 - U_3$. Then let $\tilde{Q}'$ be a new "square" in $T^2$, near $Q'$, such that $\bar{f}_0|_{\tilde{Q}'}$ is close to $\bar{f}_0|_{Q'}$ and $\tilde{Q}'$ is bounded by segments of the stable and unstable manifolds of $\bar{f}_0$. Now if $f$ is $C^2$ close to $L$, $\bar{f}_0$ will be $C^2$ close to $f_0$. Hence Theorem 3.2 gives that, for $i \geq 2$,

$$\frac{lc_{id_1(\gamma_1)}^u(\bar{f}_0)}{lc_{id_1(\gamma_2)}^u(\bar{f}_0)} \quad \text{is close to} \quad \frac{lc_{id_1(\gamma_1)}^u(f_0)}{lc_{id_1(\gamma_2)}^u(f_0)}.$$

Here we assume $c_{id_1(\gamma_1)}^u(f_0)$ and $c_{id_1(\gamma_2)}^u(f_0)$ are defined with respect to $Q'$, and $c_{id_1(\gamma_1)}^u(\bar{f}_0)$ and $c_{id_1(\gamma_2)}^u(\bar{f}_0)$ are defined with respect to $\tilde{Q}'$. Now let $\tilde{Q} = d_1^{-1}(\tilde{Q}')$ and the lemma follows.

6. Now we conclude the proof of the nondensity of Axiom A(a) by showing that for $f$ $C^2$ close enough to $L$, $\overline{W}^s(f)$ and $\overline{W}^u(f)$ have a point of tangency. Recall that such a point will belong to $\Omega(f)$ since it will be an accumulation point of homoclinic points of a fixed point of $f$.

In this section we will make the simplifying assumption that for all perturbations $f$ of $L$ which we consider, $\overline{W}_n^s(f) = \overline{W}_n^s(L)$ for $n \geq 2$. This avoids technical difficulties and indicates the main ideas needed in the proof. To make the proof rigorous we would have to enlarge each $\overline{W}_n^s(f)$ to obtain a $C^2$ foliation of $Q_2$ and proceed as below with respect to these foliations. Since, for $f$ $C^1$ close to $L$, all of these foliations can be made uniformly $C^1$ close to the natural horizontal foliation of $Q_2$, we are justified in making the estimates with respect to this foliation.

Let $f$ be close to $L$ and let $n \geq 2$. Note that $\overline{W}_n^u(f)$ has $2^{n-1}$ components and thus the boundary of $\overline{W}_n^u(f)$ consists of $2^n$ curves. Label these curves $\xi_n^1(f), \xi_n^2(f)$, ..., $\xi_n^{2^n}(f)$ so that $\xi_n^1(f)$ is below $\xi_n^2(f)$, $\xi_n^2(f)$ is below $\xi_n^3(f)$, etc. For each $i = 1, 2,$ ..., $2^n$, let $\psi_n^i(f)$ be a point of $\xi_n^i(f)$ at which $\xi_n^i(f)$ assumes its maximum with respect to the horizontal foliation of $Q_2$. Assume that $f$ is $C^1$ close enough to $L$ so that for each $i$, $\psi_n^i(f)$ is in the interior of the curve $\xi_n^i(f)$, i.e. $\psi_n^i(f)$ is not an endpoint of $\xi_n^i(f)$. Note that $\xi_n^i(f)$ is tangent to the horizontal foliation of $Q_2$ at $\psi_n^i(f)$. For $i = 1, 3, 5, ..., 2^n - 1$, let $c_n^i(f)$ be the closed rectangular strip in $Q_2$

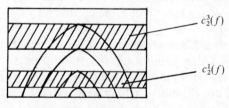

$c_2^3(f)$

$c_2^1(f)$

FIGURE 6.1

whose boundary consists of the horizontal line segment in $Q_2$ through $\psi_n^i(f)$, the horizontal line segment in $Q_2$ through $\psi_n^{i+1}(f)$ and parts of the vertical edges of $Q_2$. Then let $F_n(f) = \bigcup_{i=1,3,\ldots,2^n-1} c_n^i(f)$. Note that if $f$ is close to $L$, $\bigcap_{n \geq 2} F_n(f) \equiv F(f)$ is the product of a horizontal line segment in $Q_2$ and a Cantor set. Let $F_{\rho_0}(f) = F(f) \cap \rho_0$.

(6.1) LEMMA. *There is a $C^2$ neighborhood $N$ of $L$ such that if $f \in N$, then $F_{\rho_0}(f)$ is a $k$-thick Cantor set where $k > 1$.*

PROOF. This follows from Lemma (5.1), Lemma (5.3), and the construction of $F_{\rho_0}(f)$.

To conclude the proof of the first part of Theorem (1.1) we see that for some small $C^2$ neighborhood $N$ of $L$, if $f \in N$, then $F_{\rho_0}(f)$ and $\overline{W}_{\rho_0}^s(f)$ are $k$-thick Cantor sets for some $k > 1$. By restricting $N$ further if necessary, we may assume the hypotheses of Lemma (3.7) are satisfied by $F_{\rho_0}(f)$ and $\overline{W}_{\rho_0}^s(f)$. Thus $F_{\rho_0}(f) \cap \overline{W}_{\rho_0}^s(f) \neq \varnothing$. If $x \in F_{\rho_0}(f) \cap \overline{W}_{\rho_0}^s(f)$, then the stable manifold of $f$ through $x$ is a horizontal line segment which has a point of tangency with some unstable manifold of $f$.

7. Here we sketch a proof that for $f$ $C^2$ close enough to $L$, $f$ is not structurally stable. We also make some remarks about the $\Omega$-instability of $f$.

Let $\gamma_1$ and $\gamma_2$ be continuous arcs in the plane which have a single point $x$ of intersection. Suppose there is a disk $D$ about $x$ such that $D - \gamma_1$ has two components and $D - \gamma_2$ has two components. Say that the intersection is *one-sided* if the following is true. If $\gamma_2$ meets the component $V$ of $D - \gamma_1$, then $(\gamma_2 - \{x\}) \cap D \subset V$. Thus we have a picture as in Figure 7.1.

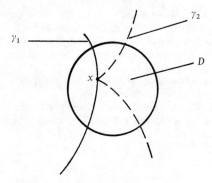

FIGURE 7.1

It is clear that if $\gamma_1$ and $\gamma_2$ are smooth arcs with nonvanishing tangent vectors, then a point of one-sided intersection is a point of tangency. In this case, we say that $\gamma_1$ and $\gamma_2$ have a point of *one-sided tangency*. Notice that if $h$ is a homeomorphism defined in a neighborhood of $x$, then $h(\gamma_1)$ and $h(\gamma_2)$ have a one-sided intersection at $h(x)$. Further, if $\gamma_1$ and $\gamma_2$ have a transversal intersection at $x$, this intersection is *not* one-sided.

These remarks together with the Kupka-Smale theorem [**6**, Theorem 6.7] imply that if the stable and unstable manifolds of a hyperbolic fixed point of a diffeomorphism $g$ have a point of one-sided tangency, then $g$ cannot be structurally stable.

Now, by the results of §6, if $f$ is $C^2$ close enough to $L$, $\overline{W}^s(f)$ has a point of one-sided tangency with $\overline{W}^u(f)$. Since $\overline{W}^s(f)$ is completely determined by the action of $f$ on $Q$, a perturbation $g$ of $f$ which agrees with $f$ except in a small neighborhood of $Q_1$ will have $\overline{W}^s(g) = \overline{W}^s(f)$. Applying techniques of Smale (I.7 of [**6**]) to $f|_Q$, we have that $f$ has a fixed point in $Q$ whose stable manifold contains a dense subset of $\overline{W}^s(f)$ and whose unstable manifold contains a dense subset of $\overline{W}^u(f)$.

Therefore, arbitrarily $C^2$ close to $f$, we can find a diffeomorphism $g$ which agrees with $f$ outside a small neighborhood of $Q_1$ and has a fixed point whose stable and unstable manifolds have a one-sided tangency. Since the set of all structurally stable diffeomorphisms is open in $\text{Diff}(S^2)$, we have that $f$ is not structurally stable.

It does not follow immediately from the one-sided tangency of the stable and unstable manifolds of a fixed point of a diffeomorphism $g$ that $g$ is not $C^2$ $\Omega$-stable. However, it is not hard to see that such a $g$ is not $C^1$ $\Omega$-stable, and that if $g_1$ is Kupka-Smale, $C^2$ close to $g$, and $\Omega$-conjugate to $g$, then the conjugating homeomorphism cannot be close to the identity. Furthermore, with a bit more work, it can be shown that, for $r \geq 2$, any $C^r$ diffeomorphism $g$ which is sufficiently $C^2$ close to $L$ is not $C^r\Omega$-stable.

## REFERENCES

1. R. Abraham and S. Smale, *Non-genericity of $\Omega$-stability*, these Proceedings, vol. 14.

2. ———, *Stable manifolds for hyperbolic sets*, Bull. Amer. Math. Soc.,**75** (1969), 149–152.

3. M. Hirsch and C. Pugh, *Stable manifolds and hyperbolic sets*, these Proceedings, vol. 14.

4. C. Pugh, *An improved closing lemma and a general density theorem*, Amer. J. Math. **89** (1967), 1010–1021.

5. S. Smale, "Diffeomorphisms with many period points", *Differential and combinatorial topology*, Princeton Univ. Press, Princeton, N.J., 1965, pp. 63–80.

6. ———, *Differentiable dynamical systems*, Bull. Amer. Math. Soc. **73** (1967), 747–817.

7. R. Williams, *The "DA" maps of Smale and structural stability*, these Proceedings, vol. 14.

UNIVERSITY OF CALIFORNIA, BERKELEY

# NONSINGULAR ENDOMORPHISMS
# OF THE CIRCLE

ZBIGNIEW NITECKI[1]

The current interest in the orbit structure of maps has been confined mainly to diffeomorphisms, and most studies have relied heavily on the invertibility of the maps in question. Two exceptions to this rule have been Shub's study of expanding maps [7] and Guckenheimer's study of rational maps [2]. Our purpose in this paper is to continue this generalization to noninvertible maps by describing the orbit structure of self-covering maps on the simplest possible manifold—the circle.

Most of the standard notions of orbit structure can be easily formulated for arbitrary maps of a manifold to itself; such formulations can be found in both of the papers cited above; for convenience, a glossary is also included at the end of this paper. We shall adopt Shub's terminology in referring to an arbitrary differentiable map of a manifold into itself as an *endomorphism* and to a diffeomorphism as an *automorphism*. We shall denote the set of $C^r$ endomorphisms of $M$ by $E^r(M)$.

The orbit structure of automorphisms of the circle is well known. Peixoto [5] has shown that the set of structurally stable automorphisms of $S^1$ is a dense, open subset of $\text{Diff}(S^1)$, and consists of the "Morse-Smale" diffeomorphisms—those whose nonwandering set consists of a finite sequence of alternating periodic sinks and sources. We wish to obtain an analogous result for a larger class of endomorphisms of the circle. A surjective endomorphism $f$ of $M$ is a *self-covering* if every point $x$ of $M$ has a neighborhood which is evenly covered under $f$, in the sense that $f$ is a homeomorphism on each component of $f^{-1}[U]$. An endomorphism $f$ of $M$ is *nonsingular* if $Df$ is injective at each point of $M$. The $C^r$ nonsingular endomorphisms of $S^1$ form a dense, open subset of the $C^r$ self-coverings ($r \geqslant 2$): the condition of nonsingularity for a self-covering is equivalent to the transversality of $Df$ to the submanifold $\{0\}$ of $R$, and such conditions are well known to be dense and open (see Abraham and Robbin, [1]). We shall determine the orbit structure of a dense, open set of nonsingular endomorphisms— which will thus also be a dense, open subset of all self-coverings.

The set of $C^r$ nonsingular endomorphisms of the circle, which will be denoted $N^r(S^1)$, divides into a sequence of components, consisting of maps of various degrees. The first component is just the set of automorphisms, and Peixoto's results apply directly to this case. However, an application of the Lefschetz trace formula to the iterates of any endomorphism of higher degree indicates that such

---

[1] Work done while the author was an NSF Graduate Fellow.

a map must have an infinite number of periodic points—so that "Morse-Smale" endomorphisms exist only in the first component of $N^r(S^1)$. Two examples considered by Shub [7] indicate what kinds of complexity must be expected to occur, even if the best conceivable situation prevails. One example is the expanding maps, for which the nonwandering set is the whole manifold. The other is an example of an $\Omega$-stable endomorphism of the circle for which the nonwandering set is a fixed sink together with a cantor set.

Our results will show that the phenomena considered above are essentially all that can occur. More precisely, we can summarize our results in a

MAIN THEOREM. *Let $\Sigma^r$ denote the set of structurally stable nonsingular $C^r$ endomorphisms of $S^1$. Then $\Sigma^r$ is dense and open in $N^r(S^1)$ for $r \geqslant 2$. Moreover, the elements of $\Sigma^r$ of degree $> 1$ are either expanding maps or else have a nonwandering set consisting of*
   (1) *a finite set of "blocks" of isolated periodic points (see §2) and*
   (2) *a cantor set of nonisolated sources.*

We shall prove this theorem in several steps. First, we state some basic approximation theorems due essentially to Shub, Peixoto and Pugh. These will define for us a certain generic set of nonsingular endomorphisms. We will study their orbit structure in detail, and then continue this study further under the assumption of Axiom A. This will lead us to the $\Omega$-stability of these endomorphisms, which we will extend to structural stability. We shall then show that Axiom A is a dense condition, via a rational approximation theorem and some results of Guckenheimer. This will establish that the conditions we first considered together with Axiom A are necessary as well as sufficient for structural stability in $N^r(S^1)$. In the final section, we shall make a few remarks as to how our methods yield a finite algorithm for deciding when two elements of $\Sigma^r$ are topologically conjugate.

1. **Preliminary approximation results.** We state here the basic results on generic subsets of $N^r(S^1)$ which we shall need in the first part of our study.

PROPOSITION (1.1) (KUPKA-SMALE THEOREM). *The set of endomorphisms for which all periodic points are generic is residual in $E^r(S^1)$.*

PROOF. This is a special case of Theorem $\beta$ in [7].

PROPOSITION (1.2) (IMPROVED CLOSING LEMMA). *If $f \in N^r(S^1)$ and $p \in \Omega(f)$, then $f$ can be uniformly $C^r$-approximated by $\tilde{f} \in N^r(S^1)$ such that $p \in \mathrm{per}\,(\tilde{f})$.*

PROOF. As C. Pugh pointed out to me, this can be proven by simply reinterpreting the proof of Lemma 4 in [5].

PROPOSITION (1.3) (GENERAL DENSITY THEOREM). *The set of nonsingular endomorphisms for which:*
   (1) $\Omega(f) = $ closure per $(f)$
   (2) *All periodic points are generic*
*is residual in $N^r(S^1)$.*

PROOF. Here, Pugh's argument in [6], §6, applies word for word (reading "$N^r(S^1)$" for "$X$" and "nonsingular endomorphism" for "flow"), substituting (1.1) and (1.2) for the corresponding results on flows.

Finally, we throw in an additional generic condition which will become useful later on: that $|Df| = 1$ for only finitely many points of $S^1$. This condition follows from the assumption that $Df \pitchfork \{\pm 1\}$, which holds on a residual set of $N^r(S^1)$ by a standard transversality theorem (Abraham-Robbin, [1]).

We summarize all of these results in

DEFINITION (1.4). Let $\mathscr{G}$ denote the set of $f \in N^r(S^1)$ satisfying:

(1) $\Omega(f) = $ closure per $(f)$

(2) all periodic points are generic

(3) $|Df| = 1$ for only finitely many points of $S^1$

then $\mathscr{G}$ is clearly residual in $N^r(S^1)$.

## 2. $\Omega$ for elements of $\mathscr{G}$.
We will now concern ourselves with developing a picture of the structure of $\Omega(f)$ for $f \in \mathscr{G}$. For simplicity, we shall assume at first that $f$ is orientation-preserving. This involves little loss of generality, since in any case $f^2$ preserves orientation and per $(f) = $ per $(f^2)$ (so that $\Omega(f) = \Omega(f^2)$). However, at the end of this study we shall stop and examine just which details change if $f$ reverses orientation.

We begin with a lemma which will prove very useful in this study. Here, as throughout this section, we take $f$ to be a given nonsingular orientation-preserving endomorphism in $\mathscr{G}$, with deg $f > 1$.

LEMMA (2.1). *Suppose $[p_1, p_2]$ is an arc of $S^1$ containing no fixed sinks, with $p_1$ and $p_2$ both fixed sources. Then $f([p_1, p_2]) = S^1$.*

PROOF. Observing that an arc with fixed endpoints is taken either into itself or onto the whole circle under an orientation-preserving map, we see that it suffices to show that the arc cannot be mapped entirely into itself.

Suppose $f([p_1, p_2]) = [p_1, p_2]$. Then by picking points in the local unstable manifolds of $p_1$ and $p_2$, we can find a slightly smaller arc $[p_1^*, p_2^*] \subset [p_1, p_2]$ which is mapped into its own interior. Then by the Brouwer fixed point theorem, there is a fixedpoint interior to $[p_1^*, p_2^*]$.

Now, consider the greatest lower bound of the fixedpoints on $[p_1, p_2]$ above $p_1$; by continuity, this is itself a fixedpoint, say $p_3$. Since $W_{\mathrm{loc}}^u(p_1)$ contains no fixedpoints besides $p_1$, and since $p_3$ is a limit of fixedpoints above $p_1$, $p_1 \neq p_3$. But, by assumption, $p_3$ is not a sink, so it must be a source, and the previous paragraph applied to $[p_1, p_3]$ shows there must be another fixedpoint interior to $[p_1, p_3]$, contradicting the choice of $p_3$. Thus, $[p_1, p_2]$ cannot be mapped into itself, and the lemma is proved.

COROLLARY (2.2). *An arc containing no periodic sinks and whose endpoints are periodic sources covers $S^1$ under some iterate of $f$.*

(Apply the theorem to $f^n$, where $n$ is the product of the periods of the endpoints.)

This result gives us, incidentally, a partial topological characterization of the nonwandering set:

COROLLARY (2.3). *If* $\Omega(f) \neq S^1$, *then it is totally disconnected.*

PROOF. Suppose $\Omega$ contains an interval, $[p_1, p_2]$. Since sinks are always isolated in $\Omega$, the sources are dense in $[p_1, p_2]$, so we can in fact assume $p_1$ and $p_2$ are sources. Since $[p_1, p_2] \subset \Omega$, it contains no sources, and hence by (2.2) $f^n[p_1, p_2] = S^1$ for some $n \geqslant 1$. But then $f\Omega \subset \Omega$ implies $\Omega = S^1$.

We now proceed to study further the structure of the collection of isolated periodic points, if there are any. Our goal here is to show that these appear in "blocks". The first step is to determine what the nonwandering points "nearest" an isolated one look like.

PROPOSITION (2.4). *Let* $p_0$ *be an isolated periodic point, and let* $(a, b)$ *be the component of* $p_0$ *in* $\{p_0\} \cup [S^1 \sim \Omega]$. *Then* $a$ *and* $b$ *are periodic, and are both sources (sinks) if* $p_0$ *is a sink (source). If* $f$ *is orientation-preserving,* $a$, $b$ *and* $p_0$ *all have the same least period.*

PROOF. We separate the two cases:

*Case* 1. $p_0$ a sink.

We can assume $p_0$ fixed. Let $J = $ component of $p_0$ in $W^s(p_0)$. Then since $fJ \subset J$ and $\deg f > 1$, we must have $\partial J = $ two points, say $p_1, p_2$. We will show these are sources, which will mean $J = (a, b)$. Since $fJ \subset J$, by continuity $fp_i \in \overline{J}$ $(i = 1, 2)$. However, if $f(p_i) \in J$, then $p_i \in W^s_{\text{loc}}(p_0)$, which is open, in which case $p_i \notin \partial J$. Hence $f\{p_1, p_2\} = \{p_1, p_2\}$. If $f$ preserves order, $fp_1 = p_1, fp_2 = p_2$. Finally, since points slightly inside $[p_1, p_2]$ contract to $p_0$, the points $p_1$ and $p_2$ can't be sinks, so must be sources.

*Case* 2. $p_0$ an isolated source.

The proof in this case is somewhat more difficult; we need to employ the following partial converse to (2.1):

LEMMA (2.5). *Any closed interval which under some iterate of* $f$ *covers* $S^1$ *contains an infinite number of sources.*

Given this lemma, the proof of Case 2 proceeds as follows:

We can assume again that $p_0$ is fixed. It is clear (since $\Omega$ is closed) that $a, b \in \Omega$. Hence, $\Omega = \overline{\text{per}(f)}$ implies the existence of a sequence $\{x_i\}$ of periodic points approaching $b$ from the right (or, resp., $a$ from the left). We shall show $b$ must be a fixed source; the argument for $a$ is similar.

Consider the interval $[p_0, b]$. We can't have $f(b) \in (p_0, b)$, since then by continuity some of the periodic points $\{x_n\}$ must also be mapped into $(p_0, b)$, so that $(a, b)$ would contain other periodic points. Suppose $f(b) > b$. Then for all $x$ in $[b, f(b)]$, $f(x) \geqslant x$. But there is some $x_n \in (b, f(b))$ which is periodic, so there is some $k > 0$ such that $f^k(x_n) = x_n$. Then $x_n > b$ and $f^k x_n < f^k b$ imply that $f^k[p_0, f(b)] = S^1$, so that $f^{k+1}[p_0, b] = S^1$. Then, by (2.5), $[p_0, b]$ contains an infinite number of periodic points; in particular, at least one of them lies in the interior of the interval, contradicting the definition of $b$. Hence, $b$ is fixed.

To see that $b$ must be a sink, we note that if it were a source, (2.1) would show that $[p_0, b]$ must cover $S^1$, so that by (2.5) again it must contain a periodic point in its interior, and this contradiction shows $b$ must be a sink.   QED.

We now return to the proof of (2.5):

Recall again that we are assuming $\deg f > 1$, so that by taking a high iterate, we can assume that under $f^k$ the interval (which we shall denote by $[e_1, e_2]$) covers itself at least twice. Thus, at least some of the components of $f^{-k}[e_1, e_2]$ lie entirely inside $[e_1, e_2]$; for $i = 1, 2$ let $F_i$ denote the component of $f^{-k}[e_1, e_2]$ nearest to $e_i$. Then, since $f^k F_i = [e_1, e_2]$, $F_i$ contains a component of $f^{-k}F_j$ ($j = 1, 2$) nearest either end of $F_i$; let $F_{i1}$ be the component nearest the left endpoint, $F_{i2}$ that nearest the right. By continuing this process, we get, for each sequence $\{i_n\}$ of 1's and 2's, a nested decreasing sequence of closed intervals $F_{i_1} \supset F_{i_1 i_2} \supset \dots$ whose intersection $F_{\{i_n\}}$ is nonempty. But $f^k(F_{\{i_n\}}) \subset F_{\{i_{n+1}\}}$, so in particular if the sequence $\{i_n\}$ is just the repetition of a block of $N$ numbers, then $f^{kN}F_{\{i_n\}} \subset F_{\{i_n\}}$, so that $F_{\{i_n\}}$ contains a fixedpoint of $f^{kN}$ by the Brouwer fixed point theorem. If this fixedpoint is a sink, we apply Case 1 of (2.4) to produce a source corresponding to it, also in $[e_1, e_2]$. Since there are an infinite number of such "periodic" sequences $\{i_n\}$ possible, $[e_1, e_2]$ contains an infinity of sources, and the lemma is proved.

We shall use a refined form of the above technique later on to determine the structure of $f$ on $\Omega$.

As a corollary of (2.4), we deduce a "finiteness" condition on uninterrupted sequences of isolated periodic points:

COROLLARY (2.6). *If $x_n \to x$ is a monotone convergent sequence of isolated periodic points, then the open interval $(x_0, x)$ contains a nonisolated periodic point.*

PROOF. Suppose not. Then by (2.4), we can find a sequence $z_0 = x_0 > z_1 > z_2 > \dots$ such that the $z_i$ are all isolated periodic points (alternately sinks and sources) of the same period, say $N$, with no periodic points whatsoever in $(z_{i+1}, z_i)$. Since $z_i \geq x_i$, the sequence is bounded below by $x$ and hence converges, say to $z \geq x$. By continuity $z$ is periodic of period $N$; but, being a generic fixedpoint of $f^N$, it cannot be a limit of fixedpoints of $f^N$, contradicting the choice of $z$.

We can now summarize the results above in a definition. A *block of isolated periodic points* is an interval $[p_1, p_2]$ whose endpoints are periodic, nonisolated sources and whose only other nonwandering points are a finite sequence $q_1, \dots, q_k$ of periodic points, say $p_1 < q_1 < q_2 \ \dots < q_k < p_2$ such that:

(1) $q_1$ and $q_k$ are sinks.
(2) $q_{i+1}$ is a sink (source) iff $q_i$ is a source (sink).
(3) period $(p_1)$ = period $(p_2)$ = period $(q_i)$ $(i = 1, \dots, k)$.

Thus, a block is a closed interval on which the map behaves just like a Morse-Smale diffeomorphism, and the block as a whole acts like one big periodic source with regard to the rest of the circle. It is clear from (2.4) and (2.6) that

COROLLARY (2.7). *Every isolated periodic point is part of a block of isolated periodic points.*

A second corollary of the above results and of (2.3) is a topological characterization of $\Omega$:

COROLLARY (2.8). $\Omega$ *is either*

or
(1) *all of* $S^1$,
(2) *a collection of isolated periodic points (appearing in blocks) together with a cantor set, K, of nonisolated nonwandering points.*

PROOF. Suppose $\Omega \neq S^1$, and let $K$ be the derived set of $\Omega$. $K$ is clearly a closed, totally disconnected subset of $\Omega$. We need to show that $K$ is perfect.

Given an element $x$ of $K$, we need to show it is a limit of points also in $K$. Suppose not. Then, by definition, there must be a sequence of isolated periodic points $y_i$ approaching $x$ as a limit; by choosing a subsequence, we can assume $\{y_i\}$ monotone. Then, applying (2.6) to the sequence $\{y_i\}_{i \geqslant n}$ for successive $n$'s, we see that $x$ is also a limit of nonisolated points of $\Omega$—i.e., elements of $K$. Thus, $K$ is a closed, perfect, totally disconnected set, so a cantor set.

All other points are isolated in $\Omega$, so must be periodic.    QED.

We note, however, that our description of $\Omega$ is thus far incomplete, since it is still conceivable that we might have an infinite number of "blocks" of isolated periodic points—although, of course, each block contains only finitely many isolated periodic points. But, under an additional hypothesis (which we shall eventually show to be generic) this complication can indeed be avoided.

Recall that an endomorphism $f$ is said to satisfy axiom A if $\Omega(f) = \overline{\operatorname{per}(f)}$ and $\Omega(f)$ has a hyperbolic structure (see glossary). Note that for members of $\mathscr{G}$, the nonisolated part of $\Omega$ (be it all of $S^1$ or a cantor set $K$) must be in $\Omega_e$, the "expanding" part of $\Omega$, since all periodic sources are in $\Omega_e$. Let $\mathscr{A}$ denote the set of all $f \in \mathscr{G}$ satisfying Axiom A.

LEMMA (2.9). *If* $f \in \mathscr{A}$, $\Omega(f)$ *has only finitely many isolated points.*

PROOF. By invariance, $\Omega_c$ consists precisely of all periodic sinks; it is also clear that $\Omega_c(f^N) = \Omega_c(f)$, $\Omega_e(f^N) = \Omega_e(f)$, where $N$ is the iterate referred to in the definition of hyperbolic structure (see glossary). Applying (2.4), it clearly suffices to show $\Omega_c(f^N)$ finite.

Suppose $\Omega_c(f^N)$ is infinite. Then we can pick a sequence $\{x_n\}$ of sinks converging to, say, $x_0$. Since $x_0 \in K$, it is in $\Omega_e$. Hence, $|Df^N(x_0)| > c\lambda^N$. But then, by the continuity of $Df^N$, for sufficiently large $n$, $|Df^N(x_n)| > c\lambda^N > c\lambda^{-N}$, contradicting the fact that $x_n \in \Omega_c$. This proves the lemma.

We can summarize our results on the topological structure of $\Omega(f)$ in a

THEOREM (2.10). *If* $f \in \mathscr{A}$, *it is either an expanding map (so that* $\Omega = S^1$) *or* $\Omega$ *consists of a finite number of finite blocks of isolated periodic points together with a cantor set $K$ of nonisolated nonwandering points.*

Let us consider now how this picture changes if we allow $f$ to reverse orientation. Since squaring the map changes neither the point-set $\Omega$ nor the quality of being a sink or source for a given periodic point, and since $f^2$ always preserves orientation, we can still apply all parts of our description to orientation-reversing maps, except for the statement that all points in a block have the same least period under $f$.

To determine the relation of periods within a block, note that if the least period of a point under $f$ is odd, it is unchanged under $f^2$, whereas if it was even, it is halved under $f^2$. Thus, if all points in a block have even periods under $f$, they must all have the same least period, whether under $f$ or $f^2$. On the other hand, suppose $p_0$ is an isolated periodic point of least period $2n + 1$ under $f$. Then we are looking at a fixedpoint of the orientation-reversing map $f^{2n+1}$. Clearly the block is taken into itself under $f^{2n+1}$ (since $f$ is monotone and $f^{2(2n+1)}$ preserves the block); it is also easy to see that a nonsingular orientation-reversing endomorphism of an interval has at most one fixedpoint (each component of the complement of a fixedpoint is taken onto another component). Hence, a given block can have at most one fixedpoint of $f^{2n+1}$, and all other periodic points in that block will have least period $2(2n + 1)$ under $f$. It is easy to see, in fact, that the point of odd period will have to be in the "middle" of the block, since each periodic predecessor will be mapped under $f^{2n+1}$ to a successor, and vice-versa. So, for orientation-reversing maps, we must modify our concept of a block to allow the "middle" periodic point to be of odd period equal to half of the period of the rest of the block. Note also that these are the only blocks with points of odd period.

3. **Orbit structure of elements of $\mathscr{A}$.** Our goal in the next three sections is to show that nonsingular endomorphisms in $\mathscr{G}$ which satisfy Axiom A are structurally stable. This will involve three basic steps: first we shall determine the structure of $f \in \mathscr{A}$ on $\Omega(f)$, then we shall show that these maps are $\Omega$-stable, and finally we shall show that an $\Omega$-conjugacy can be extended to a topological conjugacy, thus establishing structural stability. This section is concerned with the first of these steps.

To state our main structure theorem, we need first a digression to define some not entirely standard notions. Our idea here is to get a noninjective analogue of the standard "shift automorphism" of symbolic dynamics, as developed in [8]. This same notion appears in [2].

DEFINITION (3.1). The *(right) sequence space*, $C_n$, on $n$ symbols is the set of sequences $\{a_i\}_{i \geqslant 1}$ of members of an $n$-element set $S$, given the compact-open topology—or equivalently, $C_n = \Pi_{i \geqslant 1} S_i$, where $S_i = S$ is regarded as a discrete set.

DEFINITION (3.2). The *semishift* on $n$ symbols is the map $s_n: C_n \to C_n$ given by $s_n(\{a_i\}_{i \geqslant 1}) = \{a_{i+1}\}_{i \geqslant 1}$.

The following properties of $s_n$ are easily checked:

PROPERTIES OF THE SEMISHIFT (3.3):

(1) $s_n$ is an $n : 1$ local homeomorphism
(2) per $(s_n)$ is dense in $C_n$
(3) $s_n$ is topologically transitive on $C_n$
(4) $\alpha(p) = C_n$ for any $p \in C_n$.

DEFINITION (3.4). A *subsemishift* on $n$ symbols is the restriction of $s_n$ to a subset $K_n$ of $C_n$ which is invariant and on which $s_n$ is topologically transitive. A subsemishift is said to be of *finite type* if $K_n$ can be expressed as the set of all sequences in $C_n$ in which no element of a finite family $\mathscr{F} \subset \bigcup_{k \geqslant 0} \Pi_1^k S$ appears as a sequence of consecutive symbols.

It is equally easy to see that all of the properties (3.3) apply equally well to subsemishifts of finite type.

We can now state our basic structure theorem on $\Omega$.

THEOREM (3.5). *If $f \in \mathscr{A}$, then either $f$ is an expanding map or $f|K$ is topologically conjugate to $s_{nd}: K_{nd} \to K_{nd}$, a subsemishift of finite type, where $n$ = number of blocks of isolated points in $\Omega(f)$, and $d$ = degree of $f$.*

PROOF. If $f$ has no sinks, then $S^1 = \Omega = \Omega_e$, so that $f$ is expanding by definition.

Suppose $f$ has $n$ blocks of isolated periodic points. Then the closure of the complement of all these blocks consists of $n$ intervals, say $I_1, \ldots, I_n$. On each $I_j$ there are defined $d$ single-valued branches of $f^{-1}$, $g_{1j}, \ldots, g_{dj}$. Then we can let $S$ in Definition (3.1) be the collection of $nd$ symbols, $\{(1, 1), \ldots, (d, 1), (1, 2), \ldots, (d, n)\}$. Now, call an element $\{(i_m, j_m)\}_{m > 1}$ of $C_{nd}$ admissible if the corresponding compositions $g_{i_m j_m} \circ g_{i_{m+1} j_{m+1}}$ are defined for all $m \geqslant 1$—equivalently, if $g_{i_{m+1} j_{m+1}}(I_{j_{m+1}}) \subset I_{j_m}$. Then the set $K_{nd}$ of admissible sequences is a subset of $C_{nd}$ of finite type, since the corresponding $\mathscr{F}$ in (3.4) can be taken to be the finite class of "pairs of pairs" $\{(i_m, j_m), (i_{m+1}, j_{m+1})\}$ for which $g_{i_{m+1} j_{m+1}}(I_{j_{m+1}}) \not\subset I_{j_m}$. To show that $s_{nd}$ is topologically transitive on $K_{nd}$, it suffices to show that, given any two pairs of indices $(i, j)$ and $(i', j')$, there exists some finite sequence of pairs of indices, $(i_1, j_1), (i_2, j_2), \ldots, (i_p, j_p)$, which appears in some element of $K_{nd}$ (i.e., is "admissible"), with $(i_1, j_1) = (i, j)$, $(i_p, j_p) = (i', j')$. That is, we have to find a sequence $g_{i_1 j_1} \cdots g_{i_p j_p}$ of branches of $f^{-1}$ such that $g_{i_{l+1} j_{l+1}}(I_{j_{l+1}}) \subset I_{j_l}$, $l = 1, \ldots, p - 1$. But note that each $I_k$ has sources for endpoints and contains no sinks, so that by (2.2) there exists an iterate $f^\alpha$ of $f$ such that $f^\alpha I_k = S^1$. Hence, there is some branch of $f^{-\alpha}$ taking $g_{i' j'}(I_{j'})$ into $I_j$; any such branch can be expressed as $g_{i_2 j_2} \circ \cdots \circ g_{i_{p-1} j_{p-1}}$ (where $p = \alpha + 2$), and this gives us our desired sequence. Hence, $s_{nd}: K_{nd} \to K_{nd}$ is a subsemishift of finite type on $nd$ symbols.

Now, to establish a conjugacy between $f|K$ and $s_{nd}|K_{nd}$, we shall adapt a technique originated by Newhouse [4]. For any admissible sequence $\Lambda = \{(i_m, j_m)\}$, let $F_\Lambda = \bigcap_{l \geqslant 1} \mathrm{im}\,(g_{i_1 j_1} \circ \cdots \circ g_{i_l 1_l})$. Since these successive "images" are nested closed intervals, $F_\Lambda$ is either a closed interval or a single point— but in any case, it is nonempty (for $\Lambda \in K_{nd}$) by the finite intersection property.

These facts concerning the $F_\Lambda$'s are clear:

(1) $F_{\Lambda_1} \cap F_{\Lambda_2} = \varnothing$ unless $\Lambda_1 = \Lambda_2$.

(2) $f(F_\Lambda) = F_{s_{nd}(\Lambda)}$.

(3) $K \subset \bigcup_{\Lambda \in K_{nd}} F_\Lambda$ (this because any point not in one of the $F_\Lambda$'s eventually is mapped into one of the blocks).

(4) The endpoints of any $F_\Lambda$ are in $K$. (Any endpoint is an $\alpha$-limit of points of some $I$.)

Thus, we can define a map $h: \bigcup F_\Lambda \to K_{nd}$ by $h(F_\Lambda) = \Lambda$. Then, by (2) above, $h$ is a conjugacy of $f|K$ and $s_{nd}|K_{nd}$ if it is a homeomorphism and $K = \bigcup F_\Lambda$.

We first claim: $F_\Lambda$ is a single point for every admissible $\Lambda$.

If $\Lambda$ is periodic, then the endpoints of $F_\Lambda$ are periodic sources; by definition, $F_\Lambda$ contains no sinks, so, if $F_\Lambda$ is an interval, it eventually covers all of $S^1$. But it follows immediately from the definition of $F_\Lambda$ that $f^k F_\Lambda \subset I_{j_k} \neq S^1$. Hence, $F_\Lambda$ is a single point for $\Lambda$ periodic.

Second, the above argument shows that no interval can even be eventually periodic (since images of intervals have interior). Hence, if $F_\Lambda$ is an interval, $f^i F_\Lambda \cap f^j F_\Lambda = \varnothing$ for all $i \neq j$. Thus $\lim_{K \to \infty}$ length $(f^k F_\Lambda) = 0$. On the other hand, since $f \in \mathcal{G}$, there are only finitely many points where $|Df| = 1$, so that for sufficiently high $k$, points on $f^k F_\Lambda$ are all expanding or all contracting. Since $f$ satisfies Axiom A and the endpoints of $f^k F_\Lambda$ are in $K$, $|Df| > 1$ on all of $f^k F_\Lambda$ (for high $k$). This means the length of $f^k F_\Lambda$ is eventually increasing, contradicting the above limit argument. Hence, $F_\Lambda$ cannot have interior, and the claim is proved.

One can conclude immediately from this claim that $h$ is injective; on the other hand $h$ is surjective by definition (for every $\Lambda \in K_{nd}$, $F_\Lambda$ is defined and nonempty). To see that $h^{-1}$ is continuous, note that it follows from the claim that $\lim_{k \to \infty}$ length $(\text{im } g_{i_1 j_1} \circ \dots \circ g_{i_k j_k}) = 0$ for any sequence $\Lambda = \{(i_m, j_m)\}$ in $K_{nd}$. On the other hand, $\Lambda = \lim \Lambda_i$ in $K_{nd}$ iff eventually the initial $k$ terms of $\Lambda_i$ agree with those of $\Lambda$—which is to say that $F_\Lambda$ and $F_{\Lambda_i}$ are both contained in

$$\text{im} (g_{i_1 j_1} \circ \dots \circ g_{i_k j_k}).$$

Thus $\Lambda_i \to \Lambda$ implies distance $(F_{\Lambda_i}, F_\Lambda) \to 0$, so $h^{-1}$ is continuous. Being a continuous bijection of a compact and a Hausdorff space, it is a homeomorphism.

Finally, the claim together with the properties of $F_\Lambda$ clearly implies that $K = \bigcup F_\Lambda$, and hence $h$ indeed establishes a conjugacy of $f|K$ with $s_{nd}|K_{nd}$, completing the proof of the theorem.

Theorem (3.5) gives us a description of the behavior of $f \in \mathcal{A}$ on $\Omega(f)$. We now proceed to study the behavior of such an $f$ off its nonwandering set; the results obtained here will prove useful in establishing $\Omega$-stability and, even more, in extending an $\Omega$-conjugacy to a topological conjugacy.

Suppose $f \in \mathcal{A}$. If $f$ is not an expanding map, $S^1 \sim \Omega$ consists of a countable disjoint union of open intervals, say $J_1, \dots$. The following result shows that the behavior of the map on all of these intervals is in some sense determined by its behavior on a finite family of them:

PROPOSITION (3.6). $f$ as above, then every $J_i$ is eventually mapped into one of the blocks of isolated periodic points.

PROOF. As a first step, we wish to show that $J_i$ must eventually contain two periodic points. Toward this end, we claim: $f^j J_i \cap f^k J_i \neq \varnothing$ for some $j \neq k$.

Suppose not. Then, since $J_i, f J_i, f^2 J_i, \dots$ are all disjoint intervals of positive length with bounded total length, $\lim_{k \to \infty}$ length $(f^k J_i) = 0$. On the other hand, the endpoints $a_1, a_2$ of $J_i$ are in $\Omega$. If either of them is isolated, then $J_i$ is already part of a block. If not, then both are in $K$, the cantor subset of $\Omega$, so for sufficiently high $n$, $|Df^n(a_i)| > c\lambda^n > 1$ ($i = 1, 2$) and by taking possibly even higher iterates, we can assume that $|Df| \neq 1$ on $f^k \bar{J}_i$ for all large $k$. Thus on $f^k \bar{J}_i$, $|Df|$ is on the same side of 1 over the whole interval; and since the endpoints are expanding, the whole interval is expanding; since this is true for all iterates (of $f^k \bar{J}_i$), the length is increasing, contradicting the fact that it converges to zero.

So, the iterates of $J_i$ are not all disjoint; take $j \neq k$ such that $f^j J_i \cap f^k J_i \neq \varnothing$. If $f^j J_i = f^k J_i$, then the endpoints are both periodic, and we are done with our first step. If not, then an endpoint of one of the intervals is interior to the other. Since all of the endpoints are nonisolated elements of $\Omega$, some sequence of periodic points must also be interior to this interval, completing our first step in either case.

Now, for our second step, we show that the iterate above cannot contain two sources without a sink between; for if it did, then by (2.2) it would eventually cover $S^1$—so that $\bar{J}_i$ itself would eventually cover $S^1$ and hence by (2.5) would have to contain an infinity of periodic points, contradicting the definition of $J_i$ as a component of the complement of $\Omega$.

Finally, we note that this means every periodic point interior to $f^k \bar{J}_i$ is isolated—which means, in turn, that $f^k \bar{J}_i$ is contained in some block. Thus, $\bar{J}_i$ is eventually mapped into a block, proving the proposition.

We obtain, at no extra cost, four further results concerning the behavior of $f$:

COROLLARY (3.7). *If $\bar{J}_i$ is not already in a block, it is actually eventually mapped onto one.*

(The endpoints of $J_i$, and hence of any iterate, are not isolated in $\Omega$ and hence can not be interior to a block.)

COROLLARY (3.8). *A component of the (iterated) inverse image of a block coincides with the closure of a component of $S^1 \sim \Omega$.*

COROLLARY (3.9). *If $f \in \mathscr{A}$, then $S^1 = \bigcup_{p \in \mathrm{per}(f)} W^s(p)$.*

COROLLARY (3.10). *The cantor subset $K$ of $\Omega$ is both $\alpha$- and $\omega$-invariant.*

PROOF. Any point of $f^{-1} K$ is either in $K$ or in one of the $J_i$; but the $J_i$ are all eventually out of $K$, so $fK \subset K$ implies $f^{-1}K \cap J_i = \varnothing$.

4. $\Omega$- **Stability of elements of** $\mathscr{A}$. The results of the last section allow us to prove two basic results: that the elements of $\mathscr{A}$ are $\Omega$-stable, and that $\mathscr{A}$ is an open set. Since the proofs of these two facts are very closely intertwined, we combine them into a single

THEOREM (4.1). *Suppose $f \in \mathscr{A}$. Then there is a neighborhoood $\mathscr{U}$ of $f$ in $N^r(S^1)$, such that, if $\tilde{f} \in \mathscr{U}$,*
  (1) *$\tilde{f}$ is $\Omega$-conjugate to $f$.*
  (2) *$\tilde{f}$ satisfies Axiom A.*

It follows from the results of Shub [7] that the expanding maps form an open set of structurally stable maps in $N^r(S^1)$; hence we need only prove (4.1) for the case where $f$ has sinks. Recall that by (2.10), there is a finite number of blocks, $B_1, \ldots, B_k$, and by hyperbolicity, for some $\alpha$, $|Df^\alpha| > 1$ on $K$, the cantor subset of $\Omega(f)$.

The proof of the theorem will be embodied in a series of lemmas. Our general plan is first, to show that the blocks themselves are stable under perturbation, then to apply the construction of (3.5) to nearby endomorphisms and construct an $\Omega$-conjugacy. Our methods give us Axiom A for free.

The basis for the first of these steps is an adaptation of the "$\Gamma(a)$-Stability Theorem" of Abraham and Robbin (24.4 in [1]).

LEMMA (4.2). *Suppose $M$ is a compact $C^{r+2}$ manifold ($r \geqslant 1$), $a \in \mathbf{Z}^+$, and $f \in \mathscr{G}(M)$. Let $p_1, \ldots, p_m$ be the periodic points of $f$ with least period $\tau_i \leqslant a$ ($i = 1, \ldots, m$). Suppose $U_i$ is an open neighborhood of $p_i$ in $M$ ($i = 1, \ldots, m$) such that the sets $U_i$ are disjoint. Then there exists an open neighborhood $N$ of $f$ in $N^r(M)$ such that:*

(1) *for $\tilde{f}$ in $N$, there is precisely one periodic point. $\tilde{p}_i$, of least period $\leqslant a$, in each $U_i$.*

(2) *the least period of $\tilde{p}_i$ is precisely $\tau_i$.*

(3) *there are no other periodic points of $\tilde{f}$ with period $\leqslant a$.*

The proof of this lemma is just a matter of appropriately reinterpreting each step of the proof of ([1], 24.4) for endomorphisms.

We apply (4.2) to prove the stability of the blocks.

LEMMA (4.3). *For $\tilde{f}$ sufficiently near $f$, $\tilde{f}$ has blocks $\tilde{B}_1, \ldots, \tilde{B}_k$ such that $\tilde{B}_i \sim B_i$ has small measure.*

PROOF. Between any two adjacent periodic points, pick some other point $q$; also, pick additional points $q$ within a small distance beyond the end of each block. Then the complement of the $q$'s in $S^1$ forms a family of $U_i$'s as in (4.2), so that by (4.2) for $\tilde{f}$ sufficiently near $f$ there is a unique periodic point of $\tilde{f}$ between any two adjacent $q$'s, of the same period as the original "$p$" there (for $f$). Furthermore, these new points will be hyperbolic in the same sense (sink or source) as the old ones (possibly for a closer approximation), so that the $q$'s themselves cannot be periodic for $\tilde{f}$—if a $q$ were periodic, then it would have to be of the same period as the adjacent $p$'s, and analysis of nearby points would show it could be neither a sink nor a source (note that genericity of periodic points with bounded period is an open condition). This completes the proof of the lemma.

Now, as the second step in our proof of (4.1), we wish to find an analogue to the intervals $I_j$ in the proof of Theorem (3.5), to which we can then apply the process sketched in that proof simultaneously to $f$ and to any $\tilde{f}$ sufficiently near $f$.

LEMMA (4.4). *We can pick a finite collection $\{K_1, \ldots, K_m\}$ of closed intervals such that:*

(1) $K(f) \subset \bigcup_i K_i$;

(2) $|Df^\alpha|K_i| > 1$;

(3) *For $\tilde{f}$ sufficiently near $f$, $\tilde{f}$ eventually takes $S^1 \sim \bigcup_i K_i$ into its blocks, $\tilde{B}_i$;*

(4) *For $f$ sufficiently near $f$, $|D\tilde{f}^\alpha|K_i| > 1$.*

PROOF. By assumption, $|Df^\alpha| = 1$ at only finitely many points of $S^1$, so that $S^1$ divides into intervals $A_i$, on each of which either $|Df^\alpha| > 1$ or $|Df^\alpha| < 1$. In particular, $K(f)$ is interior to the union of those $A_i$ on which $|Df^\alpha| > 1$, and we can pick a closed interval $K_i$ interior to each $A_i$ so that $A_i \cap K(f) \subset \mathring{K}_i$. Then, noting that by (3.6) some iterate $f^j$ takes $S^1 \sim \bigcup \mathring{K}_i$ into the interior of the blocks, we can require $\tilde{f}^j$ to be $C^0$-$\varepsilon/3$-close to $f^j$, where $\varepsilon$ is the distance of $f^j(S^1 \sim \bigcup K_i)$ from the boundary of the $B_i$'s.

Now, note that by (4.3) we can assume, since $f^j$ takes $S^1 \sim \bigcup \mathring{K}_i$ into its blocks, that $\tilde{f}^j$ likewise takes $S^1 \sim \bigcup \mathring{K}_i$ into the $\tilde{B}_i$, so that the only elements of $\Omega(\tilde{f})$ outside the $K_i$'s are the isolated periodic points inside the $\tilde{B}_i$. Note also that, if $\delta = \min_{K_i} |Df^\alpha - 1| > 0$, then taking $\tilde{f}^\alpha$ to be $C^1$-$\delta/3$-near $f^\alpha$ insures that $|D\tilde{f}^\alpha| K_i| > 1$, as well. This proves (4.4).

CONCLUSION OF PROOF OF (4.1). Let $\tilde{f}$ be sufficiently near $f$ to satisfy the conclusions of (4.3) and (4.4). For reasons which will become apparent in the next section, we want to be careful to construct an $\Omega$-conjugacy between $f$ and $\tilde{f}$ which is order-preserving. We shall do this by mimicking the process in the proof of (3.5) in a clever way.

We again pick $g_i$ to be branches of $f^{-1}$ on the $K_j$'s, but ordered this time so that $\operatorname{im} g_i$ lies between $\operatorname{im} g_{i-1}$ and $\operatorname{im} g_{i+1}$. We note that $\operatorname{im} g_i \subset \bigcup_j K_j$ for all $i$. We can pick $\tilde{g}_i$ to be branches of $\tilde{f}^{-1}$ near the $g_i$; we are thus assured that $\operatorname{im} \tilde{g}_i \subset K_j$ iff $\operatorname{im} g_i \subset K_j$, as well as that the order of the $\operatorname{im} \tilde{g}_i$ is the same as that of the $\operatorname{im} g_i$.

We now use these $g_i$ and $\tilde{g}_i$ to define $F_\Lambda$ and $\tilde{F}_\Lambda$ as in (3.5), for $f$ and $\tilde{f}$, respectively. We note that the set of $\Lambda$'s admissible for $\tilde{f}$ will be the same as that for $f$. Also, since $|Df^\alpha| > 1$ on $K_j$, any $\alpha$-string of compositions $\tilde{g}_{i_1} \circ \ldots \circ \tilde{g}_{i_\alpha}$ will be a contraction (if defined), so that the $\tilde{F}_\Lambda$'s as well as the $F_\Lambda$'s will each be a single point, if admissible. By the same arguments as in (3.5), $K(\tilde{f}) = \bigcup_\Lambda \tilde{F}_\Lambda$, and the correspondence $F_\Lambda \leftrightarrow \tilde{F}_\Lambda$ clearly establishes an order-preserving conjugacy of $f$ and $\tilde{f}$ on $K$; furthermore, by (4.3), the rest of $\Omega(f)$ and $\Omega(\tilde{f})$ is also (order-preserving) conjugate.

Finally, the fact that all of the $\tilde{F}_\Lambda$'s are contained in $\bigcup_i K_i$ implies by construction that $|Df^\alpha| > 1$ on $\bigcup_\Lambda F_\Lambda$, so that $\tilde{f}$ satisfies Axiom A.   QED.

5. **Extension of an $\Omega$-conjugacy.** Our goal in this section is to complete the proof that elements of $\mathscr{A}$ are structurally stable, by extending the $\Omega$-conjugacies established in (4.1) to topological conjugacies. Our result leans heavily on the latter half of §3.

PROPOSITION (5.1). *If $f, g \in \mathscr{A}$, then an order-preserving conjugacy $h$ between $f|\Omega(f)$ and $g|\Omega(g)$ extends to a conjugacy of $f$ and $g$ on all of $S^1$.*

PROOF. We construct an extension in several steps.

*Step* 1. $h$ extends to any block.

It will suffice to extend $h$ to the space between two adjacent periodic points. We will, for reasons that become apparent in Step 2, construct $h$ to conjugate $f^N$ and $g^N$ on a block of period $N$. But this is just the problem of conjugating two diffeomorphisms of a closed interval, with fixed endpoints and everything in the interior moving to its right (or left). This can clearly be conjugated.

*Step* 2. $h$ extends to the orbit of any block.

Given $h$ conjugating $f^N$ and $g^N$ on the blocks $B$ and $B'$ (resp.), we have a commuting diagram of homeomorphisms:

$$B \xrightarrow{f} fB \xrightarrow{f} f^2B \xrightarrow{f} \cdots \xrightarrow{f} f^{N-1}B \xrightarrow{f} f^NB = B$$
$$h \downarrow \qquad\qquad\qquad\qquad\qquad\qquad\qquad\qquad \downarrow h$$
$$B' \xrightarrow{g} gB' \xrightarrow{g} g^2B' \xrightarrow{g} \cdots \xrightarrow{g} g^{N-1}B' \xrightarrow{g} g^NB' = B'$$

which can clearly be filled in with vertical arrows where they are now missing. (For example, let $h|fB = g \circ (h|B) \circ (f|B)^{-1}$, since $f|B$ is a homeomorphism onto $fB$.) Note that this leaves $h$ consistently defined on the whole orbit, since the outer rectangle was constructed so as to commute.

*Step* 3. $h$ extends to all of the (iterated) inverse images of any block.

By (3.8), the endpoints of any (iterated) inverse image of $B$ (or $B'$, resp.) are in $\Omega$; hence, $h$ is already defined there, and so sets up a one-one correspondence between connected (iterated) inverse images of $B$ and of $B'$. If $\tilde{B}$ and $\tilde{B}'$ are two such corresponding ones, the extension of $h$ to $\tilde{B}$ is just a matter of filling in the diagram of homeomorphisms

$$B \xleftarrow{f^k} \tilde{B}$$
$$h \downarrow$$
$$B' \xleftarrow{g^k} \tilde{B}'$$

*Step* 4. Finally, we see by (3.6) that this defines $h$ on all of $S^1$ as an order-preserving bijection conjugating $f$ and $g$; but, being order-preserving, it is a homeomorphism.  QED.

COROLLARY (5.2). $\mathscr{A} \subset \Sigma$—i.e., *elements of $\mathscr{A}$ are structurally stable in $N^r(S^1)$.*

PROOF. This is immediate from (4.1) and (5.1), since the conjugacy in (4.1) preserves order.

6. **Density of $\mathscr{A}$.** Our goal in this section is to prove that $\mathscr{A}$ is dense in $N^r(S^1)$, so that $\Sigma \subset \mathscr{A}$ (since all generic properties must in fact be possessed by any structurally stable system). We do this by first proving a rational approximation theorem and then applying an argument of Guckenheimer to show that a dense set of rational maps is in $\mathscr{A}$.

We digress to explain just what we mean by "rational maps". The "abstract" circle can be embedded as any one of many subcircles of the Riemann sphere C, all equivalent under the group of linear fractional transformations. For the moment, we shall find it useful to regard $S^1$ as the unit circle $\{x : |x| = 1\}$ in C.

DEFINITION (6.1). An endomorphism of $S^1$ is *rational* if it can be expressed as the restriction to the unit circle of a rational endomorphism of C (i.e., a map of the form $f(z) = P(z)/Q(z)$, $P$ and $Q$ polynomials) preserving $S^1$.

The exponential map $\exp(t) = e^{it}$ gives a universal covering of $S^1$ by the real line. This covering induces a "flat" metric on $S^1$ which allows us to express the derivative, $Df$, of an endomorphism $f \in E^r(S^1)$ as a real-valued function on $S^1$, given locally as

$$Df(z) = \frac{d}{dt}\Big|_{t=\arg z}[-i \arg(f(e^{it}))]$$

where arg is any local single-valued branch of the argument. It is important to notice, in the case where $f$ is the restriction of a function $F$ holomorphic on a neighborhood of $S^1$, that $Df(z)$ is quite different from the complex derivative of $F$ at $z$. This is crucial when we are considering holomorphic $C^r$-approximations of $C^r$ maps, as in our main theorem.

THEOREM (6.2). *The rational endomorphisms are dense in $E^r(S^1)$.*

PROOF. This is in essence an adaptation of a lemma due to Helson and Sarason [3].

First, we reduce the problem to that of approximating any endomorphism of degree zero by a rational one. Suppose $f$ is an endomorphism of degree $N$. Then the endomorphism of $S^1$ given by $g(z) = z^{-N} \cdot f(z)$ is of degree zero, and if $\{r_n(z)\}$ is a sequence of rational endomoprphisms whose $C^r$ limit is $g$, then $\{\tilde{r}_n(z) = z^N \cdot r_n(z)\}$ is a sequence of rational endomorphisms whose $C^r$ limit is $f$.

Now, if $f$ is a $C^r$ endomorphism of degree zero, $g(z) = \arg[f(z)]$ is a well-defined real-valued $C^r$ function on $S^1$. Our approximation is in two steps. First, using Fourier series, we $C^r$-approximate $g$ by trigonometric polynomials; these can be written in the form $\text{Im}[2Q_n(z)]$, where $Q_n$ is a (complex) polynomial whose real part vanishes at the origin. Then we approximate $Q_n$ by $R_k^{(n)}(z) = 2 \arg P_k^{(n)}(z)$, where $P_k^{(n)}$ are the partial sums of the Taylor series for $\exp Q_n$. For $n$ sufficiently large, these $R_k^{(n)}$ are, by Helson-Sarason's argument, the lifts of functions of the form $z^{-k} \cdot B(z)$, where $B(z)$ is a finite Blaschke product. We note that this means $R_k^{(n)}$ are the lifts of rational maps on $S^1$; we note also that $P_k^{(n)}(z)$ uniformly $C^r$-approximate $\exp Q_n$, so that $R_k^{(n)}$ $C^r$-approximate $2 \text{Im}[Q_n] = \text{Im}[2Q_n]$, and thus $R_n^{(n)} \to g$ in the $C^r$ sense. Finally, the fact that the exponential map is a $C^\infty$ inverse to arg shows that we have indeed $C^r$-approximated $f$ by rational functions. QED.

We remark here that while this theorem might at first glance seem to make much of our previous work unnecessary, since Guckenheimer has completely described the structure of rational endomorphisms (of C) in [2], this description cannot be regarded as complete for our case, since the rational maps form a set of only first category in $E^r(S^1)$ (being a countable union of finite-dimensional manifolds—see [2]). We can really use this theorem only to help us establish the density of $\mathscr{A}$—to establish its genericity, we needed all of the theory in §§1–5 to show that elements of $\mathscr{A}$ are structurally stable and hence form an open set.

We now proceed to the density of $\mathscr{A}$.

PROPOSITION (6.3). *The rational endomorphisms in $\mathscr{A}$ are dense in $N^r(S^1)$.*

PROOF. Guckenheimer shows in [2] that rational endomorphisms in $\mathscr{G}$ automatically satisfy Axiom A, and that a real, rational endomorphism can be approximated by a rational endomorphism in $\mathscr{G}$ which still preserves the real line. Given $f \in N^r(S^1)$, we first approximate $f$ by a rational endomorphism, $g$; then we consider $\omega \circ g \circ \omega^{-1}$, where $\omega$ is a linear fractional transformation taking $S^1$ onto $\mathbf{R} \cup \{\infty\}$; this is a real, rational endomorphism. We approximate it by a rational member of $\mathscr{G}$ (hence of $\mathscr{A}$). Then if $r_n \to \omega \circ g \circ \omega^{-1}$, $\omega^{-1} \circ r_n \circ \omega \to g$.

ADDED IN PROOF. The argument here is somewhat more involved than was at first realized. Details will appear in a subsequent paper.

COROLLARY (6.4). *$\mathscr{A}$ is a dense open set in $N^r(S^1)$.*

COROLLARY (6.5). *$\Sigma = \mathscr{A}$.*

PROOF. $\mathscr{A} \subset \Sigma$ by (5.2); $\Sigma \subset \mathscr{A}$ by (6.4), since dense properties belong to any structurally stable map.

We now simply point out that (6.5), together with (2.10) and (3.5) comprise the following restatement of the "Main Theorem" of the introduction:

MAIN THEOREM. *The set $\Sigma^r$ of all structurally stable elements of $N^r(S^1)$ is an open dense set. Its members are the Morse-Smale diffeomorphisms and the expanding maps together with maps for which the nonwandering set consists of*

(1) *a finite family of finite blocks of isolated periodic points;*

(2) *a cantor set of nonisolated nonwandering points, on which the map acts as a subsemishift of finite type, and on which the map has a hyperbolic structure.*

7. **Conjugacy classes in $\Sigma$.** We conclude our study with a few remarks on how our methods give us a way of deciding from a finite amount of data whether two elements of $\Sigma$ are conjugate to each other.

The following example shows that for this question we need to be somewhat careful, since the location of sinks alone cannot suffice, nor can knowing something about a higher iterate of the maps. Consider the following diagram:

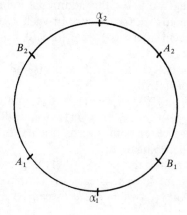

We consider two maps, $f$ and $g$. Both maps will be oreintation preserving, of degree 3. For both maps, each of the intervals $[A_1, B_1]$ and $[A_2, B_2]$ will contain a single sink ($\alpha_i$) of period 2, whose orbit consists of precisely the two points, $\alpha_i$. The two pairs $\{A_1, A_2\}$ and $\{B_1, B_2\}$ will each comprise an orbit of sources of period 2. Under $f$, each of the intervals $[A_1, B_2]$ and $[A_2, B_1]$ wraps around the circle once (that is, each point of $[A_1, B_2]$ will have two inverse images in $[A_2, B_1]$ and one in $[A_1, B_2]$). Under $g$, $[A_1, B_2]$ is mapped one-one onto $[A_2, B_1]$, and $[A_2, B_1]$ wraps around the circle twice. These might be sketched as follows:

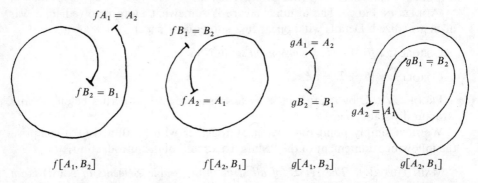

$f[A_1, B_2]$                $f[A_2, B_1]$        $g[A_1, B_2]$            $g[A_2, B_1]$

It is clear that $f$ and $g$ can be constructed so as to both be in $\mathscr{A}$ and such that even $f^2 = g^2$; but it is also clear that $f$ and $g$ themselves are not topologically conjugate. This example shows, by the way, the need for assuming our original $\Omega$-conjugacy in (5.1) to be order preserving, since in this case $f$ and $g$ *are* $\Omega$-conjugate, but not respecting order.

The example also shows that knowing the "blocks" and their arrangement (and periods) around the circle does not suffice to determine the conjugacy class of the map. However, the only respect in which the two maps $f$ and $g$ above differ is in their "degree" on each of the intervals $[A_1, B_2]$ and $[A_2, B_1]$—that is, each one "wraps around" the circle once under $f$, while under $g$ one wraps around twice and the other not at all.

Suppose, however, that we know this additional information, as well. That is, suppose we have two maps, $f_1, f_2 \in \mathscr{A}$, and for each $f_i$ we have a partition of $S^1$ into a sequence of adjacent intervals $\alpha_1(i), \beta_1(i), \alpha_2(i), \beta_2(i), \ldots, \beta_n(i), \alpha_1(i)$ $(i = 1, 2)$ (in this order) such that:

(1) each $\alpha_j(i)$ is a "block" containing $k_j$ sinks (same for $\alpha_j(1)$ and $\alpha_j(2)$);

(2) each $\beta_j(i)$ is free of sinks;

(3) the "degree" of $f_i$ on $\beta_j(i)$ is $d_j$ (same for $\beta_j(1)$ and $\beta_j(2)$);

(4) $f_1\alpha_j(1) = \alpha_k(1)$ iff $f_2\alpha_j(2) = \alpha_k(2)$; $f_1\beta_j(1) = \beta_k(1)$ iff $f_2\beta_j(2) = \beta_k(2)$;

(5) $f_i$ either both preserve, or both reverse, orientation.

Then we claim that $f_1$ is conjugate to $f_2$.

To see this, we just mimic the proof of (4.1). Suppose for simplicity that the maps preserve orientation. Then it is easy to see that each $\beta_j(i)$ will contain $d_j$ inverse images $g_{jk}\alpha_l(i)$ of $\alpha_l(i)$, where the $g_{jk}$'s can be numbered so that all $g_{j1}$'s precede all $g_{j2}$'s, etc. Then between $g_{jk}\alpha_l(i)$ and $g_{jk}\alpha_{l+1}(i)$ will be an inverse image of $\beta_l(i)$; call it $g_{jk}\beta_l(i)$. This sets up a one-one order-preserving correspondence between branches of $f_j^{-1}|\beta_l(j)$, which then by the process of (4.1) sets up an order-preserving $\Omega$-conjugacy between $f_1$ and $f_2$. Hence, such a partition together with the information (1)–(5) above in fact determines the conjugacy class of a member of $\Sigma^r$.

As a result, we see that we can effectively exhibit all of the stable conjugacy classes in $N^r(S^1)$ by writing down an example (similar to the ones at the beginning of this section) for each distinct set of information of the form (1)–(5) above.

**Appendix.** *Glossary.* Alternate formulations (for endomorphisms) of most of the notions below can be found in [7] and [2].

Let $M$ denote a compact manifold.

An *endomorphism* of $M$ is an arbitrary $C^r$ map $f: M \to M$.

$E^r(M) = \{C^r$ endomorphisms of $M\} = C^r(M, M)$.

An *automorphism* of $M$ is a diffeomorphism of $M$.

$\text{Diff}^r(M) = \{C^r$ automorphisms of $M\}$.

$f \in E^r(M)$ is *nonsingular* iff $Df(x)$ injective $(\forall x \in M)$.

$N^r(M) = \{f \in E^r(M)$ nonsingular$\}$.

$f \in E^r(M)$ is a *self-covering* iff $f: M \to M$ is a covering space.

Let $f \in E^r(M)$, $x \in M$.

*Orbit* of $x = \{f^n(x): n \in Z_+\}$.

$x$ is *periodic* iff $\exists n > 1.\exists.f^n(x) = x$. $n =$ a *period* of $x$.

per $(f) = \{x$ periodic under $f\}$.

$x \in$ per $(f)$, $n =$ period $(x)$, is *generic* iff $Df^n(x)$ is *hyperbolic*—i.e. $|\zeta| \neq 1$

                                    a *source* iff $Df^n(x)$ is *expanding* $(|\zeta| > 1)$

                                      a *sink* iff $Df^n(x)$ is *contracting* $(|\zeta| < 1)$

                                    —for all eigenvalues $\zeta$ of $Df(x)$

(dim $M = 1$ implies all generic periodic points are sinks or sources.)

$x$ is *eventually periodic* iff orbit $(x)$ is finite—i.e., $f^k x \in$ per $(f)$, some $k$.

$x$ is *nonwandering* iff $(\forall U$ nbhd of $x) \bigcup_{k \geqslant 1} U \cap f^k U \neq \varnothing$.

$\Omega(f) = \{$nonwandering points of $f\}$—clearly a closed set.

Let $p \in M$

$$\omega(p) = \{\omega\text{-limits of } p \text{ under } f\} = \{x : \exists n_i \to \infty. \exists. x = \lim f^{n_i} p\}$$

$$\alpha(p) = \{\alpha\text{-limits of } p \text{ under } f\} = \left\{x : \exists \{x_n\}_{n>1}. \exists. \begin{matrix} (1) & (\forall n > 1) fx_n = x_{n-1} \\ (2) & \text{some subsequence converges} \\ & \text{to } x \end{matrix}\right\}.$$

Clearly, $\omega(p) \cup \alpha(p) \subset \Omega(f)$.

$S \subset M$ is $(\omega)$-*invariant* $(=preserved)$ under $f$ iff $fS \subset S$.

    $S$ is $\alpha$-*invariant* iff $f^{-1}S = \{x : fx \in S\} \subset S$.

Clearly, (1) $\Omega(f)$ is $\omega$-invariant, (2) $S$ $\alpha$- and $\omega$-invariant $\Rightarrow fS = S$.

$f$ is *topologically transitive* on $S$ iff: (1) $S$ $\omega$-invariant (2) $\exists x \in S.\exists.S = \overline{\text{orbit } (x)}$.

$f$ has a *hyperbolic structure* for $S$, $\omega$-invariant, iff $TS = T_eS \oplus T_cS$ and $\exists c > 0, \lambda > 1, N \in \mathbf{Z}_+ . \exists .(\forall n > N)$ :

(1) $\|Tf^n(x)\| > c\lambda^n \|x\|$  for $x \in T_eS$,

(2) $\|Tf^n(x)\| < c\lambda^{-n} \|x\|$  for $x \in T_cS$.

If $M = S^1$, this reduces to: $S = S_e \cup S_c$ (disjoint) $.\exists.(\forall n > N)$ :

(1) $|Df^n(x)| > c\lambda^n$  for $x \in S_e$,

(2) $|Df^n(x)| < c\lambda^{-n}$  for $x \in S_c$.

$f \in E^r(M)$ is *expanding* iff $S = M$, $T_eS = TM$. $f$ expanding $\Rightarrow \Omega(f) = M([7])$.

$p \in \text{per}(f)$ generic, $n = \text{period } p$, then

$$W^s(p) = \text{stable manifold of } p = \{x : p \in \omega(x) \text{ under } f^n\}$$
$$W^u(p) = \text{unstable manifold of } p = \{x : p \in \alpha(x) \text{ under } f^n\}.$$

If $p$ is a sink (source), then $W^s(p)(W^u(p))$ is a neighborhood of $p$. Any open connected neighborhood of $p$ contained in $W^s(p)(W^u(p))$ is called a *local stable (unstable) manifold* for $p$, denoted $W^u_{\text{loc}}(p)(W^u_{\text{loc}}(p))$.

$f_i \in E^r(M_i)$ $(i = 1, 2)$, then

$$f_1 \sim f_2 \text{ (topologically conjugate) iff } \exists h : M_1 \to M_2 \text{ homeo } \exists hf_1 = f_2 h.$$
$$f_1 \underset{\Omega}{\sim} f_2 \text{ (}\Omega\text{-conjugate) iff } f_1|\Omega(f_1) \sim f_2|\Omega(f_2).$$

Let $\mathscr{S}$ be a subspace of $E^r(M)$.

$f \in \mathscr{S}$ is *structurally ($\Omega$-) stable* in $\mathscr{S}$ iff $\exists$ nbhd $\mathscr{U}$ of $f$ in $\mathscr{S}.\exists.$
$$g \in \mathscr{U} \Rightarrow g \sim f(g \underset{\Omega}{\sim} f).$$

## REFERENCES

**1.** R. Abraham and J. Robbin, *Transversal mappings and flows*, Benjamin, New York, 1967.

**2.** J. Guckenheimer, *Endomorphisms of the Riemann sphere*, these Proceedings, vol. 14.

**3.** H. Helson and D. Sarason, *Past and future*, Math. Scand. **21** (1967), 5–16.

**4.** S. Newhouse, *Nondensity of axiom A(a) on $S^2$*, these Proceedings, vol. 14.

**5.** M. Peixoto, *Structural stability on two-dimensional manifolds*, Topology, **1** (1962), 101–120.

**6.** C. Pugh, *An improved closing lemma and a general density theorem*, Amer. J. Math. **89** (1967), 1010–1021.

**7.** M. Shub, *Endomorphisms of compact differentiable manifolds*, Amer. J. Math. **91** (1969).

**8.** S. Smale, *Differentiable dynamical systems*, Bull. Amer. Soc. **73** (1967), 747–817.

University of California, Berkeley

# A NOTE ON Ω-STABILITY

J. PALIS

Recently the following important result in Dynamical Systems has been proved by Smale [4]: if a diffeomorphism $f$ of a compact $C^\infty$ manifold $M$ satisfies Axiom A and $\Omega(f)$ has the no cycle property then $f$ is $\Omega$-stable. We show in this note that if $f$ satisfies Axiom A and is $\Omega$-stable then $\Omega(f)$ has the no cycle property.

Here $\Omega(f)$ stands for the nonwandering set of $f$. We denote by Diff($M$) the set of $C^r$ diffeomorphisms of $M$ with the uniform $C^r$ topology, $0 < r$. We call $f \in \mathrm{Diff}(M)$ $\Omega$-stable if there exists a neighborhood $N$ of $f$ in Diff($M$) such that for each $g \in N$, $g/\Omega(g)$ is topologically conjugate to $f/\Omega(f)$.

(1) DEFINITION. We say that $f \in \mathrm{Diff}(M)$ satisfies Axiom A if

(i) $\Omega(f)$ has a hyperbolic structure,

(ii) the set of periodic points is dense in $\Omega(f)$.

We refer the reader to [3] for a detailed presentation of these concepts.

From the Spectral Decomposition Theorem [3], if $f$ satisfies Axiom A then $\Omega = \Omega(f)$ can be written as $\Omega = \Omega_0 \cup \ldots \cup \Omega_k$, where the $\Omega_i$ are closed invariant sets and $f$ is topologically transitive on each of them. These $\Omega_i$ are called basic sets for $f$. For any closed hyperbolic set $K$ one can define [1] its stable and unstable manifolds, denoted by $W^s(K)$ and $W^u(K)$.

(2) DEFINITION. Let $f \in \mathrm{Diff}(M)$ satisfy Axiom A. An $n$-cycle ($n \geq 1$) on $\Omega(f)$ is a sequence of basic sets $\Omega_0, \ldots, \Omega_n, \Omega_{n+1}$ (reordering the indices if necessary) such that $W^s(\Omega_i) \cap W^u(\Omega_{i+1}) \neq \varnothing$, $\Omega_0 = \Omega_{n+1}$ and otherwise $\Omega_j \neq \Omega_k$ for $j \neq k$. When there is no $n$-cycle ($n \geq 1$) on $\Omega(f)$ we say that $\Omega(f)$ has the no cycle property.

(3) THEOREM. *Let $f \in \mathrm{Diff}(M)$ satisfy Axiom A and be $\Omega$-stable. Then $\Omega = \Omega(f)$ has the no cycle property.*

PROOF. The theorem will follow by showing that if there is an $n$-cycle on $\Omega$, $n \geq 1$, then $f$ is not $\Omega$-stable.

We prove first a weaker assertion: if there is an $n$-cycle on $\Omega$ for $n > 1$ then $f$ is not $\Omega$-stable or there exists $g$ near $f$ in Diff($M$) such that $g/\Omega = f/\Omega$ and $\Omega$ is the union $\Omega_0 \cup \ldots \cup \Omega_k$ of hyperbolic sets for $g$ with an $(n-1)$-cycle. We then show that this last alternative also implies the non-$\Omega$-stability of $f$.

Let $\Omega_0, \ldots, \Omega_n, \Omega_{n+1}$ be a cycle on $\Omega = \Omega(f)$. From $W^s(\Omega_i) = \bigcup_{x \in \Omega_i} W^s(x)$ and $W^u(\Omega_i) = \bigcup_{x \in \Omega_i} W^u(x)$ [4], we have $W^s(x_i) \cap W^u(x_{i+1}^*) \neq \varnothing$ for some $x_i$, $x_i^* \in \Omega_i$ and $0 \leq i \leq n$. We claim that at least for one of the indices $i$, $\dim W^s(x_i) + \dim W^u(x_{i+1}^*) \geq \dim M$. For otherwise, if $r = \sum_{i=0}^n (\dim W^s(x_i) + \dim W^u(x_{i+1}^*))$ then $r < n \dim M$. But since $\Omega_{n+1} = \Omega_0$ and $\dim W^s(x_i) + \dim W^u(x_i^*) = \dim M$ we have $r = n \dim M$, thus reaching a contradiction. Let $y \in W^s(x_j) \cap W^u(x_{j+1}^*)$, the index $j$ chosen as above. Clearly $y$ is a wandering point for $f$. Thus after a small

221

local pertubation of $f$ near $y$ we get $f_i$ near $f$ in $\text{Diff}(M)$ so that $f_1/\Omega = f/\Omega$, $W^s(x_j, f_1)$ and $W^u(x^*_{j+1}, f_1)$ have a point of transversal intersection and $W^s(x_i, f_1) \cap W^u(x^*_{i+1}, f_1)$ $\neq \varnothing$ for $0 \leq i \leq n$. From the density of the periodic points in $\Omega$ we may assume that $x_j$, $x^*_{j+1}$ are periodic points of $f$ (and thus of $f_1$). Now, $f_1/\Omega_{j+1} = f/\Omega_{j+1}$ is topologically transitive and so we may further assume $x^*_{j+1} = x_{j+1}$. Furthermore if $z \in W^s(x_{j+1}, f_1) \cap W^u(x^*_{j+2}, f_1)$ is nonwandering for $f_1$ then $f_1/\Omega(f_1)$ is not conjugate to $f/\Omega$, which implies that $f$ is not $\Omega$-stable. Suppose $z$ is wandering for $f_1$. Then, since Closure $W^s(x_j, f_1) \supset W^s(x_{j+1}, f_1)$ [2], by a small local pertubation of $f_1$ near $z$ we get $g$ near $f$ in $\text{Diff}(M)$ with $g/\Omega = f/\Omega$ and $W^s(x_j, g) \cap W^u(x^*_{i+2}, g)$ $\neq \varnothing$, thus producing an $(n-1)$-cycle on $\Omega \subset \Omega(g)$ and hence proving our assertion.

The proof of the theorem goes now by induction on the length of the cycle and the fact that if $\Omega$ has a 1-cycle then $f$ is not $\Omega$-stable. For as above after a small local pertubation of $f$, say to $f_1$, $W^s(x_1, f_1)$ and $W^u(x^*_2, f_1)$ has a point $y$ of transversal intersection. Then $y$ is nonwandering for $f_1$ and using that $f/\Omega_i$ is topologically transitive we conclude that $f$ and $f_1$ are not $\Omega$-conjugate and thus $f$ is not $\Omega$-stable. This finishes the proof of the theorem.

From [2] and this Theorem we get

(4) COROLLARY. *Let $f \in \text{Diff}(M)$ and $\Omega(f)$ be finite. Then $f$ is $\Omega$-stable iff $\Omega(f)$ is hyperbolic and has the no cycle property.*

For the general case, Smale posed in [5] the following

(5) CONJECTURE. *Let $f \in \text{Diff}(M)$. Then $f$ is $\Omega$-stable iff $f$ satisfies Axiom A and $\Omega(f)$ has the no cycle property.*

From the $\Omega$-stability Theorem [4] and the above Theorem it follows that the remaining question to consider in the above conjecture is whether $\Omega$-stability implies Axiom A.

## REFERENCES

1. M. Hirsch and C. Pugh, *Stable manifolds and hyperbolic sets*, these Proceedings, vol. 14.
2. J. Palis, *On Morse-Smale dynamical systems*, Topology (to appear).
3. S. Smale, *Differentiable dynamical systems*, Bull. Amer. Math. Soc. 73 (1967), 747–817.
4. ———, *The $\Omega$-stability theorem*, these Proceedings. vol. 14.
5. ———, *Global stability questions in dynamical systems*, Symposium on Global Analysis, Washington, D.C., April 1968.

UNIVERSITY OF CALIFORNIA, BERKELEY
IMPA, RIO DE JANEIRO

# STRUCTURAL STABILITY THEOREMS

J. PALIS AND S. SMALE

1. We prove here that a class of diffeomorphisms, i.e. differentiable automorphisms, introduced in [3] and studied in [2] are structurally stable. Since every compact manifold has such diffeomorphisms, this implies a positive answer to the well-known question as to whether every manifold has a structurally stable diffeomorphism.

These results are proved in [2] for manifolds with dimension $\leq 3$.

We show also that the similar theorems for differential equations are true.

The precise results are as follows. Let $M$ be a compact $C^\infty$ manifold without boundary, and for $r \geq 1$ let Diff $(M)$ be the set of $C^r$ diffeomorphisms of $M$ with the uniform $C^r$ topology. For $f \in$ Diff $(M)$ we denote by $\Omega(f)$ the set of nonwandering points of $f$ (see [5] for elementary definitions and a general background). Consider $f \in$ Diff $(M)$ which satisfies

(1.1)
- (a) $\Omega(f)$ is finite,
- (b) $\Omega(f)$ is hyperbolic,
- (c) Transversality Condition.

Condition (a) implies that $\Omega = \Omega(f)$ consists of periodic points. Then (b) means that if $x \in \Omega$ and $f^m(x) = x$, for some $m > 0$, then the derivative $Df^m(x): T_x(M) \to T_x(M)$ has its eigenvalues not equal to one in absolute value. From condition (b) one has defined stable and unstable manifolds for each $x \in \Omega$ denoted by $W^s(x)$, $W^u(x)$ respectively. Then condition (c) means that for each $x, y \in \Omega$, $W^s(x)$, $W^u(y)$ have transversal intersection in $M$.

A diffeomorphism $f$ Diff $(M)$ is said to be *structurally stable* if there exists a neighborhood $N(f) \subset$ Diff $(M)$ such that if $g \in N(f)$, then $f$ and $g$ are topologically conjugate, i.e. there exists a homeomorphism $h: M \to M$ satisfying $hf = gh$.

(1.2) MAIN THEOREM. *If $f \in$ Diff$(M)$ satisfies (1.1) then $f$ is structurally stable.*

Using [4], we have the following corollaries:

(1.3) COROLLARY. *The gradient dynamical systems on a compact Riemannian manifold $M$ which are structurally stable form a dense open set among all gradient dynamical systems on $M$. Moreover, a gradient dynamical system is structurally stable iff its induced diffeomorphism at time $t = 1$ satisfies (1.1).*

(1.4) COROLLARY. *Any compact manifold admits a structurally stable diffeomorphism.*

This paper is mainly devoted to proving (1.2). §5 gives the corresponding results for differential equations and poses a question related to the main result in this paper.

Conversations with M. Hirsch and C. Pugh have been very helpful in preparing this paper.

2. In this section we first review the needed background results from [2] and [3]. We then construct for $f$ satisfying (1.1) tubular families of its stable manifolds, as defined below.

Let $f$ satisfy (1.1) and $\Omega = \Omega(f)$. If $\Omega_k$ is a periodic orbit of $\Omega$, let $W^s(\Omega_k) = \bigcup_{p \in \Omega_k} W^s(p)$ and $W^u(\Omega_k) = \bigcup_{p \in \Omega_k} W^u(p)$. Then from [3] (see also [2] and [5]):

(2.1) THEOREM. *The $\Omega_k$ are partially ordered by*: $\Omega_i \leq \Omega_k$ *iff* $W^s(\Omega_i) \cap W^u(\Omega_k) \neq \varnothing$. *For all $1 \leq k \leq m$, $W^s(\Omega_k)$ and $W^u(\Omega_k)$ are properly imbedded submanifolds of $M$.*

We now choose once and for all a simple ordering consistent with this partial ordering, so that $\Omega_1 \leq \Omega_2 \leq \Omega_3 \leq \ldots \leq \Omega_{m-1} \leq \Omega_m$, the $\Omega_k$ being all the periodic orbits of $f$. Thus $\Omega_1$ (and perhaps $\Omega_2$) is a sink (or attractor) and $\Omega_m$ is a source.

Submanifolds are always supposed to be properly imbedded and without boundary unless so stated. By the general theory for $p \in \text{Per}\,(f)$, $W^s(p)$ is a submanifold of $M$.

(2.2) DEFINITION (TUBULAR FAMILY OF $W^s(p)$). A *tubular family* $T$ of $W^s(p) = T_0$ is a collection of disjoint $C^r$ submanifolds $\{T_y\}$ of $M$ indexed by $y$ in an open neighborhood $N$ of $p$ in $W^u(p)$ with the following properties.

(1) $V = V(T_0) = \bigcup_{y \in N} T_y$ is an open set of $M$ containing $T_0$.
(2) $T_0 = T_p$.
(3) $T_y$ intersects $N$ transversally in the single point $y$.

Finally

(4) the map $V \to N$ which sends $T_y$ into $y$ is continuous; the section $s$ which sends $x \in T_y$ into the tangent space of $T_y$ at $x$ is a continuous map from $V$ into the Grassmann bundle over $V$.

Note that $s$ in (4) is not required to be smooth.

If $\Omega_k$ is a periodic orbit of $f$ satisfying (1.1), then we will speak of a tubular family for $W^s(\Omega_k)$. This is simply a tubular family $T = \{T_x\}$ for $W^s(p)$, some $p \in \Omega_k$ together with the submanifolds $\{f^i(T_x)\}_{i=0,\ldots,n-1}$ $n = \text{period } \Omega_k$.

A tubular family $T = \{T_y\}$ of $W^s(p)$ is called *invariant* if $f^{-n}T_y = T_{f^{-n}(y)}$ where $n$ is the period of $p$. In this case the corresponding tubular family of $W^s(\Omega_k)$ will also be called invariant.

A *system* of tubular families for $f$ is a set of tubular families $T^k = \{T_x^k\}$, one for each $\Omega_k$. The system will be called *compatible* when it meets the condition, if $T_x^i \cap T_y^j \neq \varnothing$, then one submanifold contains the other.

The main goal of this section is to prove that $f \in \text{Diff}\,(M)$ satisfying (1.1) has a compatible system of invariant tubular families.

A standard result in Differential Topology is:

(2.3) LEMMA. *Let $N$ be a $C^r$ manifold possibly with boundary and $I = [0, 1]$. There exist a neighborhood $U(\Delta)$ of the diagonal $\Delta$ in $N \times N$ and a $C^r$ function $\theta : U(\Delta) \times I \to N$ such that $\theta(x, y, 0) = x$, $\theta(x, y, 1) = y$, $\theta(x, x, t) = x$ for all $t \in I$, and if $x, y \in \partial N$, then $\theta(x, y, t) \in \partial N$.*

PROOF. By a theorem of Whitney, choose a $C^\infty$ structure on $N$ compatible with its $C^r$ structure and define $\theta$ via geodesics in some $C^\infty$ Riemannian metric, for which $\partial N$ is totally geodesic. Thus $\theta(x, y, t)$ is the point on the unique geodesic joining $x$ and $y$ at "time $t$ from $x$."

A $C^r$ retraction $r \geq 0$ of an open set $U \subset M$ into a closed submanifold $B$ of $U$ is a $C^r$ map $p : U \to B$ which is the identity on $B$.

(2.4) LEMMA ($C^r$ RETRACTION LEMMA). *Let $M$ be a $C^\infty$ manifold, $B$ a closed $C^r$ submanifold. Let $A$ be a compact set in $B$, $U_0$ a neighborhood of $A$ in $M$ and $r_0 : U_0 \to B$ a $C^r$ retraction onto $U_0 \cap B$. Then there exists a neighborhood $U$ of $B$ and a retraction $r : U \to B$ such that $r/U'_0 = r_0/U'_0$, $U'_0$ a neighbourhood of $A$.*

*Furthermore if $B$ has boundary, the above holds if $A \subset \text{int } B$ and $U$ is a tubular neighborhood "with boundary" $\partial U$, fibered by $r$ over $\partial B$.*

PROOF. The tubular neighborhood theorem yields a $C^r$ retraction $\pi : U_1 \to B$ where $U_1$ is a neighborhood of $B$. Let $\phi$ be a $C^\infty$ "bump function", i.e. $\phi$ is 1 on a neighborhood of $A$, with support of $\phi$ contained in $U_0$ and $0 \leq \phi \leq 1$.

Let $\theta$ be the "geodesic function" on $B$ by the previous lemma. Then define $r(x) = \pi(x)$ if $x \notin U_0$ and $\theta(\pi(x), r_0(x); \phi(x))$ if $x \in U_0$. This proves (2.4).

Having seen this basic extension process, we now prove a parameterized version of it in the next lemma, which we explicitly use in constructing tubular families.

(2.5) LEMMA *Let $T = \{T_x\}$ be a tubular family for $W^s(\Omega_i)$ and $T_0 = \bigcup_{x \in \Omega_i} T_x$. Let $W$ be a $C^r$ submanifold meeting $T_0$ transversally and $A$ be a compact set in $W$. Suppose that $r_0 : U_0 \to W$ is a given continuous retraction of a neighborhood $U_0$ of $(\partial T_0) \cap A$ ($\partial T_0$ the point set boundary) which is*

(i) *$C^r$ on each $T_x$,*

(ii) *$y \in T_x \cap W \cap U$ implies $r_0^{-1}(y) \subset T_x$ and*

(iii) *the family $\{r^{-1}(y)\}$ satisfies the Grassmannian continuity condition of (2.2)–4.*

*Then there exists a continuous retraction $r : U \to W$, $U$ a neighborhood of $\overline{T}_0 \cap A$ such that (i), (ii) and (iii) hold for $r$ and $r$ restricted to some neighborhood of $(\partial T_0) \cap A$ is $r_0$.*

(Here $\overline{T}_0$ is the closure of $T_0$ in $M$.)

We indicate how we apply this crucial lemma in the construction of tubular families. We take $W = W^u(p)$, $p \in \Omega_k$, period $\Omega_k = n$. A fundamental domain $D$ for $f^{-n}$ on $W^u(p)$ is a subset of the form closure $(N - f^n(N))$ where $N$ is a compact ball in $W^u(p)$ such that $f^{-n}(N) \subset \text{int } N$. Then $\partial D = S_E \cup S_I$, a disjoint union of two spheres, an exterior one $S_E$, an interior one $S_I$ with $f^{-n}(S_E) = S_I$. In the applications, $A$ will be either $S_E$ or $D$, and $i$ will be less than $k$.

This will be carried out in (2.6). Now we prove (2.5).

PROOF. First assume $\Omega_i$ is a point $p$. Since the question is restricted to a neighborhood of $A$, we may assume $W$ to be compact and $A \subset$ int $W$. Let $U_1$ be an open neighborhood of $\partial T_0 \cap W$ such that $\bar{U}_1 \cap A \subset U_0 \cap A$. Then choose a $C^r$ retraction $F: U(T_0) \to T_0$ defined off $U_1$ and such that $F(W) \subset W \cap T_0$. This can be done using a tubular neighborhood of $T_0$ which extends a tubular neighborhood of $W \cap T_0$ in $W$. Let $F_x = F/T_x$ so $F_x: T_x \to T_0$ is $C^r$ for each $x \in N$, where $N$, the index set for $T$, is a neighborhood of $\Omega_i$ in $W^u(\Omega_i)$. Then $N$ can be chosen small enough so that for each $x \in N$, $F_x$ is invertible on neighborhood $T_0 \cap A$.

Choose a $C^r$ retraction $\pi_0: T_0 \cap V \to T_0 \cap W$ with $V$ some neighborhood of $T_0 \cap A$. Then define for $x \in N$ a $C^r$ retraction $\pi_x: T_x \cap V \to T_x \cap W$ by $\pi_x = F_x^{-1} \circ \pi_0 \circ F_x$. This makes sense off $U_1$.

Also, off $U_1$ define via (2.3) a "geodesic function" $\mathcal{O}_0$ for $T_0 \cap W$ and then let $\mathcal{O}_x$ be the geodesic function on $T_x \cap W$ off $U_1$ induced by the map $F_x: T_x \to T_0$.

Now we follow the proof of (2.4) to obtain our desired retraction $r$. Thus let $\phi$ be a $C^\infty$ bump function which is 1 on $U_1$, has support in $U_0$ and $0 \leq \Phi \leq 1$. Define $r = r_0$ on $U_1$, $r(z) = \pi_x(z)$ if $z \in T_x$-supp $\Phi$ znd $r(z) = \theta_x(\pi_x z, r_0 z, \Phi(z))$ otherwise.

Thus $r$ is defined in some neighborhood $U$ of $T_0 \cap A$ and one checks directly that $r$ has the desired properties.

In the general case $T_0 = W^s(\Omega_i)$, proceed as above for each component of $T_0$. This proves (2.5).

(2.6) TUBULAR FAMILY THEOREM. *Let* $f \in \text{Diff}(M)$ *satisfy* (1.1) *and let* $\Omega(f) = \bigcup \Omega_k$, *where the* $\Omega_k$ *are the periodic orbits of* $\Omega(f)$ *simply ordered as above. There exists a compatible, invariant system of tubular families* $\{T^k\}$, *each* $T^k$ *being a tubular family of* $W^s(\Omega_k)$. *Thus* $T_0^k = W^s(\Omega_k)$.

*Moreover, the map* $\pi: \bigcup_{x \in N_k} T_x^k \to W^u(\Omega_k)$, *defined by* $\pi(T_x^k) = x = T_x^k \cap W^u(\Omega_k)$ *is* $C^r$ *on* $T_x^i$, *each* $x$ *and* $1 \leq i \leq k$.

PROOF. We sometimes let $T^k$ be the union $\bigcup_{x \in N_k} T_x^k$, restricting $N_k$ when necessary. Proceeding inductively define $T^1$ as simply $\{W^u(p) | p \in \Omega_1\}$. Now assume $T^1, \ldots, T^{k-1}$ constructed as in the statement of the theorem. Let us construct $T^k$. For $p \in \Omega_k$ of period $m$, let $D$ be a fundamental domain with $\partial D = S_E \cup S_I$. We will define a continuous retraction $\pi: U(D) \to W^u(p)$ using a second induction on $T^i \cap U(D)$ for $i = k - 1, \ldots, 1$, such that $\pi f^{-n} = f^{-n} \pi$ on a neighborhood of $S_E$, conditions (i), (ii) of (2.5) are met for each $T^i$, $1 \leq i \leq k - 1$ and $\{\pi^{-1}(y)\}$ satisfy (2.5)–(iii).

Once this is done, the desired $T^k$ is obtained as follows. Let $N$ be the neighborhood of $p \in \Omega_k$ in $W^u(p)$, with $\partial N = S_E$. For each $x \in N$ there is a unique point $y \in D - S_E$ and an integer $j \geq 0$ so that $x = f^{-nj}(y)$. We then set $T_x^k = f^{-nj}(\pi^{-1}(y))$ and $T_p^k = W^s(p)$. $\{T_x^k\}_{x \in N}$ is a tubular family of $W^s(p)$. This follows from Lemma 1.1 of [2]. It amounts to seeing that $\bigcup_{x \in N} T_x^k$ is open (which also follows from Hartman's Theorem) and the continuity condition (2.2)–4 is met. This continuity is also a consequence of basic known linear estimates.

We now construct the retraction $\pi: U(D) \to W^u(p)$ as required above. To give the idea, the retraction is constructed in the following order. From

$U((\partial T_0^{k-1}) \cap S_E)$ it is extended to $U(T_0^{k-1} \cap S_E)$. This induces via $f^{-n}$ the retraction on $U(T_0^{k-1} \cap S_I)$, which is now extended from $U((T_0^{k-1}) \cap S_E) \cup U(T_0^{k-1} \cap S_I)$ to $U(T_0^{k-1} \cap D)$. Then we extend it from $U((\partial T_0^{k-2}) \cap S_E)$ to $U(T_0^{k-2} \cap S_E)$. This induces the retraction on $U((T_0^{k-2}) \cap S_E)$ to $U(T_0^{k-2} \cap S_E)$. This induces the retraction on $U(T_0^{k-2} \cap S_I)$ and we extend it from $U(T_0^{k-1} \cap D) \cup U(T_0^{k-2} \cap S_E)$ $\cup U(T_0^{k-2} \cap S_I)$ to $U(T_0^{k-2} \cap D)$. We continue with $T_0^{k-3}$ in the same manner.

More precisely let $\pi: T^{k-1} \cap U(S_E) \to W^u(p)$ given by (2.5). For that we take in (2.5) $A$ to be $S_E$, $W = W^u(p)$, $T$ to be $T^{k-1}$, using the fact that $T^{k-1} \cap D$ is compact and $\partial T_0 \cap A = \varnothing$. Since $T^{k-1}$ is invariant and $f^{-n}(S_E) = S_I$, the composition $f^{-n}\pi$ defines a retraction of $T^{k-1} \cap U(S_I)$ into $W^u(p)$, denoted also by $\pi$. Here $U(S_I) = f^{-n}(U(S_E))$. Now apply again (2.5) with $A = D$ to extend $\pi$ to the desired retraction of $T^{k-1} \cap U(D)$ into $T^{k-1} \cap D$.

We proceed in the same way to extend $\pi$ to $T^{k-2} \cap U(D)$, using again (2.5) and the fact that $(\partial T_0^{k-2}) \cap D \subset (T_0^{k-1} \cup T_0^k) \cap D$. Continuing this process through $T^{k-2}, T^{k-3}, \ldots, T^1$ yields the desired $\pi$ and hence (2.6).

(2.7) REMARK. Via iteration by $f$, the system $\{T^k\}$ obtained in (2.6) can be extended to be also $f$ invariant. Such a system will be denoted by $\{T^{k,s}\}$, while $\{T^{k,u}\}$ will denote a similar system of tubular families of the $W^u(\Omega_k)$. $T^{k,s}(x)$, $T^{k,u}(x)$ will stand for the element of the family $T^{k,s}$, $T^{k,u}$ containing the point $x$.

(2.8) REMARK. The set of diffeomorphisms satisfying (1.1) is open in Diff$(M)$ [2]. Let $\{T^{k,s}\}$, $\{T^{k,u}\}$ be as above. There exists a neighborhood $N$ of $f$ such that for each $g \in N$ we can construct compatible, $g$ and $g^{-1}$ invariant systems of tubular families $\{T_g^{k,s}\}$, $\{T_g^{k,u}\}$ of the $W^s(\Omega_k(g))$, $W^u(\Omega_k(g))$ where $\{T_f^{k,s}\} = T^{k,s}$ and $\{T_f^{k,u}\} = \{T^{k,u}\}$. Moreover we may assume that the maps $\phi_1(g) = S_g^{k,s}$ and $\phi_2(g) = S_g^{k,u}$, $g \in N$, are continuous, at $f$ on compact parts of the $T^{k,s}$, $T^{k,u}$ where $S_g^{k,s}$, $S_g^{k,u}$ are the sections on the Grassman bundle of (2.2)–4 defined for $T_g^{k,s}$, $T_g^{k,u}$. This can be checked at each stage of the proof of the Tubular Family Theorem, using the same uniform estimates on linear maps mentioned there. The tubular families that appear in the following sections are the ones given by the maps $\phi_1$, $\phi_2$ above.

3. Let $N$ be a neighborhood of $f$ in Diff$(M)$. For each $g \in N$, $gf^{-1}$ is called a perturbation of $f$. The perturbation is said to be localized at an open set $U$ of $M$ if it is 1 on $M - U$, where 1 stands for the identity map.

The main purpose of this section is to show that perturbations of $f$ near 1 can be decomposed into simple ones. That is, into localized perturbations leaving invariant the tubular family structure introduced in the previous section.

We denote by supp $f$ the closure of the set where $f \neq 1$. The following lemma was detailed to us by M. Hirsch.

(3.1) LEMMA. Let $\{U_k | 1 \leq k \leq n\}$ be an open cover of $M$ and $N$ be a neighborhood of 1. There exists a neighborhood $N_1$ of 1 such that if $f \in N_1$ then $f$ can be factored as $f = f_n \circ \ldots \circ f_1$, where $f_i \in N$, and supp $f_i \subset U_i$.

PROOF. First choose $N_1$ so that if $f \in N_1$ then there is an isotopy $F: M \times I \to M$, $F_0 = 1$ and $F_1 = f$. Let now $\{\lambda_i : M \to R | \text{supp } \lambda_i = \overline{W}_i \subset U_i\}$ be a partition of unity subordinate to $\{U_i\}$ and define $\mu_i : M \to M \times I$ as $\mu_i = (1, \lambda_1 + \ldots + \lambda_i)$.

For $N_1$ small, $F$ is close to the projection map $\pi: M \times I \to M$ and the maps $g_i = F_0 \mu_i$ are close to 1. Thus $g_i \in \text{Diff}(M)$, $g_i \circ g_{i-1}^{-1} \in N$ and $g_{i-1}^{-1}(M - V_i) \subset M - \bar{W}_i$. Here $V_i$ is open, $\bar{W}_i \subset V_i$ and $\bar{V}_i \subset U_i$. Since $g_i = g_{i-1}$ on $M - W_i$ and $g_n = f$, setting $f_i = g_i \circ g_{i-1}^{-1}$ we have $f = f_n \circ \dots \circ f_1$. The lemma is proved.

(3.2) LEMMA. *Let $f \in \text{Diff}(M)$ satisfy* (1.1). *There exist open neighborhoods $N$ of $f$, $U_k = U_k(\Omega_k)$ and $J > 0$ such that*

(a) $\{f^j(U_k)\}_{0 \le j < J, 1 \le k \le m}$ *is a covering of $M$,*

(b) *for each $g \in N$, $k$ as above and $0 \le j \le J$, $f^j(U_k) \subset T_f^{k,s} \cap T_f^{k,u} \cap T_g^{k,s} \cap T_g^{k,u}$,*

(c) *$T^{k,s}(x)$ and $T^{k,u}(y)$ intersect transversally in a unique point for each $x, y \in f^j(U_k)$.*

*Here $T^{k,s}$ stands either for $T_f^{k,s}$ or $T_g^{k,s}$ and $T^{k,u}$ for $T_f^{k,u}$ or $T_g^{k,u}$.*

REMARK. We assume that $N$ is such that $T_g^{k,s}$ is defined by (2.8).

PROOF. $T_f^{k,s}$, $T_f^{k,u}$ are tubular families of $W^s(\Omega_k)$, $W^u(\Omega_k)$. Thus for each $k$ we can choose $U_k = U_k(\Omega_k)$ so that (b) and (c) are true for $j = 0$, $\bar{U}_k$ and $T_f^{k,s}$, $T_f^{k,u}$. Since $\Omega(f) = \bigcup \Omega_k$ and $M$ is compact, there exists $J > 0$ so that $\{f^j(U_k)\}_{0 \le j < J, 1 \le k \le m}$ is a covering of $M$. $T_f^{k,s}$, $T_f^{k,u}$ being invariant (a) and (b) are true for $0 \le j \le J$, $f^j(\bar{U}_k)$ and $T_f^{k,s}$, $T_f^{k,u}$. Now $\bigcup_{0 \le j \le J} f^j(\bar{U}_k)$ being compact, (2.8) implies the existence of a neighborhood $N$ as required by the lemma. This proves (3.2).

(3.3) LEMMA. *Let $\{f^j(U_k)\}$ be the covering of $M$ and let $N$ be the neighborhood of $f$ as in* (3.2). *If $g \in N$ and supp $gf^{-1} \subset f^j(U_k)$ for fixed $j$, $k$ then there are homeomorphisms $\eta^s$, $\eta^u: M \to M$ so that for all $1 \le i \le m$ and $x \in M$*

(a) $\eta^s(T_f^{i,s}(x)) = T_f^{i,s}(x)$ *and $\eta^s = 1$ on $M - U$.*

(b) $\eta^u(T_g^{i,u}(x)) = T_g^{i,u}(x)$ *and $\eta^u = 1$ on $M - U$.*

(c) $\eta^u \eta^s f(x) = g(x)$.

*Here $U = f^j(U_k) \cup f^{j+1}(U_k)$.*

PROOF. Since the systems of tubular families considered here are compatible, it is enough to construct $\eta^s$, $\eta^u$ satisfying (a) to (c) above for $i = k$. In this case (3.2) implies that the map $\eta^s: U \to U$, $\eta^s(x) = T_f^{k,s}(x) \cap g T_g^{k,u}(f^{-1}(x))$, is well defined and is, in fact, a homeomorphism. The same is true for $\eta^u: U \to U$, defined by $\eta^u(x) = g T_f^{k,s}(f^{-1}(x)) \cap T_g^{k,u}(x)$. Since supp $gf^{-1} = \bar{V} \subset f^j(U_k)$, we have $g = f$ $g^{-1} = f^{-1}$ on $M - W$, where $W = V \cup f(V)$. Consequently $\eta^s, \eta^u = 1$ on $U - W$ and so they can be extended to be 1 on $M - U$.

From the definitions of $\eta^s$ and $\eta^u$, $\eta^s(T_f^{k,s}(x)) = T_f^{k,s}(x)$ and $\eta^u(T_g^{k,u}(x)) = T_g^{k,u}(x)$ for all $x \in M$.

For $x \in U$, $\eta^u \eta^s f(x) = g(T_f^{k,s}(x) \cap T_g^{k,u}(x))$ and so $\eta^u \eta^s f(x) = g(x)$ on $U$. Since this is true on $M - U$, we have $\eta^u \eta^s f = g$ on $M$.

Thus conditions (a) to (c) of the lemma are met by $\eta^s$, $\eta^u$ and hence (3.3) is proved.

4. In this section we prove the structural stability of diffeomorphisms satisfying (1.1).

(4.1) THEOREM. *Let $f \in \text{Diff}(M)$ satisfy* (1.1). *Then $f$ is structurally stable.*

PROOF. Let $\{f^j(U_k)\}$ be the covering of $M$ as in (3.2). If $g$ is near $f$, then from

(3.1) $gf^{-1} = f_n \circ \dots \circ f_1$, each $f_i$ near 1 and $\text{supp} f_i \subset f^j(U_k)$ for some $j, k$ where $0 \leq j < J$, $1 \leq k \leq m$. Thus from the openness of the diffeomorphisms satisfying (1.1), we have that $f_i \circ \dots \circ f_1 \circ f$ also satisfy (1.1). The result will follow by proving the conjugacy between $f_i \circ \dots \circ f_1 \circ f$ and $f_{i-1} \circ \dots \circ f_1 \circ f$. Thus we may assume that $g$ is of the form $g = f_i \circ f$, $f_i$ as above. But for $g$ in this form, we have $g = \eta^u \circ \eta^s \circ f$, $\eta^s$ and $\eta^u$ being the homeomorphisms given by (3.2). Since $\eta^s$ and $\eta^u$ preserves, respectively, each element of $\{T_f^{k,s}\}$ and $\{T_g^{k,u}\}$, it is enough to prove that $\eta^s f$ and $f$ are conjugate.

Let $d = \eta^s f$. We define a conjugacy $h$ between $f$ and $d$ as follows:
(4.2) $h(x) = \lim_{n \to \infty} d^n f^{-n}(x)$.

This amounts to

(a) $h(x) = x$ for $x \in M - \bigcup_{m \geq 0} f^m(U_k)$,
(b) $h(x) = d^n f^{-n}(x)$ for $x \in \bigcup_{m \geq 0} f^m(U_k)$ and $f^{-n}(x) \in M - \bigcup_{m \geq 0} f^m(U_k)$,
(c) $h(x) = T^{k,s}(x) \cap W^u(\Omega_k(g))$ for $x \in W^u(\Omega_k)$.

To prove this assertion it is enough to show that $h(x) = T^{k,s}(x) \cap W^u(\Omega_k(g))$ for $x \in W^u(\Omega_k) \cap U_k$. Since $d = \eta^s f$ and $\eta^s(T^{k,s}(x)) = T^{k,s}(x)$, $d^n f^{-n}(T^{k,s}(x)) = T^{k,s}(x)$ and thus $h(T^{k,s}(x)) = T^{k,s}(x)$. Also, since $d = (\eta^u)^{-1} g$ the set of limit points of the the sequences $\{d^n(y) | y \in U_k\}$ is contained in $\bigcup_{i \leq k} W^u(\Omega_i(g))$. So if $K$ is the set of limit points of the sequences $\{d^n f^{-n}(y) | y \in U_k\}$ then $K \subset \bigcup_{i \leq k} W^u(\Omega_i(g)) \cap U_k = W^u(\Omega_k(g)) \cap U_k$. This implies $\lim_{n \to \infty} d^n f^{-n}(x) = T^{s,k}(x) \cap W^u(\Omega_k(g))$ and hence the assertion.

We have to show that $h$ is well defined, $hf(x) = dh(x)$ for all $x \in M$ and $h$ is a homeomorphism of $M$.

That $h$ is well defined is clear, since it is well defined on $M - W^u(\Omega_k)$ and for each $x \in W^u(\Omega_k)$ there is a unique point of intersection $T^{k,s}(x) \cap W^u(\Omega_k(g))$.

For all $x \in M$, $dhf^{-1}(x) = \lim_{n \to \infty} d^{n+1} f^{-(n+1)}(x) = h(x)$. This proves $dh = hf$.

The continuity of $h$ follows from its continuity on $W^u(\Omega_i)$, $i \leq k$. Let us show that $h$ is continuous on $W^u(\Omega_k)$. It is enough to do so on $W^u(\Omega_k) \cap U_k$. Let $x_m \in U_k$ be a sequence so that $x_m \to x \in W^u(\Omega_k) \cap U_k$. As we showed above $h(T^{k,s}(x_m)) = T^{k,s}(x_m)$. Consider now the following two cases. If $x_m \in W^u(\Omega_k)$ then $f^{-n}(x_m) \in U_k$ for all $n \geq 0$. If $x_m \notin W^u(\Omega_k)$ there exist $y_m \in f^{-1}(U_k) - U_k$, $n_m \geq 0$ so that $f^{-n_m}(x_m) = y_m$. Thus $n_m \to \infty$ as $m \to \infty$ and $h(x_m) = d^{n_m}(y_m)$. As before the set of limit points of the sequences $\{d^n(y) | y \in U_k \cup f^{-1}(U_k)\}$ is contained in $\bigcup_{i \leq k} W^u(\Omega_i(g))$. Thus by combining the two cases if necessary, the set of limit points of $h(x_m)$ is contained in $\bigcup_{i \leq k} W^u(\Omega_i(g)) \cap U_k = W^u(\Omega_k(g)) \cap U_k$.

Therefore $h(x_m) \to T^{k,s}(x) \cap W^u(\Omega_k(g))$ or $h(x_m) \to h(x)$ and thus $h$ is continuous on $W^u(\Omega_k)$.

The proof of the continuity of $h$ on $W^u(\Omega_i)$, $i < k$ is similar. For from the compatibility of $\{T^{i,s}\}_{1 \leq i \leq m}$, $h(T^{i,s}(x)) = T^{i,s}(x)$ for $i \leq k$ since this is true for $T^{k,s}$. Also if $x_m \to x \in W^u(\Omega_i)$, then the set of limit points of $h(x_m)$ is contained in $W^u(\Omega_i(g))$. From these two facts the assertion follows. Thus $h$ is continuous.

Finally, to show that $h$ is a homeomorphism we consider the map $h_1 = \lim_{n \to \infty} f^n d^{-n}(x)$. As before $h_1$ is well defined and continuous. Since $f^n d^{-n} d^n f^n = 1$ for all $n$, $h_1 h = 1$. Similarly $hh_1 = 1$ and hence $h$ is a homeomorphism of $M$. This concludes the proof of the theorem.

5. We consider here flows or vector fields on $M$ and extend the previous results to this case.

Let $\chi$ be the set of $C^r$ vector fields on $M$, with the uniform $C^r$ topology. This makes $\chi$ into a Banach space. For each $X \in \chi$ we denote by $X_t$ its induced flow. A fixed (critical) point of $X$ is called hyperbolic if $x$ is a hyperbolic fixed point for $X_{t=1}$. In a neighborhood of a closed orbit $\gamma$, $X$ is the suspension of a diffeomorphism $f$ defined in a cross section $S$ of $\gamma$ through $x \in \gamma$. We call $\gamma$ hyperbolic if $f$ has $x$ as a hyperbolic fixed point.

We now define the analogue of the diffeomorphisms satisfying (1.1) as the vector fields $X$ such that

(5.1)
    (a) $\Omega(X)$ is the union of a finite number of fixed points $x_1, \ldots, x_m$ and a finite number of closed orbits $\gamma_1, \ldots, \gamma_n$ of $X$,

    (b) the $x_i$, $\gamma_j$ are all hyperbolic,

    (c) the stable and unstable manifolds of the $x_i$, $\gamma_j$ have transversal intersection.

$X, Y \in \chi$ are called *equivalent* if there is a homeomorphism of $M$ sending trajectories of $X$ into trajectories of $Y$. $X$ is called structurally stable if there exists a neighborhood $N$ of $X$ in $\chi$ such that each $Y \in N$ is equivalent to $X$.

(5.2) THEOREM. *If $X \in \chi$ satisfies* (5.1) *then $X$ is structurally stable.*

To prove (5.2), we first indicate how to carry over to this case the corresponding concepts and results of §§2 and 3. The only novelty here is the concept of tubular families for the stable and unstable manifolds of the closed orbits of the flow.

Let $X \in \chi$ satisfy (5.1). For the fixed points of $X$, we define tubular families of their stable and unstable manifolds as in (2.2). Let now $\gamma$ be a closed orbit of $X$ with period $\pi$ (not necessarily $3.14 ..!$) and $S$ be a cross-section of $\gamma$ at $x \in \gamma$. $S$ is called invariant if $X_{t=\pi}(U) \subset S$, where $U$ is a neighborhood of $x$ in $S$. In this case we define a tubular family of $W^s(\gamma)$ as follows. Let $f: U \to S$ be the restriction of $X_\pi$ and let $\{T_f^s(y)\}_{y \in U}$ be an invariant (under $f^{-1}$) tubular family of $W_f^s(x)$. The tubular family of $W^s(\gamma)$ is now defined by $T^s(X_t(y)) = X_t(T_f^s(y))$, for $y \in U$ and all $t$.

We show that in proving (5.2) we may assume $X$, and in fact all vector fields in some neighborhood of $X$, to have invariant cross-sections for its closed orbits. Let $\gamma$ be a closed orbit of $X$ with period $\pi$, $S$ be a cross-section of $\gamma$ at $x \in \gamma$ and $z \in \gamma$, $z \neq x$. From simple properties of flows near a closed orbit (e.g. [1, p. 251]), for each $Y$ in some neighborhood $N$ of $X$ we get a positive $C^r$ function $\rho(Y): M \to R$ so that $(\rho(Y)Y)_{t=\pi}(U) \subset S$ and $\rho(Y) = 1$ on $M - V$, where $U$ is a neighborhood of $x$ in $S$ and $V$ a neighborhood of $z$ in $M$. Moreover, the correspondence $Y \in N \to \rho(Y)$ is continuous. We now repeat the same procedure for each closed orbit of $X$ and notice that the set of vector fields satisfying (5.1) is open in $\chi$ [2], this yielding in particular a natural one-one correspondence between the fixed points and closed orbits of $X$ and those of $Y$, $Y$ near $X$. Thus, taking the neighborhood $N$ of $X$ small enough, we get a continuous map $\mu: N \to \chi$ such that for each $Y \in N$, $\mu(Y)$ satisfies (5.1), it is equivalent to $Y$ with the same trajectories as $Y$

and its closed orbits have the same period and invariant cross-sections as the correspondent ones for $\mu(X)$.

The definition of a compatible, invariant system of tubular families is the same as before, the invariance being considered for the maps $X_t$, for all $t$. As in (2.6) one can prove the existence of compatible, invariant systems of tubular families of the $W^s(x_i)$, $W^s(\gamma_j)$ and $W^u(x_i)$, $W^u(\gamma_j)$. Again, it is enough to do so for vector fields equivalent to $X$ and we point out that instead of the neighborhood of the fundamental domain considered in (2.6), we take here a cross-section $S$ to the flow whose trajectories together with $W^s(x_i)$ (or $W^s(\gamma_j)$) form a neighborhood of $W^s(x_i)$ (or $W^s(\gamma_j)$).

The translation to flows of the results of §3 is immediate, once we consider flows corresponding to the maps $\eta^s f$ and $(\eta^u)^{-1} g$ in Lemma (3.3).

The proof of (5.2) becomes now very much the same as that of (4.1). We just remark that when $X_t$ has no closed orbits, (5.2) provides for $Y_t$ near $X_t$ a homeomorphism that not only sends trajectories of $X_t$ into those of $Y_t$, but it is in fact a conjugacy between the flows. This is the case for gradient flows satisfying (5.1).

Using (4.1) (or (5.2) for the corresponding case for flows) and a converse known for some time, we have:

(5.3) COROLLARY. *Let $f \in \text{Diff}(M)$, with $\Omega(f)$ being finite. Then $f$ is structurally stable if and only if $f$ satisfies* (1.1).

This gives a good characterization of structural stability in the $\Omega$-finite case and leads us to the question of finding in the general case a similar characterization.

Referring the reader to [5] for the concepts appearing here, we pose as a possible answer to this question the following

CONJECTURE. $f \in \text{Diff}(M)$ is structurally stable if and only if $f$ satisfies

(a) AXIOM A. $\Omega(f)$ is hyperbolic and the set of periodic points of $f$ is dense in $\Omega(f)$.

(b) STRONG TRANSVERSALITY CONDITION. For all $x, y \in \Omega(f)$, $W^s(x)$ and $W^u(y)$ have a transversal intersection.

## REFERENCES

1. P. Hartman, *Ordinary differential equations*, Wiley, New York, 1964.
2. J. Palis, *On Morse-Smale dynamical systems*, Topology (to appear).
3. S. Smale, *Morse inequalities for a dynamical system*, Bull. Amer. Math. Soc. **66** (1960), 43–49.
4. ———, *On gradient dynamical systems*, Ann. of Math. (2) **74** (1961), 199–206.
5. ———, *Differentiable dynamical systems*, Bull. Amer. Math. Soc. **73** (1967), 747–817.

I.M.P.A., RIO DE JANEIRO
UNIVERSITY OF CALIFORNIA, BERKELEY

# A GLOBAL APPROXIMATION THEOREM
# FOR HAMILTONIAN SYSTEMS

R. CLARK ROBINSON[1]

**I. Basic definitions and statement of the main theorem.** By a theorem of Kupka and Smale, for a residual subset of all ordinary differential equations all closed orbits and critical points are hyperbolic (residual in the sense of Baire). This property is not true for a residual subset of Hamiltonian systems. The definitions and theorem below give a suitable weakening of this property for differential equations such that it is satisfied for a residual subset of Hamiltonian systems and it still gives geometric information about the set of closed orbits.

Now the basic definitions for Hamiltonian systems will be given. For a more complete treatment of the approach used here for Hamiltonian systems see Abraham and Marsden [1] especially Chapter three. A symplectic manifold is a pair $(M, \omega)$ where $M$ is a $C^\infty$ manifold of dimension $2n$ and $\omega$ is a nondegenerate closed two form on $M$. $\omega$ induces an isomorphism between $T^*M$ and $TM$ and thus between the spaces of $C^r$ sections $C^r(T^*M)$ and $C^r(TM)$. If $H \in C^{r+1}(M, R)$ is a $C^{r+1}$ function on $M$, then $dH \in C^r(T^*M)$ and thus there corresponds a section $X_H \in C^r(TM)$ called a Hamiltonian vector field or Hamiltonian system. Given a fixed symplectic manifold and $1 \leq r \leq \infty$ the set of all such $C^r$ Hamiltonian vector fields on the manifold with the $C^r$ Whitney topology is denoted by $\mathscr{X}^r$. If $M$ is compact the topology is just the $C^r$ sup topology. $C^{r+1}(M, R) \to \mathscr{X}^r$ is an open projection which if $M$ is connected is just formed by making two functions equivalent if they differ by a constant. Thus $\mathscr{X}^r$ is Baire space which implies a residual subset of $\mathscr{X}^r$ is dense. An energy surface for $X_H \in \mathscr{X}^r$ is a subset of $M \ H^{-1}(e) = \{m \in M : H(m) = e\}$ where $H$ is the energy function inducing $X_H$.

If $m$ is a critical point of a Hamiltonian vector field $X$, then $\exp(DX(m))$ evaluated in local symplectic coordinates is a symplectic matric, so if $\lambda$ is an eigenvalue then $1/\lambda$, $\bar{\lambda}$, and $1/\bar{\lambda}$ are also. Define the principal characteristic multipliers of the critical point to be those $n$ eigenvalues of $\exp(DX(m))$ with modulus strictly greater than one or modulus one and imaginary part greater than or equal zero (take half the eigenvalues equal to 1 or $-1$). Similarly for a close orbit with prime period $b$ $D\phi_b(m)$ is symplectic where $\phi_t$ is the flow of the vector field and $m$ is a point on the orbit. The eigenvalue one will have multiplicity at least two corresponding to the directions along the orbit and increasing energy. Taking half of the remaining $2n - 2$ eigenvalues as above, call these $n - 1$ eigenvalues the principal characteristic multipliers of the closed orbit. Note these eigenvalues are associated with the prime period and no higher periods.

[1] The author held a fellowship from the National Science Foundation at the University of California at Berkeley during the time this research was done.

A critical point or a closed orbit is called generic ($N$-generic) if the principal characteristic multipliers are multiplicatively independent over the integers (between $-N$ and $N$), i.e. if $\Pi\lambda_i^{p_i} = 1$ with $p_i \in Z\,(\cap[-N, N])$ then $p_i = 0$ for all $i$. This definition implies the following conditions on the principal characteristic multipliers: (i) none is a root of unity, and (ii) all have multiplicity one. Also note that all the characteristic multipliers can not be multiplicatively independent over the integers since $\lambda$ and $1/\lambda$ are not.

A closed orbit $\gamma$ of $X$ is called elementary if $T\theta - T\,id : T\Sigma \to T\Sigma$ spans the tangent space to the energy surface in $\Sigma$ through $\gamma$, where $\theta : \Sigma \to \Sigma$ is a Poincaré map for $X$ on some transverse section to $\gamma$, $\Sigma$ (dimension $2n - 1$). If a closed orbit is elementary then the closed orbit theorem of Abraham holds (Abraham and Marsden [1, p. 178]) and the closed orbit lies in a one parameter family of closed orbits (not necessarily parameterized by energy). Below in the local lemma this fact is proved using transversality instead of the implicit function theorem but the idea is the same.

Let $\gamma$ be a closed orbit of $X$, $\Sigma$ be a transverse section to $\gamma$, and $\theta : \Sigma \to \Sigma$ a Poincaré map. A property for closed orbits is said to be true locally on $\Sigma$ or at $\gamma$ if it is true for the set of closed orbits corresponding to the fixed points of $\theta$ on $\Sigma$. The point of this definition is that only the set of closed orbits corresponding to fixed points and not higher iterations of $\theta$ are required to have the property.

A Hamiltonian vector field has property H1 (H1 $- N$) if every critical point is generic ($N$-generic). Let $\mathscr{H}^r(N) = \{X \in \mathscr{X}^r : X \text{ has property } H1 - N\}$ and $\mathscr{H}^r = \bigcap_N \mathscr{H}^r(N)$. A vector field has property H2 (H2 $- N$ for orbits of period $\leq b$) if (i) it has property H1 (H1 $- N$), (ii) every closed orbit (of period $\leq b$) is elementary and has a transverse section $\Sigma$ such that locally on $\Sigma$ at most a countable (finite) number of closed orbits are not generic ($N$-generic), and (iii) the nongeneric closed orbits occur "generically", i.e. given by a transverse intersection with a manifold as in the local lemma. Let $\mathscr{G}^r(b, N) = \{X \in \mathscr{H}^r(N) : X \text{ has property } H2 - N \text{ at all closed orbits of period } \leq b\}$, and $\mathscr{G} = \mathscr{G}^r = \bigcap_{b,N} \mathscr{G}^r(b, N)$.

THEOREM 1. *Let $(M, \omega)$ be a symplectic manifold.*

(i) *If $1 \leq r \leq \infty$ then $\mathscr{H}^r(N)$ is a dense open subset of $\mathscr{X}^r$ and $\mathscr{H}^r$ is a residual subset.*

(ii) *If $M$ is compact, $2 \leq r \leq \infty$, $b > 0$, and $N \in Z$, then $\mathscr{G}^r(b, N)$ and $\mathscr{G}^r$ are residual subsets of $\mathscr{X}^r$.*

The first half of this theorem is given as a problem in Abraham and Marsden [1, p. 182]. The proof will not be given here but it follows using the hint given for the problem and a normal form for infinitesimally symplectic matrices similar to the one given below for symplectic matrices. A statement similar to the second half of the theorem is also conjectured in that book although the actual conjecture that all closed orbits can be made generic is false. The fact that all critical points and only most closed orbits can be made $N$-generic is the reason $r$ is only needed to be greater than one for the first half and greater than two for the second half of the theorem. The proof below does not prove $\mathscr{G}^r(b, N)$ is open but only that it is residual. The problem is caused by closed orbits with characteristic multipliers which are rational rotations as discussed below in the proof of the global theorem. Also the second half of the theorem appears to be true when $M$ is noncompact by a method similar to the one used in Peixoto [3] but there is a problem proving an appropriate version of Lemma 3 below.

Two ways closed orbits can fail to be generic for $X \in \mathscr{G}$ will now be discussed. First, one will be a principal characteristic multiplier for a closed orbit $\gamma$ (multiplicity at least four as an eigenvalue of $D\phi_t(m)$) iff the family of closed orbits is tangent to the energy surface through $\gamma$. This fact is true since if the family is not tangent then in order to span the tangent space to the energy surface one can not be a principal characteristic multiplier. Conversely if the family is tangent then the direction of tangency is an eigenvector whose eigenvalue is one. K. Meyer has an example where such tangency does occur and can not be perturbed away. (See [4].) Thus it is false that generically for Hamiltonian vector fields all closed orbits do not have one as a principal characteristic multiplier. However locally the orbits with one as a principal characteristic multiplier will be isolated by condition (ii) in the definition of 1-generic. Similar examples can be constructed with other numbers of modulus one being principal characteristic multipliers of multiplicity two.

Second, if the closed orbits on a one parameter family have two principal multipliers of modulus one (or both with imaginary part zero) $\lambda$ and $\mu$ then generically $\lambda$ and $\mu$ will be multiplicatively dependent at a countable number of closed orbits of the family. This is caused by the fact that if they have multiplicity one then they can not be perturbed off the unit circle because of the pairing of eigenvalues necessary for a symplectic matrix. An example would be if $\lambda$ had increasing argument (as a complex number) and $\mu$ had decreasing argument as functions of the parameter of the closed orbits. The arguments will then be additively dependent at a countable number of orbits and the situation is stable.

## II. A local lemma.

LEMMA 2. Let $X_H \in \mathscr{X}^r$, let $\gamma$ be a closed orbit of $X_H$, let $m \in \gamma$, and let $\Sigma$ be a transverse section to $\gamma$ at $m$. Then there exist neighborhoods $\mathcal{O}$ of $X_H$ in $\mathscr{X}^r$ and $U$ of $m$ in $M$ such that (i) if $1 \leq r \leq \infty$ then there is a dense open subset $R_1 \subset \mathcal{O}$ such that locally on $\Sigma_{\bar{U}} = \Sigma \cap \bar{U}$ all closed orbits for $X \in R_1$ are elementary, and (ii) if $2 \leq r \leq \infty$ and $N \in Z$ then there is a dense open subset $R_2^N \subset R_1$ such that each $X \in R_2^N$ has property $H2 - N$ locally on $\Sigma_U$.

PROOF OF 2(i). Let $x_1, \ldots, x_{2n}$ be Hamiltonian flow box coordinates for $X_H$ at $m$, i.e. coordinates such that $X_H = \partial/\partial x_1, x_{n+1} = H$, and $\omega = \sum_{i=1}^n dx_i \wedge dx_{i+n}$. Let $\Sigma$ be the transverse section to $\gamma$ at $m$ formed by setting $x_1$ equal a constant. If $S$ is a subset of $M$ let $\Sigma_S = \Sigma \cap S$. Let $\pi: \Sigma \to R^{2n-2}$ be the $C^r$ submersion defined by $\pi(q) = (x_2(q), \ldots, x_n(q), x_{n+2}(q), \ldots, x_{2n}(q))$. Then, there exist contractible neighborhoods $\mathcal{O}$ of $X_H$ in $\mathscr{X}^r$ and $V$ of $m$ in $M$ such that (i) $V$ is contained in the above Hamiltonian flow box coordinate chart and is small enough so it can be given $C^\infty$ symplectic coordinates (given by Darboux's theorem), (ii) there is a $C^r$ Poincaré map $\theta : \mathcal{O} \times \Sigma_V \to \Sigma$ defined by the flow of each $X \in \mathcal{O}$, and (iii) for each $X_K \in \mathcal{O}$ all the $K$ energy surfaces restricted to $V$ are regular and

$$\pi : \Sigma_{V \cap K^{-1}(e)} \to R^{2n-2}$$

is a diffeomorphism into. Condition (i) is obviously possible. For condition (ii) the implicit function theorem can be used to define a $C^r$ function $\tau : \mathcal{O} \times \Sigma_V \to R$ for any $V$ small enough such that $\phi^X_{\tau(X,q)} q \in \Sigma$ where $\phi^X_t(q)$ is the flow for $X$ through $q$. Then setting $\theta(X, q) = \phi^X_{\tau(X,q)}(q)$, $\theta$ is $C^r$ as a function of both $X$ and $q$. For condition (iii), $\pi : \Sigma_{V \cap K^{-1}(e)} \to R^{2n-2}$ can be thought of as the projection from a $K$ energy surface to the $H$ energy surface through $m$. The condition is satisfied by $X_H$ at $m$ and hence will be true in a neighbourhood of $m$ for $XC^1$ close to $X_H$.

Let $U \subset \bar{U} \subset V$ be a neighborhood of $m$. Define $\rho : \mathcal{O} \to C^r(\Sigma_V, \Sigma \times \Sigma)$ by $ev(\rho)(X, q) = (q, \theta(X, q))$. Let $W = \{(p, q) : \pi p = \pi q\} \subset \Sigma \times \Sigma$. Then $\rho(X_K)q \in W$ iff $\theta(X_K, q) = q$, since $K$ is preserved on the trajectories of $X_K$ and $\pi : \Sigma_{K^{-1}K(q)} \to R^{2n-2}$ is a diffeomorphism into. Also $\rho(X) : \Sigma_V \to \Sigma \times \Sigma$ is transverse to $W$ at points of $\Sigma_{\bar{U}}$ iff locally on $\Sigma_{\bar{U}}$ all closed orbits are elementary. Let $s = \max \{2, r\}$. Claim (a) below proves that $\rho$ is $C^s$ pseudotransverse to $W$, so it follows by Theorem 6 that $R_1 = \{X \in \mathcal{O} : \rho(X) \, \rho\infty$ is transverse to $W$ at points $\Sigma_{\bar{U}}\}$ is dense and open in $\mathcal{O}$. (See §IV for the terminology.) This completes the proof of 2(i) except for claim (a).

CLAIM (a). $\rho$ is $C^s$ pseudotransverse to $W$.

PROOF. Since the Poincaré map $\theta : \mathcal{O} \times \Sigma_V \to \Sigma$ is $C^r$ and $r \geq 1$, $\rho$ is a $C^1$ pseudorepresentation, i.e. $ev(\rho^{(1)}) : \mathcal{O} \times T\Sigma_V \to T(\Sigma \times \Sigma)$ defined by $ev(\rho^{(1)})(X, v) = T^1(\rho X)v$ is $C^0$. For $m \in Z$ let $\mathcal{O}^m = \{X \in \mathcal{O} : X$ is $C^m\}$. $\mathcal{O}^{s+1}$ is dense in $\mathcal{O}$. For $X_K \in \mathcal{O}^{s+1}$ let $B = \mathcal{O}^s$ with $C^s$ norm and $\psi : \mathcal{O}^s \to \mathcal{O}$ the inclusion. $ev(\rho \circ \psi) : B \times \Sigma_V \to \Sigma \times \Sigma$ is $C^s$ since $\theta$ is $C^s$ when restricted to $\mathcal{O} \times \Sigma_V$. Thus we need only prove $ev(\rho \circ \psi)$ is transverse to $W$ on $X_K \times \Sigma_{\bar{U}}$. For $q \in \Sigma_V$ such that $q = \theta(X_K, q)$, we will show below that $T\theta(X_K, q) \cdot T_{XK}\mathcal{O}^s$ spans $T_q\Sigma_{K^{-1}K(q)}$. Then since $\pi : \Sigma_{K^{-1}K(q)} \to R^{2n-2}$ is a diffeomorphism into, $0 \times T_q\Sigma_{K^{-1}K(q)}$ forms a complement to $T_qW$ in $T_q(\Sigma \times \Sigma)$. Thus $ev(\rho \circ \psi)$ is transverse to $W$ at $(X_K, q)$ and hence at all points of $X_K \times \Sigma_{\bar{U}}$.

To prove the image spans $T_q\Sigma_{K^{-1}K(q)}$ take a $C^{s+1}$ Hamiltonian flow box for $X_K$ that ends at $q$ with coordinates $(t, x_2, ..., x_n, K, y_2, ..., y_n)$ and range $Y = [0, a] \times [-b, b]^{2n-1}$ such that $X_K = \partial/\partial t$, $q = (a, 0)$, and the two form $\omega = dt \wedge dK + \sum dx_i \wedge dy_i$. Let $L : [-b, b]^{2n-2} \to R$ be a $C^{s+1}$ function that is zero on the boundary and when all coordinates are zero. Let $g : [0, a] \to R$ be a $C^{s+1}$ function such that $\int_0^a g(t)dt = 1$ and supp $g \subset (0, a)$ so all derivatives of $g$ vanish at $0$ and $a$. Let $0 \leq \lambda \leq 1$ be a parameter and define $\lambda g L : Y \to R$ by $\lambda g L(t, x_i, K, y_i) = \lambda g(t)L(x_i, y_i)$. Then $X_{K + \lambda g L}$ is $C^s$ on the manifold (the coordinates need to be $C^{s+1}$ for this to be true). Let $\phi^\lambda_t$ be its flow. Then

$$(d/d\lambda)\phi_a^\lambda(0,0,0,0) = \int_0^a X_{gL}(a - t,0,0,0)dt$$

$$= \int_0^a \begin{pmatrix} 0 \\ g\dfrac{\partial L}{\partial y_i} \\ -L\dfrac{\partial g}{\partial t} \\ -g\dfrac{\partial L}{\partial x_i} \end{pmatrix}_{(a-z,0)} dt = \int_0^a \begin{pmatrix} 0 \\ g\dfrac{\partial L}{\partial y_i} \\ 0 \\ -g\dfrac{\partial L}{\partial x_i} \end{pmatrix}_{(a-z,0)} dt = \begin{pmatrix} 0 \\ \dfrac{\partial L}{\partial y_i} \\ 0 \\ -\dfrac{\partial L}{\partial x_i} \end{pmatrix}_{(0,0)}$$

See Abraham and Robbin [2, p. 107] for a proof of the first equality. Since these derivatives of $L$ at $(0,0)$ are arbitrary, this proves the desired result.

PROOF OF 2(ii). To prove this part of the lemma the eigenvalues of $D\phi_{\tau(X,q)}^X(q)$ must be controlled. Two of the eigenvalues corresponding to the directions of increasing energy and $X$ must be equal to one. Thus the other $2n - 2$ eigenvalues must be controlled.

Let $V$ and $\mathcal{O}$ be contractible neighborhoods as used above. Fix an arbitrary set of $C^\infty$ symplectic coordinates on $V$. Let $\mathcal{O} \times TV$ be the $C^\infty$ bundle over $\mathcal{O} \times V$. Using the $C^\infty$ local coordinates on the fiber induced by those on $V$ define $Sw:\mathcal{O} \times TV \to \mathcal{O} \times TV$ by $Sw(X, q, y_1, ..., y_{2n}) = (X, q:y_{n+1}, ..., y_{2n}, -y_1, ..., -y_n)$. A $C^r$ basis for the fibers of this bundle will be constructed by induction. Define $X_1:\mathcal{O} \times V \to TV$ by $X_1(X_K, q) = X_K(q)$. Thus $X_1$ is a $C^r$ cross section of $\mathcal{O} \times TV$, $X_1 \in C^r(\mathcal{O} \times TV)$. Define $Y_1':\mathcal{O} \times V \to TV$ by $Y_1'(X_K, q) = -(SwX_K)q = \text{grad } K(q)$. Thus $Y_1' \in C^r(\mathcal{O} \times TV)$. Since $dK \neq 0$ on $V$ for $X_K \in \mathcal{O}$, $\omega(X_1(\cdot), Y_1'(\cdot)) \neq 0$ on $\mathcal{O} \times V$. Thus by a $C^r$ scalar multiplication we can construct $Y_1 \in C^r(\mathcal{O} \times TV)$ such that $\omega(X_1(\cdot), Y_1(\cdot)) = 1$ on $\mathcal{O} \times V$. Let $S_1 = \{X \in \mathcal{O} \times TV: \omega(X_1, X) = \omega(Y_1, X) = 0\}$. $S_1$ is a $C^r$ bundle over $\mathcal{O} \times V$ since it is the kernel of two bundle maps into $R$. $\omega$ is also nondegenerate when restricted to $S_1$ because of the way $S_1$ was constructed. Assume by induction $X_i, Y_i \in C^r(\mathcal{O} \times TV)$ have been constructed for $i \leq k$ such that $\omega(X_i, X_j) = \omega(Y_i, Y_j) \equiv 0$ and $\omega(X_i, Y_j) \equiv \delta_{ij}$. Let $S_k = \{X \in \mathcal{O} \times TV: \omega(X, X_i) = 0 \text{ for } i \leq k\}$. Assume $\omega$ is nondegenerate when restricted to $S_k$. $S_k$ is a $C^r$ bundle over $\mathcal{O} \times V$ since it is the kernel of bundle maps. Now since $\mathcal{O} \times V$ is contractible $S_k$ will have a global section, $X_{k+1} \in C^r(S_k)$. Let $Y_k' = -SwX_{k+1} \in C^r(\mathcal{O} \times TV)$. Project $Y_{k+1}'$ into $S_k$ by subtracting $\sum_{i=1}^k \omega(X_i, Y_{k+1}')Y_i - \omega(Y_i, Y_{k+1}')X_i$ and use scalar multiplication to construct $Y_{k+1} \in C^r(S_k)$ such that $\omega(X_{k+1}, Y_{k+1}) \equiv 1$. Similarly define $S_{k+1}$ and it will follow that $\omega$ restricted to $S_{k+1}$ is nondegenerate. Thus by induction $2n$ cross sections of $\mathcal{O} \times TV$ are constructed that span each fiber of $\mathcal{O} \times TV$ and in terms of which

$$\omega = \begin{pmatrix} 0 & I \\ -I & 0 \end{pmatrix}$$

in matrix form. Also note by taking the $X_i$ above to be $C^\infty$ and then projecting into $S_i$ we can assume that when we restrict to $C^s$ vector fields the cross sections are $C^s$.

Let Sym($TV$, $TV$) be the bundle over $V \times V$ whose fiber over $(q, p)$ is the space of linear maps from $T_q V$ to $T_p V$ which preserve the two form $\omega$. Let $\mathcal{O} \times$ Sym be the corresponding bundle over $\mathcal{O} \times V \times V$. The $C^r$ basis for $\mathcal{O} \times TV$ constructed above gives a $C^r$ trivialization for the new bundle $m: \mathcal{O} \times$ Sym $\to \mathcal{O} \times V \times V \times J$ where $J$ is the set of all $2n \times 2n$ symplectic matrices with respect to the two form

$$\begin{pmatrix} 0 & I \\ -I & 0 \end{pmatrix}$$

Let $J'$ be the subset of $J$ formed by imposing the restriction that the first column have all zero entries except for one in the $(1, 1)$ position and the restriction that the $n + 1$ row have all zero entries except for one in the $(n + 1, n + 1)$ position. Let $E = m^{-1}(\mathcal{O} \times V \times V \times J')$. $B$ is a $C^r$ subbundle of $\mathcal{O} \times$ Sym. For $(X, q) \in \mathcal{O} \times \Sigma_V$, $D\phi^X_{\tau(X,q)}(q)$ preserves the two form $\omega$. Also because the flow preserves energy and its derivative preserves $X$, it follows that $D\phi^X_{\tau(X,q)}(q) \in J'$. Thus the map $\psi: \mathcal{O} \times \Sigma_V \to E$ given by $\psi(X, q) = m^{-1}(X, q, \theta(X, q), D\phi^X_{\tau(X,q)}(q))$ is $C^{r-1}$ and $\psi \in C^{r-1}(E)$. Define $\tilde{p}: \mathcal{O} \to C^{r-1}(\Sigma_V, E)$ by $\mathrm{ev}(\tilde{p}) = \psi$. Any matrix in $J'$ will have the eigenvalue one with multiplicity at least two. Let $S^N$ be the set of all matrices in $J'$ with principal eigenvalues (as in principal characteristic multipliers) nontrivially multiplicatively dependent over the integers between $-N$ and $N$. $S^N$ is closed and the finite union of submanifolds $S^{Nj}, j = 1, \ldots, k_M$ of codimension at least one in $J'$. The fact that $S^N$ is the union of submanifolds is proved similarly to Abraham and Robbin [2, 30.4]. The codimension can be proved by showing that $S^N$ has empty interior in $J'$. For $A \in S^N$ put it in normal form as in §V below. The eigenvalues of $A$ are then given by entries in blocks on the diagonal. Perturbing only these entries the matrix can be kept symplectic but have multiplicatively independent principal eigenvalues. Thus $A$ can be perturbed off $S^N$ into $J'$.

Now let $W^{Nj} = m^{-1}(\mathcal{O} \times W \times S^{Nj}) \cap B$ where $W$ is as in the proof of 2(i). $W^{Nj}$ has codimension at least $2n - 1$. Claim (b) below proves $\tilde{p}$ is $C^{r-1}$ pseudotransverse to $W^{Nj}$. Let $R_2^{Nj} = \{X \in \mathcal{O}: \tilde{p}(X)$ is transverse to $W^{Nj}$ at points of $\Sigma_U\}$ $\cap R_1$ and $R_2^N = \bigcap_{j=1}^{k_N} R_2^{Nj}$. Theorem 6 then proves each $R_2^{Nj}$ is residual in $\mathcal{O}$ so $R_2^N$ is. Since $W^N$ is closed in $E$ and the union of submanifolds and $\tilde{p}$ is a $C^1$ pseudo-representation $R_2^N$ is open.

For $X \in R_2^N$, $\Sigma_U \cap \bigcup_{j=1}^{k_N} \tilde{p}(X)^{-1} W^{Nj}$ is a submanifold of dimension at most zero since each of the sets in the union has codimension at least $2n - 1$. But these points are the closed orbits of $X$ which are not $N$-generic. Thus locally on $\Sigma_U$ there are a finite number of closed orbits which are not $N$-generic, or locally on $\Sigma_U$, $X$ has property $H2 - N$. This completes the proof of 2(ii) except for claim (b).

CLAIM (b). $\tilde{p}$ is $C^{r-1}$ pseudotransverse to $W^{Nj}$.

PROOF. Proceeding as in Claim (a), it suffices to prove that for $X \in \mathcal{O}^{r+1}$, $\mathrm{ev}(\tilde{p}): \mathcal{O}^r \times \Sigma_V \to E$ is transverse to $W^{Nj}$ on $X \times E_U$. Since $\mathrm{ev}(\rho)$ is transverse to $W$ on $X \times \Sigma_U$, it is sufficient to prove given $q \in \Sigma_U$ such that $\theta(X, q) = q$ one can perturb the vector field to make this one closed orbit $N$-generic, or more formally that the image of $T\tilde{p}$ acting on perturbations of the vector field which keep this one orbit closed spans the tangent space to the $(2n - 2) \times (2n - 2)$ subblock of the matrix formed by eliminating the first and $n + 1$ rows and columns. It is this block which determines the principal eigenvalues.

Let $X_K \in \mathcal{O}^{r+1}$ and $q \in \Sigma_U$ such that $\theta(X_K, q) = q$. Take a $C^{r+1}$ Hamiltonian flow box $Y$ for $X_K$, $L: [-b, b]^{2n-2} \to R$, $g: [0, a] \to R$, $0 \leq \lambda \leq 1$, $\lambda g L: Y \to R$ a $C^{r+1}$ function, and $\phi_t^\gamma$ as in Claim (a). Now also assume that $DL = 0$ as well as $L = 0$ along $\gamma = \{\phi_{-a+t}(q)\}$ so $X_{K+\lambda g L} = X_K$ along $\gamma$, and $\phi_a^\lambda(0) = q$ for all $\lambda$. Thus $(d/d\lambda)\phi_a^\lambda(0) = 0$. Then evaluating in the flow box coordinates using the special form of $DX_{K+\lambda g L}$ on $\gamma$,

$$(d/d\lambda)D\phi_a^\lambda(0) = \int_0^a DX_{gL}(t, 0)dt$$

$$= \int_0^a \begin{pmatrix} 0 & 0 & 0 & 0 \\ 0 & g(t)\dfrac{\partial^2 L}{\partial x_j \partial y_i} & 0 & g(t)\dfrac{\partial^2 L}{\partial y_j \partial y_i} \\ 0 & 0 & 0 & 0 \\ 0 & -g(t)\dfrac{\partial^2 L}{\partial x_j \partial x_i} & 0 & -g(t)\dfrac{\partial^2 L}{\partial y_j \partial x_i} \end{pmatrix} dt = \begin{pmatrix} 0 & 0 & 0 & 0 \\ 0 & \dfrac{\partial^2 L}{\partial x_j \partial y_i} & 0 & \dfrac{\partial^2 L}{\partial y_j \partial y_i} \\ 0 & 0 & 0 & 0 \\ 0 & -\dfrac{\partial^2 L}{\partial x_j \partial x_i} & 0 & \dfrac{\partial^2 L}{\partial y_j \partial x_i} \end{pmatrix}_{(0,0)}$$

By making the second derivative of $L$ arbitrary in $x_i$ and $y_i$ at $(0, 0)$ all the infinitesimally symplectic matrices in the corresponding $(2n - 2) \times (2n - 2)$ matrix can be assumed by this type of perturbation. A conjugacy between the form of the matrix in the flow box coordinates and the constructed basis completes the proof of the claim.

III. **Proof of the Global Theorem.** The difficulty in proving the global result arises from the fact that the local lemma gives information only about closed orbits of approximately the same period as $\gamma$. If $\gamma$ has a characteristic multiplier of modulus one, $\lambda$ such that $\lambda^m = 1$, then the family of closed orbits probably will bifurcate at $\gamma$ giving other families of closed orbits of period approximately $m$ times that of $\gamma$ about which the local lemma applied at $\gamma$ gives no information. Thus below the set $C$ is used to bound away from orbits of lower period. Thinking of these orbits of lower periods as points at infinity the proof is similar to the one given in Peixoto [3] for the theorem of Kupka and Smale when the manifold is noncompact.

More notation is needed to prove the global result. Fix $2 \leq r \leq \infty$, $N \in Z$, and let $\mathcal{X}^r = \mathcal{X}$. Let $d$ be a distance between points of $M$. Also let $d$ stand for the induced minimum distance from a point to a compact set and the nonsymmetric distance between two compact sets $d(A, B) = \sup\{d(a, B): a \in A\}$. For $X \in \mathcal{X}$ and $b \geq 0$ let $\Gamma(b, X) = \{m \in M: m$ is a critical point of $X$ or lies on a closed orbit of period $\leq b\}$. For $X \in \chi$, $j \in Z$, and using induction on $b \geq 0$, define $\gamma(b, X, j) = $ closure $\{m \in \Gamma(b, X): m$ is a critical point of $X$ or lies on a closed orbit $\delta$ of period $\tau \leq b$ such that $d(\delta, \gamma(5\tau/6, X, j)) \geq 2^{-j}\}$. $\gamma(b, X, j)$ is well defined because $X$ has a minimum period $\mu$ for closed orbits which meet $\{m: d(m, \Gamma(0, X)) \geq 2^{-j}\}$, so for $0 \leq b < \mu$, $\gamma(b, X, j) = \Gamma(0, X)$. Let $C(b, X, j) = \{m \in M: d(m, \Gamma(b, X)) \geq 2^{-j}\}$ and

$D(b, X, j) = \{m \in M : d(m, \gamma(b, X, j)) \geq 2^{-j}\}$. Let $\mathcal{G}(b, N, j) = \{X \in \mathcal{G}(b, N) : X$ satisfies property $H2 - N$ locally at all closed orbits of period $\leq 3b/2$ which meet $C(b, X, j)\}$. Note that $\mathcal{G}(3b/2, N) = \bigcap_j \mathcal{G}(b, N, j)$.

LEMMA 3. *Each* $X \in \mathcal{G}(b, N)$ *has a neighborhood* $\mathcal{O} \subset \mathcal{H}(N)$ *and a dense open subset* $R(b, X, j) \subset \mathcal{O}$ *such that* $\mathcal{G}(b, N) \cap R(b, X, j) \subset \mathcal{G}(b, N, j)$.

PROOF. Let $X \in \mathcal{G}(b, N)$. $\mathcal{H}(N)$ is open so there exists a neighborhood $\mathcal{O}_1 \subset \mathcal{H}(N)$ such that each $Y \in \mathcal{O}_1$ has a critical point within a distance $2^{-j-1}$ of each critical point of $X$. Let $Z_1 = M - C(0, X, j + 1)$. Let $\gamma \subset \gamma(5b/6, X, j + 1)$ be a closed orbit. Since these orbits for $X$ are elementary there exist neighborhoods $Z \subset M - D(5b/6, X, j + 1)$ of $\gamma$ and $\mathcal{O}$ of $X$ such that each $Y \in \mathcal{O}$ will have a closed orbit of period $\leq b$ within a distance $2^{-j-1}$ of each point of $Z \cap \gamma(5b/6, X, j + 1)$. Since $\gamma(5b/6, X, j + 1)$ is compact there are a finite number $Z_1$ (as above), $Z_2, \ldots, Z_k$ which cover it. Let $\mathcal{O}_i$ be the set associated with $Z_i$. For $Y \in \bigcap_{i=1}^k \mathcal{O}_i$, $C(b, Y, j) \subset D(5b/6, X, j + 1)$.

Let $Q = D(5b/6, X, j + 1) \cap \Gamma(3b/2, X)$. For each closed orbit of $X$ meeting $Q$, $\gamma$, let $m \in \gamma$, let $\Sigma$ be a transverse section, and let the neighborhoods $U \subset \bar{U} \subset V$ of $m$ and $\mathcal{O}$ of $X$ be as in Lemma 2. Take $\mathcal{O}$ and $V$ smaller if necessary to insure all trajectories of $Y$ take at least $5b/6$ units in time to pass from $\Sigma_V$ to $\Sigma$, thus all trajectories of $Y$ that are closed by two applications of the Poincaré map have period $> 3b/2$. Take neighborhoods $Z$ of $\gamma$ and $\mathcal{O}$ of $X$ (perhaps smaller) such that any trajectory of $Y \in \mathcal{O}$ that meets $Z$ passed through $\Sigma_U$ at an earlier time and it occurred the last time the trajectory met $\Sigma$. A finite number $Z_{k+1}, \ldots, Z_m$ cover $Q$. Let $\mathcal{O}_i$ be the set associated with $Z_i$. By Lemma 2 there exists a dense open subset $R_i \subset \mathcal{O}_i$ such that each $Y \in R_i$ has property $H2 - N$ locally at all closed orbits of period $\leq 3b/2$ that meet $Z_i$. Let $L = D(5b/6, X, j + 1) - \bigcup_{i=k+1}^m Z_i$. $L$ is compact and $X$ has no closed orbits of period $\leq 3b/2$ that meet $L$ so there exists a neighborhood $\mathcal{O}_{m+1}$ of $X$ such that each $Y \in \mathcal{O}_{m+1}$ has the same property on $L$.

Let $\mathcal{O} = \bigcap_{i=1}^{m+1} \mathcal{O}_i$ and $R = \mathcal{O} \cap \bigcap_{i=k+1}^m R_i$. Since $R$ is dense and open in $\mathcal{O}$ all that remains to show is that $\mathcal{G}(b, N) \cap R \subset \mathcal{G}(b, N, j)$. But for each $Y \in \mathcal{G}(b, N) \cap R$, $C(b, Y, j) \subset D(5b/6, X, j + 1)$ and $D(5b/6, X, j + 1)$ is covered by $L$ and $Z_{k+1}, \ldots, Z_m$. Thus all closed orbits of period $\leq 3b/2$ that meet $C(b, Y, j)$ locally have property $H2 - N$. This completes the proof of the lemma.

Let $b > 0$, $j$ be an integer, and $R(b, j) = \bigcup\{R(b, X, j) : X \in \mathcal{G}(b, N)\}$.

COROLLARY 4. *Let* $P$ *be an open subset of* $\mathcal{X}$. *If* $P \cap \mathcal{G}(b, N)$ *is residual in* $P$ *then* $P \cap R(b, j)$ *is dense and open in* $P$.

LEMMA 5. $R(b, j)$ *is dense and open in* $\mathcal{X}$.

PROOF. Openness follows from the definition. It is enough to prove density at points of $\mathcal{H}(N)$. For $X \in \mathcal{H}(N)$ there is an open neighborhood $P \subset \mathcal{H}(N)$, a $c > 0$, and an integer $k$ such that $(3/2)^k c > b$ and all closed orbits for $Y \in P$ have period $> c$. Such a $c$ exists since all critical points of $X$ do not have one as a characteristic multiplier. Thus $P \cap \mathcal{G}(c, N) = P$. Using Corollary 4 repeatedly and the fact that

$\mathcal{G}(3\tau/2, N) = \bigcap_j \mathcal{G}(\tau, N, j)$ it follows that $\mathcal{G}(b, N) \cap P$ is residual in $P$ and hence dense at $X$. $R(b, j)$ is dense at points of $\mathcal{G}(b, N)$ so the proof is complete.

Theorem 1 now follows from the following lemma.

LEMMA 6. *Let* $\tau > 0$. *Then* $R(\tau) = \bigcap_{j,k=1}^{\infty} R((2/3)^k \tau, j)$ *is residual in* $\mathcal{X}$ *and* $R(\tau) \subset \mathcal{G}(\tau, N)$.

PROOF. $R(\tau)$ is residual by Lemma 5. Since $R(\tau) \subset \mathcal{H}(N)$ and using an argument with minimum period of closed orbits it follows that $R(\tau) \subset \mathcal{G}(b, N)$.

IV. **A transversality theorem of Abraham.** Let $A$ be a topological space with the Baire property, $M$ and $N$ second countable finite dimensional manifolds, $K \subset M$ a compact subset, $V \subset N$ a submanifold, and $F: A \to C^1(M, N)$ a point set map. $F$ is said to be $C^r$ pseudotransverse to $V$ on $K$ if (a) $F$ is a $C^1$ pseudorepresentation, i.e. $\mathrm{ev}(F^{(1)}): A \times TM \to TN$ defined $\mathrm{ev}(F^{(1)})(a, q) = T^1(Fa)q$ is $C^0$, and (b) there exists a dense subset $D \subset A$ such that for each $a \in D$ there exists an open set $B$ in a separable Banach space and $\psi: B \to A \quad C^0$ such that (i) $\psi(a') = a$ and (ii) the evaluation map $\mathrm{ev}(F \circ \psi): B \times M \to N$ is $C^r$ and transverse to $V$ on $a' \times K$.

THEOREM 6. *With the above assumptions, also assume* $F: A \to C^r(M, N)$ *is* $C^r$ *pseudotransverse to* $V$ *on* $K$ *with* $r \geq \max\{1, 1 + \dim M - \mathrm{codim}\, V\}$. *Let* $R = \{a \in A: F(a)$ *is transverse to* $V$ *at points of* $K\}$. *Then* $R$ *is residual in* $A$. *If* $V$ *is a closed submanifold then* $R$ *is dense and open in* $A$.

PROOF. If $V$ is closed then $R$ is open because $F$ is a $C^1$ pseudorepresentation. See Abraham and Robbin [2, p. 47]. If $V$ is not closed represent $V$ as the union of compact sets $V_n$. Being transverse to $V$ at points of intersection in $V_n$ is open as above. To prove density let $a \in D$ and $\psi: B \to A$ be the map given above. The set of points where a map is transverse is open so there exists an open neighborhood of $a'$ $U \subset B$ such that $\mathrm{ev}(F \circ \psi)$ is transverse to $V$ on $U \times K$. By the density theorem for representations, Abraham and Robbin [2, p. 48], $\{x \in U: F \circ \psi(x)$ is transverse to $V$ at points of $K\}$ is dense in $U$ and thus its image in $A$ by $\psi$ is dense at $a$. Thus $R$ is dense at all points of $D$ and hence of $A$.

V. **A normal form for symplectic matrices.** The existence of a normal form for symplectic matrices will be proved by showing that given an even dimensional vector space $V^{2n}$ with two form $\omega$ which equals

$$\begin{pmatrix} 0 & I \\ -I & 0 \end{pmatrix}$$

in terms of the basis for $V$, and given a symplectic matrix $A$, the question of the existence of the normal form for $A$ can be reduced to the question of a normal form for a symplectic matrix on a lower dimensional subspace of $V$. The actual normal form will not be written out explicitly but is the combination of the three types of reductions given below for different types of eigenvalues and eigenvectors. The important feature used in the proof of the main theorem is that the eigenvalues are determined by entries in blocks on the diagonal. Eigenvalues below will be represented by $\lambda + i\mu$ where $\lambda$ and $\mu$ are real. Also a matrix is symplectic with respect to the two form

$$\omega = \begin{pmatrix} 0 & I \\ -I & 0 \end{pmatrix}$$

iff $A_1{}^t A_3$ and $A_2{}^t A_4$ are symmetric and ${}^t A_1 A_4 - A_2{}^t A_3 = I$.

First, assume $\lambda + i0$ is an eigenvalue. Then there exists $a \in V$ such that $Aa = \lambda a$. Let $b \in V$ be such that $\omega(a, b) = 1$. Using $a$ and $b$ complete the basis for $V$ such that $\omega$ has the form

$$\begin{pmatrix} 0 & I \\ -I & 0 \end{pmatrix}$$

Using the fact $\omega(\cdot, \cdot) = \omega(A\cdot, A\cdot)$, it follows that $A$ has the form

$$\begin{pmatrix} \lambda & * & * & * \\ 0 & A_1 & * & A_2 \\ 0 & 0 & \alpha & 0 \\ 0 & A_3 & * & A_4 \end{pmatrix}$$

where $\alpha = 1/\lambda$.

$$\begin{pmatrix} A_1 & A_2 \\ A_3 & A_4 \end{pmatrix}$$

satisfies the condition to be a symplectic matrix on the subspace $(a, b)^\perp = \{v \in V: \omega(a, v) = \omega(b, v) = 0\}$.

Second, assume $\lambda + i\mu$ is an eigenvalue for $A$ with $\mu \neq 0$. Then there exist two nonzero vectors $a, b \in V$ such that $Aa = \lambda a - \mu b$ and $Ab = \mu a + \lambda b$. Assume $\omega(a, b) = 0$. Then there exist $c, d \in V$ such that $\omega(a, c) = \omega(b, d) = 1$ and $\omega(a, d) = \omega(b, c) = 0$. Let $\alpha + i\beta = 1/(\lambda + i\mu)$. Complete the basis for $V$ using $a$, $b$, $c$, and $d$ such that

$$\omega = \begin{pmatrix} 0 & I \\ -I & 0 \end{pmatrix}$$

$A$ has the form

$$\begin{pmatrix} \begin{pmatrix} \lambda & \mu \\ -\mu & \lambda \end{pmatrix} & * & * & * \\ 0 & A_1 & * & A_2 \\ 0 & 0 & \begin{pmatrix} \alpha & -\beta \\ \beta & \alpha \end{pmatrix} & 0 \\ 0 & A_3 & * & A_4 \end{pmatrix}$$

$$\begin{pmatrix} A_1 & A_2 \\ A_3 & A_4 \end{pmatrix}$$

satisfies the condition to be symplectic on the subspace $(a, b, c, d)^\perp$.

Third, assume $\lambda + i\mu$ and $a, b \in V$ are as in case two but with $\omega(a, b) \neq 0$. By scalar multiplication we can insure $\omega(a, b) = 1$. Complete the basis for $V$ as above. $A$ will have the form

$$\begin{pmatrix} \lambda & 0 & \mu & 0 \\ 0 & A_1 & 0 & A_2 \\ -\mu & 0 & \lambda & 0 \\ 0 & A_3 & 0 & A_4 \end{pmatrix}$$

Again

$$\begin{pmatrix} A_1 & A_2 \\ A_3 & A_4 \end{pmatrix}$$

will be symplectic on $(a, b)^\perp$. It will also follow that $\lambda^2 + \mu^2 = 1$.

See Robinson [5] where the case with $M$ noncompact is proven and more geometric interpretation and motivation is given.

## BIBLIOGRAPHY

**1.** R. Abraham and J. Marsden, *Foundations of mechanics*, Benjamin, New York, 1967.

**2.** R. Abraham and J. Robbin, *Transversal mappings and flows*, Benjamin, New York, 1967.

**3.** M. Peixoto, *On an approximation theorem of Kupka and Smale*, J. Differential Equations 3 (1967), 214–227.

**4.** K. Meyer and J. Palmore, *A generic phenomenon in conservative Hamiltonian systems*, these Proceedings, vol. 14.

**5.** C. Robinson, *Generic properties of conservative systems*, Amer. J. Math. (to appear).

UNIVERSITY OF CALIFORNIA, BERKELEY

# STRONGLY MIXING TRANSFORMATIONS

RICHARD SACKSTEDER

1. **Introduction.** Anosov and Sinai [1], [2], [7], have given very general criteria for a transformation of a manifold to be strongly mixing. Their results contain most of the classical theorems of this type. On the other hand, L. Auslander has given an example (Example 3, §7 below) of a strongly mixing diffeomorphism which does not seem to be readily treatable by the Anosov-Sinai method.

Our purpose here is to develop a theory which includes the Auslander example and as much of the Anosov-Sinai theory as is possible without entering into the complicated technical problems of "the absolute continuity of the expanding and contracting distributions." Our treatment turns out to be rather simple and elementary and it does not require "exponentially fast expansion and contraction." Moreover, it is possible that some of its by-products such as Corollary 5.1 might be useful in other connections.

2. **Mixing modulo** $K$. Here $(X, \Omega, \mu)$ will denote a probability space, where $X$ is a set, $\Omega$ a $\sigma$-algebra of subsets of $X$, and $\mu$ a measure such that $\mu(X) = 1$. If $T$ is an automorphism of $(X, \Omega, \mu)$, $U : L^2(\mu) \to L^2(\mu)$ will denote the unitary operator defined by $(Uf)(x) = f(T(x))$. If $K$ is a closed subspace of $L^2 = L^2(\mu)$, $T$ will be said to be *strongly mixing modulo* $K$ if

$$(2.1) \qquad \lim(U^j f, g) = (f, 1)(1, g) \quad \text{as } j \to \infty$$

for every $f$ in $L^2$ and $g$ in $L^2 \ominus K$. Note that strong mixing modulo $\{0\}$ is just strong mixing in the usual sense, as is mixing modulo the constant functions.

PROPOSITION 2.1. *Suppose that $K$ reduces $U$. Then $T$ is strongly mixing modulo $K$ if and only if $T^{-1}$ is.*

PROOF. Since $K$ reduces $U$, to show that $T$ is strongly mixing modulo $K$ it suffices to verify (2.1) for $f$ in $L^2 \ominus K$. Then the desired result follows from $(U^j f, g) = (f, U^{-j}g)$.

We shall be concerned with the case where $X$ is a compact metric space with metric $d(x, y)$, $\Omega$ is the $\sigma$-algebra of Borel sets of $X$, $T$ is a map such that $T$ and $T^{-1}$ are $\Omega$-measurable and leave a Borel measure $\mu$ with $\mu(X) = 1$ invariant. We call this set of conditions *Hypothesis A*.

Under Hypothesis A a group $G_+$ (or $G_-$) of $\Omega$-measurable transformations of $X$ will be said to satisfy the *Hypothesis $B_+$* (or $B_-$) if for every element $a$ of $G_+$ (or $G_-$) there is a constant $M_a$ such that for every $E$ in $\Omega$

(2.2)                                    $\mu(a^{-1}(E)) \leq M_a^2 \mu(E),$

and

(2.3)                              $\lim(d(T^j(x), T^j(a(x)))) = 0$   (pointwise)

as $j \to +\infty$ (or $j \to -\infty$).

The condition (2.2) implies that if $f_a$ is defined by $f_a(x) = f(a(x))$, then $f_a$ is in $L^2$ whenever $f$ is, and if $\Delta_a f = f_a - f$, $\Delta_a$ is a bounded linear operator on $L^2$. The kernel of $\Delta_a$ will be denoted by $K_a$ and if $G$ is any group of $\Omega$-measurable transformations of $X$, $K(G) = \{K_a : a \in G\}$.

PROPOSITION 2.2. *Let* $(X, \Omega, \mu, T)$ *satisfy Hypothesis A and suppose that $G$ is a group of $\Omega$-measurable transformations of $X$ whose elements satisfy*

(2.4)                   $a$ *is in $G$ if and only if $TaT^{-1}$ and $T^{-1}aT$ are in $G$.*

*Then $K(G)$ reduces $U$.*

PROOF. Let $b = TaT^{-1}$. Then $(\Delta_b f)(T(x)) = (\Delta_a U f)(x)$, hence $K_a = UK_b$. Therefore (2.4) implies that $UK(G) = K(G)$.

## 3. Mixing theorems.

THEOREM 3.1. *Suppose that $(X, \Omega, \mu, T)$ satisfies Hypothesis A and that $G_+$ satisfies $B_+$. Then $T$ is strongly mixing modulo $K(G_+)$.*

The proof requires a lemma.

LEMMA 3.1. *Let $f$ be an element of $L^2$. Then for every $a$ in $G_+$, $\|\Delta_a U^j f\| \to 0$ as $j \to \infty$. (Here $\| \quad \|$ denotes the $L^2$ norm.)*

PROOF. First suppose that $g$ is continuous and let $\|g\|_0 = \text{Sup}\{|g(x)| : x \in X\}$. Since $(\Delta_a U^j g)(x) = g(T^j(a(x))) - g(T^j(x))$, (2.3) and the uniform continuity of $g$ imply that $\Delta_a U^j g \to 0$ pointwise as $j \to \infty$. But $|(\Delta_a U^j g)(x)|^2 \leq 4\|g\|_0^2$, so the desired result follows from the bounded convergence theorem.

Now if $f$ is in $L^2$, there is for any $\varepsilon > 0$ a continuous $g$ such that $\|f - g\| < \varepsilon$ because $\mu$ is a Borel measure. Then

$$\|\Delta_a U^j f\| \leq \|\Delta_a U^j g\| + \|U^j(g - f)\| + \|(U^j(f - g))_a\|.$$

It has already been shown that the first term approaches zero and the second is less than $\varepsilon$. Then (2.2) implies that the third is less than $M_a \varepsilon$ and the proof is complete.

Now to prove the theorem, note that $K(G_+)$ obviously contains the constant functions so the right side of (2.1) reduces to zero. Let $a$ be in $G_+$, $f$ in $L^2$, $g$ in $L^2 \ominus K_a$, and denote the adjoint of $\Delta_a$ by $\Delta_a^*$. Then for any $\varepsilon > 0$, $g = g_1 + g_2$ where $\|g_1\| < \varepsilon$ and $g_2 = \Delta_a^* g_3$. Now $|(U^j f, g)| \leq |(U^j f, g_1)| + |(\Delta_a U^j f, g_3)|$. Hence $\lim |(U^j f, g)| = 0$ by Lemma 3.1 and the Schwarz inequality. Now the desired result follows for $g$ in $L^2 \ominus K$ by approximating such a $g$ by linear combinations of elements each of which is in $L^2 \ominus K_a$ for some $a$ in $G$.

Now if $(X, \Omega, \mu, T)$ satisfy the Hypothesis A and $G$ is a group of $\Omega$-measurable transformations of $X$, $G$ will be said to satisfy the *Hypothesis* B if (i) $G$ is generated by the elements of groups $G_+$ and $G_-$ where $G_+$ $(G_-)$ satisfies $B_+$ $(B_-)$, (ii) the group $G_-$ satisfies (2.4). In this definition either $G_+$ or $G_-$ can be trivial.

THEOREM 3.2. *Let* $(X, \Omega, \mu, T)$ *satisfy Hypothesis* A *and let* $G$ *satisfy* B. *Then* $T$ *is strongly mixing modulo* $K(G)$.

PROOF. By Theorem 3.1, $T$ is strongly mixing modulo $K(G_+)$, and $T^{-1}$ is strongly mixing modulo $K(G_-)$. Then Propositions 2.1 and 2.2 show that $T$ is strongly mixing modulo $K(G_-)$. But one easily verifies that $K(G_+) \cap K(G_-) \subset K(G)$ hence $L^2 \ominus K(G)$ is contained in the subspace generated by $L^2 \ominus K(G_+)$ and $L^2 \ominus K(G)$, and $T$ is strongly mixing modulo $K(G)$.

4. **Differentiation in a metric space.** In this section we shall digress to indicate how a theory of differentiation in a metric space can be constructed. Let $X$ and $\mu$ be as above and let $S(v, x) = \{y \in X : d(x, y) < v\}$ for $x$ in $X$ and $v > 0$. It will be necessary to assume here that

$$(4.1) \qquad \mu(S(2v, x)) \leq k\mu(S(v, x))$$

holds for some constant $k$ independent of $v$ and $x$ and that

$$(4.2) \qquad 0 < \omega(v) \leq \mu(S(v, x))$$

where $\omega$ is nondecreasing and depends only on $v$. Clearly (4.1) and (4.2) hold if $\mu$ is defined by a volume element ($=$ $n$-form with a positive continuous coefficient) on a compact manifold.

Here it will be sufficient to consider the problem of differentiation of a functional of the form $\Phi(E) = \int_E \phi(x)\mu(dx)$ with respect to $\mu$. Of course, the Radon-Nikodym derivative $d\Phi/d\mu$ which is essentially $\phi$ provides one answer to this problem, but for our purposes it will be necessary to have a derivative which is defined by difference quotients. Define a sequence $E_1, E_2, \ldots$ of measurable subsets of $X$ to be a *regular sequence tending to* $x$ with parameter $\alpha > 0$ if the following conditions are satisfied: $x$ belongs to every $E_i$, $E_i \subseteq S(v_i, x_i)$ for some $x_i, v_i$ where $v_i \to 0$, and $\mu(E_i) \geq \alpha\mu(S(v_i, x_i))$. Now let $(D_+\Phi)(x)$ (or $(D_-\Phi)(x)$) denote the supremum (or infinum) of the numbers $l$ such that

$$\Phi(E_i)/\mu(E_i) \to l \quad \text{as } i \to \infty,$$

for some regular sequence tending to $x$. Now the arguments given in Saks [6, pp. 106–118], apply to $X$ almost word for word and show that $(D_+\Phi)(x) = (D_-\Phi)(x) = \phi(x)$ except for a set of $\mu$-measure zero.

5. **Invariance under** $G$. Let $X$ and $\mu$ be as before including (4.1) and (4.2) and let $G$ be a group of homeomorphisms of $X$. An element of $L^1(\mu)$ represented by the function $f$ will be called *almost invariant* under $G$ if for every $a$ in $G$ there is a $\mu$-null set $N_a$ such that

$$(\Delta_a f)(x) = 0 \quad \text{for } x \text{ in } X - N_a.$$

An almost invariant element will be called *invariant* if it is possible to choose $f$ so that $N_a$ is empty. Von Neumann [8] and Mackey [5] have given conditions which imply that every almost invariant element is invariant. Here we shall need another result of this type.

The condition (2.2) must now be strengthened to

(5.1)                    every $a$ in $G$ has a continuous Jacobian $J_a$.

This means that there is a continuous function $J_a$ such that for every measurable set $E$, $\mu(a^{-1}(E)) = \int_E J_a(x)\mu(dx)$. It is also necessary to assume that every $a$ in $G$ is Lipschitzian, that is, that there exists a constant $C(a)$ such that for all $x, y$ in $X$

(5.2)                              $d(a(x), a(y)) \leq C(a)d(x, y)$.

THEOREM 5.1. *Let $X$ be a compact metric space and $\mu$ a finite Borel measure on $X$ satisfying* (4.1) *and* (4.2). *Let $G$ be a group of homeomorphisms of $X$ satisfying* (5.1) *and* (5.2). *Then every almost invariant element of $L^1(\mu)$ is invariant.*

PROOF. Let $\phi$ represent an almost invariant element of $L^1(\mu)$ and let $\Phi$ be defined as in §4. Then the results of §4 show that if $f$ is defined by $f(x) = (D_+\Phi)(x) = (D_-\Phi)(x)$ whenever the latter two functions agree and are finite and $f(x) = 0$ otherwise, then $f$ and $\phi$ represent the same element of $L^1(\mu)$. It remains to show that $f$ is invariant under $G$. It can be assumed that $f \geq 0$.

Let $x$ be any point of $X$ and let $E_1, E_2, \ldots$ tend to $X$ with parameter $\alpha$. Then for any $a$ in $G$, $a(E_1), a(E_2), \ldots$ is a sequence of sets which will be shown to tend to $a(x)$ with parameter $\beta > 0$. If $x \in E_i \subseteq S(v_i, x_i)$, then $a(x) \in a(E_i) \subseteq S(C(a)v_i, a(x_i))$ by (5.2). Also

(5.3)                              $\mu(E_i) \leq J_1\mu(a(E_i))$,

where $J_1 = \text{Sup}\{J_a(x) : x \in X\}$,

(5.4)                              $\mu(a(S(v_i, x_i))) \leq J_2\mu(S(v_i, x_i))$,

where $J_2 = \text{Sup}\{J_a^{-1}(x) : x \in X\}$,

(5.5)                              $S(C^{-1}(a^{-1})v_i, a(x_i)) \subseteq a(S(v_i, x_i))$,

and

(5.6)                              $\mu(S(C(a)v_i, a(x_i))) \leq k^p\mu(S(C^{-1}(a^{-1})v_i, a(x_i)))$,

where $k$ is the constant in (4.1) and $p$ is any integer larger than $\text{Log}_2 C(a)C(a^{-1})$.

Now (5.3)–(5.6) together with $\mu(E_i) \geq \alpha\mu(S(v_i, x_i))$ imply that $\mu(a(E_i)) \geq \beta\mu(S(C(a)v_i, a(x_i)))$, where $\beta = \alpha(k^p J_1 J_2)^{-1}$.

Now suppose that $x$ is a point of $X$ where $(D_+\Phi)(x) = (D_-\Phi)(x) = f(x)$. Then if $E_1, E_2, \ldots$ tends to $x$ with parameter $\alpha$, then $\Phi(E_i)/\mu(E_i) \to f(x)$ as $i \to \infty$. Moreover it has just been shown that for every $a$ in $G$, $a(E_1), a(E_2), \ldots$ tends to $a(x)$ with parameter $\beta > 0$. Now $\mu(E_i) = \int_{a(E_i)} J_a(y)\mu(dy)$ and $\Phi(E_i) = \int_{a(E_i)} f(x)J_a(y)\mu(dy)$. For any $\varepsilon$, $0 < \varepsilon < J_a(a(x))$, $|J_a(y) - J_a(a(x))| < \varepsilon$ if $y$ is in $a(E_i)$, for large $i$, hence

$$D^{-1}(\varepsilon)\Phi(a(E_i))/\mu(a(E_i)) \leq \Phi(E_i)/\mu(E_i) \leq D(\varepsilon)\Phi(a(E_i))/\mu(a(E_i)),$$

where $D(\varepsilon) \to 1$ as $\varepsilon \to 0$. It follows that $(D_+\Phi)(a(x)) = (D_-\Phi)(a(x)) = f(a(x))$ $= f(x)$. This shows that $f(a(x)) = f(x)$ for any $x$ such that $(D_+\Phi)(x) = (D_-\Phi)(x)$ and that the set of such $x$ is invariant under $G$. Since $G$ is a group, the complement of this set is invariant under $G$. Since $f(x) \equiv 0$ on the complement, this proves that $f$ is strictly invariant.

Theorem 5.1 has the following corollary, in which $X_1 \subseteq X$, $\mu(X_1) = 1$.

COROLLARY 5.1. *If in Theorem 5.1, $G$ acts transitively on $X_1$, then any $L^1$ function which is almost invariant under $G$ is constant almost everywhere.*

It would be interesting and would strengthen our later results if the conditions (5.1) and (5.2) can be weakened in Theorem 5.1 or Corollary 5.1.

6. **Mixing diffeomorphisms.** In this section $X$ denotes a connected compact $C^k$ manifold ($k \geq 1$) and $T: X \to X$ a $C^k$ diffeomorphism. It will be supposed that $\mu$ is a measure corresponding to a volume element (cf. §4) and that $T$ leaves $\mu$ invariant. In particular, Hypothesis A is satisfied. Suppose that there are modules $A_+$ and $A_-$ of $C^k$ vector fields on $X$ which are invariant under the derivative $DT$ of $T$ and under $DT^{-1}$. The group of $C^k$ diffeomorphisms generated by one parameter subgroups corresponding to elements of $A_+$ (or $A_-$) will be denoted by $G_+$ (or $G_-$) and the group of diffeomorphisms generated by $G_+$ and $G_-$ will be called $G$. Clearly the elements of $G$ will satisfy (2.2), (2.4), (5.1) and (5.2).

In order to assure that the elements of $G_+$ satisfy (2.3) (where $j \to +\infty$) we assume that for every $\xi$ in $A_+$

$$(6.1) \qquad \operatorname{Sup}_{x\varepsilon X}|((DT^j)\xi)(x)| \to 0 \quad \text{as } j \to \infty.$$

Similarly for every $\xi$ in $A_-$ we assume

$$(6.2) \qquad \operatorname{Sup}_{x\varepsilon X}|((DT^{-j})\xi)(x)| \to 0 \quad \text{as } j \to \infty,$$

which implies that (2.3) holds (where $j \to -\infty$) for the elements of $G_-$. Therefore if (6.1) and (6.2) hold, Hypothesis B is satisfied.

Under the above assumptions, $T$ is strongly mixing modulo $K(G)$ by Theorem 3.2. In view of Theorem 5.1 we then have proved

THEOREM 6.1. *Let $X, T, \mu, A_+, A_-, G$ be as described above. Then $T$ is strongly mixing modulo $K(G)$ and an element of $L^2(\mu)$ is in $K(G)$ if and only if it can be represented by a function which is everywhere invariant under $G$.*

The most interesting special case is:

COROLLARY 6.1. *Let $X, T, \mu, A_+, A_-,$ and $G$ be as described above and suppose that $G$ acts transitively on $X$. Then $T$ is strongly mixing.*

For the proof one only need observe that when $G$ is transitive every element of $K(G)$ is represented by a constant function, by Corollary 5.1 and then apply the remark after (2.1).

It is not always easy to determine if $G$ acts transitively. Lemma A below gives a simple criterion which combined with Corollary 5.1 gives our main result:

THEOREM 6.2. *Let $X$, $T$, $\mu$, $A_+$, and $A_-$ be as described above, including (6.1) and (6.2). Suppose that $A_+$ and $A_-$ are finitely generated and that every tangent vector to $X$ can be realized from the elements of $A_+$ and $A_-$ by a finite combination of the bracket operation and linear sums. Then $T$ is strongly mixing.*

REMARK. The hypotheses can be weakened slightly. The elements of $A_+$ and $A_-$ can fail to generate all of the tangent space on a set of $\mu$-measure zero if this set does not disconnect $X$, cf. Corollary 5.1.

To apply Lemma A to obtain the above result one takes (in the notation of the Appendix) $A_1 = A_+ \vee_k A_-$ and the basis $B_1$ to consist of elements each of which is either in $A_+$ or in $A_-$. Then the hypotheses of Theorem 6.2 assert that $A_{k+1}$ is the module of all appropriately smooth vector fields on $X$, hence $G$ acts transitively. Then Corollary 6.1 gives the desired conclusion.

7. **Examples.** Here we shall give a few examples showing how the theorems can be applied.

EXAMPLE 1. (BAKER'S TRANSFORMATION). Let $X = \{(x, y) : 0 \leq x, y \leq 1\}$ and let $T(x, y)$ be $(\frac{1}{2}x, 2y)$ if $0 \leq y \leq \frac{1}{2}$ and $(\frac{1}{2}(x + 1), 2y - 1)$ if $\frac{1}{2} < y \leq 1$. Take as $G_+$ the group of $C^1$ diffeomorphisms of $X$ of the form $a(x, y) = (b(x, y), y)$. Then it is easy to check that $G_+$ satisfies Hypothesis $B_+$. Let $G_-$ be the set of $C^1$ diffeomorphisms of the form $a(x, y) = (x, b(x, y))$ and $G_-$ satisfies $B_-$. Clearly the group $G$ generated by $G_+$ and $G_-$ acts transitively on the interior of $X$ and satisfies (5.1) and (5.2). Therefore Theorem 3.2 and Corollary 5.1 show that $T$ is strongly mixing.

EXAMPLE 2. (ANOSOV DIFFEOMORPHISMS WITH SMOOTH SPLITTING). Let $X$ be a compact manifold and $T$ a diffeomorphism of $X$. Suppose that the tangent bundle $\tau(X) = \tau_+(X) \oplus \tau_-(X)$ where the splitting is $C^1$, is invariant under $DT$, the fiber dimensions of $\tau_+(X)$ and $\tau_-(X)$ are constant, and for any $\varepsilon > 0$ there is a $J(\varepsilon)$ such that for any $\xi$ in $\tau_+(X)$ (or in $\tau_-(X)$),

$$\|(DT^j\xi)\| \leq \varepsilon \|\xi\| \text{ if } j \geq J(\varepsilon) \qquad (\text{or } j \leq -J(\varepsilon)).$$

(This is weaker than the usual "exponential contraction" condition, which requires that $J(\varepsilon) \leq \alpha \operatorname{Log} \varepsilon + \beta$ for some constants $\alpha$ and $\beta$). Then let $A_+$ (or $A_-$) be the set of $C^1$ vector fields which are sections of $\tau_+(X)$ (or of $\tau_-(X)$). It is easy to check that Theorem 6.2 applies and shows that $T$ is strongly mixing.

EXAMPLE 3. (DUE TO L. AUSLANDER). Let $N$ denote the nilpotent Lie algebra of $(3 \times 3)$-matrices with zeros on and below the diagonal. The elements of $N$ are in one-to-one correspondence with the points of $R^3 = \{(x, y) : x \in R^2, y \in R^1\}$, where the elements of $R^2$ correspond to the entries just above the diagonal. Let $N_0$ be the subalgebra whose elements correspond to $(x, \frac{1}{2}y)$ such that $x$ and $y$ have integral components. If $L : R^2 \to R^2$ is a linear map defined by an element of $SL(2, Z)$, $S(x, y) = (L(x), y)$ is an automorphism of $N$ leaving $N_0$ fixed.

Let $M = \exp N$ and $M_0 = \exp N_0$. $S$ induces an automorphism $T_0$ of $M$ which leaves $M_0$ fixed, hence defines a diffeomorphism $T$ of $X = M/M_0$. Now suppose that $L$ has no eigenvalue of absolute value one and let $x_+$ (or $x_-$) be an eigenvector belonging to the eigenvalue inside (or outside) the unit circle. It can be verified that corresponding to the basis $\{(x_+, 0), (x_-, 0), (0, 1)\}$ of $N$, the tangent bundle $\tau(X) = \tau_+(X) \oplus \tau_-(X) \oplus \tau_0(X)$ has a splitting into line bundles which are invariant under $DT$. Then if $A_+$ (or $A_-$) in the set of smooth sections of $\tau_+(X)$ (or $\tau_-(X)$), it can be verified that every tangent vector can be formed from the elements of $A_+$ and $A_-$ by taking linear combinations and brackets. (In fact this is just the reflection of the fact that all of $N$ is generated by $(x_+, 0)$ and $(x_-, 0)$.) Theorem 6.2 now applies.

This construction can be generalized. The method is implicit in the paper of L. Auslander and J. Scheuneman in these Proceedings.

### APPENDIX: INTEGRAL MANIFOLDS

If $A$ and $B$ are sets of $C^i$ vector fields on a compact connected manifold $X$, the smallest $C^i$ module of vector fields containing them will be denoted by $A \vee_i B$. Suppose that $A_1$ is a finitely generated module of $C^k$ vector fields ($k \geq 1$) on $X$ and define $A_{i+1} = A_i \vee_{k-i} [A_i, A_i]$ for $i = 1, \ldots, k$, where the obvious conventions are made for $k = \infty$. Let $B_1 = \{X_i^j : j = 1, \ldots, j_1\}$ be a basis for $A_1$, let $\{a^j(t) : j = 1, \ldots, j_1\}$ be the corresponding one parameter groups of diffeomorphisms, and let $G$ be the group of diffeomorphisms generated by

$$\{a^j(t) : -\infty < t < +\infty, j = 1, \ldots, j_1\}.$$

LEMMA A. *Suppose that $A_{k+1}$ is equal either to $A_k$ or to the module of all $C^0$ (or $C^\infty$ if $k = \infty$) vector fields on $X$ and that the dimension $d$ of the subspace of the tangent space spanned by the elements of $A_{k+1}$ is constant over $X$. Then $A_{k+1}$ defines a completely integrable distribution whose maximal integral manifolds are exactly the orbits of $G$. In particular, if $d = \dim X$, $G$ acts transitively.*

PROOF. The assertion about complete integrability is clear. Define $B_2$ to be the set of vector fields of the form $X_2^j = [X_1^r, X_1^s]$ so that $B_1 \cup B_2$ spans $A_2$. Similarly define $B_i$ for $i \leq k + 1$ as the set of vector fields formed by applying the bracket operation successively to $i$ elements of $B_1$ so that $\bigcup \{B_j : 1 \leq j \leq i\}$ spans $A_i$.

The interpretation of the bracket of two vector fields as a Lie derivative shows that the elements of $B_2$ can be $C^{k-1}$ approximated by elements of the form $(1/t) \cdot (Da_1^r(t))X_1^s$. Similarly the elements of $B_3$ can be $C^{k-2}$ approximated by elements of the form $(1/t)(1/u)(D(a_1^r(t)a_1^q(u)))X_1^s$ and so on for all $B_i$. Moreover, the complete integrability of $A_{k+1}$ implies that the integral manifolds of $A_{k+1}$ are invariant under $a_1^r(t)$, so the approximating vector fields also lie in $A_{k+1}$. Since $X$ is compact, the elements of $B_1$ together with a suitably chosen set of approximating elements will span $A_{k+1}$. Call such a set of elements $E$.

If $x$ is a point of $X$ let $E_x$ be some minimal subset of $E$ such that its elements span a subspace of the tangent space of dimension $d$ at (hence near) $x$. Applying Chow's form of the Frobenius Theorem [3, Satz A] to $E_x$ shows that the group $\Gamma_x$ of diffeomorphisms generated by one parameter subgroups corresponding to elements of $E_x$ acts transitively on the components of the intersections of the integral manifolds of $A_{k+1}$ with a neighborhood of $x$. If $\Gamma$ is to $E$ as $\Gamma_x$ is to $E_x$ it is then clear that $\Gamma$ acts transitively on the integral manifolds of $A_{k+1}$. But $\Gamma \subset G$ (cf. the remarks in Chow [3] after Satz B), so $G$ acts transitively on each integral manifold. This completes the proof.

REMARKS. We only need the special case $d = \dim X$ of the lemma here. A somewhat more general result than the one stated above follows by the method of Herman [4].

## REFERENCES

**1.** D. V. Anosov, *Ergodic properties of geodesic flows on closed Riemannian manifolds with negative curvature*, Soviet Math. Dokl. **3** (1962), 1153–1156.

**2.** D. V. Anosov and Ja.G. Sinai, *Some smooth ergodic systems*, Russian Math. Surveys **22** (1967), 103–167.

**3.** W. L. Chow, *Über systeme von linearen partiellen Differentialgleichungen erster Ordnung*, Math. Ann. **117** (1940–1941), 98–105.

**4.** R. Hermann, *The differential geometry of foliations. II*, Math. and Mech. **11** (1962), 303–315.

**5.** G. W. Mackey, *Point realizations of transformation groups*, Illinois J. Math. **6** (1962), 327–335.

**6.** S. Saks, *The theory of the integral*, Hafner, New York.

**7.** Ja. G. Sinai, *Dynamical systems with countably multiple Lebesgue spectrum*, Amer. Math. Soc. Transl. (2) **68** (1968), 34–88.

**8.** J. von Neumann, *Zur Operatormethode in der klassischen Mechanik*, Ann. of Math. **33** (1932), 587–642.

CITY UNIVERSITY OF NEW YORK

# THE DEPTH OF THE CENTER
# OF 2-MANIFOLDS

A. J. SCHWARTZ[1] AND E. S. THOMAS[2]

1. **Introduction.** We consider a flow $h: X \times R \to X$ where $X$ is a metric space, $R$ is the additive group of real numbers and $h$ is a continuous function satisfying

(a) $h(x, t_1 + t_2) = h(h(x, t_1), t_2)$, and

(b) $h(x, 0) = x$,

for all $x$ in $X$ and $t_1$, $t_2$ in $R$. Each real number $t$ determines a homeomorphism, $h_t: X \to X$ according to $h_t(x) = h(x, t)$.

We will investigate a weak form of recurrence due to G. D. Birkhoff [1], which we now define:

1.1. DEFINITION. Let $Y$ be an invariant ($h_t(Y) = Y$ for all real $t$) subset of $X$. We say a point $y$ in $Y$ is *nonwandering relative to $Y$* in case for every neighborhood, $U$, of $y$ relative to $Y$, there is a $t \geq 1$ such that $h_t(U) \cap U = h_t(U) \cap U \cap Y \neq \varnothing$; otherwise, we say $y$ is *wandering relative to $Y$*. We denote by $Y'$ the set of points nonwandering with respect to $Y$.

Note that if $y$ is in $Y'$, given any neighborhood, $U$, of $y$, and any real number $T$ there can be found $t_1 \geq |T|$ and $t_2 \leq -|T|$ such that $h_{t_i}(U) \cap U \cap Y \neq \varnothing$ for $i = 1, 2$.

$Y'$ is a closed, invariant subset of $Y$ (possibly empty). If $Y$ is nonempty and compact then $Y'$ is nonempty. By taking $Y = X$ and iterating the process of forming sets of nonwandering points we get a nested sequence of closed invariant sets $X \supseteq X' \supseteq (X')' \supseteq \ldots$ which leads to the concept of the central sequence which we now define.

1.2. DEFINITION. $X^{(0)} = X$. If $\alpha$ is an ordinal and $X^{(\alpha)}$ has been defined, we set $X^{(\alpha + 1)} = (X^{(\alpha)})'$. If $\beta$ is a limit ordinal and $X^{(\alpha)}$ has been defined for each $\alpha < \beta$, set $X^{(\beta)} = \bigcap_{\alpha < \beta} X^{(\alpha)}$. The transfinite sequence $\alpha \mapsto X^{(\alpha)}$ is called the *central sequence*. If $\alpha_0$ is the least ordinal such that $X^{(\alpha_0 + 1)} = X^{(\alpha_0)}$, we say the central sequence has *depth* $\alpha_0$.

The central sequence of every flow must have some depth. The central sequence of a flow in a separable metric space must have countable depth.

1.3. DEFINITION. If $h: X \to X$ is a flow whose central sequence has depth $\alpha_0$, $X^{(\alpha_0)}$ is called the *center of the flow*.

$Y \subset X$ is the center of a flow on $X$ if and only if $Y$ is the largest invariant subset of $X$ such that $Y' = Y$. Another characterization of the center of a flow may be obtained using the following concepts:

---

[1] This work was partially supported by National Science Foundation grant GP–7445.

[2] This work was partially supported by National Science Foundation grant GP–5935.

1.4. DEFINITION. $y$ is called an $\omega(\alpha)$-*limit point of* $x$ in case there is a sequence of real numbers $t_k \to \infty$ $(-\infty)$ such that $h_{t_k}(x) \to y$.

The set of $\omega(\alpha)$-limit points of $x$, $\Omega(x)$ $(A(x))$, is called the $\Omega(A)$-*limit set of* $x$.

1.5. DEFINITION. If $x \in \Omega(x)$ $(A(x))$ we call $x$ $P^+$ $(P^-)$-*stable*. If $x$ is both $P^+$ and $P^-$ stable we say $x$ is *Poisson stable*.

Note that if $x$ is a fixed point $(h_t(x) = x$ for all $x)$ or a periodic point $(h_t(x) = x$ for some $t \neq 0)$ then $x$ is Poisson stable. Also note that if $x$ is $P^+$ or $P^-$ stable then $x$ is nonwandering relative to $O(x)$ where $O(x)$ denotes the orbit of $x$, $\{h_t(x) | t \in \mathbf{R}\}$. Thus if $x$ is $P^+$ or $P^-$ stable, $x$ is in the center of the flow. Conversely, we have the following:

1.6. PROPOSITION [6]. *The center of a flow is the closure of the set of points which are Poisson stable.*

Maïer has shown that in a 3-manifold, given an arbitrary countable ordinal $\alpha_0$, one can construct a flow whose central sequence has depth $\alpha_0$ [3], [4]. Recently, P. Frederickson has clarified this result [2]. In §2 we shall construct a flow on the Moebius strip whose central sequence has depth 3 and which may be embedded in flow on any nonorientable 2-manifold. A central result of this paper is the following:

THEOREM. *If* $h: M \times \mathbf{R} \to M$ *is a flow on a compact, orientable, 2-manifold, the central sequence has depth* $\alpha_0 \leq 2$.

In §3 we will prove the theorem for the case $M = S^2$, the 2-sphere. The proof for the general case will involve an argument by induction for which we need the following concept:

1.7. DEFINITION. We say a connected, orientable 2-manifold (not necessarily compact) is of *type* 1 in case $X - S$ is disconnected by any simple closed curve, $S$, contained in the interior of $X$. We say $X$ is of *type* $n + 1$ in case $X - S$ is of type $n$ for any simple, closed curve, $S$, contained in the interior of $X$.

Note that a sphere with $k$ handles is of type $k + 1$. In general, $X$ is of type $k + 1$ if and only if $X - \bigcup S_i$ is disconnected for *any* $k + 1$, disjoint closed curves $S_1, ..., S_{k+1}$ in the interior of $X$ while *there exist* $k$ disjoint closed curves $S_1, ..., S_k$ in the interior such that $X - \bigcup S_i$ is connected.

In §4 we will state and prove some topological propositions about manifolds of type $n$ which will be used in the proof of the main theorem.

We now frame two families of propositions which will be proved inductively in §5. All manifolds are assumed separable metric.

A($n$): If $M$ is a sphere with $n$ handles, then for any flow $h: M \times \mathbf{R} \to M$ on $M$, the depth $\alpha_0$ of the central sequence of the flow is $\leq 2$.

B($n$): If $X$ is an orientable, connected, 2-manifold of type $n$, without boundary, then for any flow $h: X \times \mathbf{R} \to X$ on $X$, the depth $\alpha_0$ of the central sequence of the flow is $\leq 2$.

The theorem quoted above is equivalent to the truth of A($n$) for all $n$. The validity of B($n$) for all $n$ extends this result.

In §3 we prove A(0). In §5 we prove $A(n) \Rightarrow B(n + 1)$ and $B(n + 1) \Rightarrow A(n + 1)$ making use of the topological results of §4. This will complete the proof. In §6 the results are generalized.

FIGURE 1

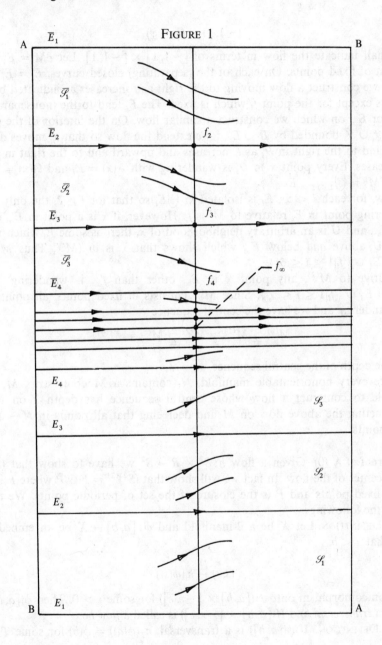

2. **A nonorientable example** (Figure 1). Consider the Moebius strip

$$M = [-1, 1] \times [-1, 1]/\sim$$

where

$$(-1, y) \sim (1, -y).$$

We shall indicate the flow in terms of $[-1, 1] \times [-1, 1]$. Let $\partial M = E_1 \cup \bar{E}_1$ consist of fixed points. On each of the (separating) closed curves, $E_i = \underline{E}_i \cup \bar{E}_i$, $i \geq 2$, we construct a flow moving to the right as $t$ increases as indicated by the arrows except for the point $f_i$ which is fixed. The $E_i$ tend to the (nonseparating) equator $E_\infty$ on which we construct a similar flow. On the interior of the strips $\mathscr{S}_i = \underline{\mathscr{S}}_i \cup \bar{\mathscr{S}}_i$ bounded by $E_i \cup E_{i+1}$ we extend the flow so that $x_t$ moves downward and to the right in $\bar{\mathscr{S}}_i$ as $t$ increases and upward and to the right in $\underline{\mathscr{S}}_i$ as $t$ increases. Every point $x$ in $\mathscr{S}_i$ is wandering with $A(x) = E_i$ and $\Omega(x) = E_{i+1}$. Thus, $M' = \bigcup_{1 \leq i \leq \infty} E_i$.

Now, for each $i < \infty$, $E_i$ is isolated in $\bigcup E_i$ so that for $i \geq 2$, the only nonwandering point in $E_i$ relative to $M'$ is $f_i$. However, if $x$ is a point in $E_\infty$ other than $f_\infty$, and $U$ is an arbitrary neighborhood of $x$, there is some $E_i$ which intersects $U$ above *and* below $E_\infty$ which shows that $x$ is in $(M')'$. Thus $M^{(2)} = E_\infty \cup E_1 \cup \{f_i | 1 \leq i \leq \infty\}$.

Relative to $M^{(2)}$, any point $x$ in $E_\infty$ other than $f_\infty$, is wandering. Thus $M^{(3)} = E_1 \cup \{f_i | 1 \leq i \leq \infty\}$. Since $M^{(3)}$ consists of fixed points, all points are nonwandering and we have the central sequence

$$M^{(0)} \supsetneqq M^{(1)} \supsetneqq M^{(2)} \supsetneqq M^{(3)} = M^{(4)}$$

and the depth of the central sequence, $\alpha_0 = 3$.

Since every nonorientable manifold, $N$, contains a Moebius strip, $M$, it is possible to construct a flow whose central sequence has depth 3 on $N$ by constructing the above flow on $M$ and decreeing that all points in $N - M$ be fixed points.

3. **Proof of A** (0). Given a flow $h: S^2 \times R \to S^2$ we have to show that $(S^2)^{(2)}$ is the center of the flow. In fact we will show that $(S^2)^{(2)} = F \cup \bar{P}$ where $F$ is the set of fixed points and $\bar{P}$ is the closure of the set of periodic points. We make use of the following:

3.1. DEFINITION. Let $X$ be a 2-manifold and $\phi: [a, b] \to X$ be an embedding such that $a < b$

$$(s, t) \overset{H}{\mapsto} h_t(\phi(s))$$

is a homeomorphism onto $H([a, b] \times [-\varepsilon, \varepsilon])$ for some $\varepsilon > 0$. Then $\phi([a, b])$ is called a *transversal* and $H([a, b] \times [-\varepsilon, \varepsilon])$ is called a *flow box*.

3.2. DEFINITION. If $\phi([a, b])$ is a transversal, $h_T(\phi(a)) = \phi(b)$ for some $T > 0$,

FIGURE 2

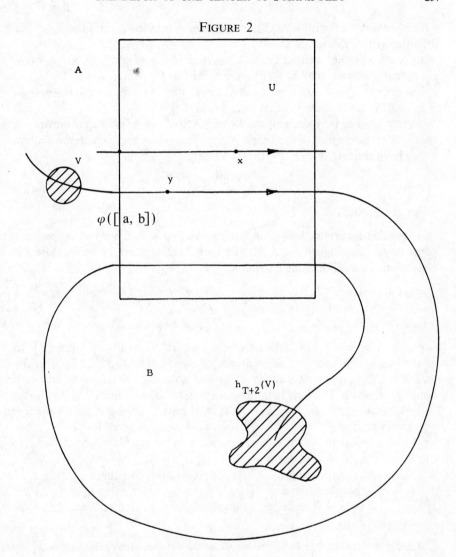

and $\gamma = \phi([a, b]) \cup \{h_t(\phi(a))|0 \le t \le T\}$ is a simple closed curve, then $\gamma$ is called a *circuit*.

Now let us prove A(0).

Let $x$ be in $(S^2)^{(2)}$. If $x$ is a fixed point then we are finished. If $x$ is not a fixed point, we may construct an arbitrarily small flow box $U$, around $x$, which contains a point $y$ in $(S^2)^{(1)}$ (possibly $x$ itself) such that $h_t(y)$ leaves $U$ as $t$ increases and, for some $t > 0$, returns to $U$. (See Figure 2.)

We assert that $y$ is periodic. For if not we may form a circuit

$$\gamma = \{h_t(y)|0 \le t \le T\} \cup \phi([a, b])$$

which separates $S^2$ with $h_{-1}(y)$ lying in one component, $A$, and $h_{T+1}(y)$ lying in the other component $B$ of $S^2 - \gamma$.

By virtue of the continuity of $h$, there exists a neighborhood, $V$, of $h_{-1}(y)$ contained in $A$ such that $h_{T+2}(V)$ is contained in $B$.

But for $t \geq T + 2$, $h_t(V)$ must remain in $B$, for no orbit can cross $\gamma$ going from $B$ to $A$. This contradicts the fact that $y$ is in $(S^2)^{(1)}$.

Thus $y$ must be periodic and it follows, since $U$ can be chosen arbitrarily small, that $x$ is in the closure of the union of periodic orbits and thus in the center.

We have shown $(S^2)^{(2)} \subset \bigcap_\alpha (S^2)^{(\alpha)}$. Clearly

$$\bigcap_\alpha (S^2)^{(\alpha)} \subset (S^2)^{(2)}$$

so we are finished.

4. **Topological preliminaries.** Before proceeding with the proof we pause to list some useful topological facts. The first of these appears to be folklore but for convenience we will include a proof.

4.1. PROPOSITION. *Let $X$ be a connected orientable, 2-manifold of type $n$. Then there is a homeomorphism of $X$ onto $S^2 \cup h_1 \cup \ldots \cup h_{n-1} - F$ where $F$ is a closed, totally disconnected set. ($S^2 \cup h_1 \cup \ldots \cup h_m$ denotes the sphere with $m$ handles.)*

PROOF. For $n = 1$, this follows easily from a theorem of Zippin [7]. In the general case, choose $n - 1$ disjoint simple closed curves, $S_1, \ldots, S_{n-1}$ in $X$ such that $X - \bigcup S_i$ is connected. Since $X$ is orientable, each $S_i$ has a closed neighborhood $A_i$ in which $X$ which is homeomorphic to the annulus $S^1 \times I$. Choose the $A_i$ to be disjoint. Then $Y = X - \bigcup A_i$ is of type 1, so, by Zippin's theorem, is homeomorphic to $S^2 - F'$ where $F'$ is totally disconnected. For $i = 1, \ldots, n - 1$, let $d_i$ and $e_i$ be the points of $F'$ corresponding to the simple closed curves forming the boundary of $A_i$ and let $F = F' - \{e_i, d_i | i = 1, \ldots, n - 1\}$. For each $i$, let $D_i$, $E_i$ be disks containing $d_i$, $e_i$ (respectively) and no other point of $F'$. Then $Y$ is homeomorphic to $S^2 - [(D_1 \cup E_1) \cup \ldots \cup (D_{n-1} \cup E_{n-1})] - F$. Then $S^2 \cup h_1 \cup \ldots \cup h_n - F$, obtained by attaching each $h_1$ along $\partial D_i$ and $\partial E_i$, is clearly homeomorphic to $X$.

4.2. PROPOSITION. *Let $\Gamma$ be a continuum (i.e., a compact connected set) properly contained in $X = S^2 \cup h_1 \cup \ldots \cup h_n$. Either $\Gamma$ lies in an open cell or else each component of $X - \Gamma$ has type $\leq n$.*

PROOF. For $n = 0$, any $\Gamma$ lies in a cell. Suppose $n \geq 1$ and some component $Z$ of $X - \Gamma$ has type $\geq n + 1$. There exist $n$ simple closed curves $S_1, \ldots, S_n$ in $Z$ such that $Z - \bigcup S_i$ is connected. Then $X - \bigcup S_i$ is connected and of type 1 so by Proposition 1 there is a homeomorphism $h$ of $X - \bigcup S_i$ onto $S^2 - F$ where $F$ is totally disconnected. $F$ lies in a component of $S^2 - h(\Gamma)$; hence there is a closed cell $D$ in $S^2$ such that $F \subset D \subset S^2 - h(\Gamma)$. Then $S^2 - D = U$ is an open cell containing $h(\Gamma)$ and $h^{-1}(U)$ is the desired cell in $X$ containing $\Gamma$.

The next proposition will be used in §6 but for the sake of continuity in that section we prove it here. In what follows, $X = S^2 \cup h_1 \cup \ldots \cup h_n$. We begin with a lemma.

**4.3. LEMMA.** *If $\Gamma$ is a continuum in $X$ and $W$ is a component of $X - \Gamma$ then cl $W$ has finitely many boundary components (here "cl" denotes closure). Moreover the quotient space, $\tilde{W}$, of cl $W$ obtained by collapsing each boundary component to a point is homeomorphic to $S^2 \cup h_1 \cup \ldots \cup h_k$ where $k \leq n$.*

**PROOF.** $W$ is of type $k$ where $k \leq n + 1$, so there is a homeomorphism $h$ taking $W$ onto $S^2 \cup h_1 \cup \ldots \cup h_k - F$ where $F$ is totally disconnected. It is easily verified that points of $F$ correspond to boundary components of cl $W$ in the sense that a sequence $w_i$ in $W$ has its cluster set in a boundary component if and only if $h(w_i)$ converges to a point of $F$. Now suppose $F$ has at least $n + 2$ points. Choose $n + 1$ disjoint simple closed curves $S_1, \ldots, S_{n+2}$ in $h(W)$ such that every component of $h(W) - \bigcup S_i$ contains a point of $F$. Then each component of $W - \bigcup h^{-1}(S_i)$ has a point of $\partial W$ in its closure. Then $[W - \bigcup h^{-1}(S_i)] \cup \Gamma$ is connected, whence $X - \bigcup h^{-1}(S_i)$ is connected. Since $X$ is of type $n + 1$ this is impossible; so, in fact, $F$ has at most $n + 1$ points and the boundary of cl $W$ has at most $n + 1$ components. The second assertion follows from the above remarks and Proposition 4.1.

**4.4. PROPOSITION.** *Let $\Gamma$ be a continuum in $X$ which meets a closed arc $T$ in a totally disconnected set. Let $W$ be a component of $X - \Gamma$ and let $\mathcal{T}$ be the collection of components of $T - \Gamma$ lying in $W$. If $\mathcal{T}$ is infinite, $\mathcal{T} = \{T_i | i \in Z\}$, then, for almost all $i$, there is an open cell $W_i \subset W$ such that the boundary of $W_i$ is contained in $T_i \cup \Gamma$.*

**PROOF.** Using the notation of the preceding lemma, suppose $\tilde{W}$ is homeomorphic, via say $h$, to $Y = S^2 \cup h_1 \cup \ldots \cup h_k$ and let $q$ be the quotient map of cl $W$ onto $\tilde{W}$. The boundary components $K_1, \ldots, K_s$ of cl $W$ are mapped by $h \circ q$ to a finite set $F = \{k_1, \ldots, k_s\}$. Let $U$ be the union of $s$ pairwise disjoint cells in $Y$ each containing one $k_i$. Thus $q^{-1}(h^{-1}(U))$ is a neighborhood of $\bigcup K_i$ in cl $W$. Almost all $T_i$ lie in this neighborhood. For each such $i$, $h \circ q(T_i)$ is a loop in $U$, hence bounds an open cell $V_i$. Then letting $W_i = q^{-1}(h^{-1}(V_i))$ we obtain an open cell in $X$ whose boundary lies in $T_i \cup \Gamma$.

**5. A(n) implies B(n + 1) for $n \geq 0$.** Assume $A(n)$ and suppose we are given a flow $\{h_t\}$ on a manifold $X$ satisfying the hypotheses of $B(n + 1)$. By Proposition 4.1 there is a homeomorphism $g$ of $X$ onto $Y = S^2 \cup h_1 \cup \ldots \cup h_n - F$. Then the family $\{f_t = gh_t g^{-1} | t \in R\}$ is a flow on $Y$ and it is easily seen that it extends to a flow, also denoted $\{f_t\}$, on all of $S^2 \cup h_1 \cup \ldots \cup h_n$ which leaves $F$ pointwise fixed. By $A(n)$, the central sequence of this flow terminates in two steps. It follows readily that the same is true of the original flow $\{h_t\}$ on $X$.

**6. Completion of the main theorem.** We now assume that the propositions $A(k)$, $B(k)$ have been established for $k = 0, \ldots, n - 1$ and that flow $\{h_t\}$ has been given on $X = S^2 \cup h_1 \cup \ldots \cup h_n$. We wish to show $X'' = X'''$.

We first observe that if there is a $P^+$ or $P^-$ stable point which is not periodic then we are done. For suppose that $x$ is $P^+$ stable. Let $\Gamma = \mathrm{cl}\, O(x)$; if $\Gamma$ lies in a cell then by a simple modification of the argument used to prove A(0) it is easily seen that $x$ is periodic.

Assume then that $\Gamma$ does not lie in a cell. Let $\mathscr{W}$ be the collection of components of $X - \Gamma$. Each $W \in \mathscr{W}$ is a separable, orientable 2-manifold without boundary and by Proposition 4.2 has type $\leq n$. By the inductive hypotheses, for each $W \in \mathscr{W}$, $W''$ is the center of the flow obtained by restricting $\{h_t\}$ to $W$. $\Gamma$ itself lies in $X''$ and in the center of $\{h_t\}$. Thus $X'' = \Gamma \cup \{W'' | W \in \mathscr{W}\}$ = the center of $\{h_t\}$. q.e.d.

Thus, for the remainder of the proof we shall assume that *every $P^+$ or $P^-$ stable point is periodic.*

Let us suppose $X'' \neq X'''$ and fix a point $x$ in $X'' - X'''$. Ultimately we shall reach a contradiction. If $T$ is an arc transverse to the flow and $x \in T$ is a non-endpoint of $T$ then we say $T$ is a transversal *through* $x$.

Let $T$ be a transversal through $x$ and let $\{y_i\}$ be a sequence of points in $X' - X''$ whose orbits make a sequence of circuits with endpoints lying on $T$ and converging to $x$. The $y_i$ and their circuits may be visualized as in Figure 3 (in fact a circuit may cross $T$ several times).

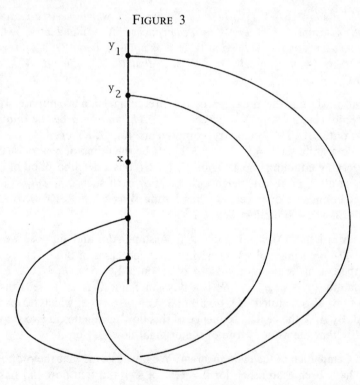

FIGURE 3

We now define a continuum $\Gamma$ associated with $\{y_i\}$. If infinitely many of the $y_i$ lie on a single orbit, say $O(y)$ where necessarily $y \in X' - X''$, then $x$ belongs to $A(y)$ or to $\Omega(y)$. (It cannot lie in $O(y)$ because $x$ is in $X''$ and $y$ is not.) We assume that $x \in \Omega(y)$ and put $\Gamma = \Omega(y)$; $\Gamma$ is an invariant continuum containing $O(x)$.

The other possibility is that only finitely many $y_i$ lie on any given orbit. Choose a subsequence, also denoted $\{y_i\}$, all of whose orbits are distinct and set $\Gamma = \limsup O(y_i) = \bigcap_{n=1}^{\infty} (\text{cl} \bigcup_{i=n}^{\infty} O(y_i))$. In this case also, $\Gamma$ is an invariant continuum containing $O(x)$.

We now want to show that the sequence $\{y_i\}$ could have been chosen so that no $y_i$ belongs to $\Gamma$. To make this precise let us introduce the following notation.

If $\sigma$ is any subsequence of $\{y_i\}$, let $\Gamma(\sigma)$ denote the continuum defined in the same way as $\Gamma$ was defined for $\{y_i\}$. We assert that there is a subsequence $\sigma$ of $\sigma_0 = \{y_i\}$ such that no term of $\sigma$ belongs to $\Gamma(\sigma)$.

There is nothing to prove if $\Gamma(\sigma_0) = \Omega(y)$, for if some $y_i$ were in $\Gamma$ then $y$ itself would be in $\Omega(y)$. Thus $y$ would be $P^+$ stable, therefore periodic and $x$ would be in $X'''$.

Let us assume then that the orbits of the $y_i$ are distinct. We may assume that infinitely many (hence, for convenience, all) of the $y_i$ belong to $\Gamma(\sigma_0)$; otherwise we have a subsequence with the desired property.

Now through $y_1$ there is a transversal $T$ and a subsequence $\sigma(y_1)$ of $\sigma_0$ such that the orbit of every term of $\sigma(y_1)$ misses $T$. If not, then for any such transversal, almost all $O(y_i)$ meet $T$. Fix a transversal $T$ through $x$ and choose $y_k$ such that $O(y_k)$ meets $T$. Let $T'$ be a transversal through $y_1$ such that $O(y_k)$ misses the arc of $O(y_k)$ from $y_k$ to $T$. The orbit of every $y_i$ sufficiently close to $y_k$ must meet $T - T'$. But almost all $O(y_i)$'s meet $T'$ (since $y_1 \in \Gamma$); hence some $O(y_i)$ meets $T$ twice. The situation is pictured in Figure 4.

But if every transversal through $y_1$ meets some $O(y_i)$ twice then $y_1$ lies in $X''$, a contradiction. So, after all there is a transversal $T$ through $y_1$ and a subsequence $\sigma(y_1)$ with the desired properties.

We now define a collection of subsequences of the original sequence $\sigma_0$ as follows:

$\sigma_1$ is the sequence whose first term is $y_1$ and whose remaining terms are those of $\sigma(y_1)$. Thus the first term of $\sigma_1$ does not belong to $\Gamma(\sigma_1)$.

Applying essentially the same argument to $\sigma_1$ as we did to $\sigma_0$ we may obtain a subsequence $\sigma_2$ whose first two terms are those of $\sigma_1$ and do not belong to $\Gamma(\sigma_2)$.

Inductively we obtain sequences $\{\sigma_i\}$ such that $\sigma_i$ agrees with $\sigma_{i-1}$ through the first $i$ terms and the first $i$ terms of $\sigma_i$ do not belong to $\Gamma(\sigma_i)$.

The sequence $\sigma$ whose $i$th term is that of $\sigma_i$ has the desired property, namely, no term of $\sigma$ belongs to $\Gamma(\sigma)$. Rather than change notation we shall assume the original sequence $\{y_i\}$ had this property.

We are now ready to make the key observation in this proof. We assert that *if $T$ is any transversal to the flow then $\Gamma \cap T$ is finite.*

FIGURE 4

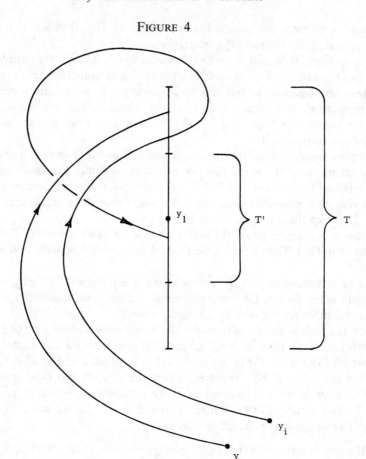

For suppose $\Gamma \cap T$ is infinite. We first treat the case that $\Gamma = \Omega(y)$; let $W$ be the component of $X - \Gamma$ containing $O(y)$. Certainly $W$ contains infinitely many components of $T - \Gamma$ (since $O(y)$ meets $T - \Gamma$ infinitely often). By Proposition 4.4, there is an open cell $W$ and a component $T_1$ of $\Gamma - T$ such that the boundary of $W$ lies in $T_1 \cup \Gamma$ and $O(y)$ crosses into $W$ through $T_1$. Since $T_1$ is transverse to the flow it follows as in the proof of $A(0)$ that $y$ cannot be in $X'$, a contradiction.

The other case is that $\Gamma = \lim \sup O(y_i)$. Let $\mathscr{W}$ be the collection of components of $X - \Gamma$ which contain some $O(y_i)$. Suppose $\mathscr{W}$ were infinite; by an obvious argument almost all $W \in \mathscr{W}$ are of type 1 and therefore by Proposition 4.1 homeomorphic to punctured spheres. Of these, at most $n + 1$ can fail to be cells—otherwise one finds $n + 2$ disjoint simple closed curves in $X$ whose union fails to separate $X$. Thus we have infinitely many, and therefore at least one, open invariant cell $W$ which contains some $O(y_i)$. Now $T \cap W$ is a collection of transverse arcs which separate $W$, and $O(y_i)$ crosses one of these arcs. It follows that $y_i$ is not in $W'$ and hence not in $X'$, a contradiction. So we have shown that $\mathscr{W}$ is finite.

Let $\mathcal{T}$ be the collection of components of $T - \Gamma$ which contain a point of some $O(y_i)$. $\mathcal{T}$ is infinite since $\Gamma \cap T$ is infinite. Hence some $W \in \mathcal{W}$ contains infinitely many elements of $\mathcal{T}$. Now apply Proposition 4.4; there is a cell $W'$ lying in $W$ whose boundary is contained in $\Gamma \cup T'$ where $T'$ is in $\mathcal{T}$. But some $O(y_i)$ lying in $W$ crosses $T'$ and so cannot be in $X'$. This final contradiction shows that $\Gamma \cap T$ must be finite, as asserted.

Since $\Gamma$ meets any transversal only finitely many times each orbit in $\Gamma$ is relatively open in $\Gamma$. We now assert that $\Gamma$ has a neighborhood which is either a cell or an annulus whose boundary components are separated by $\Gamma$.

First notice that $\Gamma - O(x)$ has either one or two components— since each such component meets one of $A(x)$, $\Omega(x)$. Now if $K$ is a component of $\Gamma - O(x)$ and $K$ does not lie in a cell then by Proposition 4.2 and the inductive hypotheses applied to the component $W$ of $X - K$ which contains $O(x)$ we find that $W'' = W'''$ and, therefore, $x \in X'''$—a contradiction. Thus each component of $\Gamma - O(x)$ lies in a cell.

Suppose $\Gamma - O(x)$ has one component, $K$. Let $U$ be an open cell containing $K$ but not containing all of $O(x)$. Let $\hat{K}$ be the continuum obtained by adding to $K$ all components of $U - K$ whose closures in $U$ are compact; then $\hat{K}$ is an invariant subcontinuum of $X$ missing $O(x)$; $\hat{K}$ does not separate $U$ and therefore does not separate $X$. The quotient space obtained by shrinking $\hat{K}$ to a point is again homeomorphic to $X$. (This is a special case of a theorem of R. L. Moore, [5].) The image $\Gamma'$ of $\Gamma$ in the quotient space is a simple closed curve. Let $V$ be an annulus whose boundary components are separated by $\Gamma'$, the preimage of $U$ under the quotient map is an annulus $U$ in $X$ whose boundary components are separated by $\Gamma$.

Let $T$ be a small transversal through $x$ lying in $U$. Since $U$ is a neighborhood of $\Gamma$ almost all $O(y_i)$ lie entirely in $U$. One of these meets $T$ twice and by essentially the same argument used to prove $A(0)$ any such orbit is periodic.

In case $\Gamma - O(x)$ has two components a similar construction shows that $\Gamma$ has a cell neighborhood and consequently some $y_i$ must be periodic.

Thus in either case we have reached a final contradiction, $X'' - X'''$ is empty and the main theorem is proved.

## 6. A generalization of the main theorem.

The most general result of this paper is now a rather easy consequence of the main theorem.

6.1. THEOREM. *Let $X$ be a connected orientable 2-manifold of finite type; then for any flow on $X$, $X''$ is the center.*

PROOF. Boundaries will be denoted by $\partial$. If $\partial X = \varnothing$, the result follows from some $B(n)$. Suppose then that $\partial X \neq \varnothing$ and let $\mathcal{K}$ be the collection of boundary components of $X$. Each $K \in \mathcal{K}$ is an open arc or a simple closed curve. In the first case let $U(K) = \{(x, y) \in \mathbf{R}^2 | x^2 + y^2 < 1, 0 \leq y\}$ and in the second case let $U(K) = \{(x, y, z) \in \mathbf{R}^3 | x^2 + y^2 = 1, 0 \leq z < 1\}$. In either case let $h(K)$ be a homeomorphism of $\partial U(K)$ onto $K$.

Let $\tilde{X}$ be the manifold obtained by attaching each $U(K)$ to $X$ via $h(K)$. Then $\tilde{X}$ is separable, orientable, has the same type as $X$ and $\partial \tilde{X} = \varnothing$.

Suppose a flow is given on $X$. We can extend this to a flow on $\tilde{X}$ with the property that points in the interior of any $U(K)$ are wandering points. Since $\partial \tilde{X} = \varnothing$, the central sequence for the extended flow on $\tilde{X}$ terminates in two steps and by construction the same is true for the original flow on $X$.

Our final result is not directly concerned with the length of the central series but rather with the behavior of alpha and omega limit sets.

6.2. THEOREM.  *If $X$ is a connected orientable 2-manifold of finite type and $x \in X'$ but $x$ is not $P^+$ or $P^-$ stable then $A(x)$ and $\Omega(x)$ contain no regular points.*

PROOF. We first observe that it suffices to prove the theorem for the case that $\partial X = \varnothing$. The general case then follows by the construction and argument of the preceding theorem.

Thus we assume $\partial X = \varnothing$. If $X$ is of type 1 and $x \in X'$ then the argument used to prove $A(0)$ shows·that if some point of $\Omega(x)$ (or $A(x)$) is not fixed then $x$ is periodic. Assuming the result true for manifolds of type $\le n$, suppose $X$ has type $n + 1$. Let $\Gamma = \Omega(x)$ where $x \in X'$ and $x$ is not $P^+$ or $P^-$ stable; by the same argument as in the main theorem, every regular orbit in $\Gamma$ is relatively open in $\Gamma$. If $\Gamma$ contains a regular orbit then as in the main theorem it has an annular or cell neighborhood and the usual trapping argument gives a contradiction.

## REFERENCES

**1.** G. D. Birkhoff, *Dynamical Systems*, Amer. Math. Soc., Providence, R.I., 1927.

**2.** Paul Frederickson, Informal communication.

**3.** A. Maïer, *Sur un problème de Birkhoff*, C.R. (Doklady) Acad. Sci. USSR, (N.S.) **55** (1947), 473–475.

**4.** ———, *Sur les trajectoires dans l'espace à trois dimensions*, C. R. (Doklady) Acad. Sci. USSR, (N.S.) **55** (1947), 579–581.

**5.** R. L. Moore, *Concerning upper semicontinuous collections of continua*, Trans. Amer. Math. Soc. **27** (1925), 416–428.

**6.** V. V. Nemytskii and V. V. Stepanov, *Qualitative theory of differential equations*, Princeton Univ. Press, Princeton, N.J., 1960.

**7.** L. Zippin, *On continuous curves and the Jordan curve theorem*, Amer. J. Math. **52** (1930), 331–350.

UNIVERSITY OF MICHIGAN

# SECOND ORDER ORDINARY DIFFERENTIAL EQUATIONS ON DIFFERENTIABLE MANIFOLDS

S. SHAHSHAHANI

In this paper we discuss some global aspects of the theory of second order ordinary differential equations on manifolds. Let $M$ be a compact smooth manifold with a fixed Riemannian metric, $TM$ its tangent bundle, and $K_i$ the subset of $TM$ consisting of tangent vectors of length $\leq i$. Since $TM$ is not compact, we have a choice of topology for the set $\Gamma_1^r(TM)$ of $C^r$ vector fields on $TM$. It turns out that for our purposes the *Whitney $C^r$-topology* is the proper choice. A basis for this topology is given by sets of the form

$$B(\xi, \delta) = \{\eta \in \Gamma_1^r(TM) | d_r(\xi | K_i - \text{int } K_{i-1}, \eta | K_i - \text{int } K_{i-1}) < \delta_i \text{ for all } i\}$$

where $\xi \in \Gamma_1^r(TM)$, $\delta : TM \to R$ is a continuous positive-valued function with $\delta_i = \min \delta$ on $K_i - \text{int } K_{i-1}$, and $d_r$ is the usual $C^r$-distance for vector fields on $K_i - \text{int } K_{i-1}$. The Whitney $C^r$-topology is independent of the choice of the expanding sequence $K_0 \subset K_1 \subset \ldots$ with $\bigcup K_i = TM$. We shall henceforth suppress the superscript "$r$" unless there is danger of confusion. It can be proved that $\Gamma_1(TM)$ has the Baire property, i.e., a countable intersection of open-dense sets is dense. For the proof of the above statements as well as reasons for the use of this topology see [5].

An intrinsic definition of second order ordinary differential equations is as follows. Let $\pi : TM \to M$ be the natural projection, then a tangent vector field $\xi$ on $TM$ is called a *second order ordinary differential equation on $M$* if $(D\pi)(\xi) = 1_{TM}$. This corresponds, locally, to autonomous systems of 2nd order ODE, i.e., systems of the form $\ddot{x} = f(x, \dot{x})$ where $x = (x_1, \ldots, x_n)$. In fact, let $U$ be an open subset of $R^n$ and suppose $\pi \in \Gamma_1(TU)$ is given by

(*)  $$(x, v) \to (x, v; f_1(x, v), f_2(x, v))$$

where $(x, v) \in U \times R^n = TU$ and each $f_i$ is a $C^r$ function $U \times R^n \to R^n$. Then $\xi$ is second-order on $M$ if and only if $f_1(x, v) = v$ for all $(x, v) \in TU$ (see [1]). We shall denote the set of 2nd order ODE on $M$ by $\Gamma_{II}^r(M)$ or simply $\Gamma_{II}(M)$. This set will be endowed with the Whitney $C^r$-topology as a subspace of $\Gamma_1(TM)$. The best known examples of 2nd order ODE are sprays ([4]) and (classical) mechanical systems ([1]) both of which exhibit highly nongeneric behavior. It is not known whether any tangent-bundle admits gradient vector fields ([6]) which are second-order and possess isolated singularities. One can prove, however, that if such a vector field exists, all its singularities are saddle points of type $(n, n)$ where $n = \dim M$. This is impossible in the case $M = S^1$ as later discussion will show.

265

**Some Elementary Observations.** (i) It follows from (\*) that all singularities of $\xi \in \Gamma_{II}(M)$ belong to the zero-section $M$. In fact $\xi$ is vertical at a point $P$ of the tangent-bundle (i.e., for $\pi : TM \to M$, $(D\pi)(\xi_P) = 0$) if and only if $P$ is on the zero-section. At a singularity the Jacobian matrix has the form

$$\begin{pmatrix} 0 & I_n \\ A & B \end{pmatrix}$$

where $n = \dim M$ and $I_n$ is the identity matrix of order $n$.

(ii) In the case $M = S^1$, $TM = S^1 \times R$ is a cylinder. Using again the local expression of a 2nd order vector field we see that the trajectories in the upper-cylinder run from left to right while those in the lower cylinder move from right to left. If we further suppose that the singularities are isolated, every such point displays one of the following forms:

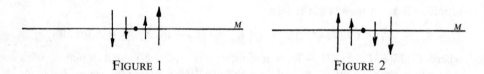

FIGURE 1                                    FIGURE 2

Figure 1 is exhibited by saddle points; the vector field points upward on the right-hand side and downward to the left. Sinks, sources, and centers are as in Figure 2 where the vector field points downward on the right and upward on the left side of the singularity. It follows that if the singularities are isolated, there are an even number of them with exactly half being saddle points.

(iii) The noncompactness of $TM$ destroys some pleasant features encountered in the global theory of first order ODE on manifolds. This is perhaps best illustrated by the fact that even under the assumption of Axiom $A'$, $TM$ is in general not exhausted by the union of stable and unstable manifolds of the components of the nonwandering set (for an exposition of these concepts see [6]). The following figure for $M = S^1$ provides an example. We shall later see that this picture represents a structurally stable second-order equation.

Just as in the first-order case, one is interested in characterizing a "large" subset of $\Gamma_{II}(M)$. The following is analogous to the Kupka-Smale Theorem (see [2], [5], or [6]).

THEOREM 1. *There is a Baire subset $\mathcal{G}$ of $\Gamma_{II}(M)$ so that if $\xi \in \mathcal{G}$ the following hold:*
(i) *Every singularity of $\xi$ is hyperbolic.*
(ii) *Every closed orbit of $\xi$ is hyperbolic.*
(iii) *The stable and unstable manifolds of singularities and closed orbits of $\xi$ are in general position.*
(iv) *If $\dim M > 1$, no closed orbit of $\xi$ meets the zero section.*

A sharper theorem can be proved in the case $M = S^1$.

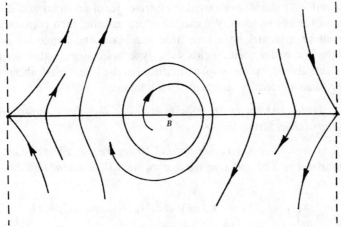

FIGURE 3

THEOREM 2 (BARETO[3]). *The structurally stable elements of* $\Gamma_{II}(S^1)$ *form an open and dense subset* $\mathscr{S}$. $\xi \in \mathscr{S}$ *if and only if the following conditions are satisfied:*

(i) *The nonwandering set contains only singularities and closed orbits.*

(ii) *The singularities are hyperbolic and hence finite in number.*

(iii) *The closed orbits are hyperbolic and countable in number. Only a finite number of these intersect the zero-section.*

(iv) *If a trajectory* $\lambda$ *has a saddle as its* $\alpha$-*limit (resp.* $\omega$-*limit), then every trajectory in some tubular neighborhood of* $\lambda$ *has the same* $\omega$-*limit (resp.* $\alpha$-*limit). In particular no trajectory joins saddle points.*

Aside from questions of genericity, one is interested in relations between $\Gamma_{II}(M)$ and $\Gamma_I(TM)$. We first remark that $\Gamma_{II}(M)$ is a closed connected subspace of $\Gamma_I(TM)$ where we use $C^r$-Whitney topology on both spaces. A more substantial result and of crucial importance in the proof of Theorem 1 is

FUNDAMENTAL LEMMA. *Let* $\xi \in \Gamma_{II}(M)$ *and* $S \subset \Gamma_I(TM)$ *satisfy:* (a) *There is a compact flow-box* $F$ *of* $\xi$ *contained in a coordinate neighborhood and disjoint from the zero-section so that every* $\eta \in S$ *agrees with* $\xi$ *outside* $F$.

(b) $\xi \in$ *closure* $S$.

*Then there is a subset* $T$ *of* $\Gamma_{II}(M)$ *so that:*

(i) *Every* $\zeta \in T$ *agrees with* $\xi$ *outside* $F$.

(ii) $\xi \in$ *closure* $T$.

(iii) *Each* $\zeta \in T$ *is differentiably conjugate to some* $\eta \in S$.

The rest of this paper will be devoted to some details of the proof of Theorem 1 including a proof of the Fundamental Lemma. Some familiarity with the proof of Kupka-Smale Theorem (particularly Peixoto's version in [5]) will be assumed.

PART (i). We have seen already that the Jacobian matrix at a singularity has the form

$$J = \begin{pmatrix} 0 & I_n \\ A & B \end{pmatrix}$$

Since $\det J = (-1)^n \det A$, nonsingular matrices form an open and dense subset of matrices of the above form. We can therefore assume, by a transversalty argument, that all singularities have invertible Jacobians and hence are isolated. To show that the 2nd order vector fields with hyperbolic singularities form a dense-open set of the above, we use approximation on the linear part about the singularities. For this we need the following two lemmas.

LEMMA 1. *Within the set of* $2n \times 2n$ *matrices of the above form those with no imaginary eigenvalues form a dense-open set.*

LEMMA 2 (*Partition of Unity for Second Order Vector Fields*). *Let* $\{W_\alpha\}$ *be a locally finite cover of* $TM$ *and suppose that on each* $W_\alpha$ *a vector field* $\xi^\alpha$ *of the local form*

$$(x_1, ..., x_n; v_1, ..., v_n) \to (v_1, ..., v_n; f_1(x, v), ..., f_n(x, v))$$

*is given. Then if* $\{\lambda_\alpha\}$ *is a partition of unity subordinate to* $\{W_\alpha\}$, *the vector field* $\sum \lambda_\alpha \xi^\alpha : TM \to T\ TM$ *is second order.*

PROOF OF LEMMA 1. First consider the space of all characteristic polynomials of degree $2n$. This is clearly in one-one correspondence with $\mathbf{R}^{2n}$ as the coefficient of $\lambda^{2n}$ is 1. Now if a characteristic polynomial has imaginary eigenvalue $\pm i\alpha$ it factors as

$$(\lambda^2 + \alpha^2)(\lambda^{2n-2} + ...)$$

and hence its $2n$ coefficients are polynomial functions of $2n - 1$ variables. It follows that the coefficients are algebraically dependent and define an algebraic variety of codimension one in $\mathbf{R}^{2n}$. The complement is then open and dense in $\mathbf{R}^{2n}$ and thus the characteristic polynomials with no imaginary eigenvalues form an open and dense set in all such polynomials.

Next identify the space of matrices considered with $\mathbf{R}^{2n^2}$ and let $p : \mathbf{R}^{2n^2} \to \mathbf{R}^{2n}$ be the map that associates to each matrix of the form $J$ its characteristic polynomial. This map is clearly smooth. The lemma will be proved once we show $p$ is regular on an open and dense subset of $\mathbf{R}^{2n^2}$. For this, form the Jacobian of the partial derivatives of the coefficients of $p(J)$ with respect to the entries of $J$. To show that the rank of this Jacobian is $2n$ on some open and dense subset of $\mathbf{R}^{2n^2}$ it suffices to show that the determinant of some $2n \times 2n$ submatrix is nonzero on such a subset. Take for example the matrix of the partial derivatives with respect to the entries of row $n + 1$. The elements of this matrix are polynomial functions of $2n^2$ variables and for it to have rank $2n$ on an open and dense subset of $\mathbf{R}^{2n^2}$ it is sufficient that the determinant does not identically vanish. A lengthy but routine computation establishes this fact.

PROOF OF LEMMA 2. This is immediate from the local expression of second-order vector fields.

PARTS (ii) AND (iii). A close examination of the proof of Kupka-Smale Theorem for 1st order flows [5] shows that all the perturbations needed can be reduced to flow-box perturbations. Moreover note that since a 2nd order vector field is vertical at the zero-section, no trajectory (except a singularity) is entirely contained in the zero-section and hence the flow-boxes may be chosen disjoint from $M$. Finally, the eigenvalues of the Poincaré map are invariants of differentiable conjugacy. Therefore the Fundamental Lemma enables us to extend the proofs of (ii) and (iii) for first-order flows to the 2nd order case.

PROOF OF THE FUNDAMENTAL LEMMA. Our aim is to find a sequence of 2nd order flows converging to $\xi$, each agreeing with $\xi$ outside $F$ and differentiably conjugate to some elements of $S$. Let the flow-box $F$ be given by

$$\theta : B^{2n-1} \times [-1, +1] \to F$$

where $\theta$ is a $C^r$ diffeomorphism, $B^{2n-1}$ is the closed unit disc in $R^{2n-1}$, and each $\theta(\{b\} \times [-1, 1])$ is a trajectory of $\xi$. By the lateral surface of $F$ we shall mean $\theta(\partial B \times [-1, 1])$. Take a trajectory

$$\lambda : [-1, +1] \xrightarrow{C^r} F$$

of some $\eta \in S$. We seek a curve $\gamma : [-1, +1] \xrightarrow{C^r} F$ so that if $\gamma(t) = (\gamma_1(t), ..., \gamma_{2n}(t))$ (in a coordinate neighborhood containing $F$),

(i) $d\gamma_i(t)/dt = \gamma_{n+i}(t)$ for $i = 1, ..., n$ and $t \in [-1, 1]$ (i.e., the tangents to $\gamma$ define a second-order flow),

(ii) $\gamma(\pm 1) = \lambda(\pm 1)$, say $= P^{\pm} = (P_1^{\pm}, ..., P_{2n}^{\pm})$, and

(iii) $d^j\gamma_i(t)/dt^j|_{t=\pm 1} = d^j\lambda_i(t)/dt^j|_{t=\pm 1}$, $i = 1, ..., 2n$ and $j \leq r$.

We consider a family $\{\phi_t\}$, $t \in [-1, 1]$ of auxiliary functions defined by

$$\phi_t(x) = 1 + t \exp(x^2/(x^2 - 1)) \qquad \text{for } x \in [-1, 1]$$
$$= 0 \qquad \text{elsewhere.}$$

There is a $C^\infty$ map $f : C^0[-1, +1] \times [-1, 1] \to R$ defined by

$$(\lambda, t) \mapsto \int_{-1}^{+1} \phi_t(x)\lambda(x)dx.$$

One can show by direct computation or by the use of the Implicit Function Theorem that if $\lambda_0 \in C^0[-1, 1]$ is nowhere-zero, there is a neighborhood $U$ of $\lambda_0$ in $C^0[-1, 1]$ and a neighborhood $V$ of $f(\lambda_0, 0)$ in $R$ so that for every $r \in V$ there exists a unique $t \in (-1, 1)$ satisfying $f(\lambda, t) = r$. Moreover, $t$ depends smoothly on $\lambda$ and $r$.

Now parameterize the trajectories of $\xi$ in $F$ as $\{\theta^x\}$ with $x \in B^{2n-1}$. Since $F \cap$ zero-section $= \varnothing$, we may assume that the coordinates of all points of $F$ are nonzero. There is a $C^r$ map

$$u = (u_1, ..., u_n) : B^{2n-1} \to R^n$$

where

$$u_i(x) = \int_{-1}^{+1} \theta_{n+i}^x(t)dt$$

and $\theta = (\theta_1, ..., \theta_{2n})$. Note that since $\xi$ is 2nd order, $\theta_{n+i}^x(t) = (\theta_i^x)'(t)$, and hence

$$\int_{-1}^{+1} \theta_{n+i}^x(s)ds = \theta_i^x(1) - \theta_i^x(-1).$$

Suppose the trajectory $\lambda^x$ of $\eta$ starts at $\theta_i^x(-1)$, i.e., $\theta_i^x(-1) = P_i^-$. Then if $\eta$ is sufficiently close to $\xi$, an $n$-fold application of the result of the previous paragraph gives a $C^r$ map

$$B^{2n-1} \to [-1, +1]^n$$

assigning an $n$-tuple $(t_1, ..., t_n)$ to each point $x \in B^{2n-1}$ so that if $\theta^x(-1) = \lambda^x(-1) = P^-$, then

$$P_i^+ - P_i^- = \int_{-1}^{+1} \phi_{t_i}(t) \lambda_{i+n}^x(t)dt, \quad i = 1, ..., n.$$

Define $\gamma^x : [-1, +1] \to TM$ by $\gamma^x(t) = (\gamma_1^x(t), ..., \gamma_{2n}^x(t))$ where

$$\gamma_i^x(t) = P_i^- + \int_{-1}^{t} \phi_{t_i}(s)\lambda_{n+i}^x(s)ds \quad \text{if } i = 1, ..., n$$
$$\gamma_{i+n}^x(t) = \phi_{t_i}(t)\lambda_{n+i}^x(t).$$

Note that if $\theta^x = \lambda^x$, the uniqueness of $(t_1, ..., t_n)$ gives $t_1 = ... = t_n = 0$, and hence $\theta^x = \lambda^x$ implies $\theta^x = \gamma^x$. This is particularly the case for the trajectories near the lateral surface of $F$. It follows that the map $h : F \to TM$ defined by:

$$(\lambda_1^x(t), ..., \lambda_{2n}^x(t)) \to (\gamma_1^x(t), ..., \gamma_{2n}^x(t))$$

has its range contained in $F$ for $\eta$ sufficiently close to $\xi$. In fact, the openness of $C^r$ diffeomorphisms in $C^r$ maps implies that by taking $\eta$ sufficiently $C^r$ close to $\xi$ we can make the map $h : F \to F$ a diffeomorphism. The vector field $h_* \eta = \zeta$ has the $\gamma$'s as its integral curves in $F$ and is easily checked to have the desired properties. Moreover our construction shows that by taking $\eta$ sufficiently close to $\xi$ we can make the $C^r$ distance between $\xi$ and $\zeta$ arbitrarily small. This finishes the proof of the Fundamental Lemma.

PART (iv). We assume dim $M > 1$. Let $\mathcal{H}$ be the set of those second order vector fields for which all singularities and closed orbits are hyperbolic. It clearly suffices to show that $\mathcal{G} \cap \mathcal{H}$ contains a Baire subset of $\mathcal{H}$ (where $\mathcal{G}$ is the set of vector fields satisfying (iv)). We shall use the following notation:

$\mathcal{H}_1$: the set of 2nd order vector fields with hyperbolic singularities,

$\mathcal{G}(t)$: the set of elements $\xi$ of $\mathcal{H}$ no closed orbit of period $\leq T$ of which intersects the zero-section.

It is sufficient to show that each $\mathcal{G}(T)$ is open and dense in $\mathcal{H}$. We shall make heavy use of the following facts the proofs of which can be found in [5].

A. Let $P$ be a singularity of $\xi \in \mathcal{H}_1$ and let $T > 0$. There exists a neighborhood $U$ of $P$ in $TM$ and a neighborhood $\mathcal{U}$ of $\xi$ in $\mathcal{H}_1$ so that every $\eta \in \mathcal{U}$ has exactly one singularity in $U$; this singularity depends continuously on $\eta$. Moreover every closed orbit of $\eta$ meeting $U$ has period $> T$.

B. Let $\gamma$ be a hyperbolic closed orbit of $\xi \in \mathcal{H}_1$ with period $\leq T$. Then there exists a tubular neighborhood $U$ of $\gamma$ and a neighborhood $\mathcal{U}$ of $\xi$ in $\mathcal{H}_1$ so that every $\eta \in \mathcal{U}$ has a unique closed orbit $\gamma(\eta) \subset U$; $\gamma(\eta)$ depends continuously on $\eta$. With the exception of $\gamma(\eta)$, every closed orbit of $\eta$ meeting $V$ has period $> T$.

C. Let $K$ be a compact subset of $TM$ no point of which belongs to singularities or closed orbits of period $\leq T$ of $\xi$. Then there exists a neighborhood $\mathcal{U}$ of $\xi$ in $\Gamma_{II}(M)$ so that every closed orbit of $\eta \in \mathcal{U}$ meeting $K$ has period $> T$.

OPENNESS OF $\mathcal{G}(T)$. Let $\xi \in \mathcal{G}(T)$ and let $P_1, \ldots, P_q$ be all the singularities of $\xi$. By (A), there are neighborhoods $U_1, \ldots, U_g$ of $P_1, \ldots, P_g$ and a neighborhood $\mathcal{U}_1$ of $\xi$ so that if $\eta \in \mathcal{U}_1$, every closed orbit of $\eta$ meeting $\bigcup U_i$ has period $> T$. Let $K = M - (\bigcup_{i=1}^q (M \cap U_i))$. $K$ is compact and therefore by (C), there is a neighborhood $\mathcal{U}_2$ of $\xi$ so that if $\eta \in \mathcal{U}_2$, no closed orbit of period $\leqq T$ of $\eta$ intersects $K$. $\mathcal{U}_1 \cap \mathcal{U}_2$ is the desired neighborhood of $\xi$.

DENSITY OF $\mathcal{G}(T)$. Let $\xi \in \mathcal{H}$, we shall $C^r$-approximate $\xi$ by some $\eta \in \mathcal{G}(T)$. $\xi$ has only a finite number of closed orbits of period $\leqq T$ that intersect the zero-section; call them $\gamma_1, \ldots, \gamma_P$. A transversality argument shows that each $\gamma_i$ intersects the zero-section in only a finite number of points. Let $\gamma$ be any of the $\gamma_i$'s and $A$ an intersection of $\gamma$ with $M$. Around $\gamma$ choose a pair of tubular neighborhoods $U$, $\hat{U}$ so that $U \subset \hat{U}$ and choose a neighborhood $\mathcal{U}$ of $\xi$ in $\mathcal{H}$ so that:

a. Every $\eta \in \mathcal{U}$ has at most one closed orbit $\gamma(\eta)$ that intersects $U$. In fact, by (C), $\mathcal{U}$ can be so chosen that if $\gamma(\eta)$ meets $\hat{U}$, then it actually meets $U$. This implies that if $\gamma(\eta)$ intersects the zero-section in $\hat{U} \cap M$, then the point of intersection belongs to $U \cap M$.

b. Near $A$, the two tubular neighborhoods are required to be representable as a pair of flow-boxes by the diffeomorphism

$$h : (\hat{B}^{2n-1}, B^{2n-1}) \times [-1, 2] \to (\hat{F}, F) \subset (\hat{U}, U)$$

where $(\hat{B}^{2n-1}, B^{2n-1})$ is a pair of concentric closed discs at $0 \in R^{2n-1}$. Suppose also that $L(\{0\} \times [-1, 2]) = \gamma \cap F$ with $h(0, 0) = A$. Further we may assume that $M \cap F$ and $M \cap \hat{F}$ are images under $h$ of the $n$-slices:

$$C = \{(x_1, \ldots, x_{2n-1}) \in B^{2n-1} | x_{n+1} = \ldots = x_{2n+1} = 0\}$$
$$\hat{C} = \{(x_1, \ldots, x_{2n-1}) \in \hat{B}^{2n-1} | x_{n+1} = \ldots = x_{2n+1} = 0\}.$$

We make a perturbation of the Poincaré transformation on $h(B^{2n-1} \times \{1\})$ so that the new fixed point is $h(P, 1)$ with $P \notin C$ (this is possible since dim $M = n < 2n - 1$ if $n > 1$). Let $Q \in B^{2n-1}$ be the point such that $h(Q, 0)$ is the next entry of the trajectory of the old flow through $h(P, 0)$ to $h(B^{2n-1} \times \{0\})$. Given $\varepsilon > 0$, we can make the perturbation so small that the Isotopy Theorem of Thom [2] provides an isotopy $\{\psi_s\}_{s \in (1,2)}$ of $\hat{B}^{2n-1}$ satisfying:

(i) $\psi_1 = $ identity,

(ii) $\psi_s$ is equal to identity in some neighborhood of $\partial \hat{B}^{2n-1}$ for all $s$,

(iii) each $\psi_s$ is arbitrarily $C^r$-close to the identity map,

(iv) $\partial^n \psi_s(x) / \partial s^n = 0$ for $x \in \hat{B}^{2n-1}$, $s = 0, 1$, and $n = 1, 2, \ldots$.

(v) The Poincaré transformation on $h(B^{2n-1}) \times \{1\})$ obtained by making the diffeomorphism $D$ of $\hat{F}_1 = h(\hat{B}^{2n-1} \times [1, 2])$, where $D(h(b, s)) = h(\psi_s(b), s)$ has its unique fixed point off $h(C \times \{1\})$; in fact $\psi_2(h(P, 1)) = h(Q, 2)$.

The diffeomorphism $D$ defines a new vector field $D_* \xi$. The new vector field has the perturbed trajectories as its integral curves and can be made arbitrarily $C^r$-close to $\xi$. By (a), if a closed orbit of period $\leqq T$ of $D_* \xi$ meets the zero-section in $\hat{U} \cap M$, the intersection actually belongs to $U \cap M$. But because of (v), a trajectory leaving $U \cap M$ will not intersect $U \cap M$ again after one turn around $U$. It follows that if

272

a closed orbit of $D_*\xi$ intersects $\hat{U} \cap M$, it has period $> T$. Finally note that since $\hat{F}_1 = h(\hat{B}^{2n-1} \times [1, 2])$ is disjoint from the zero-section, and since the perturbation is chosen to leave some neighborhood of the lateral surface of $\hat{F}_1$ unchanged (property (iii) above), we may assume by the Fundamental Lemma that $D_*\xi$ is 2nd order.

We can now repeat the process to eliminate all the (finite number of) intersections of $\gamma_i$'s with $M$. (C) shows that no new closed orbits of period $\leq T$ are generated, and the proof is complete.

The author wishes to thank Professors S. Smale and C. Pugh for their guidance.

## BIBLIOGRAPHY

1. R. Abraham and J. Marsden, *Foundations of mechanics*, Benjamin, New York, 1967.
2. R. Abraham and T. Robbin, *Transversal mappings and flows*, Benjamin, New York, 1967.
3. A. Bareto, *Thesis*, IMPA, Rio de Janeiro, 1963.
4. S. Lang, *Introduction to differentiable manifolds*, Interscience, New York, 1962.
5. M. Peixoto, *On an approximation theorem of Kupka and Smale*, J. Differential Equations 3 (1966), 214–227.
6. S. Smale, *Differentiable Dynamical Systems*, Bull. Amer. Math. Soc. 73 (1967), 747–817.

UNIVERSITY OF CALIFORNIA, BERKELEY

# EXPANDING MAPS

MICHAEL SHUB

It is the purpose of this note to give a short survey of some of the facts known about expanding maps and of one approach to the fundamental problem of expanding maps, which is stated below. Another approach to the problem is contained in an article by M. W. Hirsch in these same proceedings (vol. 14). Generalizations of some of these results, as well as some of the results themselves treated slightly differently, are contained in John Franks, *Anosov Diffeomorphisms*, also in these proceedings (vol. 14). Smale [7] and R. F. Williams [8] have used expanding maps and generalizations to branched manifolds to study basic sets of diffeomorphisms. Manifold will be used throughout to denote a connected $C^\infty$-manifold without boundary.

DEFINITION 1. Let $M$ be a complete Riemannian manifold. A $C^1$ endomorphism $f: M \to M$ is expanding if there exist constants $c > 0$ and $\lambda > 1$ such that $\|Tf^m v\| \geq c\lambda^m \|v\|$ for all $v \in TM$ and all $m > 0$.

If $M$ is compact this definition is clearly independent of the metric.

DEFINITION 2. Let $f: X \to X$ and $g: Y \to Y$ be continuous maps of topological spaces. $f$ and $g$ are said to be topologically conjugate (semiconjugate) if there exists a homeomorphism (continuous map) $h: X \to Y$ such that $hf = gh$. $h$ is called a conjugacy (semiconjugacy).

FUNDAMENTAL PROBLEM OF EXPANDING MAPS. Find all expanding maps of compact manifolds, up to topological conjugacy.

**Examples of expanding maps.** These examples are motivated by the examples of Anosov diffeomorphisms given in Smale [7].

Let $G$ be a 1-connected Lie group with a left invariant metric and let $A: G \to G$ be an expanding Lie group automorphism. It is easy to see that $G$ must be nilpotent. Now let $\Gamma \subset G$ be a uniform discrete subgroup of $G$, i.e. $G/\Gamma$ is a compact manifold called a nilmanifold. If $A(\Gamma) \subset \Gamma$, $A$ defines a map $\bar{A}: G/\Gamma \to G/\Gamma$ which is expanding, which we shall call a nil-expanding map.

EXAMPLE 1. $G = R$; $\Gamma = Z$; $R/Z = S^1$ and $A$ is multiplication by an integer $n$, with $|n| > 1$.

2. $G = R^n$; $\Gamma = Z^n$; $R^n/Z^n = T^n$ and $A$ is a matrix with integer entries and eigenvalues greater than 1 in absolute value.

3. $G$ is the group of lower triangular matrices

$$\begin{pmatrix} 1 & 0 & 0 \\ x & 1 & 0 \\ z & y & 1 \end{pmatrix}$$

$x, y, z \in R$ and $\Gamma$ the uniform discrete subgroup of matrices of the form

$$\begin{pmatrix} 1 & 0 & 0 \\ \alpha & 1 & 0 \\ \gamma & \beta & 1 \end{pmatrix}$$

where $\alpha, \beta, \gamma \in Z$. If $a, b, c \in Z$ with $|a|, |b|, |c| > 1$ and $ab = c$ then the map

$$\begin{pmatrix} 1 & 0 & 0 \\ x & 1 & 0 \\ z & y & 1 \end{pmatrix} \rightarrow \begin{pmatrix} 1 & 0 & 0 \\ ax & 1 & 0 \\ cz & by & 1 \end{pmatrix}$$

defines an expanding Lie group automorphism of $G$ which takes $\Gamma$ into itself and, hence, projects to a nil-expanding map of $G/\Gamma$.

**Modification of these examples.** Let $G$ be a 1-connected nilpotent Lie group, $C$ a compact group of automorphisms of $G$, $G \cdot C$ the semidirect product of $G$ and $C$ considering $G$ as acting on itself by left translations, and $\Gamma$ a torsion free uniform discrete subgroup of $G \cdot C$. It is a theorem of L. Auslander [1] that $\Gamma \cap G$ is a uniform discrete subgroup of $G$ and $\Gamma \cap G$ is normal in $G$ with $\Gamma/\Gamma \cap G$ finite. $G/\Gamma$ the orbit space of $G$ under the action of $\Gamma$ is a compact manifold. Let $A : G \cdot C \rightarrow G \cdot C$ be an automorphism such that $A(\Gamma) \subset \Gamma$, $A(G) = G$ and $A|G$ is an expanding Lie group automorphism. $A$ induces an expanding map $\bar{A} : G/\Gamma \rightarrow G/\Gamma$, which following the terminology of M. W. Hirsch we call an infranil-expanding map. $\bar{A} : G/\Gamma \rightarrow G/\Gamma$ is finitely covered by the nil-expanding map $\overline{A|G} : G/G \cap \Gamma \rightarrow G/G \cap \Gamma$.

$$\begin{array}{ccc} \overline{A|G} : G/G \cap \Gamma & \rightarrow & G/G \cap \Gamma \\ \downarrow & & \downarrow \\ \bar{A} : \quad G/\Gamma & \rightarrow & G/\Gamma \end{array}$$

EXAMPLE. $G = R^n$, $C = \mathcal{O}(n)$ the orthogonal group. Then the $G/\Gamma$ are the compact flat Riemannian manifolds.

THEOREM (D. B. A. EPSTEIN—SHUB [3]). *If $M$ is a compact flat Riemannian manifold, $M$ admits an infranil-expanding map.*

The problem of finding infranil-expanding maps when the group is not abelian seems much more difficult. While such examples do exist, D. B. A. Epstein has an example of a nilmanifold which does not admit an expanding map.

PROBLEM. Find all infranil-expanding maps up to topological conjugacy.

QUESTION. Are all expanding maps of compact manifolds topologically conjugate to infranil-expanding maps?

**Theorems about expanding maps.** Proofs of these theorems may be found in [6].

THEOREM 1. *Let $M$ be a compact manifold and $f : M \rightarrow M$ an expanding map. Then*

(a) *$f$ has a fixed point, $m_0$.*

(b) *The universal covering space of $M$ is diffeomorphic to $R^n$.*

(c) *The periodic points of f are dense in M.*

(d) *f has a dense orbit.*

(e) *M is a $K(\Pi_1(M, m_0), 1)$.*

(f) *$\Pi_1(M, m_0)$ is torsion free.*

(g) *$x(M) = 0$ where $x(M)$ is the Euler characteristic of M.*

THEOREM 2. *If M is compact, $f, g : M \to M$ are expanding maps and f is homotopic to g then f is topologically conjugate to g.*

COROLLARY. *An expanding endomorphism of a compact manifold is structurally stable.*

THEOREM 3. *Let N be compact and $g : N \to N$ continuous with $g(n_0) = n_0$. Let $f : M \to M$ be expanding with $f(m_0) = m_0$. Then if $A : \Pi_1(N, n_0) \to \Pi_1(M, m_0)$ is a homomorphism such that*

$$\Pi_1(N, n_0) \overset{A}{\to} \Pi_1(M, m_0)$$

$$g_\# \downarrow \qquad \qquad \downarrow f_\#$$

$$\Pi_1(N, n_0) \overset{A}{\to} \Pi_1(M, m_0)$$

*commutes, there exists a unique semiconjugacy $h : N \to M$ such that $h(n_0) = m_0$, $fh = hg$, and $h_\# = A$. Moreover, if M is compact, g is expanding, and A is an isomorphism then h is a topological conjugacy, i.e. h is a homeomorphism.*

**Applications of Theorem 3.** Let $M$ be compact, $f : M \to M$ expanding, and $m_0$ a fixed point of $f$. If $\Pi_1(M, m_0)$ has a nilpotent subgroup of finite index, it follows that the Hirsch-Plotkin radical of $\Pi_1(M, m_0)$, i.e. the maximal locally nilpotent normal subgroup of $\Pi_1(M, m_0)$, is of finite index in $\Pi_1(M, m_0)$. Recalling that $\Pi_1(M, m_0)$ is torsion free, $\Pi_1(M, m_0)$ is in the terminology of L. Auslander and E. Schenkman [2] a nil-admissible group. Thus, $\Pi_1(M, m_0)$ embeds as a uniform discrete subgroup of a nilpotent group $N$ semidirect product with a finite group of automorphisms $F$. Moreover, by L. Auslander [1] $f_\# : \Pi_1(M, m_0) \to \Pi_1(M, m_0)$ extends uniquely to an automorphism $A : N \cdot F \to N \cdot F$ with $A(N) = N$.

THEOREM 4 ( JOHN FRANKS [4]). *$A|N : N \to N$ is expanding. Thus, A induces an infranil-expanding map $\bar{A}$ of $N/\Pi_1(M, m_0)$ with $A|\Pi_1(M, m_0) = f_\#$.*

Theorem 3 now says that $\bar{A}$ and $f$ are topologically conjugate, so we have

THEOREM 5. *An expanding map $f : M \to M$ of a compact manifold M is topologically conjugate to an infranil-expanding map if and only if $\Pi_1(M, m_0)$ has a nilpotent subgroup of finite index.*

ALGEBRAIC FACTS ABOUT $\Pi_1(M, m_0)$. Let $f : M \to M$ be expanding. We state some facts about $\Pi_1(M, m_0)$ not stated above.

PROPOSITION [6]. *Let $m_0$ be a fixed point of f. Then $f_\# : \Pi_1(M, m_0) \to \Pi_1(M, m_0)$ is a monomorphism, $\mathrm{Im}(f_\#)$ is of finite index in $\Pi_1(M, m_0)$ and $\bigcap_{i=0}^{\infty} \mathrm{Im}(f_\#^i) = e$ the identity of $\Pi_1(M, m_0)$.*

DEFINITION. A finitely generated group is said to have polynomial growth if given a set of generators of the group the function $\gamma(S)$, the number of distinct elements of the group which can be written as a word of length $\leq S$ in the generators, is majorized by a polynomial.

The following Theorem is proven in J. Wolf [9] and is a combination of Theorems of J. Wolf [9] and Milnor [5].

THEOREM. *A finitely generated solvable group, either is polycyclic and has a nilpotent subgroup of finite index and is thus of polynomial growth, or has no nilpotent subgroup of finite index and is not of polynomial growth.*

THEOREM 6 ( JOHN FRANKS [4]). *Let $f : M \to M$ be an expanding map with $\Pi_1(M, m_0)$ finitely generated; then $\Pi_1(M, m_0)$ has polynomial growth and thus nilpotent may be replaced by solvable in Theorem 5.*

The question of whether an expanding map of a compact manifold $M$ is topologically conjugate to an infranil-expanding map is thus equivalent to the question of whether $\Pi_1(M, m_0)$ has a solvable subgroup of finite index.

## REFERENCES

1. L. Auslander, *Bieberbach's theorems on space groups and discrete uniform subgroups of Lie groups*, Ann. of Math. (2) **71** (1960), 579–590.

2. L. Auslander and E. Schenkman, *Free groups, Hirsch-Plotkin radicals, and applications to geometry*, Proc. Amer. Math. Soc. **16** (1965), 784–788.

3. D. B. A. Epstein and M. Shub, *Expanding endomorphisms of flat manifolds*, Topology **7** (1968), 139.

4. J. Franks, *Anosov diffeomorphisms*, these Proceedings, vol. 14.

5. J. Milnor, *Growth of finitely generated solvable groups*, J. Differential Geometry (to appear).

6. M. Shub, *Endomorphisms of compact differentiable manifolds*, Amer. J. Math. (to appear).

7. S. Smale, *Differentiable dynamical systems*, Bull. Amer. Math. Soc. **73** (1967), 747–817.

8. R. F. Williams, *One dimensional nonwandering sets*, Topology **6** (1967), 473–487.

9. J. Wolf, *The rate of growth of discrete solvable groups and its influence on the curvature of Riemannian manifolds with solvable fundamental group* (to appear).

BRANDEIS UNIVERSITY

# NOTES ON DIFFERENTIABLE DYNAMICAL
# SYSTEMS

STEPHEN SMALE[1]

A differentiable dynamical system is a smooth action of a Lie group on a manifold. Specifically, this is a homomorphism $\phi$ of a Lie group $G$ into the group $\text{Diff}^r(M)$ of $C^r$ diffeomorphisms of the manifold $M$ such that the induced evaluation map $G \times M \to M$ defined by $(g, x) \to \phi_g(x)$ is smooth. We shall assume throughout that $M$ is a compact connected $C^\infty$ manifold and $r \geq 1$. There are only two Lie groups with which we shall be concerned.

*Case* 1. $G = R$. This is the classical case. The homomorphism $\phi: R \to \text{Diff}(M)$ defines a *flow* on $M$. There is a natural 1-1 correspondence between flows on $M$ and smooth vector fields on $M$ defined as follows. To the flow $\phi_t: M \to M$ we associate the vector field $X(x) = d\phi_t(x)/dt|_{t=0}$. To go the other way start with a vector field $X$ on $M$. Now a vector field on $M$ is nothing more than an ordinary differential equation defined on $M$. Therefore, the classical existence and uniqueness theorems for differential equations produce a unique flow which "solves" $X$ in the sense that $X$ is the vector field assigned to the flow by the correspondence defined above. The flow is defined for all time because $M$ is compact.

*Case* 2. $G = Z$. This is the case of discrete time. In many ways it is conceptually easier to work with than the classical case. The action of $G$ on $M$ is determined by a generator of $\phi(G)$ in $\text{Diff}(M)$. The problem is reduced to studying the iterates of this one diffeomorphism and its inverse.

The question we are interested in dealing with is the orbit structure of $\phi$. More precisely, what are the sets of the form $\{\phi_g(x) | g \in G\} = O_x, \ x \in M? \ O_x$ is the *orbit* of $x$.

If $G = R$, the orbits are homeomorphic to
(1) $R$, or
(2) $S^1 = R/Z$: a *closed orbit*, or
(3) point: a *fixed, singular,* or *stationary* point.
If $G = Z$, the orbits are homeomorphic to
(1) $Z$,
(2) a finite set of points: *periodic* points.

A very simple kind of dynamical system is the *gradient* dynamical system. Let $M$ be a smooth, compact Riemannian manifold and let $f: M \to R$ be a differentiable function. $Df(x) \in T_x^*$ for each $x \in M$. The Riemannian structure induces a particular isomorphism between $T_x(M)$ and $T_x^*(M)$. Using this identification, $Df(x)$ becomes a vector field denoted $-\text{grad}(f)$. $\phi_t$ is the flow obtained, as above, by solving the differential equation $-\text{grad}(f)$. $\phi_t$ has the property that $f(\phi_t(x))$ is a

---

[1] Notes by John Guckenheimer.

strictly decreasing function of $t$ or a constant. It follows that there are no closed orbits of $\phi_t$. The fixed points of $\phi_t$ are the critical points of $f$.

An example of a gradient dynamical system is obtained by embedding $M$ isometrically in $R^N$ and taking $f$ to be the projection onto the "vertical" axis. The orbits of $\phi_t$ are then the lines of steepest descent along $M$.

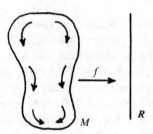

We are interested in the global geometric picture of a flow. The question arises as to how this picture changes under perturbation of the flow. We need to make precise the idea of perturbation. The set of $C^r$ vector fields on a manifold $M$ forms a Banach space $\mathfrak{X}(M)$ with the uniform $C^r$ norm. The norm is defined as the maximum of the first $r$ derivatives of a vector field with respect to some finite covering of $M$ by coordinate systems. The norm is unique up to an equivalent norm. We give $\mathfrak{X}(M)$ the norm topology. Given a vector field $X \in \mathfrak{X}(M)$, a *perturbation* of $X$ is any element of a previously specified neighborhood of $X$ in $\mathfrak{X}(M)$. In this setting, we make two definitions.

DEFINITION. Suppose $(X, \phi_t)$, $(Y, \psi_t)$ are two vector fields with their corresponding flows on $M$. $(X, \phi_t)$ and $(Y, \psi_t)$ are *topologically equivalent* if there exists a homeomorphism $h: M \to M$ which preserves orbits, as point sets, in a sense preserving way.

DEFINITION (PONTRYAGIN, ANDRONOV). $(X, \phi_t)$ is *structurally stable* if there exists a neighborhood $N(X)$ of $X$ in $\mathfrak{X}(M)$ such that if $Y \in N(X)$, $X$ is topologically equivalent to $Y$.

Now consider the question as to when a gradient dynamical system is structurally stable. Observe that a topological equivalence between flows preserves fixed points.

THEOREM (MORSE). *Any $f: M \to R$ can be approximated by one with nondegenerate critical points.*

The proof of this theorem uses transversality methods. Recall that a nondegenerate critical point is a critical point $x$ where $D^2 f(x)$ defines a nondegenerate quadratic form on $T_x(M)$. Nondegenerate critical points are isolated. Therefore, we have the following

THEOREM. *If* $X = -\text{grad } f$ *is a structurally stable gradient dynamical system,* $f$ *has nondegenerate critical points.*

We digress somewhat to consider some more properties of arbitrary flows on a manifold $M$.

DEFINITION. Suppose $x$ is a fixed point for the flow $\phi_t$. The *stable manifold* $W^s(x) = \{y \in M | \phi_t(y) \to x \text{ as } t \to \infty\}$. The *unstable manifold* $W^u(x) = \{y \in M | \phi_t(y) \to x \text{ as } t \to -\infty\}$. $x$ is a *hyperbolic* fixed point if none of the eigenvalues of $D\phi_t : T_x(M) \to T_x(M)$ lie on the unit circle (for any fixed $t > 0$).

The stable and unstable manifolds will be seen to play an important role in determining whether a dynamical system is structurally stable. For a gradient dynamical system $(-\text{grad } f, \phi_t)$, the statement that it is a hyperbolic fixed point of $\phi_t$ is equivalent to the statement that $x$ is a nondegenerate critical point of $f$.

. If $x$ is a hyperbolic fixed point of the flow $\phi_t$, we define $E^s$ to be the eigenspace of the eigenvalues of $D\phi_t$ lying inside the unit circle. Similarly, $E^u$ is defined to be the eigenspace of the eigenvalues of $D\phi_t$ lying outside the unit circle. We now state

STABLE MANIFOLD THEOREM. *If* $x$ *is a hyperbolic fixed point of the flow* $\phi_t$, *there exist manifolds* $\overline{W}^s$, $\overline{W}^u$ *such that*
(1) $\dim \overline{W}^s = \dim E^s$, $\dim \overline{W}^u = \dim E^u$,
(2) $\overline{W}^s$, $\overline{W}^u$ *are diffeomorphic to cells,*
(3) *there exist smooth injective immersions* $\overline{W}^s \to W^s(x)$, $\overline{W}^u \to W^u(x)$.

$W^s(x)$ and $W^u(x)$ are invariant under $\phi_t$ and tangent to $E^s$ and $E^u$, respectively, at $x$.

EXAMPLE. Consider the "standard" torus of revolution $R^3$ standing on end, and let $f$ be projection onto the vertical axis. Label the critical points of $f$ by $p$, $q$, $r$, $s$ from top to bottom.

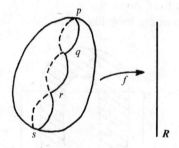

$W^s(p) = p$, $W^u(p)$ is a 2 cell; $W^s(q)$ is a 1 cell such that $W^s(q) \cup \{p\}$ is a nontrivial one cycle, $W^u(q)$ is a 1 cell such that $W^u(q) \cup \{r\}$ is a nontrivial one cycle; and so on. We obtain two cellular decompositions of the torus $T$ as

$$T = \bigcup_{x \varepsilon \{p,q,r,s\}} W^s(x) = \bigcup_{x \varepsilon \{p,q,r,s\}} W^u(x).$$

These two decompositions are essentially Poincaré duals of each other.

This particular dynamical system is *not* structurally stable. This is plausible because $W^s(r) \cap W^u(q) = S^1 - \{r, q\}$. By tilting the torus slightly in the right direction, we can make $W^u(q)$ pass below $r$ and $W^s(r)$ come from above $q$. In the tilted system $W^u(q) \cap W^s(r) = \varnothing$.

*Transversality condition for* $\phi_t$. For each $x, y$ hyperbolic fixed points of $\phi_t$, $W^s(x)$ and $W^u(y)$ have only transversal intersections; that is, if $p \in W^s(x) \cap W^u(y)$, the tangent planes of $W^s(x)$ and $W^u(y)$ at $p$ span $T_p(M)$.

THEOREM. (1) *A structurally stable gradient dynamical system satisfies the transversal intersection condition.*

(2) *Any gradient dynamical system can be approximated by one with nondegenerate singular points which satisfies the transversality condition.*

THEOREM (PALIS-SMALE). *A gradient dynamical system is structurally stable if and only if the system has hyperbolic fixed points and satisfies the transversal inter-section condition.*

In this case, the topological equivalence between a structurally stable system and a perturbation of the system can be chosen to preserve the group action of $R$ on $M$. This cannot be done for a structurally stable flow with a closed orbit because the period of the orbit is not invariant under perturbation.

At this point we shift attention from flows to discrete dynamical systems. Most of what we say is applicable to flows via a procedure described later. Nonlinearity is essential to the example we now consider.

Define a map $f$ of the unit square $I^2$ into the plane which has the following properties:

(1) $f$ is a diffeomorphism onto its image,

(2) $f(I^2) \cap I^2$ is *two* rectangles (this is where nonlinearity forces itself upon us),

(3) $f$ is linear on $f^{-1}(I^2) \cap I^2$,

(4) $f$ has a hyperbolic fixed point $p$.

A picture of such a map might look like a horseshoe

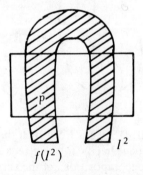

$f(I^2)$  $I^2$

If $W^u(p)$ and $W^s(p)$ have a point of intersection other than $p$, then the orbit of this point of intersection lies in $W^u(p) \cap W^s(q)$. Consequently, the stable and unstable manifolds of $p$ start to oscillate wildly.

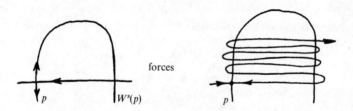

forces

By keeping track of the successive intersections of $I^2$ with itself under iteration, we find that

$$\bigcap_{m>0} f^{-m}(I^2) = \text{Cantor set} \times \text{interval},$$

$$\bigcap_{m>0} f^m(I^2) = \text{Cantor set} \times \text{interval},$$

$$\bigcap_{m\varepsilon\mathbf{Z}} f^m(I^2) = \text{Cantor set which we denote } \Lambda.$$

All of the recurrence phenomena of $f$ occur on the set $\Lambda$. We can extend $f$ to a diffeomorphism of $S^2$ as follows. "Cap" the top and bottom of the unit square by attaching disks $D_1$ and $D_2$. Denote by $D = D_1 \cup I^2 \cup D_2$. By choosing $D_1$ and $D_2$ large enough to contain $f(I^2)$, we can extend $f$ so that $f(D) \subset D$. Then extend $f$ to map the complement of $D$ onto the complement of $f(D)$. The extension of $f$ can be chosen to have one sink in $D_2$ and one source in the complement of $D$. $f$ can be chosen so that the whole geometric picture remains unchanged under $C^1$ perturbation.

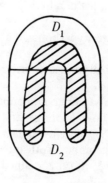

The map $f|_\Lambda$ has a particularly nice description. Consider the set of bi-infinite sequences of 0's and 1's; that is, $\{g : \mathbf{Z} \to \{0, 1\}\}$, with the compact open topology. This forms a Cantor set which we denote $\Lambda_2$. We write elements of $\Lambda_2$ as $a = \{a_i\}_{i\varepsilon\mathbf{Z}}$. There is a *shift map* $\alpha : \Lambda_2 \to \Lambda_2$ defined by $\alpha(a) = b$ where $a_{i+1} = b_i$.

PROPOSITION. $\Lambda$ *can be given coordinates from* $\Lambda_2$ *so that* $f|_\Lambda$ *is the shift on* $\Lambda_2$. *Precisely, there is a homeomorphism* $h:\Lambda \to \Lambda_2$ *such that the diagram*

$$\begin{array}{ccc} \Lambda & \xrightarrow{f|_\Lambda} & \Lambda \\ \downarrow h & & \downarrow h \\ \Lambda_2 & \xrightarrow[\alpha]{} & \Lambda_2 \end{array}$$

*commutes.*

This example describes a level of geometric complexity which we did not encounter in gradient dynamical systems.

We show how to obtain a flow with basically the same geometric complexity as the above example. This is done by a general construction for obtaining a flow from a discrete dynamical system.

Given a diffeomorphism $f: M \to M$, we define $N$ to be the compact manifold $M \times R/\sim$ where $\sim$ is the equivalence relation generated by $(x, t) \sim (f(x), t + 1)$. The unit vector field along $R$ induces a vector field on $N$ with flow $\phi_t$. $\phi_t$ has the property that $\phi_1: M \times \{0\} \to M \times \{0\}$ is naturally equivalent to $f$. The periodic points of $f$ lie on closed orbits of $f$.

In the horseshoe example, the periodic points of $f$ in $\Lambda$ are dense in $\Lambda$ because the periodic points of the shift in $\Lambda_2$ are dense. It follows that there is a flow on a 3-dimensional manifold (which is an $S^2$ bundle over $S^1$) which has an infinite number of closed orbits. Under perturbation, the flow continues to have an infinite number of closed orbits. This contrasts markedly with the behavior of a gradient dynamical system.

Now we turn to the problem of finding *generic* properties of $\text{Diff}(M)$ (or $\mathfrak{X}(M)$). A generic property is one satisfied by a Baire set (countable intersection of open, dense sets) in the space. We give a list of all known generic properties of $\text{Diff}^r(M)$, $r > 1$.

(1) The periodic points of $f$ are hyperbolic; that is, if $f^m(x) = x$, $Df^m$ has no eigenvalues on the unit circle.

(2) $W^s(x)$, $W^u(y)$ have transversal intersections for each pair of periodic points $x$, $y$ of $f$.

$W^s(x)$, $W^u(x)$ are defined in an analogous manner to the stable and unstable manifolds of a fixed point for a flow. More generally, if $f: M \to M$ is a $C^r$ diffeomorphism, we say $x \sim_s y$ if $f(f^m(x), f^m(y)) \to 0$ as $m \to \infty$. This defines an equivalence relation on $M$ which decomposes $M$ into *stable sets* $\{W^s_\alpha\}$. We write $W^s(x) =$ the $W^s_\alpha$ such that $x \in W^s_\alpha$. $W^u(x)$ is defined similarly by replacing $f$ with $f^{-1}$.

In the horseshoe example, the $W^s(x)$ contain horizontal lines in $I^2$.

Two recurrent themes appear in our attempt to understand dynamical systems. These are

(1) the smoothness of the $W^s_\alpha$,

(2) the transversality of the $W^s_\alpha$ and $W^u_\alpha$ if smoothness is satisfied.

*Question.* Are the $W^s_\alpha$ generically smooth?

The stable manifold theorem states that $W^s(x)$ is a smoothly immersed cell if $x$ is a hyperbolic periodic point.

The definition of structural stability we gave for flows has a natural analogue for discrete dynamical systems.

DEFINITION. $f \in \mathrm{Diff}^r(M)$ is *structurally stable* if there exists a neighborhood $N(f)$ of $f$ in $\mathrm{Diff}^r(M)$ such that if $g \in N(f)$, $g$ is *topologically conjugate* to $f$; that is, there is a homeomorphism $h: M \to M$ such that

$$\begin{array}{ccc} M & \xrightarrow{f} & M \\ \downarrow h & & \downarrow h \\ M & \xrightarrow{g} & M \end{array}$$

commutes.

THEOREM. *If $f$ is structurally stable, then* (1) *and* (2) *above are satisfied for $x$ and $y$ periodic.*

THEOREM. *For gradient dynamical systems,* (1) *and* (2) *are necessary and sufficient conditions for structural stability.*

There exist examples of $f \in \mathrm{Diff}(M)$ for which the $\{W_\alpha^s\}$, $\{W_\alpha^u\}$ are families of smooth, injectively immersed manifolds of constant dimension, and the two families meet transversally everywhere. Thus the stable and unstable manifolds in this case give transversal foliations of the manifold $M$.

For example, take $f_0 \in \mathrm{SL}(2, \mathbf{Z})$ with eigenvalues $\lambda$, $\mu$ such that $|\lambda| < 1 < |\mu|$. $f_0$ induces a map $f: T^2 \to T^2$ of the torus $T^2 = \mathbf{R}^2/\mathbf{Z}^2$. The stable manifolds of $f$ wrap around the torus parallel to the eigenvector belonging to $\lambda$, and the unstable manifolds wrap around parallel to the eigenvector of $\mu$. Each stable and unstable manifold is dense in the torus. The periodic points of $f$ are dense.

DEFINITION. $f: M \to M$ is *Anosov* if $T(M) = E^s + E^u$ where $E^s$ and $E^u$ are (continuous) vector bundles invariant under $Df$ such that there exist constants $0 < c, 0 < \lambda < 1$ for which

$$\begin{aligned} \|Df^m(x)(v)\| &< c\lambda^m\|v\|, & v \in E^s, \quad m \in \mathbf{Z}, \\ \|Df^{-m}(x)(v)\| &< c\lambda^m\|v\|, & v \in E^u, \quad m \in \mathbf{Z}. \end{aligned}$$

Little is known about Anosov maps. We mention the following.

*Question.* Suppose $f_0 \in \mathrm{SL}(n, \mathbf{Z})$ induces an Anosov toral diffeomorphism $f$. Does $f$ have a minimal set of dimension 1?

*Problem.* Find all Anosov diffeomorphisms up to topological conjugacy.

There are partial results with regard to this problem. Some infra-nil manifolds support Anosov diffeomorphisms and these are the only known examples. For the latest work in this direction, the reader is referred to the thesis of J. Franks.

THEOREM (ANOSOV). *Anosov diffeomorphisms are structurally stable.*

Using a toral Anosov diffeomorphism, we give an example of a diffeomorphism of $T^2$ which has a neighborhood containing no structurally stable diffeomorphisms.

This example demonstrates that structurally stable systems are not dense on the torus. It follows easily that structurally stable diffeomorphisms are not dense on any manifold with the exceptions of $S^1$ and $S^2$. Structurally stable systems are dense on $S^1$; it is open as to whether structurally stable diffeomorphisms are dense on $S^2$.

We start with an Anosov diffeomorphism of the torus having a fixed point $p$. If we choose coordinates centered at $p$ with the axes along the eigendirections then

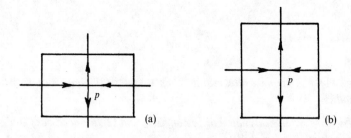

(a)                                            (b)

$f$ maps the rectangle (a) into the rectangle (b). Modify $f$ by averaging with a function that is the identity outside a small neighborhood of $p$, leaves $p$ fixed, and expands strongly at $p$. The effect of the modification is to make $p$ a source. Two saddle points are created on the stable manifold of $p$. The diagram of stable and unstable manifolds is the following:

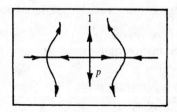

This new diffeomorphism $g$ we have obtained is called a DA (derived from Anosov) diffeomorphism. The unstable manifold of $p$ is now an immersed open 2 cell in the torus. $T^2 = W^u(p) \cup \Lambda^1$ where $\Lambda^1$ is a compact, invariant dimensional set, and $f|_{\Lambda^1}$ is topologically transitive. $\Lambda^1$ is locally the product of a Cantor set and an interval.

A further modification of $g$ gives a counter example to the denseness of structurally stable systems on $T^2$. $g$ can be chosen so that the stable manifolds of $f$ are not altered; that is, the stable manifolds of $g$ are horizontal lines in our picture. Choose a small disk $D$ about $p$ and consider its first two images under $g$. Introduce

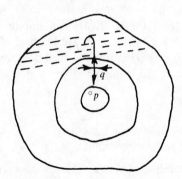

a new saddle point $q$ in $g(D) - D$ so that the unstable manifold of $q$ has a "hook" in $g^2(D) - g(D)$. The hook is a point of tangency of $W^u(q)$ with some stable manifold. Note that we have not changed $g$ outside $g^2(D)$. It follows that the stable manifolds inside $g^2(D) - g(D)$ are still horizontal lines.

The hook creates a lack of transversality between a stable and unstable manifold. The stable manifolds of periodic points are dense in the torus. Thus, in some neighborhood of the diffeomorphism we have produced, those diffeomorphisms with a tangential intersection between the stable and unstable manifolds of periodic points are dense. We have exhibited an open set of $\text{Diff}(T^2)$ with no structurally stable diffeomorphisms because structural stability implies that the stable and unstable manifolds of periodic points intersect transversally. Bob Williams found this example by modifying earlier examples of mine.

The question now arises as to whether there is not a weaker equivalence relation which might replace structural stability and have the property that a generic set is stable in the weaker sense. Such an attempt led to the concept of $\Omega$-stability.

DEFINITION. The *nonwandering set* $\Omega(f)$ of $f \in \text{Diff}(M) = \{x \in M \,|\, \text{for all neighborhoods } U \text{ of } x, \text{ there is an } n > 0 \text{ such that } f^n(U) \cap U \neq \varnothing\}$.

An $\Omega$-*conjugacy* of $f$ to $g$ is a topological conjugacy of $f|_{\Omega(f)}$ to $g|_{\Omega(g)}$; that is, a homeomorphism $h : \Omega(f) \to \Omega(g)$ such that

$$\begin{array}{ccc} \Omega(f) & \xrightarrow{f} & \Omega(f) \\ \downarrow h & & \downarrow h \\ \Omega(g) & \xrightarrow{g} & \Omega(g) \end{array}$$

commutes.

$f$ is $\Omega$-*stable* if $f$ has a neighborhood $N(f)$ in $\text{Diff}(M)$ such that $g \in N(f)$ implies $g$ is $\Omega$-conjugate to $f$.

$\Omega(f)$ is where the "action" of $f$ takes place. In this sense, $\Omega$-stability is almost as good as structural stability. The example we gave above is $\Omega$-stable but not structurally stable.

Our next goal is to describe the $\Omega$-stability theorem which gives sufficient conditions for a diffeomorphism to be $\Omega$-stable.

DEFINITION   Suppose $f : M \to M$ is a map, $\Lambda \subset M$ is closed, and $f(\Lambda) = \Lambda$. $f$ is *hyperbolic* on $\Lambda$ if $Df : T_\Lambda(M) \to T_\Lambda(M)$ is a bundle homeomorphism which splits so that $T_\Lambda(M) = E^s + E^u$ with $E^s$, $E^u$ subbundles of locally constant dimension, invariant under $Df$, and $Df$ contracts $E^s$ and expands $E^u$. Contraction and expansion here means that the estimates given in the definition of Anosov diffeomorphism are satisfied.

Let $f \in \text{Diff}^r(M)$.

AXIOM A.   (a) $\Omega(f)$ has a hyperbolic structure.
(b) *The periodic points of $f$ are dense in* $\Omega$.

Note that for $r = 1$, condition (b) is generic by the Closing Lemma. Condition (a) is *not* generic.

PROBLEM.   Find a replacement for (a) which is generic and allows the construction of smooth stable manifolds.

SPECTRAL DECOMPOSITION THEOREM. *If $f$ satisfies Axiom A, then there is a canonical decomposition of* $\Omega = \Omega_1 \cup \ldots \cup \Omega_m$; *the* $\Omega_i$ *are closed disjoint sets and* $f|_{\Omega_i}$ *is topologically transitive.*

For example, the spectral decomposition of the horseshoe is given by $\Omega(f)$ = sink $\cup$ source $\cup \Lambda_2$. The spectral decomposition of the $DA$ is given by $\Omega(g)$ = source $\cup \Lambda^1$. The whole torus is indecomposable for the Anosov diffeomorphisms on $T^2$.

*Question.* If $f$ is an Anosov diffeomorphism of $M$, is $\Omega(f) = M$?

Axiom A is not a sufficient condition for $\Omega$-stability. A phenomenon known as an $\Omega$-explosion can occur.

DEFINITION.   Assume $f$ satisfies Axiom A and has spectral decomposition $\Omega(f) = \Omega_1 \cup \ldots \cup \Omega_m$. A *cycle* of $\{\Omega_i\}$ is an ordered set $\{\Omega_{i_1}, \ldots, \Omega_{i_k}\}$ such that $W^u(\Omega_{i_j}) \cap W^s(\Omega_{i_{j+1}}) \neq \varnothing$ and $W^u(\Omega_{i_k}) \cap W^s(\Omega_{i_1}) \neq \varnothing$. The $\Omega_{i_j}$ are to be distinct and $k > 1$.

*No Cycle Property.* $f$ (assumed to satisfy Axiom A) has no cycles.

THEOREM (PALIS).   *If $f$ is $\Omega$-stable and satisfies Axiom A, $f$ has the no cycle property.*

$\Omega$-STABILITY THEOREM.   *If $f$ satisfies Axiom A and the no cycle property, then $f$ is $\Omega$-stable.*

The proof of the $\Omega$-stability theorem depends heavily on a generalized stable manifold theory. This theory states that if $\Lambda$ has a hyperbolic structure for $f$, then for all $x \in \Lambda$, there exist smooth, injectively immersed subcells $W^s(x)$ of $M$. The $W^s(x)$ vary continuously with $x \in \Lambda$ in the $C^r$ topology. If $\Omega_i$ is a component of the spectral decomposition of $f$, $W^s(\Omega_i)$ is defined to be $\bigcup_{x \in \Omega_i} W^s(x)$. The proof of the $\Omega$-stability theorem proceeds by a generalized handlebody theory. A filtration of $M$ is constructed starting at the attractors and building via stable manifolds.

Very little is known about the structure of the $\Omega_i$ that can occur in the spectral decomposition of a diffeomorphism satisfying Axiom A.

We end with a few questions and comments about dynamical systems.

PROBLEM. Find necessary and sufficient conditions for $f$ to be $\Omega$-stable (structurally stable).

An important question here is whether $\Omega$-stability implies Axiom A.

*Strong Transversality Condition.* Assuming Axiom A, $W^s(x)$ and $W^u(y)$ intersect transversally for all $x, y \in \Omega$.

THEOREM. *Structural stability implies the Strong Transversality Condition.*

CONJECTURE. Axiom A + Strong Transversality Condition implies structural stability.

This is true if $\Omega$ is a finite set (Smale-Palis).

PROBLEM. Find generic properties of Diff($M$). In particular, find generic stability properties. Is the following property generic?

DEFINITION. $f$ is *future stable* if for all perturbations $f'$, there exists $h: M \to M$ which sends stable sets of $f$ into stable sets of $f'$; that is, $h(W^s(x)) = W^s(h(x))$ for all $x \in M$.

*References.* A much more detailed survey of part of this material appears in S. Smale, *Differentiable dynamical systems*, Bull. Amer. Math. Soc., **73** (1967), 747–817. This survey contains an extensive bibliography. The survey is up to date as of about a year ago. Results obtained within the past year have not yet been published, though several preprints exist.

ADDENDUM. M. Shub and R. Williams have since pointed out that future stability is not generic using the example in these notes of R. Williams.

UNIVERSITY OF CALIFORNIA, BERKELEY

# THE $\Omega$-STABILITY THEOREM

S. SMALE

1. We give a proof here of the $\Omega$-stability Theorem, announced with the proof sketched in [4]. The paper [4] also provides a good introduction and background for this work. We recall that if $f \in \text{Diff}(M)$, the set of nonwandering points, $\Omega = \Omega(f)$, is defined as the closed invariant set of $x \in M$ such that for any neighborhood $U$ of $x$, $f^m(U) \cap U \neq \emptyset$ for some $m \in Z$, $m \neq 0$. If $f, g \in \text{Diff}(M)$, they are $\Omega$-*conjugate* if there exists an $\Omega$-conjugacy $h : \Omega(f) \to \Omega(g)$, i.e. a homeomorphism $h$ such that $gh = hf$. We assume from now on that $M$ is a compact $C^\infty$ manifold and that $\text{Diff}(M)$ has the uniform $C^r$-topology, $0 < r < \infty$. Then $f$ is $\Omega$-*stable* if there is a neighborhood $N(f)$ in $\text{Diff}(M)$ such that every $g \in N(f)$ is $\Omega$-*conjugate* to $f$.

Recently it was discovered that $\Omega$-stable diffeomorphisms are not dense in $\text{Diff}(M)$ in general [1]. On the other hand a large class of diffeomorphisms are $\Omega$-stable and it is the purpose of this paper to give a proof of this fact. (Presumably, there is no difficulty in proving similar theorems for ordinary differential equations by the same method.)

The main condition on $f \in \text{Diff}(M)$ is that $\Omega = \Omega(f)$ has a *hyperbolic structure*. This means that the tangent bundle of $M$ restricted to $\Omega$, $T_\Omega(M)$ splits into a sum of subbundles $E^s$ and $E^u$ as follows: $E^s$ is invariant under the derivative $Df : T_\Omega(M) \to T_\Omega(M)$ and contracting in that for some $c > 0$, $0 < \lambda < 1$ and all $v \in E^s$, $m \in Z^+$, $\|Df^m(v)\| \le c\lambda^m \|v\|$ (using some Riemannian metric on $M$). Finally $E^u$ is invariant under $Df$, and *expanding for $Df$* or equivalently contracting for $Df^{-1}$ (see [4] for details). Then $f$ satisfies "Axiom A" if

AXIOM A: (i) There is a hyperbolic structure on $\Omega(f)$.

(ii) The periodic points are dense in $\Omega$.

One can find in [4] (using the stable manifold theory appearing in [2]).

(1.1) THE SPECTRAL DECOMPOSITION THEOREM. *If $f \in \text{Diff}(M)$ satisfies Axiom A, then $\Omega(f)$ can be canonically written as the finite disjoint union $\Omega = \Omega_0 \cup \ldots \cup \Omega_k$ of closed invariant sets, on each of which $f$ is topologically transitive.*

These $\Omega_i$ are called *basic sets* for $f$.

Define $\Omega_i \le \Omega_j$ iff $W^s(\Omega_i) \cap W^u(\Omega_j) \neq \emptyset$ where

$$W^s(\Omega_i) = \{y \in M \mid f^m(y) \overset{m \to \infty}{\to} \Omega_i\}, \quad W^u(\Omega_j) = \{y \in M \mid f^m(y) \overset{m \to -\infty}{\to} \Omega_i\}.$$

Then $f$ satisfying Axiom A has the *no cycle property* if:

NO CYCLE PROPERTY. Under the above conditions, $\Omega_{i_1} \le \Omega_{i_2} \le \ldots \le \Omega_{i_k} \le \Omega_{i_1}$, with distinct $i_m$ and $k > 1$ is impossible. An early use of this condition in a closely related context is H. Rosenberg [3].

The goal of this paper is then to prove

(1.2) THE $\Omega$-STABILITY THEOREM. *If $f \in \mathrm{Diff}(M)$ satisfies Axiom A and has the no cycle property, then $f$ is $\Omega$-stable.*

In [4] there are plenty of examples of $f$ satisfying the conditions of the previous theorem.

The proof of (1.2) actually yields an $\Omega$-conjugacy close to the identity.

In his thesis, Jacob Palis proves (1.2) for the case $\Omega$ is finite.

In [4], the statement of the $\Omega$-stability theorem is a little weaker than here. The no cycle property is replaced by

AXIOM B. If $f \in \mathrm{Diff}(M)$ satisfies Axiom A with $\Omega_i$ as in the spectral decomposition theorem, then when $W^s(\Omega_i) \cap W^u(\Omega_j) \neq \varnothing$, there are periodic points $p \in \Omega_i$, $q \in \Omega_j$ such that $W^s(p)$, $W^u(q)$ have a point of transversal intersection.

These properties are related by the following proposition, proved in [4, (8.4 and 8.5)].

(1.3) PROPOSITION. *If $f \in \mathrm{Diff}(M)$ satisfies Axiom A, then Axiom B for $f$ implies the no cycle property.*

The main theorem above moves us toward the yet unsolved

(1.4) PROBLEM. Find necessary and sufficient conditions ("practical conditions"!) for $f \in \mathrm{Diff}(M)$ to be $\Omega$-stable.

Discussions with Bowen, Hirsch, Palis, Pugh and Shub among others have been helpful in the preparation of this paper.

We take this opportunity to remark that the example in the last paragraph of p. 763 of [4] is wrong, as has been pointed out to us by A. Borel. This was to have been an Anosov diffeomorphism of a noncontractible simply-connected manifold. Thus the possibility of such a diffeomorphism remains open. The main examples in that section are unaffected.

2. We are concerned here with neighborhoods of hyperbolic sets and in particular with some applications of stable manifold theory [2].

If $f: U \to M$ is a diffeomorphism of an open set $U$ of $M$ onto its image in $M$ let $f^m(U)$, $m > 0$, be defined inductively by $f(f^{m-1}(U) \cap U)$ and similarly for $m < 0$.

Suppose $U$ is an open set of $M$, $f: U \to M$ is a diffeomorphism onto its image and $\Lambda \subset U$ is compact, invariant with Axiom A true for $\Lambda$. Then by restricting $U$ to a smaller neighborhood of $M$, still denoted by $U$, there is a stable (and unstable) manifold theory for $f$ on $U$ [2]. A (local) stable manifold is given for each point of $x \in \Lambda$ and is denoted by $W_U^s(x)$ and $W_U^s = \bigcup_{x \in \Lambda} W_U^s(x)$. If $f$ is given globally then a global stable manifold $W^s(x)$ is defined as the set of $y$ such that $f^m(y) \in W_U^s(f^m(x))$ for sufficiently large $m$ and this is independent of $U$.

The unstable manifolds are defined similarly (as stable manifolds for $f^{-1}$) and denoted by $W^u(x)$ etc.

Following [2] let $W_\varepsilon^s(x) = W_U^s(x)$ intersected with the $\varepsilon$-ball about $x$ in an appropriate metric, and $W_\varepsilon^s = \bigcup_{x \in \Lambda} W_\varepsilon^s(x)$. Then there is $\lambda < 1$ such that for $\varepsilon$ small enough, $f(W_\varepsilon^s(x)) \subset W_{\lambda\varepsilon}^s(f(x))$, all $x \in \Lambda$. From now on we assume that $\varepsilon$ is this small and fixed. Clearly $W_{\lambda\varepsilon}^s \subset W_\varepsilon^s \subset W_U^s \subset U$, although it is not clear that the first is a proper containment.

From the definitions it follows that $\bigcap_{k>0} W^s_{\lambda^k \varepsilon} = \Omega_i$. Thus for the "dynamical system" $f: W^s_\varepsilon \to W^s_\varepsilon$, $\Lambda$ is an attractor.

Now apply Lemma (4.2) of this paper (see the remark following (4.2)). Note that this does not involve any circularity! This yields a compact subset $V$ of $W^s_\varepsilon$ with $W_{\lambda^k \varepsilon} \subset V \subset W^s_\varepsilon$, some $k \geq 1$ and $f(V) \subset \text{int } V$. We have thus constructed a "fundamental domain" $V - f(V)$ for $f: W^s_\varepsilon \to W^s_\varepsilon$, or for $f: V \to V$ (outside $\Lambda$ anyway). Let $F^s(\Lambda) = V - f(V)$ and similarly define $F^u(\Lambda)$ relative to $W^u_\varepsilon(\Lambda)$. Of course there is uniqueness of $F^s(\Lambda)$.

(2.1) PROPOSITION. *Suppose $U$ is an open set of $M$, $f: U \to M$ is a diffeomorphism onto its image and $\Lambda \subset U$ is compact, invariant and has a hyperbolic structure. Suppose also that there exists a neighborhood $U_1$ of $F^s(\Lambda)$ with*

$$(\text{Closure}(\bigcup_{m>0} f^{-m}(U_1))) \cap \Lambda = \varnothing.$$

*Then there is $\varepsilon > 0$ and a neighborhood $Q$ of $\Lambda$ in $U$ such that if $f^{-m}(x) \in Q$ for all $m > 0$ then $x \in W^u_\varepsilon(p)$ for some $p \in \Lambda$.*

(2.2) COROLLARY. *If in (2.1), $U = M$ and if $W^u(\Lambda) = \{x \mid f^{-m}(x) \to \Lambda, \text{ as } m \to \infty\}$ (as in §1) then $W^u(\Lambda) = \bigcup_{x \in \Lambda} W^u(x)$. Similarly $W^s(\Lambda) = \bigcup_{x \in \Lambda} W^s(x)$.*

(2.3) COROLLARY. *As in (2.1), $\bigcap_{m>0} f^m(U) \subset W^u_U$.*

The proof of (2.1) follows from

(2.4) LEMMA. (*Proof to be supplied by stable manifold theorists Hirsch and Pugh, to appear, see [2]*). *For any neighborhood $U_1$ of Closure $F^s$, $\bigcup_{m>0} f^m(U_1) \cup W^u_\varepsilon(\Lambda)$ contains a neighborhood $Q$ of $\Lambda$.*

3. We now concern ourselves with neighborhoods of the "basic sets" $\Omega_i$ of §1. As we have shown in [4], these basic sets possess a "local product structure" which implies that for suitable choice of $U = U(\Omega_i)$ each $W^s_U(x)$, $W^u_U(y)$ intersect in at most one point and $\Omega_i$ consists precisely of these points of intersection. Thus $W^s_U(\Omega_i) \cap W^s_U(\Omega_i) = \Omega_i$ and so by (2.2) there is a neighborhood $Q$ of $\Omega_i$ in $M$ such that if $f^m(x) \in Q$ all $m \in Z$, then $x \in \Omega_i$.

Furthermore from §7 of [4], it follows that $W^s(\Omega_i) \cap W^u(\Omega_i) \subset \Omega$ and thus $W^s(\Omega_i) \cap W^u(\Omega_i) = \Omega_i$.

We have the following proposition:

(3.1) PROPOSITION. *Let $\Omega_i$ be a basic set of $f \in \text{Diff}(M)$. Then there are neighborhoods $U = U(\Omega_i)$ of $\Omega_i$, $N(f)$ of $f$ in $\text{Diff}(M)$ such that if $g \in N(f)$, and $g$ satisfies the condition of (2.1) then there is a conjugacy $h: \Omega_i \to \Lambda$, where $\Lambda = \{x \in U \mid g^m(x) \in U$ all $m \in Z\}$.*

Thus $h$ is a homeomorphism with $gh(x) = hf(x)$ for all $x \in \Omega_i$.

After §2, (3.1) becomes essentially a consequence of the stable manifold theory [2]. From [2], in fact, for $g$ close enough to $f$, it follows that there is a canonical continuous correspondence between the local stable manifolds of $f$ relative $U$ and those of $g$ relative $U$. In fact $W^s_U(f)$ moves continuously with $f$ and similarly $Q$

of (2.4). The stable, unstable manifolds of $g$ induce a hyperbolic structure on their intersections, which is $\Lambda$ by §2. In this way the correspondence between the local stable manifolds of $f$ and those of $g$ induces a homeomorphism $h: \Omega_i \to \Lambda$. This proves (3.1).

One can also use the original argument in [4] in conjunction with §2 to prove (3.1).

4. The goal of this section is to prove the following

(4.1) LEMMA. *Suppose $\Omega_l$ is a basic set for $f \in \mathrm{Diff}(M)$ (satisfying Axiom A), $U$ a neighborhood of $\Omega_l$ and $X$ is a compact subset of $M$ such that $f(X) \subset \mathrm{int}\, X$ and $\bigcup_{m>0} f^{-m}(X) \cup \Omega_l \supset W^u(\Omega_l)$.*

*Then there is $m > 0$, compact $Y \subset M$ with $f(Y) \subset \mathrm{int}\, Y$ and $\Omega_l \subset \mathrm{int}\, Y - f^{-m}(X) \subset U$.*

We will need the following

(4.2) LEMMA. *Suppose $Q$ is a compact neighborhood of a subset $P$ of a compact manifold $M$ with $f \in \mathrm{Diff}(M)$ satisfying $\bigcap_{m \geq 0} f^m(Q) = P$. Then there is a compact neighborhood $V$ of $P$ in $Q$ such that $f(V) \subset \mathrm{int}\, V$.*

REMARK. The proof is valid for $f$ a homeomorphism of a locally compact space $M$, etc.

PROOF. Let $A_r = \bigcap_{r \geq m \geq 0} f^m(Q)$. Then $A_1 \supset A_2 \supset \ldots$ is a decreasing sequence of sets whose intersection is $P$. Choose $r$ such that $A_r \subset Q$ and let $U = Q \cap f(Q) \cap \ldots \cap f^{r-1}(Q)$. Then $f(U) \subset U$ and $\bigcap_{m \geq 0} f^m(U) = P$. Since $f^r(U)$ is a decreasing sequence of sets converging to $P$, there is an $n \geq 2$ with $f^n(U) \subset \mathrm{interior}\, U$ ($U$ is compact).

Now choose $W$ compact such that $U \subset \mathrm{int}\, W$ and $f^n W \subset \mathrm{int}\, U$. Let $E = U \cup (W \cap f^{n-1} W)$. We claim that $f^{n-1}(E) \subset \mathrm{interior}\, E$, in which case we are finished by downward induction on $n$.

For this first note

(1)     $f^{n-1}(U) \subset U \subset \mathrm{int}\, f^{n-1}(W) \subset (\mathrm{int}\, W) \cap \mathrm{int}\, f^{n-1}(W) \subset \mathrm{int}\, E$

and noting $n \geq 2$,

(2)  $f^{n-1}(W \cap f^{n-1}(W)) \subset f^{2n-2} W = f^{n-2} f^n W \subset f^{n-2} \mathrm{int}\, U \subset \mathrm{int}\, U \subset \mathrm{int}\, E$.

This finishes the proof of (4.2).

Now turning directly to the proof of (4.1), we may assume the $U$ there compact and that $\bigcap_{k \geq 0} f^k(U) \subset W^u(\Omega_l)$.

For $x \in \Omega_l$, $\varepsilon > 0$, define $W^u_\varepsilon(x)$ as the set of $q \in W^u(x)$ of distance less than $\varepsilon$ from $x$ measured in the intrinsic metric in $W^u(x)$. Let $W^u_\varepsilon(\Omega_l) = \bigcup_{x \in \Omega_i} W^u_\varepsilon(x)$ and for $\varepsilon > \delta > 0$ define

$$F^u(\varepsilon, \delta) = F^u = W^u_\varepsilon(\Omega_l) - W^u_\delta(\Omega_l).$$

Then for $\varepsilon$ small enough, $F^u$ will compact with $\bigcup_{k>0} f^{-k}(F^u) \subset U$; and if $\delta$ is small enough, for every $x \in W^u(\Omega_l)$ there will be $k \in Z$ such that $f^k(x) \in F^u$. One can think of $F^u$ as being an enlarged fundamental domain of $f$ on $W^u(\Omega_l)$.

Furthermore since $F^u$ is compact and in $\bigcup_{k>0} f^{-k}(M_{l-1})$ we can choose $m > 0$ such that $f^{-m}(X) \supset F^u$.

Let $P = W^u(\Omega_l) \cup (\bigcap_{k>0} f^k(X))$, and $Q = U \cup f^{-m}(X)$. Then $\bigcap_{k\geq 0} f^k(Q) = P$.

Choose compact $V$ by (4.2) so that $Q \supset V \supset P$, $f(V) \subset \text{int } V$ and let $Y = V \cup f^{-m}(X)$. Then clearly $Y$ satisfies (4.1).

5. We apply (4.1) to the situation at hand in the Ω-stability theorem (1.2). Thus we assume $\Omega_i$, $i = 0, ..., k$ are given by the spectral decomposition (1.1) for $f \in \text{Diff}(M)$ satisfying Axiom A. We are assuming also the no cycle property. Choose a simple ordering $<$ on the $\Omega_i$, using indices such that $\Omega_0 < \Omega_1 < \Omega_2 < ... < \Omega_k$, and with the property that if $\Omega_i < \Omega_j$ then it is false that $\Omega_j \leq \Omega_i$ (so $W^s(\Omega_j) \cap W^u(\Omega_i) = \varnothing$).

Now (for each $i$) take disjoint $U_i \supset \Omega_i$ with the property of (3.1). Thus in particular $\bigcap_{m \in Z} f^m(U_i) = \Omega_i$, $\bigcap_{m>0} f^m(U_i) \subset W^u(\Omega_i)$ and $\bigcap_{m>0} f^{-m}(U_i) \subset W^s(\Omega_i)$.

(5.1) PROPOSITION. *With the $\Omega_i$, $U_i$ as above there are compact sets $M_0 \subset M_1 \subset ... \subset M_k = M$ and positive integers $m_1, ..., m_k$ with these properties*:

$$f^{-ml}(M_{l-1}) \subset \text{int } M_l, \qquad f(M_l) \subset \text{int } M_l,$$

$$\Omega_l \subset \text{int } M_l - f^{-ml}(M_{l-1}) \subset U_l, \qquad \text{and } M_l \subset \bigcup_{i \leq l} W^s(\Omega_i).$$

PROOF.     First since $W^u(\Omega_0) \cap W^s(\Omega_i) = \varnothing$ for $i > 0$, $W^u(\Omega_0) = \Omega_0$. (We recall that $M = \bigcup_{i=0}^{k} W^s(\Omega_i)$ and that $W^u(\Omega_i) \cap W^s(\Omega_i) = \Omega_i$). Apply (4.1) with $M_{l-1} = \varnothing$, to obtain $M_0$ as needed.

Now suppose $M_{l-1}$ has been constructed as described above. To apply (4.1) note that we need only to check $\bigcup_{r>0} f^{-r}(M_{l-1}) \cup \Omega_l \supset W^u(\Omega_l)$.

But $W^u(\Omega_l) = \bigcup_{i=0}^{k}(W^u(\Omega_l) \cap W^s(\Omega_i)) = \bigcup_{i \leq l}(W^u(\Omega_l) \cap W^s(\Omega_i))$ by the no cycle assumption. We get our desired condition from this and the obvious fact that $\bigcup_{m>0} f^{-m}(M_{l-1}) \supset W^s(\Omega_i)$, each $i < l$.

Then (4.1) yields an $M_l$ which we claim satisfies the above claimed properties of (5.1). All this is immediate except for $M_l \subset \bigcup_{i \leq l} W^s(\Omega_i)$. This property follows from an induction on $l$ and the choice of $U_i$.

6. We give here the final construction of the conjugacy, proving (1.2).

From the definition of nonwandering points it follows using the notation and construction of the previous section that there is an $\varepsilon > 0$ with this property. If $g \in \text{Diff}(M)$ is $C^0$ within $\varepsilon$ of $f$ then $\Omega(g) \subset \bigcup_{l=0}^{k} \text{int } M_l - f^{-m_l}(M_{l-1})$; we write $\Omega(g) = \bigcup_{l=0}^{k} \Lambda_l$, $\Lambda_l = \Omega(g) \cap \text{int } M_l - f^{-m_l}(M_{l-1})$, $\Lambda_l$ of course being compact and invariant. Since $\Lambda_l \subset U_l$, (3.1) yields the desired conjugacy, with a possible further shrinking of $\varepsilon$. That is, the global $h: \Omega(f) \to \Omega(g)$ is defined for each $l$, $\Omega_l \to \Lambda_l$ by (3.1).

7. We add a remark to the main result. The proof of (1.2) yields also:

(7.1) PROPOSITION. *Diffeomorphisms satisfying Axiom A and the no cycle property form an open set in* Diff(*M*). *Furthermore, sufficiently small perturbations g of such f* ∈ Diff(*M*) *admit the same compatible simple ordering on the* $\Omega_i(g)$ *as on* $\Omega_i(f)$.

We emphasize that *g* and *f* will not necessarily have the same relations defined by ≤ on the $\Omega_i$ as Example 3 in the next section shows. It is easily shown that Axiom B on *f* would imply that $\Omega_i(f) \to \Omega_i(g)$ preserves ≤ relation.

8. We give here a heuristic account of 3 examples which shed some light on the main theorem (see [1] for example, or [5] for the type of mathematics needed to make these rigorous).

EXAMPLE 1. This example shows that Axiom B is actually stronger than the no cycle property. It also shows that among diffeomorphisms satisfying Axiom A, Axiom B is not a generic property.

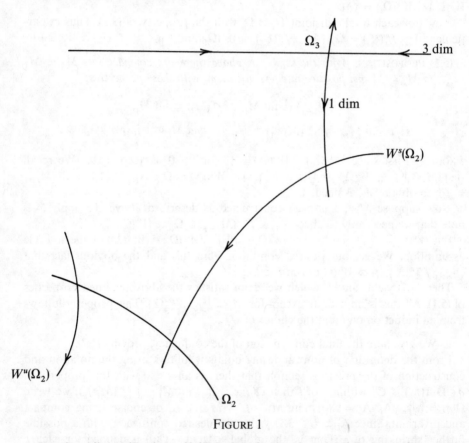

FIGURE 1

In a four-dimensional manifold $\Omega_3$ is a fixed point of saddle type with a three-dimensional stable manifold and one-dimensional unstable manifold. Then $\Omega_2$ will be a two-dimensional toral diffeomorphism (see [4, §3]) and a neighborhood

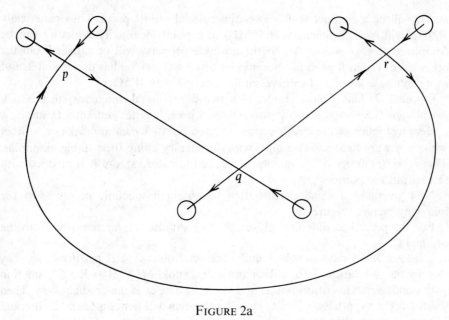

FIGURE 2a

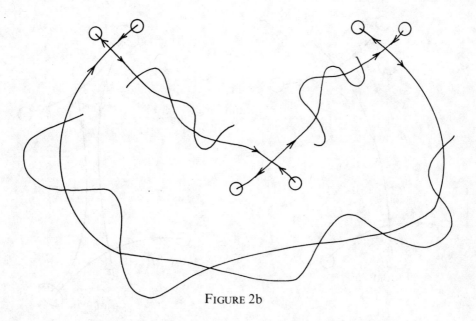

FIGURE 2b

in $M_4$ will be a product with a two-dimensional saddle point. Thus generically $W^u(\Omega_3)$ will meet 3-dimensional $W^s(\Omega_2)$ in a point locally. This picture can be completed so that Axiom A and the no cycle property will be satisfied. On the other hand Axiom B must fail because for any $p \in \Omega_2$, $W^s(p)$ has dimension 2, and so it can't have a point of transversal intersection with $W^u(\Omega_3)$.

EXAMPLE 2. This is an example of a two-dimensional diffeomorphism which permits an $\Omega$-*explosion*. It shows why Axiom A itself does not guarantee $\Omega$-stability.

Here in Figure 2a the circles represent fixed points which are sinks or sources while $p$, $q$, $r$ are fixed saddle points with degeneracy along their stable manifolds. That is, $W^u(p)$ meets $W^s(q)$ on an arc, etc., as indicated. Axiom A is satisfied with $\Omega$ a set of fixed points.

Now we make a small perturbation to obtain homoclinic points as in the following picture (Figure 2b).

For the perturbed diffeomorphism, $\Omega$ is an infinite set, far removed from the original $\Omega$.

EXAMPLE 3. We show now how under the conditions of (1.2), perturbations may destroy the $\leq$ relation. Take a Riemannian manifold $M$, $\phi : M \to R$ a $C^\infty$ function with nondegenerate critical points and $f = \psi_1$ where $\psi_t$ is the gradient flow. Then $f$ satisfies the hypotheses of the $\Omega$-stability theorem and hence is $\Omega$-stable. It could happen however that for fixed points $p$, $q \in M$ that $W^s(p)^l - p = W^u(q) - q$ and for small perturbations $g$, $W^s(p, g) \cap W^u(q, g) = \varnothing$. For example:

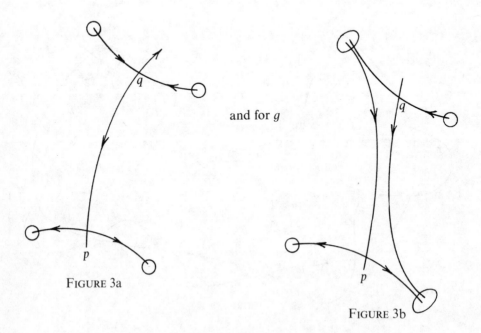

and for $g$

FIGURE 3a

FIGURE 3b

## REFERENCES

**1.** R. Abraham and S. Smale, *Nongenericity of Ω-stability*, these Proceedings, vol. 14.

**2.** M. Hirsch and C. Pugh, *Stable manifolds and hyperbolic sets*, these Proceedings, vol. 14.

**3.** H. Rosenberg, *A generalization of Morse-Smale inequalities*, Bull. Amer. Math. Soc. **70** (1964), 422–427.

**4.** S. Smale, *Differentiable dynamical systems*, Bull. Amer. Math. Soc. **73** (1967), 747–817.

**5.** ———, *Structurally stable systems are not dense*, Amer. J. Math. **88** (1966), 491–496.

UNIVERSITY OF CALIFORNIA, BERKELEY

# ANOSOV FLOWS ON INFRA-HOMOGENEOUS SPACES

PER TOMTER

0. **Introduction.** This is an investigation of Anosov flows on compact manifolds, satisfying additional symmetry conditions.

Roughly, the condition is that the manifold can be given as a double coset space $M = \Gamma \backslash G / K$; where $G$ is a Lie group, $K$ a (compact) closed subgroup, and $\Gamma$ a discrete, uniform subgroup; and the flow is given through right-multiplication by a one-parameter subgroup $(\exp t\alpha) \subseteq G: \Gamma g K \to \Gamma g (\exp t\alpha) K$. This condition means essentially that by lifting the flow to a covering manifold $M'$ one can find a transitive Lie symmetry group (i.e., sending parameterized orbits into parameterized orbits) acting on $M'$, with (compact) isotropy group $K$; and containing the group of deck transformations $\Gamma$. The relevant axioms for describing an Anosov flow in this setting (a "$(G, \Gamma)$-induced Anosov flow"), is given in Chapter 1. For basic definitions, motivation, and the role of Anosov flows in the study of general flows on compact manifolds, we refer to the survey article by Smale [41].

The two classes of examples of Anosov flows which are known at present are the geodesic flows of compact Riemannian manifolds of strictly negative curvature (restricted to the unit tangent bundle), and the suspensions of the Anosov diffeomorphisms on compact manifolds. For manifolds of constant negative curvature, it is well known that the geodesic flow can be given as a $(G, \Gamma)$-induced flow (e.g. see Gelfand and Fomin [20] for dimension 2). Our treatment seems justified in light of the following results:

(i) The geodesic flow of any compact, locally symmetric space of noncompact type and strictly negative curvature is a $(G, \Gamma)$-induced Anosov-flow.

(ii) The suspension of any Anosov diffeomorphism given by a hyperbolic automorphism of an infra-nilmanifold (see Chapter 1 and Chapter 2 for definitions) is a $(G, \Gamma)$-induced Anosov-flow. It has been conjectured that all Anosov-diffeomorphisms are of this type; and progress in this direction has been made by A. Avez, J. Franks, and M. Shub. In particular, see the work of J. Franks appearing in this volume. Thus, it is clear that the main known examples of Anosov-flows will be covered by this work. One might, in fact, conjecture that all Anosov flows on compact manifolds are topologically conjugate to these.

In §1 we discuss the relevant axioms for $(G, \Gamma)$-induced Anosov flows. Our assumptions are that $G$ is a Lie group with a finite number of components, and with Lie algebra $\mathfrak{g}$; $K$ is a compact subgroup with Lie algebra $\mathfrak{k}$; $\Gamma$ is a uniform, discrete subgroup of $G$ (i.e. $\Gamma \backslash G$ is compact), which acts freely on $G/K$ by left multiplication; $\alpha$ is an element of $\mathfrak{g}$ such that $(\exp t\alpha)$ normalizes $K$; $\forall t \in \mathbf{R}$. Then the flow $\Gamma gK \to \Gamma g(\exp t\alpha)K$ on $\Gamma \backslash G/K$ is an Anosov flow if we assume

AI: The kernel of ad$\alpha$ is $\mathfrak{k} + \boldsymbol{R}\alpha$.

AII: ad$\alpha$ has no nonzero imaginary eigenvalues.

We include a detailed proof of this result, and a discussion of some examples. All Anosov flows of the kind described here will have a finite, flow-invariant smooth measure; the periodic orbits are dense; and one can see immediately how to get the expanding and contracting leaves of the invariant foliations by using group translation.

In §2 we show that the flows we get by restricting to solvable groups $G$ are essentially the suspensions of Anosov diffeomorphisms induced by suitable hyperbolic affine transformations on infranilmanifolds; these are in fact easily shown to be differentiably conjugate to the usual diffeomorphisms given by hyperbolic automorphisms.

In §3 we study the flows for semisimple groups $G$; in this case we get exactly the geodesic flows on the unit tangent bundles of the (irreducible) Riemannian locally symmetric spaces of noncompact type and rank 1, and the flows induced by these on manifolds which are finitely covered by such unit tangent bundles. On the basis of work by J. Milnor and J. A. Wolf it also follows easily that the fundamental groups of the examples in §2 and §3 must have exponential growth, (we call a finitely generated group a group of exponential growth if the number of distinct elements of the group which can be written as words of length not greater than $n$, grows exponentially with $n$).

In §4 we study the structure of $G$-induced Anosov flows locally; i.e. on the universal covering manifold. We demonstrate a decomposition theorem for $\mathfrak{g}$ which permits us to understand the lift of a $G$-induced Anosov flow to the universal covering manifold in terms of a geodesic flow of the type investigated in §3, and the suspension of an Anosov diffeomorphism. In fact, $\mathfrak{g} = \mathfrak{N} \oplus_s \mathfrak{S}$, a semidirect product of its nilpotent radical $\mathfrak{N}$ and a Levi complement $\mathfrak{S}$ which is a real rank 1, simple Lie algebra, such that $\alpha \in \mathfrak{S}$, $\mathfrak{k} \subseteq \mathfrak{S}$. The principal tools used for this are the Levi-Malcev theorem and some of the theory of algebraic groups. In particular, in dimension 3 it is now easy to see that we get only the generalized geodesic flows corresponding to surfaces of constant negative curvature, and the suspensions of the toral Anosov diffeomorphisms. As a corollary, we get Margulis result on exponential growth of the fundamental group for the case of 3-dimensional manifolds with $(G, \Gamma)$-induced Anosov flows.

We then proceed to demonstrate that we get new examples of Anosov flows by this method; and give the detailed construction of a new Anosov flow on a compact 7-dimensional manifold, which has both a "semisimple" and a "solvable" part: it can be analyzed in terms of the geodesic flow on a compact Riemannian surface of constant negative curvature and the suspension of a 4-dimensional toral Anosov diffeomorphism. It is obtained by letting $SL(2, \boldsymbol{R})$ act suitably on $\boldsymbol{R}^4$, and taking semidirect product. Actually, we realize $SL(2, \boldsymbol{R})$ as the group of unit quaternions in the even-dimensional Clifford algebra corresponding to an aniso-tropic quadratic form on $\boldsymbol{R}^3$, (i.e. a quadratic diophantine equation without integer solutions). The problem is to find a uniform, discrete subgroup, which we do by finding a uniform discrete subgroup of SL(2, $R$) which preserves an integer lattice of $R^4$, using the theory of arithmetic subgroups. We notice that if we do not require our manifold to be compact, but only to have finite measure, this difficulty essentially disappears.

The results in this chapter are not final. Actually one can show (using Borel's density theorem) that $\Gamma$ must in fact intersect the radical $N$ in a uniform, discrete subgroup, and $\alpha$ must correspond to one of the lattice-preserving hyperbolic maps studied in Chapter 2. This will also imply that for all $(G, \Gamma)$-induced Anosov flows, the manifold $\Gamma\backslash G/K$ has fundamental group of exponential growth. Moreover, we can construct examples of analogous new Anosov flows from suitable representations of all the real rank 1, simple Lie groups of type $B_n$, $D_n$. The details of these investigations will be given in a later paper.

In §5 we use the theory of group representations to study ergodic properties of the flows. This method goes back to Gelfand and Fomin [20], and has been refined by Mautner [28], Auslander and Green [6], and C. C. Moore [31]. Most of the results are by now known in one version or another, but in our case the analysis can be made very simple; this is in fact one of the motivations for studying Anosov flows with symmetry conditions. They always have a finite, flow-invariant measure, and we get the following criteria for the spectral properties: A $(G, \Gamma)$-induced Anosov flow is always ergodic; if it is a suspended flow (i.e. $G$ is solvable), it is not weakly mixing; in all other cases it has Lebesgue spectrum, and hence satisfies all mixing properties. The proofs use an adaption of the methods quoted above, and the structure theory developed in §4. We also remark that these methods can be used to obtain information on ergodicity of Anosov-diffeomorphisms induced by hyperbolic affine transformations on infra-nilmanifolds—they are seen, by studying their suspension, to be weakly mixing. Actually, it is also possible to extend the classical method of proving Lebesgue spectrum of the "geodesic" flow by proving ergodicity of the "horocycle" flow to prove that these Anosov diffeomorphisms have Lebesgue spectrum.

I am grateful to Professor Stephen Smale and Professor Calvin C. Moore for many helpful suggestions and discussions during this investigation. I have also had stimulating contact with other members of the Department of Mathematics at Berkeley; in particular I would like to thank Professor Joseph A. Wolf and Professor Nolan R. Wallach; and finally Professor George D. Mostow for helpful conversations during the seminar on algebraic groups at Bowdoin College.

NOTATIONS AND CONVENTIONS. We use the usual symbols $Z$, $R$, $C$ for the integers, the real numbers and the complex numbers, respectively. If $A$ is a group or an algebra, $Z(A)$ denotes the center of $A$. For linear groups and Lie algebras, we follow the usual conventions, e.g. $\mathfrak{S}O(n)$ means the Lie algebra of skew-symmetric $n \times n$ matrices.

1. **G-induced Anosov-flows.** We first recall some of the fundamental notations and definitions. For more details, see Smale [41].

An Anosov-flow on a complete Riemannian manifold $M$ is a flow $(\phi_t)$ whose induced flow $(D\phi_t)$ on the tangent bundle $T(M)$ is hyperbolic in the following sense: $T(M)$ is a continuous Whitney-sum of three invariant subbundles: $T(M) = E_1 + E_2 + E_3$, where $(D\phi_t)$ is contracting on $E_1$ (i.e., $\exists c, \lambda > 0$, such that $\|D\phi_t(v)\| \leq ce^{-\lambda t}, \forall v \in E_1, \forall t > 0$), expanding on $E_2$ (i.e., $\exists c_1, \mu > 0$; such that $\|D\phi_t(v)\| \geq c_1 e^{\mu t}, \forall v \in E_2, \forall t > 0$) and $E_3$ is the 1-dimensional bundle defined by the velocity vector (which is thus assumed nonvanishing at all points).

If $M$ is compact, a flow is Anosov for some Riemannian metric iff it is Anosov for all. In this case the set of Anosov-flows form an open set among all vector-fields ($C^k$-topology); and such a flow is structurally stable in the following sense: Any sufficiently small perturbation $(\psi_t)$ of $(\phi_t)$ is topologically equivalent to $(\phi_t)$, i.e., there is a homeomorphism of $M$ sending orbits of $(\phi_t)$ into orbits of $(\psi_t)$. Finally, the Anosov-flow is ergodic for any invariant measure on $M$.

As mentioned, the known examples are some geodesic flows, and the suspensions of Anosov-diffeomorphisms. The latter construction goes as follows:

(a) Let $f \in \text{Diff}(M')$, $M'$ a compact manifold. The suspension of $f$ is a flow on a compact manifold $M$ given as follows: $g \in \text{Diff}(M' \times \mathbf{R})$ is given by $g(m, s) = (f(m), s + 1)$. Then the discrete group $(g^n)$, $n \in \mathbf{Z}$, acts freely on $M' \times \mathbf{R}$, and $M$ is the orbit space. The flow on $M$ is induced by $\phi_t(m, s) = (m, s + t)$.

(b) In analogy with the above, a diffeomorphism $f$ of a compact manifold $M'$ is an Anosov-diffeomorphism if there is a splitting of the tangent bundle $T(M)$ into a continuous, invariant Whitney-sum: $T(M) = E_1 + E_2$, with $E_1$ contracting (i.e., $\exists c > 0$, $0 < \lambda < 1$ such that $\|(Df)^m(v)\| < c\lambda^m \|v\|$, $\forall v \in E_1$, $m \in \mathbf{Z}^+$) and $E_2$ expanding.

The most general known way of constructing Anosov-diffeomorphisms is by "hyperbolic automorphisms on infra-nilmanifolds", as follows: Let $N$ be a simply-connected, nilpotent Lie-group, $S$ a finite subgroup of $\text{Aut}(N)$, $G = N \cdot_s S$ (semidirect product), and $\Delta \subseteq G$ a uniform, discrete subgroup such that $\Delta \backslash G/S$ is a manifold (i.e., $\Delta$ acts freely on $G/S$) i.e., if $\delta \in \Delta$ and a conjugate of $\delta$ lies in $S$; then $\delta$ acts trivially on $G/S$ (all conjugates of $\delta$ lie in $S$). Let $\phi$ be an automorphism of $G$ such that $\psi = \phi/N : N \to N$ is hyperbolic; (i.e., the differential $D\psi$ on the Lie-algebra of $N$ has no eigenvalues on the unit circle), $\psi$ commutes with all automorphisms in $S$, and $\phi(\Delta) = \Delta$. Then $\phi$ induces an Anosov-diffeomorphism on $\Delta \backslash G/S$ (we call these $N$-induced Anosov-diffeomorphisms); and the suspension gives an Anosov flow, on a manifold which is a bundle over the circle, with an infra-nilmanifold as fibre.

We say that a flow on a manifold $M$ is a $G$-induced flow under the following conditions: $M = \Gamma \backslash G/K$, where $G$ is a Lie-group with Lie-algebra $\mathfrak{g}$, $K$ is a closed subgroup with Lie-algebra $\mathfrak{k}$, such that $G/K$ is connected, $\Gamma$ is a group of diffeomorphisms on $G/K$ which acts freely and properly discontinuously. The flow is given by $\phi_t : \Gamma gK \to \Gamma g(\exp t\alpha)K$ where $\alpha \in \mathfrak{g}$, and the one-parameter-group $\exp t\alpha$ normalizes $K$. To obtain a well-defined flow on $M$, we also assume that $\phi_t$ normalizes $\Gamma$ in $\text{Diff}(G/K)$. We notice that in this case $\{\phi_t\}$ must in fact centralize $\Gamma$. For $t \to \phi_t(\gamma(\phi_{-t}(gK))) = \gamma_t(gK)$ is continuous, and $\gamma_0 = \gamma$. Since the action of $\Gamma$ is properly discontinuous, there is a neighborhood $V$ of $gK$ such that $A = \{\gamma' \in \Gamma; \ \gamma'(V) \cap V \neq \varnothing\}$ is finite. For $t$ in a suitable neighborhood of 0, $\gamma_t(gK) \in V$; i.e. $\gamma_t \in A$. Since $\Gamma$ acts freely, $\gamma' \neq \gamma$ implies $\gamma'(gK) \neq \gamma(gK)$. Hence, by dropping to a smaller neighborhood $V'$, we get $\gamma_t = \gamma$ for $t$ sufficiently close to 0.

We call two $G$-induced flows with groups $G_1$ and $G_2$ $G$-equivalent if
(1) There is a homomorphism $\phi: G_1 \to G_2$ (or vice versa: $G_2 \to G_1$).
(2) $\phi(K_1) \subseteq K_2$.
(3) The induced map $\bar{\phi}: G_1/K_1 \to G_2/K_2$ is a surjective diffeomorphism.
(4) $d\phi(\alpha_1) = \alpha_2$.
(5) $\phi\Gamma_1\phi^{-1} = \Gamma_2$.

It is easily checked that $G$-equivalent flows are differentiably conjugate. In fact, the relevant identifications are given by the following 3-dimensional, commutative diagram:

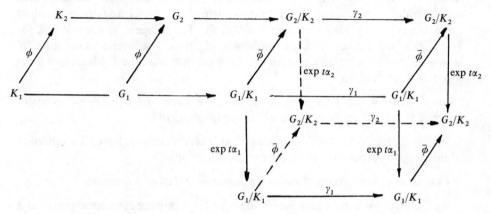

PROPOSITION 1. *Up to $G$-equivalence we can assume: No nontrivial, normal subgroup of $G$ is contained in $K$.*

PROOF. The set of elements $h$ such that $ghg^{-1} \in K$; $g \in G$; constitutes a normal, closed subgroup of $G$ contained in $K$, and maximal under those conditions. Let $G_2 = G/H$, $\phi$ be the canonical projection, $K_2 = \phi(K) = K/H$, $\Gamma_2 = \phi\Gamma_1\phi^{-1}$.

COROLLARY 1. *Up to $G$-equivalence we can assume: No nontrivial ideals of $\mathfrak{g}$ are contained in $\mathfrak{k}$.*

COROLLARY 2. *Up to $G$-equivalence we can assume: $G$ acts effectively on $G/K$.*

PROPOSITION 2. *Up to $G$-equivalence we can assume: $G$ is connected and simply connected. If we then substitute for $K$ its connected component $K_0$, we get the universal covering manifold of $M$, with the lifted Anosov flow.*

PROOF. Since $K$ meets all the components of $G$, this follows by standard type arguments.

PROPOSITION 3. *The condition for a flow $\{\phi_t\}$ on a manifold $M$ to be a $G$-induced flow is that the lift to a (universal) covering manifold $M'$ satisfies the following* *condition*: *There is a transitive Lie-symmetry group G acting on M'*; (*i.e. G centralizes* $\{\phi_t\}$ *in* Diff $(M')$; *i.e. G sends parameterized orbits into parameterized orbits*).

PROOF. It is clear that the condition is necessary (use Proposition 2). Conversely, let $p \in M'$; and let $K$ be the isotropy group. By Corollary 2 we may assume $G \subseteq$ Diff $(G/K)$. If $T$ is the one-parameter group $\{\phi_t\}$ in Diff $(G/K)$, we have $G' = G \cdot T = T \cdot G$ is a transitive Lie group of diffeomorphisms of $M = G/K = G'/K'$ containing $\{\phi_t\}$ in its center and the result will follow. In fact, it is not hard to see that $\{\phi_t\}$ must be given by right-multiplication by some one-parameter group $\{\exp t\alpha\}$ in $G : \phi_t(gK) = g(\exp t\alpha)K$.

We now wish to add the Riemannian structure; i.e. we assume that the covering space $G/K$ is a Riemannian homogeneous space and that $\Gamma$ is a group of isometries. To get $G/K$ as a Riemannian homogeneous space with $G$ acting as a group of isometries, we have to add the condition: Ad $K$ is compact. We can then orthogonalize: $\langle x, y \rangle = \int_{\text{Ad} K}(kx, ky)d\mu(k)$ where ( , ) is any inner product on $\mathfrak{g}$, and $\mu$ is a Haar-measure on Ad$(K)$. The tangent space of $G/K$ at $\bar{e} = K$ can be identified with a complement of $\mathfrak{k}$ in $\mathfrak{g}$. The metric tensor for $G/K$ is now defined by left-translation of the inner product at $e$; the $K$-invariance gives that this is well defined.

PROPOSITION 4. *Up to G-equivalence we can assume that G is the full isometry group of G/K; and hence that $\Gamma$ is a discrete subgroup of G.*

PROOF. By Corollary 2 we can assume that $G$ acts effectively, hence is a subgroup of the isometry group. Let $\phi$ be the natural injection.

COROLLARY. *Up to G-equivalence we can assume that K is compact.*

By the above, it is now reasonable to make the following assumptions: $G$ has a finite number of components, K is compact and $\Gamma$ is a discrete subgroup of $G$. We are mainly interested in compact manifolds; and $\Gamma\backslash G/K$ is now compact iff $\Gamma$ is uniform in $G$.

DEFINITION. We say that a flow on a manifold $M$ is a $(G, \Gamma)$ induced flow if it is a $G$-induced flow with $K$ compact and $\Gamma$ a discrete subgroup of $G$ acting by left multiplication. (For such a discrete subgroup; the conditions that it acts properly discontinuously and that $\{\phi_t\}$ centralizes $\Gamma$ in Diff $(G/K)$ are automatic.) A $G$-equivalence $\phi$ of $(G, \Gamma)$-induced flows is called a $(G, \Gamma)$-equivalence; in particular, $\phi(\Gamma_1) = \Gamma_2$ is a sufficient condition on the discrete subgroups.

It follows easily that Proposition 1 and its corollaries hold for $(G, \Gamma)$-induced flows.

In the case of Proposition 2, the situation is slightly more complicated. We can assume that $\Gamma$ meets all components of $G$. $\Gamma \cap G_0$ is of finite index in $\Gamma$; and $\Gamma \cap G_0\backslash G_0/K \cap G_0 \to \Gamma\backslash G/K$ is a finite covering. Thus, by restricting to connected groups $G$, we get all $(G, \Gamma)$-induced flows up to finite coverings, only. However, when considering simply connected groups $G$, we simply lift $G_0 \cap \Gamma$ to $G$; hence this is no further restriction. In this case, however, $\tilde{K}$ will no longer generally be compact.

Finally, Proposition 3 will now sound:

PROPOSITION 3'. *The condition for a flow* $\{\phi_t\}$ *on a manifold* $M$ *to be a* $(G, \Gamma)$-*induced flow is that the lift to some covering manifold* $M'$ *satisfies the following condition*: *There is a transitive Lie symmetry group* $G$ *acting on* $M'$ *such that* $G$ *contains the group of deck transformations of the covering* $M' \to M$ *and the isotropy group of a point*, $K$, *is compact.*

In that case, we can ᵈefine a Riemannian structure on $G/K$ as above; and project it down to $\Gamma \backslash G/K$.

We are now ready to give our axioms for $(G, \Gamma)$-induced Anosov flows:

Let $G$, $K$, $\Gamma$, $\alpha$ be as above; i.e. $G$ is a Lie group with a finite number of components and with Lie algebra $\mathfrak{g}$, $K$ is a compact subgroup with Lie algebra $\mathfrak{k}$ such that $G/K$ is connected, $\Gamma$ is a discrete subgroup of $G$ which acts freely on $G/K$ by left multiplication; the one-parameter group $\exp(t\alpha)$ normalizes $K$.

AI. The kernel of ad $\alpha$ is $\mathfrak{k} + \mathbf{R}\alpha$.

AII. ad $\alpha$ has no nonzero, imaginary eigenvalues.

NOTE. It then follows: $\exp t\alpha$ centralizes $K$. It centralizes $K_0$ by AI, and $\exp t\alpha$ defines a one-parameter group of automorphisms of the finite group $K/K_0$; hence it is trivial.

AIII. $\Gamma$ is uniform in $G$.

THEOREM I. *Let* $G$, $K$, $\Gamma$, $\alpha$ *be as above, and assume* AI–AII *is satisfied. Then* $\mathfrak{g}$ *has the following decomposition*: $\mathfrak{g} = \mathfrak{k} + \mathbf{R}\alpha + \mathscr{E}$ *(vector-space direct sum), where* $\mathscr{E}$ *is the space spanned by the nonzero eigenspaces of* $\mathrm{ad}(\alpha)$. $G/K$ *has a* $G$-*invariant, Riemannian structure, and the flow given by* $gK \to g(\exp t\alpha)K$ *is an Anosov flow. The Riemannian manifold* $\Gamma \backslash G/K$ *is covered by* $G/K$; *and the induced flow*: $\Gamma gK \to \Gamma g(\exp t\alpha)K$ *is an Anosov flow.* $\Gamma \backslash G/K$ *is compact iff* AIII *is satisfied.*

PROOF. First, let $\mathscr{E}^+$, $\mathscr{E}^-$ denote the invariant subspaces in the primary decomposition of ad $\alpha$ corresponding to eigenspaces of positive and negative real parts, respectively. Then $\mathfrak{g} = \mathfrak{k} + \mathbf{R}\alpha + \mathscr{E}^+ + \mathscr{E}^-$.

NOTE. $\mathscr{E}$, $\mathscr{E}^+$, $\mathscr{E}^-$, $\mathbf{R}\alpha$ are $K$-invariant.

Since $K$ centralizes $\exp(t\alpha)$; Ad $K$ commutes with $\mathrm{Ad}(\exp t\alpha)$; hence with the infinitesimal generator ad $\alpha$; and hence with the semisimple part of ad $\alpha$. By simultaneous diagonalization over the complexification we see that all full eigenspaces corresponding to $\lambda$, $\bar\lambda$-eigenvalues (i.e. the invariant subspaces in the primary decomposition) are Ad $K$-invariant. But $\mathscr{E}^+$, $\mathscr{E}^-$, $\mathscr{E}$ are spanned by such full eigenspaces. On $\mathbf{R}\alpha$, $K$ acts trivially.

We can now identify $\mathscr{E} + \mathbf{R}\alpha$ with the tangent space of $G/K$ at $\bar{e} = K$, and define the metric tensor as above. Since $\Gamma$ is a discrete group of isometries acting freely, it follows that $G/K \to \Gamma \backslash G/K$ is a normal Riemannian covering. (Wolf [**42**, pp. 99, 39, 60]), with a unique Riemannian metric on $\Gamma \backslash G/K$.

It is sufficient to show that we have an Anosov flow on $G/K$. For if $G/K \overset{\pi}{\to} \Gamma \backslash G/K$, then $T(G/K) \overset{d\pi}{\to} T(\Gamma \backslash G/K)$ projects invariant subbundles, and preserves lengths on the tangent-space part. We want to define subbundles $E_1, E_2, E_3$ by translating

$\mathscr{E}^+$, $\mathscr{E}^-$ and $\mathbf{R}\alpha$. Define $(E_1)_{gK} = Dg(\mathscr{E}^+)$, where $Dg$ is the differential of the isometry $g$. This is well defined by the following argument: $\mathscr{E}^+$ can be identified with the set of curves in $G/K$: $\exp(\beta t)K$, $\beta \in \mathscr{E}^+$. Then $Dg(\mathscr{E}^+) = \{g(\exp(\beta t)K, \beta \in \mathscr{E}^+)\}$. But if $k \in K$:

$$\{gk \exp(\beta t)K; \beta \in \mathscr{E}^+\} = \{gk \exp(\beta t)k^{-1}K; \beta \in \mathscr{E}^+\}$$
$$= \{g \exp(t\mathrm{Ad}k(\beta))K; \beta \in \mathscr{E}^+\}$$
$$= \{g \exp(\beta t)K; \beta \in \mathscr{E}^+\}$$

by $K$-invariance of $\mathscr{E}^+$. Similarly for $\mathscr{E}^-$ and $\mathbf{R}\alpha$. It is clear that $E_3$ is the subbundle corresponding to the direction of the flow. Now for $\mathscr{E}^+$ and $\mathscr{E}^-$ it is sufficient to examine the growth of tangent vectors at $\bar{e} = K$. For any tangent vector $v$ at $gK$ is given as a curve $g \exp(s\beta)K$, $\beta \in \mathbf{R}\alpha + \mathscr{E}$. The length of $v$ is determined by the length of $D(g^{-1})(v) \simeq \exp(s\beta)K$ at $\bar{e}$. $D\phi_{t_0}(v) \simeq g \exp(s\beta)\exp(t_0\alpha)K$, a curve through $g \exp(t_0\alpha)K$, with length determined by:

$$\exp(-t_0\alpha)g^{-1}g \exp(s\beta)\exp(t_0\alpha)K = \exp(-t_0\alpha)\exp(s\beta)\exp(t_0\alpha)K.$$

But $D\phi_{t_0}(Dg^{-1}(v)) \simeq \exp s\beta \exp(t_0\alpha)K$, with length determined by $\exp(-t_0\alpha)$ $\exp s\beta \exp(t_0\alpha)K$. Hence length increases at the same rate as at $\bar{e}$.

Now, let $v$ be a tangent vector at $\bar{e} = K$.

$$\langle D\phi_t(v), D\phi_t(v)\rangle_{(\exp t\alpha)K}$$
$$= \langle D\exp(-t\alpha)D\phi_t(v), D\exp(-t\alpha)D\phi_t(v)\rangle_{\bar{e}}$$
$$= \langle D(\exp(-t\alpha) \circ \phi_t)v, D(\exp(-t\alpha) \circ \phi_t)v\rangle_{\bar{e}}.$$

It is sufficient to consider the map $gK \to \exp(-t\alpha)g \exp(t\alpha)K$. Let $\beta \in \mathbf{R}\alpha + \mathscr{E}$. Then $D(\exp(-t\alpha) \circ \phi_t)(\beta) = \exp(-t \, \mathrm{ad} \, \alpha)(\beta)$. For, if $F \in C^\infty(G/K)$, then

$$D(\exp(-t\alpha) \circ \phi_t)(\beta)(F) = \beta(F \circ \exp(-t\alpha) \circ \phi_t)$$

$$= \frac{d}{ds_{s=0}}(F(\exp(-t\alpha)\exp s\beta \exp(t\alpha)K))$$

$$= \frac{d}{ds_{s=0}}(F(\exp[s \exp(-t \, \mathrm{ad} \, \alpha)(\beta)]))$$

$$= \{\exp(-t \, \mathrm{ad} \, \alpha)(\beta)\}(F).$$

It remains to show that $\exp(-t \, \mathrm{ad} \, \alpha)$ is a contracting operator on $\mathscr{E}^+$, expanding on $\mathscr{E}^-$. We note that all norms on the space of operators of $\mathscr{E}^+$ are topologically equivalent. Now choose a complex basis such that $\mathrm{ad}_{\mathscr{E}+\alpha}$ is in Jordan form:

$$\mathrm{ad}_{\mathscr{E}+\alpha} = \begin{pmatrix} A_1 & \\ & \ddots & \\ & & A_r \end{pmatrix}$$

$$A_j = \begin{pmatrix} x_j + iy_j & & 0 \\ 1 & \ddots & \\ & \ddots & \ddots \\ 0 & & 1 & x_j + iy_j \end{pmatrix}$$

Then

$$\exp(-\operatorname{ad}_{\mathscr{E}+} t\alpha) = \begin{pmatrix} e^{-tA_1} & 0 \\ 0 & e^{-tA_r} \end{pmatrix}$$

with

$$e^{-tA_j} = \begin{pmatrix} e^{-tx_j-ity_j}, & 0, & 0\ldots,0 \\ -te^{-tx_j-ity_j}, & e^{-tx_j-ity_j}, 0\ldots,0 \\ \pm\dfrac{t^{k_j-1}}{(k_j-1)!}e^{-tx_j-ity_j}, & \ldots, & \ldots, e^{-tx_j-ity_j} \end{pmatrix}$$

Now, let $\| \quad \|_m$ be the operator norm given by maximum of the matrix entries relative to this basis. For large enough $t$, this is $(t^{k_j-1}/(k_i-1)!)e^{-tx_j}$ for some $j$. Let the usual operator norm be $\| \quad \|$, then

$$\|\exp(-t\operatorname{ad}_{\mathscr{E}+}\alpha)\| \leq c\left|\dfrac{t^{k_j-1}}{(k_j-1)!}e^{-tx_j}\right|$$

for some $j$ and some constant $c$ and $t$ large enough. For large $t$:

$$c\dfrac{t^{k_j-1}}{(k_j-1)!} \leq e^{tx_j/2}$$

$$\|\exp(-t\operatorname{ad}_{\mathscr{E}+}\alpha)\| \leq e^{-tx_j/2}$$

$$\|h\| \leq e^{-tx_j/2}\|\exp(t\operatorname{ad}_{\mathscr{E}+}\alpha)(h)\|$$

$$\|\exp(t\operatorname{ad}_{\mathscr{E}+}\alpha)(h)\| \geq e^{tx_j/2}\|h\|$$

for $t >$ some constant $K > 0$. By letting $h$ vary over the compact unit ball, and $t \in [0,K]$, we find a constant $m > 0$ such that $\|(\exp t\operatorname{ad}_{\mathscr{E}+}\alpha)(h)\| \geq m\|h\|$, for $t \in [0,K]$. Let $a = \min(1, me^{-Kx_j/2})$, then $\|\exp\operatorname{ad}_{\mathscr{E}+}(t\operatorname{ad}\alpha)(h)\| \geq ae^{tx_j/2}\|h\|$ for $t > 0$. It follows: $\|\exp\operatorname{ad}_{\mathscr{E}+}(-t\alpha)(h)\| \leq \frac{1}{a}e^{-tx_j/2}\|h\|$ i.e. $\exp(\operatorname{ad}_{\mathscr{E}+}(-t\alpha))$ is contracting on $\mathscr{E}^+$. Similarly, it is expanding on $\mathscr{E}^-$.  Q.E.D.

We show how two concrete, motivating examples fit into this formalism.

(A) Let $f$ be the standard toral Anosov-diffeomorphism given by

$$\begin{pmatrix} 2 & 1 \\ 1 & 1 \end{pmatrix}$$

The eigenvalues are $(3 \pm \sqrt{5})/2$, and it follows that we can reach $f$ by a one parameter group of linear automorphisms, $\exp(t\alpha)$; $f = \exp(\alpha)$. Now, form the semi-direct product $R^2 \times_s R$:

$$(g_1, s)\cdot(g_2, t) = (g_1 + (\exp s\alpha)(g_2), s + t).$$

$Z^2$ is a subgroup, and $T^2 \times R = Z^2\backslash R^2 \times R$ is naturally identified with $Z^2\backslash(R^2 \times_s R)$. Here, the flow is described through right-multiplication by $(0, t)$. On the other hand, left multiplication by $(0, 1)$ gives the correct identification:

$$(0, 1)(g, s) = (f(g), s + 1).$$

Let $J$ be the group $\{(0, n)\}_{n\varepsilon Z}$. Since $f$ preserves the integer lattice: $J \cdot Z^2 = Z^2 \cdot J$ $= \Gamma$ is a discrete (and uniform) subgroup; and we get our suspended Anosov flow on $\Gamma \backslash R^2 \times_s R$. It is easy to check that all axioms are satisfied; $K = (e) = ((0, 0), 0)$; $\alpha = ((0, 0), 1)$; since $f$ has no eigenvalues on the unit circle, $\alpha$ has no imaginary eigenvalues.

We notice that if $f$ was given by

$$\begin{pmatrix} 1 & 2 \\ 1 & 1 \end{pmatrix}$$

then it cannot be reached by a 1-parameter group of automorphisms. However, we can write $f = J \cdot \exp B$, where $J^2 = I$. Let $K = \{I, J\}$. Then we can describe the suspension by using the semidirect product $R^2 \cdot_s K$ instead of $R^2$; and $((0, J), 1)$ instead of $(0, 1)$ in $\Gamma$.

Actually, by a careful consideration of Jordan decompositions, semisimple and unipotent parts, we can show that any element of $GL(n, R)$ can be written as $J \exp B$, where

$$J = \left(\begin{array}{c|c} -I & 0 \\ \hline 0 & I \end{array}\right)$$

relative to a basis that gives the primary decomposition, and we have $-I$ for the eigenspaces corresponding to negative, real eigenvalues.

Hence we can treat all toral Anosov diffeomorphisms in this manner. We do not give details; since we give a more general treatment at the beginning of Chapter II.

(B) Next, we give the construction for the geodesic flow on the unit tangent bundle of a compact Riemannian manifold $M$ of constant negative curvature (normalized to $-1$). For more details, see Wolf [42], Mautner [28], and Gelfand-Fomin [20].

$M$ is covered by hyperbolic $n$-space, $\Sigma^+$.

$$\Sigma = \{(x_0, ..., x_n) \in R^{n+1}; -x_0^2 + x_1^2 + ... + x_n^2 = -1\},$$

has two sheets, $\Sigma^+$ with $(+1, 0, ..., 0)$ and $\Sigma^-$ with $(-1, 0, ..., 0)$. The isometry group is the subgroup of $O(n + 1, n)$ which sends $\Sigma^+$ to $\Sigma^+$; it contains the component of the identity; is of index 2; and has Lie-algebra

$$\mathfrak{S}O(n + 1, 1) = \left\{\left(\begin{array}{c|c} 0 & x \\ \hline {}^t x & \mathfrak{S}O(n) \end{array}\right)\right\}$$

A Cartan decomposition of this is given by $\mathfrak{k}' + \mathfrak{p}$, where

$$\mathfrak{k}' = \mathfrak{S}O(n), \mathfrak{p} = \left\{\left(\begin{array}{c|c} 0 & x \\ \hline {}^t x & 0 \end{array}\right)\right\}$$

Now the geodesics are given by intersections with planes through 0; one example is: $-x_0^2 + x_1^2 = -1$; $x_j = 0$ for $j > 1$. We find the parametric representation: $x_0(t) = \cosh t, x_1(t) = \sinh t$.

We know that the isometry group $G$ of $\Sigma^+$ is transitive and also transitive on the unit tangent bundle. Hence $\Sigma^+ = G/K'$, where $K'$ is the isotropy group of $(1, 0, ..., 0)$ (with Lie-algebra $\mathfrak{k}'$); and the unit tangent bundle is $G/K$ where $K$ is the subgroup of $K'$ that fixes the above geodesic. $M = \Gamma\backslash\Sigma^+ = \Gamma\backslash G/K'$, where $\Gamma$, the fundamental group of $M$, is a discrete, uniform subgroup of $G$. It follows that the unit tangent bundle of $M$ is $\Gamma\backslash G/K$. The geodesics (lifted to the unit tangent bundle) are the orbits of the flow; and the isometry group $G$ maps geodesics to geodesics; hence is a symmetry group in the sense of Proposition 3. It follows that the flow is given as $\Gamma gK \to \Gamma g \exp(t\alpha)K$; where $\alpha \in \mathfrak{g}$ corresponds to the above geodesic; i.e. $x_0(t) = \sinh\, t$, $x_i(t) = \cosh t$, $x_j(t) = 0$ for $j > 0$ corresponds to $\exp(\alpha t)K$. Computation gives:

$$\exp(\alpha t) = \left(\begin{array}{cc|c} \cosh t & \sinh t & 0 \\ \sinh t & \cosh t & \\ \hline 0 & & I \end{array}\right); \quad \alpha = \left(\begin{array}{cc|c} 0 & 1 & 0 \\ 1 & 0 & \\ \hline 0 & & 0 \end{array}\right).$$

It is now easy to check the axioms.

$$\mathfrak{k} = \left\{\left(\begin{array}{cc|c} 0 & 0 & 0 \\ 0 & 0 & \\ \hline 0 & & \mathfrak{S}O(n-1) \end{array}\right)\right\} \mathcal{E} = \left\{\left(\begin{array}{cc|c} 0 & 0 & x \\ 0 & 0 & y \\ \hline x & {}^t\!-y & 0 \end{array}\right)\right\}$$

The only eigenvalues are $\pm 1$;

$$\text{The} + 1 \text{ eigenvectors are:} \left(\begin{array}{cc|ccccc} 0 & 0 & 0 & \ldots & 1 & \ldots & 0 \\ 0 & 0 & 0 & \ldots & 1 & \ldots & 0 \\ \hline 0 & \ldots & 0 & & & & \\ 1 & \ldots & 1 & & & 0 & \\ 0 & 0 & & & & & \end{array}\right)$$

$$\text{The} - 1 \text{ eigenvectors are:} \left(\begin{array}{cc|ccccc} 0 & 0 & 0 & \ldots & 1 & \ldots & 0 \\ 0 & 0 & 0 & -1 & \ldots & 0 \\ \hline 0 & \ldots & 0 & & & & \\ 1 & \ldots & 1 & & & 0 & \\ 0 & 0 & & & & & \end{array}\right)$$

They determine the contracting leaf and the expanding leaf respectively.

The same group also gives the geodesic flow on a certain pseudo-Riemannian manifold of constant positive curvature.

NOTE. Up to $G$, $\Gamma$-equivalence we can assume $\mathfrak{g}$ *has no center*.

PROOF.
$$k + s\alpha + e \in Z(\mathfrak{g}) \Rightarrow 0 = [k + s\alpha + e, \alpha] = [e, \alpha], \quad \text{i.e. } e = 0.$$

Also $\text{ad}_e k = - \text{ad}_e s\alpha$; since $\mathfrak{k}$ acts with imaginary eigenvalues, $s = 0$, i.e. $Z(\mathfrak{g})$ is an ideal contained in $\mathfrak{k}$, now use Proposition 1, Corollary 1.

Finally, we recall: The subalgebra $\mathcal{U}$ generated by $\mathcal{E}$ is an ideal (the unstable ideal)-see Auslander and Green. It is $\alpha$-invariant; hence $\mathbf{R}\alpha + \mathcal{U}$ is an ideal.

PROPOSITION 5. *Up to G-equivalence, we can assume* $\mathfrak{g} = \mathcal{U} + \boldsymbol{R}\alpha$.

PROOF. $G$ has only a finite number of components, and $K$ meets them all. It follows that $G_0$ must act transitively (since $G/K$ is connected). Let $U'$ be the connected, normal subgroup of $G$ corresponding to $\mathcal{U} + \boldsymbol{R}\alpha$. Then $U'$ also acts transitively, and the natural injection defines a $G$-equivalence.

AXIOM IV. The Lie algebra $\mathfrak{g}$ is spanned by $\mathcal{U}$ and $\boldsymbol{R}\alpha$.

For the remainder of this work we will assume AIV whenever convenient.

We fix the following notation: $\mathscr{E}_{\lambda,\bar{\lambda}}$ is the $\alpha$-invariant subspace of $\mathscr{E}$ corresponding to eigenvalues $(\lambda, \bar{\lambda})$ in the primary decomposition of ad $\alpha$. For $\lambda$ real, we write $\mathscr{E}_{\lambda,\lambda} = \mathscr{E}_\lambda$.

In a later chapter we will see that all $(G, \Gamma)$-induced Anosov flows have a finite, flow-invariant measure. From this it follows immediately that the nonwandering set is the whole manifold $M$. From Pugh's closing lemma, and structural stability of the Anosov flow, it follows that the periodic orbits are dense in $M$.

## 2. Anosov-flows on infrasolvmanifolds.

THEOREM II. *The suspension of an N-induced Anosov-diffeomorphism is a $(G, \Gamma)$-induced Anosov-flow with $\mathfrak{g}$ solvable.*

PROOF. Let $N, S, G, \Delta, \phi, \psi$ be given as in Chapter 1, and let $\mathfrak{N}$ be the Lie-algebra of $N$. $\mathrm{Aut}(N) = \mathrm{Aut}(\mathfrak{N})$; i.e. an algebraic linear group. Let $A(\psi)$ be the algebraic group hull of $\psi$. This constitutes the real points of an Abelian, algebraic subgroup; in particular it has a finite number of components in its Lie group topology. All automorphisms in $A(\psi)$ commute with all automorphisms in $S$ (since the set of such automorphisms form an algebraic subgroup containing $\psi$). The identity component of $A(\psi)$ is the direct product of a vector group $V$ and a compact group, hence $A(\psi)/V$ is compact; and by a well-known structure theorem, $V$ is a semi-direct factor of $A(\psi)$. (Hochschild [23, p. 39]). In this commutative case, $V$ must be a direct factor, and it follows that $A(\psi)$ is the direct product of $V$ and a compact subgroup $K'$. Hence $\psi = J'A'$, $J' \in K'$, $A' \in V$. Since $K'$ is compact, $J'^n \in (K')_0$ for some $n > 0$, hence $J'^n = \exp B$, with $B$ in the Lie-algebra of $K'$. Now let $J = J' \exp(B/n)$. Then $J^n = I$, and $\psi = J \cdot A$, where $A$ can be reached by a one-parameter group $\{\psi_t\}$ in $A(\psi)$, and $J$ commutes with all automorphisms in $S$. Hence $K = \{J^k \cdot C, C \in S\}$ is again a finite group of automorphisms of $N$, which commute with all elements of $A(\psi)$. We form the semidirect product $N \cdot_s K$. Now, $\boldsymbol{R}$ acts on $N \cdot_s K$ through the group

$$(\psi_t): \psi_t(n, k) = (\psi_t(n), k). \quad \psi_t((n_1, k_1) \cdot (n_2, k_2)) = \psi_t(n_1 \cdot k_1(n_2), k_1 k_2)$$

$$= (\psi_t(n_1) \cdot (\psi_t \circ k_1)(n_2), k_1 k_2);$$

$$\psi_t(n_1, k_1) \cdot \psi_t(n_2, k_2) = (\psi_t(n_1), k_1)(\psi_t(n_2), k_2) = (\psi_t(n_1) \cdot (k_1 \circ \psi_t)(n_2), k_1 k_2).$$

For the semidirect product:

$$(N \cdot_s K) \times_s \boldsymbol{R}: ((n_1, k_1); s_1) \cdot ((n_2, k_2); s_2) = ((n_1, k_1) \cdot \psi_{s_1}(n_2, k_2), s_1 + s_2)$$

$$= ((n_1 \cdot (k_1 \circ \psi_{s_1})(n_2), k_1 k_2); s_1 + s_2).$$

Now we have a number of natural identifications:

$N \cong N \cdot {}_s S/S \cong N \cdot {}_s K/K$ given by $n \to (n, e) \cdot S \to (n, e) \cdot K = \{(n, k), k \in K\}$. $K$ is a subgroup of $(N \cdot {}_s K) \times {}_s \mathbf{R}$; and $(N \cdot {}_s K/K) \times \mathbf{R} \cong ((N \cdot {}_s K) \times {}_s \mathbf{R})/K$ (hence $= N \times \mathbf{R} = (N \cdot {}_s S/S) \times \mathbf{R}):(n \cdot K, s) \to ((n, e); s) \cdot K$. It is easy to prove directly that $\Delta \backslash N \cdot {}_s K/K$ is a manifold. Also, we see that the action of $\Delta$ on $N \cdot {}_s K/K$ and on $N \cdot {}_s S/S$ respect the above identification. (Under the identification with $N$, the action is given as follows: let $(\delta, k) \in \Delta$, then $(\delta, k) \cdot n = \delta \cdot k(n)$.) It follows: $\Delta \backslash N \cdot {}_s S/S \cong \Delta \backslash N \cdot {}_s K/K$. Now, $M' \times \mathbf{R} = (\Delta \backslash N \cdot {}_s S/S) \times \mathbf{R} \cong (\Delta \backslash N \cdot {}_s K/K) \times \mathbf{R} \cong \Delta \backslash (N \cdot {}_s K) \times {}_s \mathbf{R}/K$. The last identification is obtained by noticing that the action of $\Delta$ on $(N \cdot {}_s K/K) \times \mathbf{R}$ and $(N \cdot {}_s K) \times {}_s \mathbf{R}/K$ respect the above identification. The flow on $M' \times \mathbf{R}$ is given by $(\Delta nS, s) \to (\Delta nS, s + t)$, i.e. on $\Delta \backslash (N \cdot {}_s K) \times {}_s \mathbf{R}/K$: $\Delta((n, k); s)K \to \Delta((n, k); s + t)K$ i.e. by right multiplication by the one-parameter-group $((e, e); t)$ (which centralizes $K$). The action of $\mathbf{Z}$ on $M' \times \mathbf{R}$ is given by the generator:

$$(\Delta(n, k)K, s) \to (\Delta(\phi(n), k)K, s + 1)$$

i.e. on $\Delta \backslash (N \cdot {}_s K) \times {}_s \mathbf{R}/K : \Delta((n, k); s)K \to \Delta((\phi(n), k), s + 1)K$. But this is precisely the left action of the cyclic group $\Sigma$ with generator $((e, J); 1)$. Now:

$$((\delta, k); 0)((e, J^s), s) = ((\delta, kJ^s); s). \quad s \in \mathbf{Z}.$$

Also $((e, J^s); s)((\delta, k); 0) = ((e, J^s)(A^s(\delta), k); s) = ((\psi^s(\delta), J^s k); s)$. Hence, since $\phi(\Delta) = \Delta : \Sigma \cdot \Delta = \Delta \cdot \Sigma = \Gamma$, which is then a uniform, discrete subgroup of $(N \cdot {}_s K) \times {}_s \mathbf{R}$; $M = \Gamma \backslash (N \cdot {}_s K) \times {}_s \mathbf{R}/K$, and the flow is given through right-multiplication by $((e, e); t)$. The Lie-algebra $\mathfrak{g} = \mathfrak{N} \oplus {}_s \mathbf{R}$, where the action of $\mathbf{R}$ by derivations on $\mathfrak{N}$ is the differential of the above action through $(\psi_t)$. Since $A = \psi_1$ has no eigenvalues on the unit circle, it follows that $\alpha \in \mathbf{R} \subseteq \mathfrak{g}$ acts with no imaginary eigenvalues on $\mathfrak{N}$. Hence all the axioms for $(G, \Gamma)$-induced Anosov flows are satisfied. $(\mathfrak{k} = (0))$.

We now assume that $G$ is a solvable Lie group, with a $(G, \Gamma)$-induced Anosov flow; and assume for simplicity ($G$-equivalence) that $G$ is connected and simply connected.

LEMMA. $\mathfrak{k}$ acts trivially on $\mathscr{E}$, i.e. is an ideal; and hence we can assume, without any loss of generality, that $\mathfrak{k} = (0)$.

PROOF. Since $\mathfrak{g}$ is solvable, $\mathfrak{g} \neq [\mathfrak{g}, \mathfrak{g}]$. It follows that $\alpha \notin \mathscr{U}$, the unstable ideal. $(= [\mathfrak{g}, \mathfrak{g}]$ in this case$)$. Let $\mathscr{U}_0$ be the 0-eigenspace of $\mathrm{ad}\, \alpha$ in $\mathscr{U} = [\mathfrak{g}, \mathfrak{g}]$. Then, for any $k \in \mathfrak{k} : k = s\alpha + u, u \in \mathscr{U}_0$. Since $[\mathfrak{g}, \mathfrak{g}]$ is nilpotent, $u$ acts with 0-eigenvalues; hence $k$ acts with the same eigenvalues as $s\alpha$. But, since $\mathfrak{k}$ is compactly imbedded in $\mathfrak{g}$, it acts semisimply with imaginary eigenvalues; and it follows $s = 0$; $\mathrm{ad}_{\mathfrak{g}} k = 0$.

Thus we can write: $\mathfrak{g} = \mathfrak{N} \oplus {}_s \mathbf{R}\alpha$; $\mathfrak{N} = \mathscr{E}$ a nilpotent Lie-algebra; and $\alpha$ an $\mathfrak{N}$-derivation without imaginary eigenvalues. Correspondingly: $G = N \times {}_s \mathbf{R}$; with $N$ the corresponding simply-connected Lie-group; and $\mathbf{R}$ acting through the corresponding hyperbolic automorphisms: $\psi_t = \exp(\mathrm{ad}\, \mathfrak{N}\alpha)$. Now $K$ is trivial; since the centralizer of $(e, t)$, $t \neq 0$, is $\mathbf{R}:((e, t)(n', t')(e, -t)) = (\psi_t(n'), t') = (n', t')$ only if $n' = e$. ($K \cap \exp(\mathbf{R}\alpha) = (e)$ by looking at Ad).

Now the limit commutator group of $G$, $G^\infty = \lim_{k \to \infty}[G, \dots [G, G] \dots ]$ is $N$. Hence it follows by a theorem of Auslander (Auslander-Green-Hahn (5)) that $\Delta = N \cap \Gamma$ is a uniform, discrete subgroup of $N$.

PROPOSITION 6. *There is a minimal positive value $t_0$ such that $(a, t_0) \in \Gamma$ for some $a \in N$.*

PROOF. Since $\Gamma$ is uniform in $G$, such values must exist. Now $\Gamma \cdot N$ is closed in $G$ by a result of Mostow [32]. $N \times {}_s R / N \cdot \Gamma$ is a compact 1-dimensional manifold (a circle), but if $t_0$ could be arbitrarily small, we would have $\Gamma \cdot N = G$. Let $Z =$ the cyclic group generated by $(a, t_0)$.

PROPOSITION 7. $\Gamma = Z \cdot \Delta = \Delta \cdot Z$.

PROOF. Let $(b, nt_0) \in \Gamma$. We have: $(a', nt_0) = (a \cdot \psi_{t_0}(a) \dots \psi_{(n-1)t_0}(a), nt_0) \in Z$. Then $(a', nt_0) \cdot (b, nt_0)^{-1} = (a', nt_0)(\psi_{-nt_0}(b^{-1}), - nt_0) = (a'b^{-1}, 0) \in \Gamma \cap N = \Delta$.

PROPOSITION 8. *Let $A = \mathrm{Ad}_N(t_0)$. Then $A(\Delta) = a^{-1}\Delta a$.*

PROOF. Let $(\delta, 0) \in \Delta$. $(a, t_0)(\delta, 0)(a, t_0)^{-1} = (a\psi_{t_0}(\delta)a^{-1}, 0)$ i.e. $\Delta = aA(\Delta)a^{-1}$.

Now; on $\Delta \backslash N \times R$ the flow is: $(\Delta n, s) \to (\Delta n, s + t)$. The identification under $\Gamma$ is: $(\Delta n, s) \to (\Delta a \cdot A(n), s + t_0)$. Hence this flow is the suspension of the diffeomorphism on $\Delta \backslash N$ induced by $n \to a \cdot A(n)$, (up to change of the time scale by a constant). Thus we are lead to examine Anosov-diffeomorphisms on nilmanifolds given by affine transformations.

Let $N$ be a simply-connected, nilpotent group, $\Delta$ a uniform discrete subgroup, $a \in N$, $A \in \mathrm{Aut}(N)$, $A$ hyperbolic. Let $\iota_a$ be the inner automorphism: $n \to ana^{-1}$, $l_a, r_a$ left- and right-translation by $a$ respectively. Then $l_a \circ A : x \to aA(x)$ is an affine transformation of $N$. Assume $A(\Delta) = \iota_a^{-1}(\Delta)$; then $l_a \circ A$ induces a well-defined diffeomorphism of $\Delta \backslash N$.

LEMMA. *The map $b \to b \cdot A(b)^{-1}$ is surjective: $N \to N$.*

PROOF. The map is injective, for $x \cdot A(x)^{-1} = y \cdot A(y)^{-1}$ implies $y^{-1}x = A(y^{-1}x)$. If $y^{-1}x = \exp z$, then $\exp(dA(z)) = \exp(z)$ i.e. $dA(z) = z$; which implies $z = 0$ by hyperbolicity of $A$. The map is given as:

$$\exp(z) \to \exp z \cdot \exp(- dA(z)) = \exp(z - dA(z) - \tfrac{1}{2}[z, dA(z)] + \dots)$$

through the Campbell-Hausdorff formula. $N$ being nilpotent, this must end with a finite number of terms; and on $\mathfrak{N}$ the map is given by $z \to z - dA(z) - \tfrac{1}{2}[z, dA(z)] + \dots$ which is an injective polynomial map. By a result of Bialynicki-Birula and Rosenlicht (10), an injective polynomial map of $R^n$ into $R^n$ is surjective.

(Actually, we could avoid using this theorem for our application. In the case of Abelian $N$ the proposition is obvious by surjectivity of $dA - I$; in the general nilpotent case one can prove surjectivity by a "ladder"-type argument as usual.)

THEOREM III. *Let $N$, $\Delta$, $a$, $A$ be as above. Then $\iota_a \circ A$ is an hyperbolic automorphism of $N$ preserving $\Delta$. The affine transformation $l_a \circ A$ induces an Anosov diffeo-*

*morphism on* $\Delta\backslash N$ *which is differentiably conjugate to the N-induced Anosov diffeomorphism from* $\iota_a \circ A$.

PROOF. Let $bA(b)^{-1} = a$. On $N$, $\iota_a \circ A$ is differentiably conjugate to $A$ through $\iota_b : (\iota_a \circ A) \circ \iota_b = \iota_b \circ A$. It follows that $\iota_a \circ A$ is a hyperbolic automorphism. Now $r_b$ induces a diffeomorphism of $\Delta\backslash N$, which gives our conjugacy: $(l_a \circ A) \circ r_b(\Delta n)$ $= (l_a \circ A)(\Delta nb) = \Delta a \cdot A(n)A(b) = \Delta a \cdot A(n)a^{-1}b = r_b(\Delta a A(n)a^{-1}) = r_b \circ (\iota_a \circ A)(\Delta n)$.

Q.E.D.

THEOREM IV. *A* $(G, \Gamma)$*-induced Anosov-flow with G solvable, connected and simply connected is, (up to change of the time scale by a constant factor), the suspension of an Anosov-diffeomorphism induced by an affine hyperbolic transformation of a simply connected nilpotent Lie-group; and is thus differentially conjugate to the suspension of an N-induced Anosov diffeomorphism. This classifies all* $(G, \Gamma)$*-induced Anosov-flows with G solvable, up to finite coverings and* $(G, \Gamma)$*-equivalence.*

As we have seen: if $G$ is disconnected, this permits suspensions of $N$-induced Anosov diffeomorphisms on infra-nilmanifolds. Conversely, any disconnected $G$ with Lie algebra $\mathfrak{g}$ is of the form $(N \times {}_sK) \times {}_s\mathbf{R}$; where $K$ is a finite group of automorphisms of $N$; $\Gamma \cap (N \times {}_sK)$ is uniform in $N \times {}_sK$; etc. We leave the details for this case as an exercise.

We remark that the question of existence of such lattice-preserving hyperbolic automorphisms of nilpotent Lie groups has recently been studied by Auslander-Scheunemann [9].

Finally, we check that the manifolds studied here have fundamental groups of exponential growths:

As we have seen, $\Gamma = Z \cdot \Delta$, where $\Delta$ is a uniform, discrete subgroup of a nilpotent group $N$; and $Z$ is the cyclic subgroup generated by $(a, 1) \in N \times {}_s\mathbf{R}$. By nilpotency we know that $\Delta$ is of polynomial growth. Now, it is clear that $\Gamma$ is polycyclic, as defined in J. A. Wolf [43]. Hence, by a theorem in the above reference: if there is no nilpotent subgroup of finite index in $\Gamma$, then $\Gamma$ must have exponential growth.

Now, assume $D$ is of finite index in $\Gamma$; then $(a, 1)^n \in D$ for some $n > 0$. Let $(a, 1)^n = (b, n)$. Let $(d, 0) \in \Delta \cap D$ with $d \neq e$. Then

$$(b, n)(d, 0)(b, n)^{-1}(d^{-1}, 0) = (b \cdot A^n(d), n)(A^{-n}(b^{-1}), -n)(d^{-1}, 0)$$
$$= (b \cdot A^n(d), n)(A^{-n}(b^{-1}) \cdot A^{-n}(d^{-1}), -n)$$
$$= ((\iota_b \circ A^n)(d) \cdot d^{-1}; 0).$$

But $\iota_b \circ A^n$ is hyperbolic; and we have seen before that $d \to (\iota_b \circ A)(d) \cdot d^{-1}$ is then injective. Hence $d_1 = (\iota_b \circ A^n)(d) \cdot d^{-1}$, and repeat the process with $(d_1, 0)$. Hence we get commutators of higher and higher orders different from the identity: this means that $D$ cannot be nilpotent.

Q.E.D.

3. **Flows with semisimple group.** The basic result of this chapter is the following theorem:

THEOREM V. (a) *Let a* $(G, \Gamma)$*-induced Anosov flow with* $\mathfrak{g}$ *semisimple be given on* *a manifold M. We can assume, up to $(G, \Gamma)$-equivalence, that $\mathfrak{g}$ is simple. We then have*: $\mathfrak{g}$ *is a real, simple Lie algebra of noncompact type with real rank* 1; *i.e. the Lie algebra of Killing vector fields of a rank* 1 *symmetric space of noncompact type, S. For dim M $\neq$ 3, it follows that the universal covering space of M is the unit tangent bundle of S, U(S), and the lift of the Anosov flow is the geodesic flow on U(S). For dim M = 3, S is the Poincaré upper half-plane with constant negative curvature. U(S) is not simply connected; but the universal covering spaces for M and U(S) coincide, and the lift of the given Anosov flow on M coincides with the lift of the geodesic flow of U(S).*

(b) *Now, specify the group G to be the isometry group of the symmetric space S. Then we can conclude that $M = \Gamma\backslash G/K$ is finitely covered by the unit tangent bundle U of a compact Clifford-Klein form of a noncompact, rank* 1 *symmetric space; and the flow is induced by the geodesic flow on U. Moreover, if the fundamental group $\Gamma$ has no torsion, then $\Gamma\backslash G/K$ is already such a unit tangent bundle. By the results of Chapter I, it then follows: All $(G, \Gamma)$-induced Anosov flows with $\mathfrak{g}$ semisimple are finitely covered by the above geodesic flows.*

PROPOSITION 9. *Up to $(G, \Gamma)$-equivalence, we can assume $\mathfrak{g}$ is simple.*

Let $\mathfrak{g} = \mathfrak{g}_1 \oplus \ldots \oplus \mathfrak{g}_n$; $\mathfrak{g}_i$ simple ideals of $\mathfrak{g}$. $\alpha = \alpha_1 + \ldots + \alpha_n$. We can assume all $\alpha_i \neq 0$. For if $\alpha_i = 0$, it is easy to see that $\mathfrak{g}_i$ is an ideal of $\mathfrak{g}$ contained in $\mathfrak{k}$. Since $\mathrm{ad}_{\mathfrak{g}_i}\alpha = \mathrm{ad}_{\mathfrak{g}_i}\alpha_i$, no $\mathfrak{g}_i$ can be of compact type. Now $\alpha_i$ is in the 0-eigenspace for $\mathrm{ad}\,\alpha$, hence $\alpha_i = k + s(\alpha_1 + \ldots + \alpha_n)$, $k \in \mathfrak{k}$. Since $\alpha_i$ acts with nonimaginary eigenvalues, $s \neq 0$. But this is a contradiction unless $n = 1$, $k = 0$.

PROPOSITION 10. $\alpha$ *acts semisimply on* $\mathfrak{g}$.

PROOF. $\alpha = \alpha_n + \alpha_s$, where $\alpha_n$, $\alpha_s$ are the nilpotent and semisimple parts of $\alpha$ respectively. Then $\alpha_n = k + s\alpha + e$, $k \in \mathfrak{k}$, $s \in \mathbf{R}$, $e \in \mathscr{E}$. Now $0 = [\alpha, \alpha_n] = [\alpha, e]$: i.e. $e = 0$, $\alpha_n = k + s\alpha$. Since $[k, s\alpha] = 0$, $\mathrm{ad}\,k$ and $\mathrm{ad}(s\alpha)$ can be simultaneously triangulated over $C$; and since $\mathrm{ad}(k + s\alpha)$ is nilpotent, the eigenvalues of $\mathrm{ad}\,k$ are the negatives of the eigenvalues of $\mathrm{ad}(s\alpha)$. Hence $k = 0$, $s = 0$, and $\alpha = \alpha_s$.

PROPOSITION 11. $\alpha$ *acts with real eigenvalues.*

PROOF. $\mathbf{R}\alpha$ is an Abelian subalgebra and acts semisimply on $\mathfrak{g}$. Let $\mathfrak{h}$ be maximal under those conditions; i.e. a Cartan subalgebra. Let $\mathfrak{g}^C = \mathfrak{k}^C + C\alpha + \mathscr{E}^C$ be the complexification of $\mathfrak{g}$. Then $\mathfrak{h}^C$ is a Cartan subalgebra of $\mathfrak{h}^C$. Now $\mathfrak{k}$ is the Lie-algebra of a compact group, i.e. $\mathfrak{k} = Z(\mathfrak{k}) + \mathfrak{k}'$, where $\mathfrak{k}'$ is semisimple of compact type. Let $'$ be a Cartan subalgebra of $\mathfrak{k}'$, then $\mathfrak{h} = \mathbf{R}\alpha + Z(\mathfrak{k}) + \mathfrak{h}'$. Take the root-space decomposition of $\mathfrak{g}^C$ with respect to $\mathfrak{h}^C$. Then $\mathscr{E}^C = \Sigma\mathfrak{g}_\gamma$, where $\gamma$ varies over the roots which are nonzero on $\alpha$. Let $(\mathfrak{h}^C)_{\mathbf{R}}$ be the real form of $\mathfrak{h}^C$ spanned by the roots (under the usual identification of $\mathfrak{h}^C$ with its dual). Then: $i(Z(\mathfrak{k}) + \mathfrak{h}') \subseteq (\mathfrak{h}^C)_{\mathbf{R}}$. By a dimension argument: $bk + c\alpha \in (\mathfrak{h}^C)_{\mathbf{R}}$ for some $k \in Z(\mathfrak{k}) + \mathfrak{h}'$; $b$, $c \in C$; $c \neq 0$: $[bk + c\alpha, e_\gamma] = (b_\gamma(k) + c_\gamma(a))e$ for $e_\gamma \in \mathfrak{g}_\gamma$, i.e. $b\gamma(k) + c\gamma(\alpha) \in \mathbf{R}$ for all roots $\gamma$, $\arg b + \pi/2 + \arg c + \arg \gamma(\alpha) = 0 \pmod \pi$. Now, the $\gamma(\alpha)$'s represent the eigenvalues of $\mathrm{ad}\,\alpha$. We know that these are closed under complex conjugation, i.e. there is a $\gamma'$ such that $\gamma'(\alpha) = \overline{\gamma(\alpha)}$ i.e. $\arg b + \pi/2 + \arg c - \arg \gamma(\alpha) = 0 \pmod \pi$. Hence $\arg \gamma(\alpha) = 0 \pmod{\pi/2}$. But by assumption $\gamma(\alpha)$ is not imaginary, hence we conclude that $\gamma(\alpha)$ is real.

COROLLARY. $\mathbf{R}\alpha$ is a maximal Abelian subalgebra under the condition that it acts semisimply with real eigenvalues; i.e. $\mathfrak{g}$ has real rank 1.

It is known that this forces $\mathfrak{g}$ to be the Lie algebra of Killing vector fields of an irreducible symmetric space of noncompact type of rank 1. Since we have not found any reference for this, we give an argument:

PROPOSITION 12. Let $\mathfrak{g}$ be a real, semisimple Lie algebra of noncompact type. Then the rank of the corresponding symmetric space equals the real rank of $\mathfrak{g}$.

PROOF. Let $\mathfrak{g} = \mathfrak{k} + \rho$ be a Cartan decomposition. By definition, the rank of the symmetric space is the dimension of a maximal, (Abelian) subalgebra of $\rho$. It is clear that this acts semisimply with real eigenvalues, and is maximal under these conditions.

Conversely, let $\mathcal{R}$ be such a subalgebra. We wish to find a Cartan decomposition such that $\mathcal{R}$ is included in the $\mathfrak{p}$-part. Extend first $\mathcal{R}$ to a Cartan subalgebra $\mathfrak{h} = \mathcal{R} + \mathfrak{S}$. In $\mathfrak{g}^C$, define a real compact form in the standard way from the root system with respect to $\mathfrak{g}^C$ (and the choice of Weyl-basis); and call the corresponding conjugation of $\mathfrak{g}^C$ for $\mu$. Also, let $\sigma$ be the conjugation of $\mathfrak{g}$ corresponding to $\mathfrak{g}$. Clearly, $\mathcal{R} \subseteq (\mathfrak{h}^C)\mathbf{R}$ is in the $(-1)$-eigenspace of $\mu$. Now the standard way to get a Cartan involution of $\mathfrak{g}$ from $\mu$ is as follows (e.g. see Helgason [22]). Let $N = \sigma\mu$. Then $N$ is Hermitian with respect to the inner product $\langle X, Y \rangle = -Tr(\text{ad } X \text{ ad } \mu Y)$. Let $N^2 = \exp(W)$, and let $\mu' = \exp(\frac{1}{4}W)\mu \exp(-\frac{1}{4}W)$. The corresponding Cartan decomposition of $\mathfrak{g}$ is $\mathfrak{g} \cap \mathfrak{g}_{\mu'} + \mathfrak{g} \cap i\mathfrak{g}_{\mu'}$. We have to show that $\mathcal{R}$ is in the $(-1)$-eigenspace for $\mu'$. Now, clearly $\mathcal{R}$ is in the 1-eigenspace for $N^2 = \sigma\mu\sigma\mu$ i.e. $\exp(W)$ is a Hermitian, positive-definite operator which is the identity on $\mathcal{R}$. Then $\mathcal{R}^C$ has an invariant, orthogonal complement on which $\exp(W)$ is also Hermitian, positive definite, so clearly the one-parameter group $\exp(tW)$ is the identity on $\mathcal{R}$, and it follows that $\mu'|_\mathcal{R} = \mu|_\mathcal{R}$, i.e. $\mathcal{R}$ is in the $(-1)$-eigenspace of $\mu'$.

We wish to study the structure of $\mathfrak{g}$ further, and in particular, relate our $(\mathfrak{k}, \alpha, \mathscr{E})$-decomposition to the usual Cartan- and Iwasawa-decompositions.

PROPOSITION 13. $\mathfrak{k}$ is contained in the compact part of a Cartan-decomposition.

LEMMA. Let $\mathfrak{g}$ be semisimple of compact type, and $\mathfrak{h}$ a Cartan subalgebra. Then, with an appropriate choice of Weyl-basis in $\mathfrak{g}^C$ relative to $\mathfrak{h}^C$; it is possible to make the canonically defined compact form of $\mathfrak{g}^C$ coincide with $\mathfrak{g}$.

This is easy to prove (for example using techniques similar to the above argument). First, notice that $Z(\mathfrak{k}) + \mathfrak{k}'$ is in the 1-eigenspace for $\mu$, i.e. in the 1-eigenspace for $N^2$, $\exp(\frac{1}{4}W)$, and for $\mu'$. Furthermore, the roots of $\mathfrak{k}'^C$ with respect to $\mathfrak{h}'^C$ correspond to roots of $\mathfrak{g}^C$ which annihilate $\alpha$. By the above lemma, we can choose the Weylbasis $e_\beta \in \mathfrak{k}_\beta'^C$ such that $e_\beta - e_{-\beta}$, $i(e_\beta + e_{-\beta})$ will be in $\mathfrak{k}' \subseteq \mathfrak{k}$. Hence it follows that $\mathfrak{k}$ is contained in $\mathfrak{g} \cap \mathfrak{g}_{\mu'}$.

Now, it is clear that the rest of the compact and noncompact part of the Cartan-decomposition are spanned by linear combinations of eigenspaces in $\mathscr{E}^+$ and $\mathscr{E}^-$. Furthermore, it follows immediately, that in the Iwasawa-decomposition of $\mathfrak{g}$, $R\alpha$ corresponds to the Abelian part, $\mathscr{E}^+$ to the nilpotent part.

Now, from Cartan's list, we specify the noncompact symmetric spaces of rank 1: These occur as follows (see Helgason [22]):

$A_n$: $\mathfrak{S}l(2, R)$; $\mathfrak{S}l(2, C)$; $\mathfrak{S}\mathfrak{U}^*(4)$ $\mathfrak{S}\mathfrak{U}(m, 1)$ where $U(m, 1)$ is the group preserving the Hermitian form: $-z_1\bar{z}_1 + z_2\bar{z}_2 + \cdots + z_{m+1}\bar{z}_{m+1}$ (complex Lorentz-group), and $SU(m, 1) = U(m, 1) \cap SL(m = 1, C)$.

$B_n$: $\mathfrak{S}O(m, 1)$ with $m$ even.

$C_n$: $\mathfrak{S}p(1, R)$
   $\mathfrak{S}p(1, m)$

$D_n$: $\mathfrak{S}O^*(4)$, $\mathfrak{S}O^*(6)$ where $SO^*(2n) = \{A \in SO(2n, C\}$; which leaves invariant the form: $-z_1\bar{z}_{n+1} + z_{n+1}\bar{z}_1 - z_2\bar{z}_{n+2} + z_{n+1}\bar{z}_2 - \cdots$
$\mathfrak{S}O(m, 1)$ with $m$ odd.

$F_4$: This exceptional structure has one real noncompact form of dim 52. The corresponding symmetric space is the Cayley plane, of dimension 16.

Now we have special isomorphisms in low dimensions:

$$\mathfrak{S}l(2, R) \cong \mathfrak{S}\mathfrak{U}(1, 1) \cong \mathfrak{S}O(2, 1) \cong \mathfrak{S}p(1, R).$$
$$\mathfrak{S}l(2, C) \cong \mathfrak{S}O(3, 1).\; \mathfrak{S}p(1, 1) \cong \mathfrak{S}O(4, 1).$$
$$\mathfrak{S}U^*(4) \cong \mathfrak{S}O(5, 1).\; \mathfrak{S}O^*(6) \cong \mathfrak{S}U(3, 1).$$
$$\mathfrak{S}O^*(4) \cong \mathfrak{S}U(2) \times \mathfrak{S}l(2, R).$$

Hence it is sufficient to consider the Lie algebras: $\mathfrak{S}O(m, 1)$, $\mathfrak{S}U(m, 1)$, $\mathfrak{S}p(m, 1)$ and the exceptional Lie algebra described above, which are the Lie algebras of Killing vector fields of real hyperbolic space, complex hyperbolic space, quaternionic hyperbolic space, and the Cayley hyperbolic plane, respectively.

From the curvature formula for symmetric spaces in terms of the Lie algebra bracket, it follows that these spaces are exactly those which have strictly negative curvature; moreover, by compactness of the Grassmannian and homogeneity, the curvature is bounded away from 0. Moreover, these are the nonflat, noncompact, two-point homogeneous Riemannian spaces; i.e. the isometry group acts transitively on the unit tangent bundle. (See Helgason [22, p. 211, 355], Wolf [43, p. 289–296].)

Now let $\mathfrak{g} = \mathfrak{f}' + \mathfrak{p}$ be a Cartan decomposition as above, i.e. $\mathfrak{f} \subseteq \mathfrak{f}'$, $\alpha \in \mathfrak{p}$. Let $G$ be the connected component of the isometry group of the corresponding symmetric space $G/K'$; where $G/K'$ must be simply connected and $K'$ connected. Since $G$ acts transitively on the unit tangent bundle, it is clear that this is $G/K''$, where $K''$ is the isotropy group of the tangent vector at $\bar{e} = K'$ given by the curve $\exp(t\alpha)K'$. We claim that $K''$ is connected. This is clear for all cases except dimension 2 for $G/K'$, by simple connectivity of the unit tangent bundle. But by inspection the exceptional case reduces to the Poincaré upper half-plane with $G = SO(2, 1)_0$. In that case we check directly that $K'' = (e)$.

PROPOSITION 14. *The Lie-algebra of $K''$ is $\mathfrak{k}$.*

PROOF. Let $k \in \mathfrak{k}''$. Then

$$\exp(sk)\exp(t\alpha)K' = \exp(t\alpha)K', \forall s, \forall t,$$

i.e. $\exp(-s\,\mathrm{Ad}(\exp t\alpha)(k)) \in K', \forall s, \forall t, \mathrm{Ad}(\exp t\alpha)(k) \in \mathfrak{k}', \forall t, [\alpha,\,k] \in \mathfrak{k}'$. But $[\alpha, k] \subseteq \mathfrak{p}$; hence $[\alpha,\,k] = 0$ and $k \in \mathfrak{k}$.

Any other choice of group $K$ with Lie-algebra $\mathfrak{k}$ will give a space $G/K$ covered by $G/K''$. By the formula for geodesics of a symmetric space, together with the methods of Chapter 1, it is also clear that the flow on $G/K''$ must be the geodesic flow.

For (b), we now assume that the $G$-induced flow is given with $G$ the full isometry group of the symmetric space. Then $G/K$; with $K$ the isotropy subgroup of $\exp(t\alpha)K'$ as before, is the unit tangent bundle, and must cover the manifold we get by different choices of $K$. Now, it is proved in Borel [11] that $G$ always has a uniform, discrete subgroup $\Gamma$. Notice that $\Gamma\backslash G/K'$ is a manifold (i.e. a compact Clifford-Klein form of $G/K'$) iff $\Gamma$ has no torsion: In fact, any finite subgroup of $\Gamma$ can be conjugated into the maximal compact subgroup $K'$; and vice versa, if $\Gamma$ intersects a conjugate of $K'$ nontrivially, the intersection must be a finite subgroup of $\Gamma$, i.e. a torsion subgroup. We now refer to a result due to Selberg and Borel (Borel [11]): Any discrete uniform subgroup $\Gamma$ of $G$ has a proper, normal, torsion-free subgroup of finite index $\Gamma_1$. Hence, if $\Gamma$ satisfies our axioms, taking $\Gamma_1$ as above, we get that $\Gamma_1\backslash G/K$ is a finite covering of $\Gamma\backslash G/K$, with $\Gamma_1\backslash G/K$ the unit tangent bundle of the compact Clifford-Klein form $\Gamma_1\backslash G/K'$.

This finishes the discussion of Theorem V.

Notice that it follows by known results of Milnor, and J. A. Wolf, that the groups $\Gamma$ as above must have exponential growth.

4. **Structure of $(G, \Gamma)$-induced Anosov flows.** Let $G$, $\mathfrak{g}$, $K$, $\mathfrak{k}$, $\Gamma$, $\alpha$ be as usual. We first notice: For $k \in \mathfrak{k}$, the flow-generator $\alpha' = \alpha + k$ defines the same flow on $\Gamma\backslash G/K$ as $\alpha$. Moreover, if $k$ is in the center of $\mathfrak{k}$, the system $G$, $\mathfrak{g}$, $K$, $\mathfrak{k}$, $\Gamma$, $\alpha + k$ will satisfy the same axioms. We call this a permissible change of flow-generator. We will need the following:

LEMMA. *Let $\mathfrak{g} = \mathfrak{k} + \mathbf{R}\alpha + \mathscr{E}$ be as usual, with $\mathfrak{g}$ simple. Then, for each eigenvalue $\lambda \neq 0$, we can choose eigenvectors $e_\lambda \in \mathscr{E}_\lambda$; $e_{-\lambda} \in \mathscr{E}_{-\lambda}$ such that $[e_\lambda, e_{-\lambda}] = 2\alpha/\lambda$; i.e. $\alpha$ can be embedded in an $\mathfrak{Sl}(2, \mathbf{R})$ subalgebra of $\mathfrak{g}$.*

This fact could be checked by a case by case inspection; one can also find a direct proof.

THEOREM VI. (a) *Let $\mathfrak{g} = \mathfrak{k} + \mathbf{R}\alpha + \mathscr{E}$ satisfy AI, AII, and Proposition 1 from Chapter 1; with $\mathfrak{k}$ compactly embedded; and let $\mathfrak{g}$ be nonsolvable. Then we can, possibly after a permissible change of flow-generator, and a reduction to the subalgebra generated by the unstable ideal and the flow-generator (i.e. assuming AIV), decompose $\mathfrak{g}$ as follows: $\mathfrak{g} = \mathscr{R} \oplus {}_s\mathfrak{S}$; where $\mathscr{R}$ is the radical of $\mathfrak{g}$, $\mathfrak{S}$ a Levi subalgebra with $\alpha \in \mathfrak{S}$, $\mathfrak{k} \subseteq \mathfrak{S}$.*

*Moreover, $\mathcal{R}$ is nilpotent and the systems* $(\mathfrak{S}, \alpha, \mathfrak{k})$, $(\mathbf{R}\alpha + \mathcal{R}, \alpha, (0))$ *satisfy* AI, AII, AIV.

(b) *From this it follows*: *For a G-induced Anosov flow on a manifold M, the lift to the universal covering manifold $\tilde{M}$ is given by*: $\tilde{M} = G/K$, $G = R \times_s S$, *where S is simple of real rank* 1, *R is simply connected, nilpotent*; *and K is a connected subgroup of S. The flow is given by* $gK \to g(\exp t\alpha)K$. *Restricting to the invariant submanifold $S/K$, we get a subflow of the type discussed in* §3.

The proof is obtained through a sequence of observations on the action of $\alpha$.

LEMMA 1. $\alpha \notin \mathcal{R}$.

PROOF. If $\alpha \in \mathcal{R}$, it follows $[\alpha, \mathfrak{g}] \subseteq \mathcal{R}$, hence $\mathscr{E} \subseteq \mathcal{R}$. If $\alpha$ and the unstable ideal are contained in $\mathcal{R}$, $\mathfrak{g} = \mathcal{R}$ (AIV), which is a contradiction.

LEMMA 2. *Let $\mathfrak{S}'$ be a Levi complement of $\mathcal{R}$, let $\pi$ be the homomorphism*: $\mathfrak{g} \to \mathfrak{g}/\mathcal{R} \cong \mathfrak{S}'$. *Then* $(\mathfrak{S}', \pi(\mathfrak{k}), \pi(\alpha))$ *satisfies* AI, AII, AIV.

PROOF. Clearly $\pi(\mathfrak{k})$ is a compactly embedded subalgebra that centralizes $\pi(\alpha)$. In $\mathscr{E}_{\lambda,\bar{\lambda}}$; choose a basis such that the first basis vectors span the $\alpha$-invariant subspace $\mathscr{E}_{\lambda,\bar{\lambda}} \cap \mathcal{R}$. Then the matrix of ad $\alpha$ is

$$\begin{pmatrix} A_{\lambda\bar{\lambda}} & * \\ 0 & B_{\lambda\bar{\lambda}} \end{pmatrix}$$

and the matrix of ad $\pi(\alpha)$ on $\pi(\mathscr{E}_{\lambda,\bar{\lambda}})$ is $B_{\lambda,\bar{\lambda}}$. It follows that $\pi(\mathscr{E}_{\lambda,\bar{\lambda}})$ is the $(\lambda, \bar{\lambda})$-eigenspace of ad $\pi(\alpha)$; and that the kernel of $\pi|_{\mathscr{E}_{\lambda,\bar{\lambda}}}$ is $\mathscr{E}_{\lambda,\bar{\lambda}} \cap \mathcal{R}$. From $\mathfrak{g}$ nonsolvable, we see $\mathscr{E} \nsubseteq \mathcal{R}$; hence $\pi(\alpha)$ acts with nontrivial, nonimaginary eigenvalues on $\pi(\mathscr{E})$; $\pi(\alpha) \notin \pi(\mathfrak{k})$; and AI, AII, AIV are verified immediately.

It now follows: $\mathfrak{S}'$ is a simple Lie algebra of real rank 1.

LEMMA 3. $\mathcal{R} \cap (\mathfrak{k} + \mathbf{R}\alpha) = \mathcal{R} \cap \mathfrak{k} \subseteq Z(\mathfrak{k})$.

PROOF. Assume $s\alpha + k \in (\mathbf{R}\alpha + \mathfrak{k}) \cap \mathcal{R}$, $s \neq 0$; and let $\lambda \neq 0$ be an eigenvalue for $\pi(\alpha)$. Then $\mathrm{ad}(s\alpha + k)$ maps $\mathscr{E}_\lambda$ bijectively onto $\mathscr{E}_\lambda \cap \mathcal{R}$, which is a contradiction. If $k \in \mathfrak{k} \cap \mathcal{R}$, $k' \in \mathfrak{k}$, we have: $[k, k'] \in \mathfrak{k} \cap [\mathcal{R}, \mathfrak{g}]$, i.e. $[k, k']$ is an element of $\mathfrak{k}$ which acts nilpotently; hence $[k, k'] = 0$.

LEMMA 4. *After a permissible change of flow generator, and reduction to the subalgebra generated by the unstable ideal and the flow-generator*; *we may assume*: $\mathfrak{g} = [\mathfrak{g}, \mathfrak{g}]$.

PROOF. *There exist* $e_\lambda \in \mathscr{E}_\lambda$, $e_{-\lambda} \in \mathscr{E}_{-\lambda}$ such that $\pi(\alpha) = [\pi(e_\lambda), \pi(e_{-\lambda})]$; i.e. $[e_\lambda, e_{-\lambda}] = \alpha + k$; $k \in \mathcal{R} \cap \mathfrak{k}$. Then $\alpha \to \alpha + k$ is a permissible change of flow-generator; and $\alpha + k$ is in the unstable ideal, $\mathscr{U}$. We can then reduce to $\mathscr{U}$; and obviously $[\mathscr{U}, \mathscr{U}] = \mathscr{U}$.

LEMMA 5. *The radical $\mathcal{R}$ is nilpotent.*

PROOF. Consider the adjoint action of $\mathcal{R}^C$ on itself; use Lie's theorem to find an invariant flag. The diagonal elements in the corresponding triangulization are the roots of $\mathscr{R}$, $\phi_1, \ldots, \phi_r$. Then the nilradical $\mathfrak{N}$ is the intersection of the kernels of the roots. Then the roots span $(\mathscr{R}/\mathfrak{N})^*$. The adjoint group $G$ acts on $\mathscr{R}/\mathfrak{N}$; and on $(\mathscr{R}/\mathfrak{N})^*$ by the contragradient representation. This action must permute the roots; since $G$ is connected, it is trivial on the set of roots; hence on $(\mathscr{R}/\mathfrak{N})^*$ and $\mathscr{R}/\mathfrak{N}$. It follows that $\mathfrak{g}$ acts trivially on $\mathscr{R}/\mathfrak{N}$; $\mathfrak{g}/\mathfrak{N}$ is a reductive Lie algebra with semisimple part isomorphic to $\mathfrak{g}/\mathscr{R}$. Hence $\mathfrak{g}/\mathfrak{N} = [\mathfrak{g}/\mathfrak{N}, \mathfrak{g}/\mathfrak{N}] \cong \mathfrak{g}/\mathscr{R}$; i.e. $\mathscr{R} = \mathfrak{N}$.

LEMMA 6. $\mathfrak{g}$ *is an algebraic Lie algebra.*

In fact, it is known that the derived algebra of any Lie algebra is algebraic (assuming characteristic 0, of course). See Chevalley [16, p. 177]. Also, one can prove that any Lie algebra with nilpotent radical is algebraic.

LEMMA 7. $\alpha$ *is a semisimple element.*

PROOF. $\alpha$ has a Jordan decomposition $\alpha = \alpha_s + \alpha_n$ with $[\alpha, \alpha_n] = 0$; i.e. $\alpha_n = s\alpha + k$; by nilpotency $\alpha_n$ must be 0 as usual.

Since $\mathscr{R}$ is the nilradical, $\mathscr{R} \cap \mathfrak{k} = (0)$. With $e_\lambda, e_{-\lambda}$ as before, we now have: $[\alpha, e_\lambda] = \lambda e_\lambda$, $[\alpha, e_{-\lambda}] = -\lambda e_{-\lambda}$, $[e, e_{-\lambda}] = \alpha$. Hence $\{\alpha, e_\lambda, e_{-\lambda}\}$ span an $\mathfrak{Sl}(2, \mathbf{R})$ subalgebra of $\mathfrak{g}$; and by a corollary of the Levi-Malcev theorem (Bourbaki [14]), it can be embedded in a Levi subalgebra.

We remark briefly on the $(G, \Gamma)$-induced Anosov flows in dimension 3. Let $\mathfrak{g} = \mathfrak{k} + \mathbf{R}\alpha + \mathscr{E}$ be as usual, with dim $\mathscr{E} = 2$. If $\mathscr{E}$ is Abelian; we have the solvable case; i.e. up to a finite covering we have the suspension of a toral diffeomorphism. In the general case, if $\mathscr{E}$ is nonabelian, the eigenvalues of ad $\alpha$ must be $\pm\lambda$, hence real. Since $\mathfrak{k}$ preserves the one-dimensional eigenspaces, it must be trivial. Hence $[\mathscr{E}_\lambda, \mathscr{E}_{-\lambda}] = \mathbf{R}\alpha$, $\mathfrak{g}$ is isomorphic to $\mathfrak{Sl}(2, \mathbf{R})$; and we have a generalized geodesic flow of a surface of constant negative curvature. In particular, we have verified Margullis' result on exponential growth of the fundamental group of 3-dimensional manifolds with an Anosov flow. Actually, a careful analysis of the discrete subgroups will show that this holds also for arbitrary $(G, \Gamma)$-induced Anosov flows.

We are now ready to demonstrate that the preceding theory leads to new classes of Anosov flows; and give the detailed construction of one such example here. The natural way to proceed is to let a simple Lie group $S$ of real rank 1 with uniform discrete subgroup $\Gamma$ and flow-generator $\alpha$ as in §3, act on Euclidean space $E$. The action should be such that $\alpha$ acts without imaginary eigenvalues on $E$; and moreover, that $\Gamma$ preserves a lattice $\Delta$ in $E$. Taking semidirect products, $G = E \times_s S$; it is then clear that $\alpha$ defines an Anosov flow as in §1; $\Gamma$ normalizes $\Delta$, and $\Gamma \cdot \Delta$ is a uniform, discrete subgroup of $G$. It is also easily checked that no conjugate of $\Gamma \cdot \Delta$ intersects $K$ nontrivially.

Thus we are led to consider the compactness criterion for arithmetic subgroups of $S$. From this theory we will need the following: Let $G$ be an affine algebraic group defined over $\mathbf{Q}$. A subgroup $\Gamma$ of $G$ is called an arithmetic subgroup if there exists a faithful $\mathbf{Q}$-representation $\rho: G \to \mathrm{GL}_n$ such that $\rho(\Gamma)$ is commensurable with $\rho(G) \cap \mathrm{GL}_n(\mathbf{Z})$. The same condition is then fulfilled for every faithful $\mathbf{Q}$-representation of $G$. Using the fact that isogenies map arithmetic groups to arithmetic groups; one can see that if $\rho$ is an isogeny of $G$ into $\mathrm{GL}_n$, defined over $\mathbf{Q}$;

and $\rho(\Gamma)$ is commensurable with $\rho(G) \cap \mathrm{GL}(n, \mathbf{Z})$, then $\Gamma$ is an arithmetic subgroup of $G$. An arithmetic subgroup $\Gamma$ of $G_{\mathbf{R}}$ is always discrete in $G_{\mathbf{R}}$. By the Godement conjecture, for reductive $G$, $\Gamma \backslash G_{\mathbf{R}}$ is compact if and only if $G$ is anisotropic over $\mathbf{Q}$; i.e. if it has no $\mathbf{Q}$-split torus $S \neq \{e\}$. One main example which we will use is $SO(F)$; where $F$ is a nondegenerate quadratic form on a $\mathbf{Q}$-vector space $V$ with rational coefficients. Then $SO(F)$ is anisotropic over $\mathbf{Q}$ if and only if $F$ does not represent 0 over $\mathbf{Q}$.

Consider the quadratic form on $M = \mathbf{R}^3 : n_0 x_0^2 - n_1 x_1^2 - n_2 x_2^2$; where $n_0, n_1, n_2$ are positive, rational integers. The rational structure is defined by the natural basis $(e_0, e_1, e_2)$. Let $S'$ be the corresponding special orthogonal group. This is an algebraic matrix group defined over $\mathbf{Q}$, whose set of real points is isomorphic to the group $SO(2; 1)$. Construct the corresponding Clifford algebra $C$. This has basis: $1, \mathbf{e}_0, \mathbf{e}_1, \mathbf{e}_2, \mathbf{e}_0\mathbf{e}_1, \mathbf{e}_0\mathbf{e}_2, -n_0\mathbf{e}_1\mathbf{e}_2, \mathbf{e}_0\mathbf{e}_1\mathbf{e}_2$, with the following multiplication table:

| | 1 | $\mathbf{e}_0$ | $\mathbf{e}_1$ | $\mathbf{e}_2$ | $\mathbf{e}_0\mathbf{e}_1$ | $\mathbf{e}_0\mathbf{e}_2$ | $-n_0\mathbf{e}_1\mathbf{e}_2$ | $\mathbf{e}_0\mathbf{e}_1\mathbf{e}_2$ |
|---|---|---|---|---|---|---|---|---|
| 1 | 1 | $\mathbf{e}_0$ | $\mathbf{e}_1$ | $\mathbf{e}_2$ | $\mathbf{e}_0\mathbf{e}_1$ | $\mathbf{e}_0\mathbf{e}_2$ | $-n_0\mathbf{e}_1\mathbf{e}_2$ | $\mathbf{e}_0\mathbf{e}_1\mathbf{e}_2$ |
| $\mathbf{e}_0$ | $\mathbf{e}_0$ | $n_0$ | $\mathbf{e}_0\mathbf{e}_1$ | $\mathbf{e}_0\mathbf{e}_2$ | $n_0\mathbf{e}_1$ | $n_0\mathbf{e}_2$ | $-n_0\mathbf{e}_0\mathbf{e}_1\mathbf{e}_2$ | $n_0\mathbf{e}_1\mathbf{e}_2$ |
| $\mathbf{e}_1$ | $\mathbf{e}_1$ | $-\mathbf{e}_0\mathbf{e}_1$ | $-n_1$ | $\mathbf{e}_1\mathbf{e}_2$ | $n_1\mathbf{e}_0$ | $-\mathbf{e}_0\mathbf{e}_1\mathbf{e}_2$ | $n_0n_1\mathbf{e}_2$ | $n_1\mathbf{e}_0\mathbf{e}_2$ |
| $\mathbf{e}_2$ | $\mathbf{e}_2$ | $-\mathbf{e}_0\mathbf{e}_2$ | $-\mathbf{e}_1\mathbf{e}_2$ | $-n_2$ | $\mathbf{e}_0\mathbf{e}_1\mathbf{e}_2$ | $n_2\mathbf{e}_0$ | $-n_2\mathbf{e}_1$ | $-n_2\mathbf{e}_0\mathbf{e}_1$ |
| $\mathbf{e}_0\mathbf{e}_1$ | $\mathbf{e}_0\mathbf{e}_1$ | $-n_0\mathbf{e}_1$ | $-n_1\mathbf{e}_0$ | $\mathbf{e}_0\mathbf{e}_1\mathbf{e}_2$ | $n_0n_1$ | $-n_0\mathbf{e}_1\mathbf{e}_2$ | $n_0n_1\mathbf{e}_0\mathbf{e}_2$ | $n_0n_1\mathbf{e}_2$ |
| $\mathbf{e}_0\mathbf{e}_2$ | $\mathbf{e}_0\mathbf{e}_2$ | $-n_0\mathbf{e}_2$ | $-\mathbf{e}_0\mathbf{e}_1\mathbf{e}_2$ | $-n_2\mathbf{e}_0$ | $n_0\mathbf{e}_1\mathbf{e}_2$ | $n_0n_2$ | $-n_0n_2\mathbf{e}_0\mathbf{e}_1$ | $-n_0n_2\mathbf{e}_1$ |
| $-n_0\mathbf{e}_1\mathbf{e}_2$ | $-n_0\mathbf{e}_1\mathbf{e}_2$ | $-n_0\mathbf{e}_0\mathbf{e}_1\mathbf{e}_2$ | $-n_0n_1\mathbf{e}_2$ | $n_0n_2\mathbf{e}_1$ | $-n_0n_1\mathbf{e}_0\mathbf{e}_2$ | $n_0n_2\mathbf{e}_0\mathbf{e}_1$ | $-n_0^2n_1n_2$ | $n_0n_1n_2\mathbf{e}_0$ |
| $\mathbf{e}_0\mathbf{e}_1\mathbf{e}_2$ | $\mathbf{e}_0\mathbf{e}_1\mathbf{e}_2$ | $n_0\mathbf{e}_1\mathbf{e}_2$ | $n_1\mathbf{e}_0\mathbf{e}_2$ | $-n_2\mathbf{e}_0\mathbf{e}_1$ | $n_0n_1\mathbf{e}_2$ | $-n_0n_2\mathbf{e}_1$ | $n_0n_1n_2\mathbf{e}_0$ | $-n_0n_1n_2$ |

The even-dimensional part of the Clifford algebra, $C^+$, is a quaternion algebra. In the notation of O'Meara [36], this is the quaternion algebra $(\alpha, \beta)$, with $\alpha = n_0 n_1$, $\beta = n_0 n_2$. The pure quaternions, $(\alpha, \beta)^0$, (spanned by $\mathbf{e}_0\mathbf{e}_1$, $\mathbf{e}_0\mathbf{e}_2$, $-n_0\mathbf{e}_1\mathbf{e}_2$) is the quadratic space $\langle n_0 n_1 \rangle \perp \langle n_0 n_2 \rangle \perp \langle -n_0^2 n_1 n_2 \rangle$. Then the following statements are equivalent (O'Meara, 57.9).

(i) $(\alpha, \beta)$ is not a division algebra;

(ii) $(\alpha, \beta)$ is isotropic;

(iii) The pure quaternions, $(\alpha, \beta)^0$, are isotropic;

(iv) $\langle \alpha \rangle \perp \langle \beta \rangle$ represents 1.

Here we get $\langle n_0 n_1 \rangle \perp \langle n_0 n_2 \rangle \perp \langle -n_0^2 n_1 n_2 \rangle$ is isotropic if and only if $\langle -n_0 n_2 \rangle \perp \langle -n_0 n_1 \rangle \perp \langle 1 \rangle$ is isotropic. This is equivalent to the condition that $\langle 1 \rangle \perp \langle -n_0 n_2 \rangle$ represent $n_0 n_1$; i.e. $\exists \xi, \eta$ such that $n_0 n_1 = \xi^2 - n_0 n_2 \eta^2$; i.e. $-n_1 - n_2 \eta^2 + n_0(\xi/n_0)^2 = 0$.

$S'$ is anisotropic over $\mathbf{Q}$ if and only if the quadratic diophantine equation $n_0 x_0^2 - n_1 x_1^2 - n_2 x_2^2$ has no integer (rational) solutions. It is well known from number theory that such triples $(n_0, n_1, n_2)$ exist; and we choose one (e.g. $n_0 = 3$, $n_1 = 1$, $n_2 = 1$). From the above calculation, it then follows that $(\alpha, \beta)^0$ is anisotropic over $Q$; i.e. $C^+$ is a division algebra over $Q$.

We now refer to some results in Chevalley [18]. The special Clifford group $\Gamma^+$ is the group of invertible elements $s$ of $C^+$ such that $sMs^{-1} \subseteq M$. In our case, $\Gamma^+$ is the group of all invertible elements in $C^+$ (e.g. if $N$ is the quaternion algebra norm, $s = \xi_0 + \xi_1 \mathbf{e}_0\mathbf{e}_1 + \xi_2 \mathbf{e}_0\mathbf{e}_2 - \xi_3 n_0\mathbf{e}_1\mathbf{e}_2$ with $Ns \neq 0$; we have: $se_0 s^{-1}$ $= (Ns)^{-1}(\xi_0 + \xi_1\mathbf{e}_0\mathbf{e}_1 + \xi_2\mathbf{e}_0\mathbf{e}_2 - \xi_3 n_0\mathbf{e}_1\mathbf{e}_2)\,(\xi_0\mathbf{e}_0 - \xi_1 n_0\mathbf{e}_1 - \xi_2 n_0\mathbf{e}_2$ $+ \xi_3 n_0\mathbf{e}_0\mathbf{e}_1\mathbf{e}_2$, where the $\mathbf{e}_0\mathbf{e}_1\mathbf{e}_2$-terms will cancel). It is easily checked that the norm, defined in Chevalley [18, p. 52], when restricted to $C^+$, is the usual quaternion norm. Thus, the reduced Clifford group $\Gamma_0^+$ is the group of unit quaternions. It acts on the space $M$ by inner conjugation, and this representation $\chi$ maps $\Gamma_0^+$ onto a subgroup of $S'$. If we specify the field to be the reals; it follows that $S/\chi(\Gamma_0^+)$ $= Z_2$; i.e. $\chi(\Gamma_0^+) = S_0' = \mathrm{SL}(2, R)/\{\pm I\}$, and $\Gamma_0^+ = \mathrm{SL}(2, R)$. Since $C^+$ is a division algebra over $Q$, $\Gamma_0^+$ is anisotropic over $Q$ (alternatively: $\chi$ is an isogeny onto $S_0'$). Also, in this case the invertible elements of $C$, $C^*$, form a real Lie group with Lie algebra $C$; in fact, we can view these as a linear Lie group, resp. Lie algebra on $C$, where the action is defined by left multiplication. Here, the Lie product in $C$ is the commutator product.

The Lie algebra of $\Gamma^+$ is $C^+$; and in this case we can restrict the action to $C^+$. Then the Lie algebra of $\Gamma_0^+$ is the space of pure quaternions; which is isomorphic to $\mathfrak{Sl}(2, R)$. The element $\mathbf{e}_0\mathbf{e}_1$ spans a real Cartan subalgebra:

$$[\mathbf{e}_0\mathbf{e}_1, \mathbf{e}_0\mathbf{e}_2] = -2n_0\mathbf{e}_1\mathbf{e}_2, \quad [\mathbf{e}_0\mathbf{e}_1, -n_0\mathbf{e}_1\mathbf{e}_2] = 2n_0\mathbf{e}_0\mathbf{e}_2;$$

hence the eigenvalues of ad $\mathbf{e}_0\mathbf{e}_1$ on the space of pure quaternions are: $0, \pm 2(n_0 n_1)^{1/2}$. It follows that we can take $\mathbf{e}_0\mathbf{e}_1$ as the flow-generator; and it remains only to check the eigenvalues on $C^+$.

The matrix of $\mathbf{e}_0\mathbf{e}_1$ acting on $C^+$, expressed in the basis $1$, $\mathbf{e}_0\mathbf{e}_1$, $\mathbf{e}_0\mathbf{e}_2$, $-n_0\mathbf{e}_1\mathbf{e}_2$ is

$$\begin{pmatrix} 0 & n_0 n_1 & 0 & 0 \\ 1 & 0 & 0 & 0 \\ 0 & 0 & 0 & n_0 n_1 \\ 0 & 0 & 1 & 0 \end{pmatrix}$$

Hence the eigenvalues are $\pm(n_0 n_1)^{1/2}$, and this completes the proof. Thus, we have constructed an Anosov flow of "mixed type" on a compact 7-dimensional manifold. It is also clear that by taking multiples of this representation, we get examples on manifolds of dimension $3 + 4n$.

5. **Spectral properties of $(G, \Gamma)$- induced Anosov flows.** We let $(G, \mathfrak{g}, K, \mathfrak{k}, \Gamma, \alpha)$ be as in §1; i.e. AI–AIV are satisfied. For simplicity, we also assume that $\Gamma$ acts effectively on $G/K$ (see Proposition 1). Then the map $\Gamma\backslash G \to \Gamma\backslash G/K$ is a fibration with fibre $K$.

PROPOSITION 15. *G is unimodular.*

PROOF. Since $G/G_0$ is finite, we may assume $G$ connected, using the following result in Nachbin [35]:

LEMMA. *Let $G$, $H$ be locally compact groups, and $\pi: G \to H$ a continuous, open homomorphism of $G$ onto $H$. If $H$ has only the trivial real continuous character, then $G$ is unimodular if and only if the kernel of $\pi$ is unimodular.*

Now, in the nonsolvable case, we then have (§4): $G = [G, G]$; and hence the modular function is the trivial character.

For solvable $G$, unimodularity is proved by the following result:

LEMMA. *If $G$ is solvable, $|\det \mathrm{Ad}(g)| = 1$ for all $g \in G$.*

PROOF. This is clear if $g$ is in the nilradical $N$. Now, by results of L. Auslander and C. C. Moore, there exists a subgroup $\Gamma' \subseteq \Gamma$ such that $\Gamma' = \exp C$, where $C$ is a lattice in the Lie algebra $\mathfrak{N}$ of $N$, and the original hyperbolic automorphism $A$ preserves $\Gamma'$ (see §2); i.e. Ad $A$ preserves $C$, and hence has an integral matrix relative to a basis that spans $C$ over the ring of integers. It follows that $|\det \mathrm{Ad}\, A| = 1$. Now, $\det(\exp(t \, \mathrm{ad}\, \alpha))$ is a real one-parameter group which takes values $\pm 1$ for integer $t$; i.e. $\det(\exp(t \, \mathrm{ad}\, \alpha)) \equiv 1$. Since $G = N \times_s R$, the lemma follows, and this finishes the proof of the proposition.

From standard theory on invariant integrals on coset spaces, it now follows that $\Gamma\backslash G$ has a right $G$-invariant integral for any discrete subgroup $\Gamma$. When $\Gamma$ is uniform, as here, this integral must be finite. Actually, for the work in this chapter we could replace AIII by the requirement of finite measure on $\Gamma\backslash G$. In the solvable case this is equivalent to compactness; but in the semisimple case we have already noticed that all arithmetic subgroups give finite measure coset spaces.

The fibration $\Gamma\backslash G \to \Gamma\backslash G/K$ now gives a finite, flow-invariant measure on $\Gamma\backslash G/K$; and the space of square integrable functions on $\Gamma\backslash G/K$ corresponds to the closed subspace of $K$-invariant square-integrable functions on $\Gamma\backslash G$.

In general, let $\{\phi_t\}$ be a measure-preserving flow on a finite measure space $M$. We recall briefly the connection between ergodic properties and representation theory. The flow is ergodic if and only if $M$ is indecomposable; i.e. for any flow-invariant, measurable set $P$, the measure of $P$ or its complement is 0. It is weakly mixing if

$$\lim_{T \to \infty} T^{\perp} \int_0^T \left| \mu(\phi_t(A) \cap B) - \frac{\mu(A)\mu(B)}{\mu(M)} \right| dt = 0$$

for all measurable sets $A$, $B$; and strongly mixing if, similarly,

$$\lim_{t \to \infty} \mu(\phi_t(A) \cap B) = \frac{\mu(A)\mu(B)}{\mu(M)}.$$

Define the strongly continuous one-parameter group $U_t$ of unitary operators on the Hilbert-space $L^2(M): U_t(f)(x) = f(\phi_{-t}(x))$. By the Hille-Yosida theory $U_t$ has a skew-adjoint, infinitesimal generator $A$. The spectrum of the flow is the spectrum of $A$, and can be defined through Stone's theorem: $U_t = \int_{-\infty}^{\infty} e^{it\lambda} dE(\lambda)$; where $E(\lambda)$ is a projection-valued spectral measure. We recall that if $\psi$ is in the domain of $A$: $A_\psi = \lim_{t\to 0}(U_t\psi - \psi)/t$. It follows that $A_\psi = i\lambda\psi$ if and only if $U_t\psi = e^{it\lambda}\psi$.

We have the following criteria for ergodic properties in terms of spectral invariants: The flow is ergodic if and only if the eigenvalue 0 has multiplicity one in the spectrum; i.e. the representation $t \to U_t$ contains the trivial representation exactly once. The flow is weakly mixing if and only if the point spectrum contains only 0 with multiplicity one; (alternatively: if and only if all finite-dimensional, invariant subspaces of $L^2(M)$ reduce to the constant functions). By definition, the flow has absolutely continuous spectrum if and only if the spectrum, except for the value 0, is absolutely continuous with respect to Lebesgue measure. Then absolutely continuous spectrum implies strong mixing.

We notice that all these properties of the spectrum are inherited from the flow on $\Gamma\backslash G : \Gamma g \to \Gamma g(\exp t\alpha)$ to the flow on $\Gamma\backslash G/K : \Gamma gK \to \Gamma g(\exp t\alpha)K$. Hence, we study only the flow on $\Gamma\backslash G$. The crucial advantage we obtain by this is that we have extra symmetry: the representation $U_t$ is the restriction of a strongly continuous unitary representation of $U$ of $G$; and $G_0$ acts transitively on $\Gamma\backslash G$. From this we can immediately conclude

PROPOSITION 16. *If* $\psi \in L_2(\Gamma\backslash G)$ *and* $U(g)\psi = \psi$ *for all* $g \in G_0$, *then* $\psi$ *is a constant function.*

As a side remark we notice that, in the nonsolvable case, since $\mathfrak{g}$ is algebraic; the corresponding group $G$ can be seen to be of type I. We conjecture that this is also true for the solvable groups $N \times_s R$ treated here. Type I reduces, via direct integral decompositions, many questions about general unitary representations to the irreducible unitary representations.

Now, the key lemma for proving ergodicity and weak mixing is the following: (Auslander and Green [6])

LEMMA. *If* $U(\exp t\alpha)\psi = e^{i\lambda t}\psi$, *then* $U(\exp Y)\psi = \psi$ *for all* $Y$ *in the unstable ideal of* $\mathfrak{g}$.

We notice that the proof can be simplified, using Appendix 2 of Auslander and Green [6]; in particular, we do not need to consider the "Mautner subgroups".

In fact, let $Y$ be in the eigenspace of ad $\alpha$ corresponding to an eigenvalue with positive real part. Then $\langle U(\exp Y)\psi, \psi \rangle = \langle U(\exp n\alpha)U(\exp(-n\alpha)\exp Y\exp n\alpha)U(\exp(-n\alpha))\psi, \psi \rangle = \langle U(\exp[\mathrm{Ad}(\exp(-n\alpha))(Y)]U(\exp(-n\alpha))\psi, U(\exp(-n\alpha))\psi \rangle = \langle U(\exp[\exp(-n\,\mathrm{ad}\,\alpha)(Y)])\psi, \psi \rangle \to_{n\to\infty} \langle \psi, \psi \rangle$. It follows that $U(\exp Y)\psi = \psi$. Similarly for $Y$ in any eigenspace corresponding to an eigenvalue with negative real part. Now, the strongly continuous, unitary representation $U$ of $G$ gives rise to a "$C^\infty$ Lie algebra" of skew-adjoint operators on a dense Gårding domain of $C^\infty$-vectors (e.g. the subspace of $L^2(M)$ consisting of the $C^\infty$-functions) such that the operator $dU(X)$ is the infinitesimal generator of the one-parameter group $U(\exp tX)$. As noted above, the 0-eigenvectors for $dU(X)$ then correspond to the

1-eigenvectors for $U(\exp tX)$. Then $dU(Y)\psi = 0$ for $Y \in \mathscr{E}$, hence for $Y$ in the unstable ideal. The lemma follows.

PROPOSITION 17. *If $\mathfrak{g}$ is nonsolvable, the flow is ergodic and weakly mixing.*

PROOF. This is now trivial. By §4, the unstable ideal is $\mathfrak{g}$, hence by the last lemma and Proposition it follows that the only eigenfunctions for $U(\exp t\alpha)$ are the constant functions.

PROPOSITION 18. *If $\mathfrak{g}$ is solvable, the flow is ergodic, but not weakly mixing.*

PROOF. Let $N$ be the normal subgroup of $G$ corresponding to the unstable ideal $\mathfrak{N}$. Then $\Gamma \cdot N$ is closed (Mostow [**32**]), let $p$ be the natural map $\Gamma\backslash G \to \Gamma \cdot N\backslash G$. We use the following lemma from Auslander and Green [**6**]:

LEMMA. *Every eigenvector of the flow on $\Gamma\backslash G$ is of the form $\psi \cdot p$, where $\psi$ is an eigenvector of the corresponding flow on $\Gamma \cdot N\backslash G$. In particular, the flow on $\Gamma\backslash G$ is ergodic (weakly mixing) if and only if the flow on $\Gamma \cdot N\backslash G$ is.*

In §2 it was shown that $\Gamma \cdot N\backslash G_1$ is the circle; and the flow is seen to be the flow around the circle with uniform speed. The one-parameter group $U_t$ is the translation group of the circle, the infinitesimal generator is the differentiation operator $f \to df/dt$; the eigenfunctions are the trigonometric functions $e^{int}$, $n \in \mathbf{Z}$. Consequently, the flow is ergodic, but not weakly mixing.                                    Q.E.D.

We now digress briefly to demonstrate that we can also extend these methods to give information about ergodic properties of $N$-induced Anosov diffeomorphisms.

PROPOSITION 19. *Let $\phi$ be an $N$-induced Anosov diffeomorphism of an infranilmanifold $M'$. Then $\phi$ is weakly mixing (and hence ergodic).*

PROOF. Let $M$ be the manifold with the suspended flow. By looking at the flow orbit of a $\phi$-invariant subset of $M'$, it is seen that ergodicity is equivalent for the diffeomorphism $\phi$ and the suspended flow. For the latter, it has already been proved.

Let $\psi_t$ be the one-parameter group of automorphisms constructed in §2, such that $\phi^n = \psi_n$. Consider the following class of functions on $M' \times \mathbf{R}: \{f\,|\,f(x, t + s) = f(\psi_t(x), s)\}$. These define a $U_t$-invariant subspace of functions on $M$, and are in one-to-one correspondence with the functions on $M'$. Now there is a corresponding connection between ergodic properties of $\phi$ and spectral properties of the unitary operator $V$ on $L^2(M')$: $V(f(x)) = f(\phi^{-1}(x))$. Assume $\xi'$ is an eigenfunction for $V$; i.e. $V\xi' = e^{i\lambda}\xi'$. Construct the function $\xi$ on $M$ as above: $\xi(x, t) = \xi'(\psi_t(x))$. Then $U_n(\xi)(x, t) = \xi(x, t - n) = \xi'(\psi_{-n+t}(x)) = \xi'(\phi^{-n}(\psi_t(x))) = V^n(\xi')(\psi_t(x)) = e^{in\lambda}\xi'(\psi_t(x)) = e^{in\lambda}\xi(x, t)$. By Stone's formula again, we see that $\xi$ is in the $e^{it\lambda}$ eigenspace of $U_t$, (i.e. in the $(i\lambda)$-eigenspace of the infinitesimal generator).

Using the lemma by Auslander and Green again, it follows that $\xi$ is invariant under the action of $N$, corresponding to the unstable ideal. But $N$ acts transitively on the nilmanifold, and $\xi$, by construction, is determined by its values on that fibre.

Hence $\xi$ (and $\xi'$) is constant; it follows that $N$-induced Anosov diffeomorphisms are weakly mixing.

Finally, we prove that the spectrum of the flow is absolutely continuous in the nonsolvable case; actually the proof gives that it is Lebesgue; i.e. the spectral measure $E$ is equivalent to Lebesgue measure on $\mathbf{R}$, except for the eigenvalue 0. This follows methods used by C. C. Moore; the generalization to nonsemisimple groups is fairly straightforward for the special case of interest here, using the structural information obtained in §4.

PROPOSITION 20. *If $\mathfrak{g}$ is not solvable, the flow has Lebesgue spectrum.*

PROOF. As in §4, we can assume that $G_0 = R \times {}_s S$, with $S$ semisimple, $R$ nilpotent. We then have an Iwasawa-decomposition of $S$, $S = K'AN$, with $A = \exp(\mathbf{R}\alpha)$, $N = \exp(\mathscr{E}^+)$; where $A$, $N$, and $A \cdot N$ are closed, simply connected subgroups of $G_0$ diffeomorphic to the Lie algebras $\mathbf{R}\alpha$, $\mathscr{E}^+$, $\mathbf{R}\alpha + \mathscr{E}^+$ by the exponential map.

Choose $\lambda$ the highest eigenvalue of $\mathrm{ad} \in \alpha$, and $e_\lambda \in \mathscr{E}_\lambda$, $e_{-\lambda} \in \mathscr{E}_{-\lambda}$. $U(\exp te_\lambda) = \int_{-\infty}^{\infty} e^{it\mu} dP(\mu)$. We use a logarithmic change of variable, and define a new spectral measure $Q$ by $Q(\sigma) = P(\sigma^*)$, where $\sigma^* = \{x \in \mathbf{R} | \log|x| \in \sigma\}$. To check that $Q$ is a spectral measure, we must verify that $Q(\mathbf{R}) = 1$; for this it is sufficient that $P(\{0\}) = 0$; i.e. there are no eigenvectors of eigenvalue 1 for $U(\exp te_\lambda)$. Now, $\exp(te_\lambda)$ is a normal subgroup of the simply connected group $N$, hence it is closed and noncompact. Taking the restriction of $U$ to $S$; and using Theorem 2 in C. C. Moore [31], we conclude that if $U(\exp te_\lambda)\psi = \psi$, then $U(s)\psi = \psi$, for all $s \in S$. In particular $U(\exp t\alpha)\psi = \psi$. As we have already seen, it then follows that $\psi$ is constant. (The essential argument for this is contained in a result by Sherman on representations of groups with Lie algebra $\mathfrak{Sl}(2, \mathbf{R})$, see C. C. Moore [31] for discussion and references.)

We have $U(\exp se_\lambda) = \int_{-\infty}^{\infty} e^{is\mu} dP(\mu)$; hence

$$U(\exp(-t\alpha))U(\exp se_\lambda)U(\exp t\alpha) = \int_{-\infty}^{\infty} e^{is\mu} d\{U(\exp(-t\alpha))P(\mu)U(\exp t\alpha))\}.$$

But

$$U(\exp(-t\alpha))U(\exp se_\lambda)U(\exp t\alpha) = U(\exp(\mathrm{Ad}(\exp(-t\alpha))(se_\lambda)))$$
$$= U(\exp(se^{-t\lambda}e_\lambda) = \int_{-\infty}^{\infty} e^{is \, \exp \, (-t\lambda)\mu} dP(\mu) = \int_{-\infty}^{\infty} e^{is\mu} dP(e^{t\lambda} \cdot \mu).$$

It follows that $U(\exp(-t\alpha))P(\mu)U(\exp t\alpha) = P(e^{t\lambda} \cdot \mu)$. Also,

$$Q(t\lambda + \sigma) = P((t\lambda + \sigma)^*) = P(e^{t\lambda} \cdot \sigma^*) = U(\exp(-t\alpha))P(\sigma^*)U(\exp t\alpha)$$

$$= U(\exp(-t\alpha))Q(\sigma)U(\exp t\alpha).$$

Now we define another one-parameter group $V_s$: $V_s = \int_{-\infty}^{\infty} e^{-ivs/\lambda} dQ(v)$. Then $V_s U(\exp t\alpha) = e^{ist} U(\exp t\alpha)V_s$. But these are the Weyl commutation relations; and from their standard model (Mackey [25]), one sees that $U(\exp t\alpha)$ has Lebesgue spectrum. Q.E.D.

We remark that we can also extend this method (using ergodicity of the "horo-cycle" flow to prove Lebesgue spectrum of the "geodesic" flow) to prove that $N$-induced Anosov diffeomorphisms have Lebesgue spectrum.

Finally, we sum up the results for $(G, \Gamma)$-induced Anosov-flows proved in this chapter:

THEOREM VII. *Let* $(G, \mathfrak{g}, K, \mathfrak{k}, \Gamma, \alpha)$ *satisfy the conditions from* §1 (AI–AIV). *Then there are two possibilities for the ergodic behavior of the Anosov flow induced by* $\alpha$ *on* $\Gamma \backslash G/K$:
*Either*: (1) *The flow has Lebesgue spectrum, or*
(2) *The flow is ergodic, but not weakly mixing.*

*In this latter case, the flow is, up to a change of the time scale by a constant factor; the suspension of an N-induced Anosov diffeomorphism. The manifold is a bundle over the circle; the eigenfunctions of the flow are constant on fibres, and correspond to the characters of the circle group.*

ADDED IN PROOF: Proposition 11 should be modified as follows: By changing the flow generator from $\alpha$ to $\alpha + k$, $k \in \mathfrak{k}$ we can always assume: $\alpha$ acts with real eigenvalues. The proof is immediate by embedding $\alpha$ in a Cartan sunalgebra, since the compact part of that must have codimension 1.

## REFERENCES

**1.** American Mathematical Society, *Lecture notes prepared in connection with the Summer Institute on Algebraic Groups and Discontinuous Subgroups held at the University of Colorado, Boulder, Colorado, July 5–August 6*, 1965.

**2.** D. V. Anosov, *Roughness of geodesic flows on compact Riemannian manifolds of negative curvature*, Soviet Math. Dokl. **3** (1962), 1068–1070.

**3.** ———, *Ergodic properties of geodesic flows on closed Riemannian manifolds of negative curvature*, Soviet Math. Dokl. **4** (1963), 1153–1156.

**4.** ———, *Geodesic flows on compact Riemannian manifolds of negative curvature*, Trudy Mat. Inst. Steklov. **90** (1967).

**5.** L. Auslander, L. Green and F. Hahn, *Flows on homogeneous spaces*, Ann. of Math. Studies No. 53, Princeton Univ. Press, Princeton, N.J., 1963.

**6.** L. Auslander and L. Green, *G-induced flows*, Amer. J. Math. **88** (1966), 43–60.

**7.** L. Auslander, *Modifications of Solvmanifolds and G-induced flows*, Amer. J. Math. **88** (1966), 615–625.

**8.** ———, *On a problem of Philip Hall*, Ann. Math. **86** (1967), 112–117.

**9.** L. Auslander and J. Scheunemann, *On certain automorphisms of nilpotent Lie groups*, these Proceedings, vol. 14.

**10.** A. Bialynicki-Birula and M. Rosenlicht, *Injective morphisms of real algebraic varieties*, Proc. Amer. Math. Soc. **13** (1962), 200–203.

**11.** A. Borel, *Compact Clifford-Klein forms of symmetric spaces*, Topology **2** (1963), 111–122.

**12.** ———, *Ensembles fondamentaux pour les groupes arithmetiques et formes automorphes*, Secr. Math. Ecole Norm. Sup., Paris, 1967.

**13.** A. Borel and Harish-Chandra, *Arithmetic subgroups of algebraic groups*, Ann. of Math. (2) **75** (1962), 485–535.

**14.** N. Bourbaki, *Groupes et algebres de Lie*, Paris, 1966.

**15.** C. Chevalley, *Theory of Lie groups*, Princeton Univ. Press, 1946.

**16.** ———, *Theorie des groupes de Lie, II*, Hermann, Paris, 1951.

**17.** ———, *Theorie des groupes de Lie, III*, Hermann, Paris, 1955.

**18.** ———, *The algebraic theory of spinors*, Columbia Univ. Press, New York, 1954.

**19.** J. Franks, *Anosov diffeomorphisms*, these Proceedings, vol. 14.

**20.** I. M. Gelfand and S. V. Fomin, *Geodesic flows on manifolds of constant negative curvature*, Amer. Math. Soc. Transl. (2) **1** (1955), 49–66.

**21.** L. Gårding, *Note on continuous representations of Lie groups*, Proc. Nat. Acad. Sci. U.S.A. **33** (1947), 331–332.

**22.** S. Helgason, *Differential geometry and symmetric spaces*, Academic Press, New York, 1953.

**23.** G. Hochschild, *The structure of Lie groups*, Holden-Day Inc., San Francisco, Calif., 1965.

**24.** N. Jacobson, *Lie algebras*, Interscience Publishers, New York, 1962.

**25.** G. W. Mackey, *A theorem of Stone and von Neumann*, Duke Math. J. **16** (1949), 313–326.

**26.** A. Malcev, *On a class of homogeneous spaces*, Amer. Math. Soc. Transl. (1) **9** (1962), 276–307 (originally **39** (1951)).

**27.** G. A. Margulis, *Anosov flows on three-dimensional manifolds*, (Appendix to the article *Some smooth ergodic systems* by D. V. Anosov and Ya. G. Sinai in Uspehi Mat. Nauk **22** (1967)).

**28.** F. Mautner, *Geodesic flows on symmetric Riemann spaces*, Ann. of Math. (2) **65** (1957), 416–431.

**29.** J. Milnor, *A note on curvature and fundamental group*, J. Differential Geometry **2** (1968), 1–9.

**30.** C. C. Moore, *Decomposition of unitary representations defined by discrete subgroups of nilpotent groups*, Ann. of Math. (2) **82** (1965), 146–182.

**31.** ———, *Ergodicity of flows on homogeneous spaces*, Amer. J. Math. **88** (1966), 154–178.

**32.** G. D. Mostow, *Factor spaces of solvable groups*, Ann. of Math. **60** (1954), 1–27.

**33.** ———, *Homogeneous spaces with finite invariant measure*, Ann. of Math. **75** (1962), 17–38.

**34.** G. D. Mostow and T. Tamagawa, *The compactness of arithmetically defined homogeneous spaces*, Ann. of Math. **76** (1962), 446–463.

**35.** L. Nachbin, *The Haar integral*, Van Nostrand, Princeton, N.J., 1965.

**36.** O. T. O'Meara, *Introduction to quadratic forms*, Springer, Berlin, 1963.

**37.** C. C. Pugh, *The closing lemma*, Amer. J. Math. **89** (1967), 956–1009.

**38.** ———, *An improved closing lemma and a general density theorem*, Amer. J. Math. **89** (1967), 1010–1021.

**39.** Seminaire S. Lie, *Groupes et algebres de Lie*, Paris, 1955.

**40.** A. Selberg, *On discontinuous groups in higher dimensional symmetric spaces*, Contributions to function theory, Tata Institute of fundamental research, Bombay, 1960.

**41.** S. Smale, *Differentiable dynamical systems*, Bull. Amer. Math. Soc. **73** (1967), 747–817.

**42.** J. A. Wolf, *Spaces of constant curvature*, McGraw-Hill, New York, 1967.

**43.** ———, *The rate of growth of discrete solvable groups and its influence on the curvature of Riemannian manifolds with solvable fundamental group*, to appear.

UNIVERSITY OF CALIFORNIA, BERKELEY

# THE "DA" MAPS OF SMALE AND
# STRUCTURAL STABILITY

R. F. WILLIAMS[1]

The purpose of this paper is to describe in detail the "derived from Anosov" maps of the title (due to S. Smale [2, p. 789]) and to show how these examples allow one to lower by 1 the dimension in the

THEOREM (SMALE [1]). *There is an open set U in the space of $C^r$ vector fields* $(r > 0)$ *on a 4-dimensional manifold such that no $X \in U$ is structurally stable.*

The resulting theorem is definitive in that M. Peixoto has shown [3] that structurally stable systems on compact two manifolds are dense. The present construction is based on a one-dimensional attractor (generalized solenoid [5]) whereas Smale's is based on a two-dimensional attractor (a torus). Otherwise this construction is just like that of Smale's. The paper of Peixoto-Pugh [4] contains another variation of Smale's construction for noncompact 2-manifolds and has a good discussion as to how all of these examples work.

The DA maps are formulated as Theorem A in §0. In §1 a DA map is used to show (Theorem B) that structurally stable systems on a certain 3-manifold are not dense. The DA maps are described in detail in §2. In §3 we point out further structure of the DA maps including computation of the invariant

$$g_*: \pi_1(K) \to \pi_1(K)$$

used in [6] to classify 1-dimensional attractors. As a by-product of §3 we obtain the following curious number theoretic

THEOREM C. *Every $A \in G1(2, \mathbf{Z})$ whose eigenvalues are off the unit circle is similar over the integers to a matrix B, all of whose entries have the same sign (0 allowed).*

*Conversely, such a B has eigenvalues off the unit circle, with the exceptions*

$$\pm \begin{pmatrix} 1 & n \\ 0 & 1 \end{pmatrix}, \quad \pm \begin{pmatrix} 1 & 0 \\ n & 1 \end{pmatrix},$$

as one can see by computing its eigenvalues. Daniel Zelinsky has shown me an elementary proof of this result in case $A$ has determinant $+ 1$. Were a direct proof of this result available, it would aid in the description in §§2, 3 instead of being a consequence. Conversations with M. Peixoto, S. Smale and R. Thom have been very helpful in the preparation of this paper.

---

[1] Supported in part by NSF Grant 5591.

0. **Formulation of DA maps.** Let $A \in Gl(2, Z)$ have eigenvalues off the unit circle, say $\lambda$, $\mu$. As $|\lambda\mu| = |\det A| = 1$, we may take $|\lambda| < 1 < |\mu|$ and note that $\lambda$ and $\mu$ are real. Let $\mathcal{L}$, $\mathcal{M}$ be the families of all lines in $R^2$ in the eigen directions corresponding to $\lambda$, $\mu$, respectively. These lines have irrational slope and thus can be thought of as on the torus $T^2 = R^2/(Z \oplus Z)$; $A$ induces a diffeomorphism also called $A: T^2 \to T^2$ and leaves these $\mathcal{L}$, $\mathcal{M}$ invariant, as families. These are the generalized stable and unstable manifolds [2, p. 781] of $A$ as $A$ contracts (expands) distances in the directions of the lines of $\mathcal{L}$ ($\mathcal{M}$). Let $L_0$ be the line of $\mathcal{L}$ which contains the origin, $\theta$.

THEOREM A (SMALE [2, p. 789]). *Corresponding to A, there is a diffeomorphism* $f: T^2 \to T^2$ *such that*

(a) *$f$ is smoothly isotopic to $A$;*

(b) *the nonwandering set $\Omega(f) = \{\theta\} \cup \Lambda$ where $\theta$ is a (point) source and $\Lambda$ is a one-dimensional attractor;*

(c) *$\Omega(f)$ has a hyperbolic structure;*

(d) *the generalized stable manifolds [2, p. 781] of $f|\Lambda$ are the lines of $\mathcal{L}$ except $L_0$, which divided by $\theta$ forms two generalized stable manifolds.*

In his presentation, Smale does not mention part (d).

1. **Proof of Theorem B.** The diffeomorphism $f$ of Theorem A is altered in two standard ways. First the source $\theta$ is replaced by two sources $a$, $b$ and one hyperbolic point $\theta$ in such manner that the line $L_0$ becomes the unstable manifold $W^u(\theta)$ of $\theta$. Call the resulting map $g$ and note that $g = f$ except near $\theta$, $a$ and $b$. Second, choose two intervals $[c, d]$, $[d, e]$ of $W^u(\theta)$ so that $g([c, d]) = [d, e]$ and add a "bump function" to $g$, at $[c, d]$. That is, let

$$h(x) = g(x) + r(x) \cdot \overline{m},$$

where $\overline{m}$ is a vector in the direction of the lines of the family $\mathcal{M}$. For $r$ smooth, small enough and vanishing except near the midpoint of $[c, d]$, where it is positive, the resulting $h$ is a diffeomorphism of $T^2$ and has the following structure:

(1) $\Omega(h) = \Lambda \cup \{\theta, a, b\} = \Omega(g)$.

(2) The generalized stable manifolds $W^s(x, h)$ for $x \in \Lambda$ are curves which simply cover $T^2 - \theta$.

Thinking of the lines of $\mathcal{L}$ as horizontal and $\overline{m}$ pointing up,

(3) there is a unique $x \in \Lambda$ so that $W^s(x, h)$ contains the local maximum of $W^u(\theta, h)$, near $[d, e]$.

Furthermore these properties persist for all $h' \in U$, a sufficiently small neighborhood $U$ of $h$ ($C^r$ topology, $r > 0$). The corresponding $\Lambda'$, $\theta'$, $a'$, $b'$, $x' \in \Lambda'$, $W^s(x', h')$ and $W^u(\theta', h')$ may of course be different from those of $h$, but are nearby. Note that $x'$ is either a periodic orbit of $h'|\Lambda'$ or not.

Now for $t \in [0, 1]$ and $s$ another small "bumping function," define

$$h_t(x) = h'(x) + t \cdot s(x) \cdot \overline{m}.$$

Then the structure $\Sigma'$, $\theta'$ still applies to $h_t$. But the point $x_t$ such that $W^u(\theta', h_t)$

has its local maximum at a point $W^s(x_t, h_t)$ varies with $t$ and can be arranged to be either a periodic or nonperiodic point of $h_t|\Lambda' = h'|\Lambda'$. As one of these is different from the behavior of $h$, $h$ is not structurally stable.

Then $\{h^i : i \in \mathbf{Z}\}$ is an action of $\mathbf{Z}$ on $T^2$ and can be turned into an action of $\mathbf{R}$ on a 3-manifold by a familiar procedure—called suspension in [2, p. 797]. This is the required example and proves

THEOREM B. *There is an open set $U$ in the space of $C^r$ vector fields ($r > 0$) on a compact 3-manifold such that no $X \in U$ is structurally stable.*

2. **That DA maps exist.** We use the notation $A$, $\mathcal{L}$, $\mathcal{M}$ of §0. $L_0 \in \mathcal{L}$, $M_0 \in \mathcal{M}$ are the lines through the origin $\theta$. As before, we can think of $A$ as acting either on $\mathbf{R}^2$ or the torus $T^2$; similarly the lines of $\mathcal{L}$ and $\mathcal{M}$ as being in either $\mathbf{R}^2$ or $T^2$. Choose an open line interval $D_0 \subset M_0$, centered at the origin.

Define $\mathcal{C} = \{C : C$ is a component of $L - D_0$ for some $L \in \mathcal{L}\}$. Then $\mathcal{C}$ is a partition of $T^2 - D_0$; let $K$ be the corresponding quotient space and $q : T^2 - D_0 \to K$ the quotient map.

2.1 LEMMA. *There is a diffeomorphism $\phi : T^2 \to T^2$ such that*

(a) *$\phi$ is smoothly isotopic to the identity;*

(b) *$\phi(A(C)) \subset A(C)$ for each $C \in \mathcal{C}$;*

(c) *$\phi \circ A$ is expanding at each point of $D_0$;*

(d) *for some disks $D^2 \supset E^2 \supset \bar{D}_0$, $\phi \circ A$ takes $T^2 - D^2$ into itself, is a contraction on each $C \cap (T^2 - D^2)$, an expansion on each $C \cap E^2$ and $\phi \circ A$ takes $D^2 - E^2$ into $T^2 - D^2$.*

PROOF. By using a smooth conjugacy on a neighborhood $N$ of $f(D_0)$ we may assume that $f(D_0) = D_1$ is on the $x$-axis and that the intervals $C \cap N$, for all $C \in \mathcal{C}$ are vertical. In this notation we define

$$\phi(x, y) = (x, p(x, y) \cdot y)$$

where $p(x, y) \geq 0$, $p(x, y) = 1$ except near $D_1$ and $p(x, 0) > 1$ is sufficiently large to counteract the shrinking that $A$ does in this "vertical" direction. Finally small disks $D^2$, $E^2$ are found and $p$ chosen to satisfy (d) as well.

2.2 PROOF OF THEOREM A. We claim that $f = \phi \circ A$ is as required. Part (a) is clear as $\phi$ itself is smoothly isotopic to the identity. Note that $\theta \in \Omega(f)$ and that $\Omega(f) - \theta \subset T^2 - D^2$ as the origin is fixed and, by Lemma 2.1(d), points of $D^2 - \theta$ eventually leave $D^2$ under iteration by $f$. Note also that $f$ has a hyperbolic structure on $T^2 - D^2$ as it is contracting on the interval $C \cap (T^2 - D^2)$ and expanding in directions (roughly) corresponding to the lines of $\mathcal{M}$. This proves part (c) of Theorem A.

To compute $\Omega(f)$, we proceed as in [5, p. 483] with the diagram

$$
\begin{array}{ccc}
\vdots & & \vdots \\
\downarrow & & \downarrow \\
f(T^2 - D^2) \xleftarrow{\ f\ } T^2 - D^2 \\
\downarrow i & & \\
T^2 - D^2 & & \Big\downarrow q \\
\downarrow q & & \\
K \xleftarrow{\ g\ } K \xleftarrow{\ g\ } \cdots
\end{array}
$$

in which $i$ is the inclusion and $g$ is defined as $q \circ i \circ f \circ q^{-1}$. This is well defined as $f[C \cap (T^2 - D^2)]$ lies in some $C' \in \mathscr{C}$ for each $C \in \mathscr{C}$. Define $\Lambda = \bigcap_{i>0} f^i(T^2 - D^2)$. It follows as in [5] from (**) that $(\Lambda, f|\Lambda)$ is conjugate to the shift map $h : \Sigma \to \Sigma$ where $\Sigma$ is the inverse limit of the sequence

$$ K \xleftarrow{\ g\ } K \xleftarrow{\ g\ } K \xleftarrow{\ g\ } \cdots, $$

and the shift map $h$ is defined by $h(x_0, x_1, x_2, \ldots) = (gx_0 \cdot gx_1, gx_2, \ldots) = (gx_0, x_0, x_1, \ldots)$, for $(x_0, x_1, \ldots) \in \Sigma$.

To see that all points of $\Lambda$ are nonwandering it is perhaps easiest to point out that $g : K \to K$ satisfies the axioms 1–3 of [5, p. 476] which in turn follow from the criterion of [6]. $K$ is a smooth branched 1-manifold as shown in [5], and is orientable as both $T^2$ and the family $\mathscr{C}$ are. It has the homotopy type of $T^2 - D^2$ and thus of a figure 8. Thus $K$ is the union of two smooth 1-spheres which are either tangent at a point or coincide along a line interval; as this latter case can always be arranged by choice of $D_0$, we suppose that it attains. That $g$ is expanding is clear. Finally $g$ is onto: $f(D_0)$ intersects each $C \in \mathscr{C}$ so that for each $C \in \mathscr{C}$, some $C_x$ with end point $x \in D_0$ satisfies $f(C_x) \subset C$. The axioms 1–3 easily follow.

The structure of the generalized stable manifolds (part d of Theorem A) is seen as follows. First, for each $C \in \mathscr{C}$, $C \cap (T^2 - D^2)$ is a local generalized stable manifold by construction and by 2.1(d) so is $C$ itself. For $x \in T^2 - D_0$, let $C(x)$ be the element of $\mathscr{C}$ containing $x$. Then for $x \in \Lambda$, the global generalized stable manifold, $W^s(x)$ containing $x$ is

$$ \bigcup_{i>0} f^{-i} C(f^i x). $$

Let $a$ be an end point of $C(x)$ where $a \neq \theta$. There is an integer $n$ such that $f^n(a) \notin D_0$, so that $C(f^n(x))$ contains $f^n(a)$ together with $f^n(C(x))$ and some $f^n(C^n(y))$ where $C(x) \cap C(y) = \varnothing$ and $a$ is a common end point. Thus $f^{-1} C(f^n(x)) \supset C(x) \cap C(y)$. By repeating this process, it follows that $W^s(x)$ is either an entire line of $\mathscr{L}$ or one of the half lines which $\theta$ divides $L_0$ into.

3. **Classification of $\Sigma$ and a result in $G1(2, \mathbf{Z})$.** In this section we round out the description of the DA map $f$ by showing

(3.1) $g_* : H_1(K, \mathbf{Z}) \to H_1(K, \mathbf{Z})$ is a (fairly) well-determined integral matrix $B$;

(3.2) $B \in G1(2, \mathbf{Z})$;

(3.3) $B$'s entries are all of the same sign;

(3.4) $B$ is similar *over the integers*, to $A$.

This last proves Theorem C. Finally, we compute $g_*\pi_1(K) \to \pi_1(K)$ because it is in terms of this map that generalized solenoids are classified in [6].

(3.1) is just the fact that $K$ is orientable so that choosing one generator, say $x \in H_1(K, \mathbf{Z}) \approx \mathbf{Z} + \mathbf{Z}$ determines the other, say $y$. As the choice $(-x, -y)$ would not change the matrix $B$, the only alternative would be $(y, x)$, which would change $B$ only by conjugation with the matrix

$$\begin{pmatrix} 0 & 1 \\ 1 & 0 \end{pmatrix}$$

To see (3.2) note that $q: T^2 - D^2 \to K$ has a section $p$, embedding $K$ in $T^2 - D^2$. Then $p(K)$ is essentially a 1-skeleton of $T^2$, as is $f \circ p(K)$. The determinant of $B$ can be viewed as the intersection number of the "two" 1-spheres in $f \circ p(K)$ which is $\pm 1$.

The fact (3.3) that $B$'s entries are all the same sign follows from the fact that $g$ is an immersion and thus does no "folding back". It immerses the 1-spheres $x$ and $y$ in $K$ in (necessarily) the same orientations. This last also allows us to define $\pi_1(K)$ without using base points: $\pi_1(K)$ is the free group on the generators $x, y$ corresponding to the two embedded 1-spheres (with a chosen orientation). Words in $\pi_1(K)$ are determined up to cyclic permutation.

Then $g_*: \pi_1(K) \to \pi_1(K)$ (up to cyclic permutation, of course) is given by the two words $g_*(x) = w_1$, $g_*(y) = w_2$, and these words are the most evenly spaced words which abelianize to the columns of $B$. Thus if

$$B = \begin{pmatrix} 3 & 1 \\ 2 & 1 \end{pmatrix}$$

then $w_1 = xyxyx$, $w_2 = xy$ (instead of $x^3y^2$, etc.). More precisely

(3.5) The exponents in $w_1$ and $w_2$ are all of the same sign; any subword of $w_i$ beginning in $x$ and ending in $y$ has as near as possible (i.e. within one) the same ratio of $x$'s and $y$'s as the whole word $w_i$, $i = 1, 2$.

PROOF. Let $b_{ij}$ be the entries of $B$ and take the case that $b_{ij} \geq 0$. Then according to the order in which the line interval $I$ joining $(0, 0)$ to $(b_{11}, b_{21})$ crosses the grid lines demarking the fundamental domains of $T^2$ in $\mathbf{R}^2$, write the word $w_1$ down. Thus, if $I$ first crosses $\alpha_1$ vertical lines then $\beta_1$ horizontal lines, then $\alpha_2$ vertical lines, etc., then $w_1 = x^{\alpha_1} y^{\beta_1} x^{\alpha_2} \dots$. If $b_{ij} \leq 0$, the exponents are taken to be negative.

To prove (3.4) we use the following easy and well known

REMARK. If $A, B \in \mathrm{Gl}(2, \mathbf{Z})$ are topologically conjugate as maps on $T^2$, i.e. if there is a homeomorphism $h: T^2 \to T^2$ such that

$$\begin{array}{ccc} T^2 & \overset{A}{\Rightarrow} & T^2 \\ h\downarrow & & \downarrow h \\ T^2 & \underset{B}{\Rightarrow} & T^2 \end{array}$$

is commutative, then there is a $P \in \mathrm{Gl}(2, \mathbf{Z})$ such that $BP = PA$.

To apply this result, note first that $B$ being positive (or negative) has a real eigenvalue $\delta > 1$ (or $< -1$) and hence also an eigenvalue $\gamma$ with $|\gamma| < 1$. Thus $T^2$, which we take as a square $S$ (fundamental domain) with center the origin, is "fibered" by eigenlines like $\mathscr{L}$ and $\mathscr{M}$ above. Now embed $K$ in $S$ as follows: $K$ is the union of the upper edge ( = lower edge) of $S$ and an arc spanning the upper and lower edges, and quite close to the sides, but tangent to the top and bottom. This is done so that the eigenlines $\mathscr{L}'$ corresponding to $\gamma$, $|\gamma| < 1$, are transverse to $K$ at all points.

At this stage, note that the two structures, $p(K)$ ($p$ is the embedding of $K$ into $T^2$ chosen as a section of $q$) and the lines of $\mathscr{L}$ away from $\theta$ are topologically just like those of $K$ and $\mathscr{L}'$ away from $\theta$. Similarly, the $A \circ p(K)$ is just like $B(K)$, as this is essentially the way in which $B$ was defined. Thus we can define $h$ so that

$$T^2 - \text{(a disk containing } \theta) \xrightarrow{A} T^2 - \text{(a disk containing } \theta)$$
$$\downarrow h \qquad\qquad\qquad\qquad\qquad\qquad \downarrow h$$
$$T^2 - \text{(a disk containing } \theta) \xrightarrow[B]{} T^2 - \text{(a disk containing } \theta)$$

is commutative. This is enough to apply the remark above, as we only need to know what $h$ does on the 1-skeleton of $T^2$ (i.e. basis for $R^2$) to define $P$.

As a final remark, a large part of these results apply as well to DA maps on the $n$-torus. The role of generalized solenoids is played by $m$-dimensional attractors, $1 \le m < n - 1$. Though these latter are not as well understood in general, they are not complicated in the DA case.

## BIBLIOGRAPHY

**1.** S. Smale, *Structurally stable systems are not dense*, Amer. J. Math. **88** (1966), 491–496.

**2.** ———, *Differentiable dynamical systems*, Bull. Amer. Math. Soc. **73** (1967), 747–817.

**3.** M. M. Peixoto, *Structural stability on 2-dimensional manifolds*, Topology **2** (1962), 101–102.

**4.** M. M. Peixoto and C. C. Pugh, *Structurally stable systems on open manifolds are never dense*, Ann. of Math. (to appear).

**5.** R. F. Williams, *One dimensional nonwandering sets*, Topology **6** (1967), 473–478.

**6.** ———, *Classification of one-dimensional attractors*, these Proceedings, vol. 14.

NORTHWESTERN UNIVERSITY

# ZETA FUNCTION IN GLOBAL ANALYSIS

R. F. WILLIAMS [1]

This is a survey of most of the known results concerning zeta functions in global analysis; in cases in which other proofs have appeared or will shortly appear we only sketch the proof.

If $X$ is a set and $f: X \to X$ a function then infinitely many integers $N_i$ are determined by

$$N_i = \text{cardinality of Fix } f^i.$$

Here Fix $f$ means the fixed point set of $f$. If $X$ is also endowed with a topology, then we only count the isolated fixed points. Following Weil and Artin-Mazur [1], we combine this information in a single "zeta" function:

$$\zeta(f) = \zeta_f(t) = \exp\left(\sum_{i=0}^{\infty} \frac{N_i t^i}{i}\right).$$

Note that this is an invariant of the topological conjugacy class of a map $f$ because if

$$
\begin{array}{ccc}
 & f & \\
X & \to & X \\
h \downarrow & & \downarrow h \\
Y & \to & Y \\
 & g &
\end{array}
$$

is commutative where $h$ is a homeomorphism, then

$$N_i(f) = N_i(g) \quad \text{for all } i.$$

Smale has raised the question as to whether this function is generically rational for diffeomorphisms of a smooth manifold. In particular it is reasonable to

CONJECTURE (SEE S. SMALE [7]). *If* $f \in \mathscr{S} \subset \text{Diff}(M)$ *where* $\mathscr{S}$ *is the set of all diffeomorphisms of* $M$ *which satisfies Smale's Axioms* A *and* B, *then* $\zeta_f$ *is rational.*

So far, the four main results on this question are:

THEOREM 1 (K. MEYER [5]). *If* $f$ *satisfies Axiom* A *then* $\zeta(t)$ *has a positive radius of convergence.*

THEOREM 2. *If* $f \in \mathscr{S}$, $\Omega_i$ *is a basic set of* $f$ *having only finitely many components and if the bundle* $E^u$ *given by Axiom* A *for* $\Omega_i$ *is orientable, then* $\zeta_{f/\Omega_i}$ *is rational.*

[1] Research supported in part by NSF Grant GP 5591.

THEOREM 3 [9]. *If $\Lambda$ is an attractor and $f|\Lambda$ satisfies Axiom A (first part) then $\zeta_{f|\Lambda}(f)$ is rational. In particular, $\zeta_f$ is rational for any Anosov diffeomorphism f.*

THEOREM 4 (BOWEN-LANGFORD [4] AND BOWEN [2]). *If $\Lambda$ is a zero dimensional basic set and $f|\Lambda$ satisfies (both parts of) Axiom A, then $\zeta_{f|\Lambda}$ is rational.*

Concerning Theorem 1, Artin-Mazur [1] prove a parallel theorem by quite different techniques. Meyer's proof uses "standard" estimates from analysis. M. Shub has proved the stronger

THEOREM $1^+$ (M. SHUB). *If $X$ is compact metric and $f : X \to X$ is an expansive homeomorphism, then $\zeta_f$ has a positive radius of convergence.*

PROOF. Let $\varepsilon$ be an expansive constant for $f$, and let $U_1, ..., U_n$ be a cover of $X$ of mesh $< \varepsilon/2$. For $x \in X$, choose $a_i(x)$ to be the least integer $n$ such that $f^i(x) \in U_n$. Now as $f$ is expansive, no two points $x, y$ could correspond to the same sequence $... a_{-1}a_0a_1a_2 ....$. Also, if $x$ is of period $m$, the symbol $a(x)$ is periodic of period $m$, so that $N_m \leq n^m$, which proves $1^+$.

Theorems 2 and 3 are based on the Lefschitz fixed point formula which states

$$\sum_{p \,\varepsilon\, \mathrm{Fix}\, f} L(p) = \sum (-1)^i \mathrm{trace}\, f_{*i},$$

where $f_{*i}$ is the induced map on $H_i(X; R)$, $i = 1, 2, ..., n$, and $L(p) = \pm 1$ is the Lefschitz index of the fixed point $p$. The two principal facts relating the Lefschitz formula to rationality of the zeta function are

A. If one has control on the $L(p)$—i.e., can decide uniformly whether $L(p) = +1$ or $-1$—then the zeta function can be computed by this formula and is rational.

B. Conversely, if $\zeta$ is rational, it is so by virtue of some Lefschitz formula; that is, there are integral matrices, $A, B$, such that

$$N_j = \mathrm{tr}\, A^j - \mathrm{tr}\, B^j.$$

(See [8] or Bowen-Lanford [4].)

The control on $L(p)$ given by Axiom A occurs as follows. First, if $p$ is a hyperbolic fixed point of a diffeomorphism, then (essentially by definition)

$$L(p) = \mathrm{sign\ of\ det}\,(I - Df(p)).$$

Hence, we have the

LEMMA (SMALE [7]). *If $x$ is a hyperbolic fixed point of $f \in \mathrm{Diff}(M)$, then*

$$L(x) = (-1)^u \Delta$$

*where $u$ is the dimension of the eigenspace $E^u$ corresponding to eigenvalues $\mu$ with $|\mu| > 1$ and $\Delta = \pm 1$ according as to whether $Df|E^u$ preserves or reverses the orientation of $E^u$.*

PROOF. Let $(I - Df(x)) = \pi(1 - \lambda_i)\pi(1 - \mu_j)\pi(1 - v_k)$, where $\lambda_i, \mu_j$ are the real eigenvalues of $Df(p)$, $|\lambda_i| < 1 < |\mu_j|$ and $v_k$ are the complex ones. Now as the

$(1 - \lambda_i)$ are all positive, and the $v_k$ occur in conjugate pairs [thus $(1 - v_k)(1 - \bar{v}_k)$ $= |1 - v_k|^2$], $L(x) = \text{sign } \pi(1 - \mu_j)$. Thus $L(x) = (-1)^p$ where $p$ is the number of positive eigenvalues larger than 1. If $n$ is the number of negative eigenvalues $< -1$, then $L(x) = (-1)^{p+2n} = (-1)^u(-1)^n$.

This last because the $v_k$ occur in pairs. Now $(-1)^n = \Delta$ because (a) a negative eigenvalue reverses a direction in a vector space and (b) the complex eigenvalues correspond to rotation and hence are not involved.

PROOF OF THEOREM 2 (after Smale). From the spectral decomposition theorem of Smale [7] we have the nonwandering set $\Omega(f) = \bigcup_{i=1}^m \Omega_i$, where $\Omega_1 \le \Omega_2 \le \ldots \le \Omega_m$. This is the partial ordering introduced by Smale [7], who also shows that there are manifolds with boundary $M_1 \subset M_2 \subset \ldots \subset M_m = M$ such that $\Omega_i \subset M_i - M_{i-1}$ and $f(M_i) \subset M_i$. As we have assumed that $f|\Omega_i$ has a hyperbolic structure with $E^u$ orientable, the sign $(-1)^u$ of the lemma above is determined.

Now if $\Omega_i$ has $q$ components, then $f$ must permute them cyclically. Hence $\zeta_{f|\Omega_i}(t) = \zeta'(t^q)$ where $\zeta'$ is the zeta function of $(f|\Omega_i)^q$. But the derivative of $(f|\Omega_i)^q$ either preserves the orientation of every fiber or reverses all of them, so that the sign $\Delta$ of the lemma is determined.

Thus it will suffice to show that the following formula

$$(*) \qquad N_n = (-1)^u \sum_j (-1)^j \left[ \text{tr}(f|M_i)_{*j}^{qn} - \text{tr}(f|M_{i-1})_{*j}^{qn} \right]$$

holds for $\zeta'$. But the contribution of a point in $\text{Fix } (f/M_i)^{qn}$ is $+1$ whereas the contribution of a point in $\text{Fix } (f/M_{i-1})^{qn}$ is zero as it enters twice, with opposite signs.

PROOF OF THEOREM 3. As [9] will appear shortly, we only sketch this proof. $\Omega_0$ has a nice neighborhood $N$; $N$ is a manifold with boundary, $f(N) \subset N$, and the splitting $E^u + E^s$ extends to $N$. Also the map $Df|(E^u/\Omega_0)$ extends (not as the derivative, of course) to $N$ as $f^u : E^u \to E^u$.

There is a double cover $\tilde{N} \to N$ so that the induced bundle $\tilde{E}^u \to E^u$ is oriented. The map $f$ lifts to $\tilde{f} : \tilde{N} \to \tilde{N}$ as does $f^u$ to $\tilde{f}^u : \tilde{E}^u \to \tilde{E}^u$. Let $T$ be the involution on $\tilde{N}$ which interchanges the two fibers: $T$ is the deck transformation of the cover. Then $T\tilde{f}$ covers $f$ also and induces $(T\tilde{f})^u : \tilde{E}^u \to \tilde{E}^u$. One of $\tilde{f}^u$, $(T\tilde{f})^u$ preserves, the other reverses, the orientation of this bundle (note: $T^u : \tilde{E}^u \to \tilde{E}^u$ necessarily reverses this orientation). We suppose $\tilde{f}^u$ preserves this orientation.

Let $\tilde{N}_n =$ the number of fixed points of $\tilde{f}^n$,
$N'_n =$ the number of fixed points of $T\tilde{f}^n$.

LEMMA. $N_n = (\tilde{N}_n + N'_n)/2$.

PROOF. Above a point $x \in N$ fixed under $f^n$ are two points $y$ and $Ty$. One of the maps $\tilde{f}^n$, $T\tilde{f}^n$ fixes both of them, the other reverses them. Then using Smale's Lemma,

$$\tilde{N}_n + N'_n = (-1)^u \left[ \sum_i (-1)^i \text{tr}(\tilde{f}^n)_{*i} - \sum_i (-1)^i \text{tr}(T\tilde{f}^n)_{*i} \right].$$

Now as $T^2 = 1, (T_{*i})^2 = 1$, so that we may take bases so that $T_{*i}$ is diagonal with entries $\pm 1$, say $B_i$. Let $A_i$ be the matrix of $\tilde{f}_{*i}$. Then

$$2N_n = \tilde{N}_n + N'_n = (-1)^u[\sum(-1)^i(\text{tr}A_i^n - \text{tr}B_iA_i^n)]$$
$$= (-1)^u\sum(-1)^i\text{tr}A_i^n(I_i - B_i).$$

As $I_i - B_i$ has the form

$$\begin{pmatrix} 2I_i & 0 \\ 0 & 0 \end{pmatrix}$$

this reduces to $2N_n = 2(-1)^u\sum(-1)^i\text{tr}C_i^n$.

Theorem 4 is proved by first, characterizing zero-dimensional basic sets (Bowen [2]) as being sub-shifts of finite type. Next (actually previously!) Bowen and Lanford [4] computed the zeta-function of sub-shifts of finite type. Finite type is crucial here, as these authors also show that some (actually most) sub-shifts have irrational zeta functions.

DEFINITIONS. Let $S$ be a finite set with $N \geq 2$ elements and consider the Cantor set $X$ of bi-infinite sequences $(\dots a_{-1}a_0a_1a_2 \dots)$ topologized in the obvious way. Define the shift $\alpha : X \to X$ by $(\alpha a)_n = a_{n+1}$ for $a \in X$. For a finite sequence $\sigma = \sigma_0$, $\sigma_1, \dots, \sigma_m$, set $X_\sigma = \{a \in X : a$ nowhere contains the string $\sigma\}$. Then $X_\sigma$ is closed and invariant under $\alpha$. A subset $Y \subset X$ is of *finite type* provided $Y = X_{\sigma 1} \dots X_{\sigma r}$ for some finite array $\sigma^1, \dots, \sigma^r$ of finite sequences.

The author produces a square matrix $T$ such that $N_n = \text{tr}T^n$; the order of $T$ is the number of sequences of length $n_0$ which are permitted in $Y$ where $n_0 + 1$ is the length of the longest sequence among $\sigma^1, \dots, \sigma^r$. Note that in this case $\zeta$ is the reciprocal of a polynomial; this seems to correspond to zero-dimensionality. In this connection there is

THEOREM $3^-$ [8]. *If* $\Omega_1$ *is a 1-dimensional attractor which satisfies Axiom* A *(part 1) then* $\zeta_f$ *is rational.*

This of course is implied by Theorem 3; of interest here is that it was proved without resort to double covers. One can interpret the machinery of [8] used in the proof as a homology theory which makes the Lefschitz trace formula correct and not just correct up to the index $L(p)$. The novel feature is that the zero-dimensional chains are oriented, in such a manner that the chain map $f_\#$ respects this structure.

In conclusion, it seems to the author at least that a proof of this main conjecture may well arise by combining several of these proofs. In particular, a fortuitous "box-structure" based on the tubular family theory of Palis-Smale would seem to provide a homology theory as alluded to in the above paragraph.

## REFERENCES

1. M. Artin and B. Mazur, *On periodic points*, Ann. Math. **81** (1965), 82–89.
2. R. Bowen, *Topological entropy and Axiom* A, these Proceedings, vol. 14.

**3.** R. Bowen and O. E. Lanford III, *Zeta functions of restrictions of the shift transformation* (to appear).

**4.** K. Meyers, *Periodic points of diffeomorphisms,* Bull. Amer. Math. Soc. **73** (1967), 615–617.

**5.** M. Shub, *Periodic orbits of hyperbolic diffeomorphisms and flows,* Bull. Amer. Math. Soc. **75** (1969), 57–58.

**6.** S. Smale, *Differential dynamical systems,* Bull. Amer. Math. Soc. **73** (1967), 747–817.

**7.** R. Williams, *One dimensional nonwandering sets,* Topology **6** (1967), 473–487.

**8.** ———, *The zeta function of an attractor,* to appear in the Proceedings of the Michigan State Topology conference (Spring 1967). (Mimeographed notes, Northwestern University.)

UNIVERSITY OF GENEVA,
  GENEVA, SWITZERLAND

# CLASSIFICATION OF ONE DIMENSIONAL ATTRACTORS

R. F. WILLIAMS[1]

This is a sequel to a paper [3] in which one dimensional attractors are characterized. Some familiarity with [3] is assumed (see §1, below). *Attractor* is meant in the sense of S. Smale; for a general introduction to this subject, see the important paper *Differential dynamical systems* (DDS) [2] of Smale. As this is written, the axioms A and B of Smale [DDS, pp. 777–8] have been shown to be nongeneric by Abraham-Smale [0]. Thus the formulation of Smale (sometimes called Anosov-Smale systems or diffeomorphisms) will have to be varied again. But for technical reasons, the example of Abraham-Smale does not directly affect work done on attractors. Moreover, the analysis carried out here and in [3] does concern an open set of diffeomorphisms, and thus will be pertinent to any formulation of this theory. Conversations with S. Smale, L. Evens and M. Shub have been helpful in the preparation of this paper.

0. **Statement of results.** Let $K$ be a branched 1-manifold and $g: K \to K$ an immersion satisfying axioms 1–3 (see §1 below or [3]). This determines a *generalized solenoid* $\Sigma$, as inverse limit of the sequence

$$K \underset{g}{\leftarrow} K \underset{g}{\leftarrow} K \underset{g}{\leftarrow} \dots$$

and the shift map $h: \Sigma \to \Sigma$, by $h(x_0, x_1, \dots) = (gx_0, x_0, x_1, \dots)$; $h$ is a homeomorphism. We say $g: K \to K$ is a *presentation* of $\Sigma$ and $h: \Sigma \to \Sigma$.

Given two presentations, $g: K \to K$ and $g': K' \to K'$ of $h: \Sigma \to \Sigma$ and $h': \Sigma' \to \Sigma'$, a map $f: K \to K'$ such that

$$
\begin{array}{ccc}
K & \overset{g}{\to} & K \\
f \downarrow & & \downarrow f \\
K' & \underset{g'}{\to} & K'
\end{array}
$$

is commutative, induces a map $F_i: \Sigma \to \Sigma'$, for $i \in Z$, by

$$F_i(x_0, x_1, \dots) = \begin{cases} (fx_i, fx_{i+1}, \dots) & \text{for } i \ge 0, \\ (g'^{-i}fx_0, g'^{-i}fx_1, \dots) & \text{for } i \le 0. \end{cases}$$

DEFINITION. By a *ladder map* from $\Sigma$ to $\Sigma'$ is meant such an $F_i$, $i \in Z$, for such an $f$.

These maps are interrelated; in particular, $F_i = F_0 h^i = h'^i F_0$ and $F_i h = h' F_i$.

---

[1] Research supported in part by NSF Grant 5591.

THEOREM A. *The only maps between generalized solenoids which commute with their shift maps are the ladder maps* (3.1).

That is, if $\phi: \Sigma \to \Sigma'$ such that $\phi h = h'\phi$, then there is an $f: K \to K'$ such that $fg = g'f$, and an $i \in Z$, such that $\phi = F_i$ (notation as above).

COROLLARY. *A necessary and sufficient condition that the shift maps presented by $g: K \to K$ and $g': K' \to K'$ be topologically conjugate is that there exist maps $r: K \to K'$, $r': K' \to K$ and a positive integer $m$ such that the following are commutative* (3.3):

$$(**) \qquad \begin{array}{ccc} K \xrightarrow{g} K & K \xrightarrow{g} K & K \xrightarrow{g^m} K \\ r\downarrow \quad \downarrow r & r'\uparrow \quad \uparrow r' & r \searrow r'\nearrow \quad \searrow r \\ K' \xrightarrow{g} K' & K' \xrightarrow{g} K' & K' \xrightarrow{g'^m} K' \end{array}$$

DEFINITION. If two maps $g, g'$ satisfy (**) for some $r, r'$ and $m$, we say $g$ is *shift equivalent* to $g'$. Notation: $g \sim_s g'$.

This is an equivalence relation and is useful in many categories. In particular, if $K$ and $K'$ are finite dimensional vector spaces and the maps are linear, we have, via the Weil Zeta function (see [3] or [4]),

THEOREM B. *If two linear maps are shift equivalent then their characteristic polynomials differ only by a factor of the form $\pm t^j$* (4.8).

Let $g: K \to K$ be a presentation of $h: \Sigma \to \Sigma$ and assume $g$ has a fixed point $x_0 \in K$, or what is equivalent, $h$ has a fixed point $(x_0, x_0, \ldots) \in \Sigma$. Then $g$ induces a map

$$g_*: \pi_1(K, x_0) \to \pi_1(K, x_0)$$

and by Theorem A the shift equivalence class of $g_*$ is an invariant of $h$, called the *shift class* of $h$. Notation: $S(h) = [g_*]$. More accurately, $S(h, x_0)$, as all of this depends upon the choice of the base point. Also from Theorem A we have

THEOREM C. *If $h$ and $h'$ are topologically conjugate shift maps of generalized solenoids which have fixed points, then $S(h) = S(h')$* (7.2).

The converse to Theorem C is not known and probably not true; the available $K$'s are just too diverse. We formulate a partial converse by introducing *elementary presentations* $g: K \to K$ as being presentations in which $K$ is a (smooth) wedge of one-spheres and $g$ leaves the base point $b \in K$ fixed. Then

THEOREM D. *Two shift maps which have elementary presentations are topologically conjugate iff their shift classes are equal* (7.3).

Theorem D does not cover all shift maps of generalized solenoids of course; but it covers a lot:

THEOREM E. *If $h$ is a shift map of a generalized solenoid then there is an integer $n$ such that $h^n$ has an elementary presentation* (5.2).

In the course of proving Theorems A and E we obtain

THEOREM F. *If $\Sigma$ is an oriented solenoid, then $\Sigma$ is the total space of a fiber bundle $C \to \Sigma \to S^1$, where $C$ is a Cantor set* (2.1).

The formulation of generalized solenoids given by Theorem F may have other applications; it would seem to be a good version to generalize to higher dimensions. As stated in [3] it is felt that a large part of this material will generalize to higher dimensions.

The proof of Theorem D makes use of a natural measure on $K$:

THEOREM G. *If $g: K \to K$ is a presentation of a connected generalized solenoid then there is a unique measure on $K$ relative to which $g$ is uniformly expanding* (6.3).

This type of result is probably true in much greater generality.

1. **Axioms for g: K → K and a criterion.** Differentiable, branched 1-manifolds are defined [3] just as the unbranched variety except that (finite) ramification points are allowed as long as the "angles" at a ramification point are either $0°$ or $180°$. Branched manifolds have tangent bundles and thus one defines as usual differentiable maps and immersions between them. For a compact branched 1-manifold $K$ and an immersion $g: K \to K$ we list the following:

Axiom 0. $g: K \to K$ is indecomposable (see below);

Axiom 1. $g$ is an expansion;

Axiom 2. $\Omega(g) = K$; and

Axiom 3°. each point of $K$ has a neighborhood $N$ such that for some $n$, $g^n(N)$ is a 1-cell.

Axiom 3° is weaker than the two versions of it which were given in [3]:

Axiom 3. Each point of $K$ has a neighborhood $N$ such that $g(N)$ is a 1-cell.

Axiom 3'. $g(B) \cap B = \varnothing$. ($B$ is the set of all branch points of $K$.)

In [3; 3.5] we show that if $g$ satisfies Axioms 1–3 it is shift equivalent (see §4) to $g_0$ which satisfies Axioms 1–4 and 3'. The same arguments can be used to show that if $g$ satisfies Axioms 1, 2, 3°, then $g$ is shift equivalent to $g$, which satisfies Axioms 1–3. The weakest version, 3°, allows the *elementary* presentations of (§5).

For completeness we also mention

Axiom 4. There is a finite set $A \subset K$ such that $g(A \cup B) \subset A$.

Axiom 0 means $K$ cannot be split into two nonempty, closed, invariant sets (see part 3 of the definition of *basic sets*, above). It was not assumed in [3]. Clearly if $g: K \to K$ satisfies Axioms 1–3, then there are disjoint $K_1, \ldots, K_n \subset N$ such that $g|K_i$ satisfies Axioms (0–3), $i = 1, \ldots, n$, and $K = K_1 \cup \ldots \cup K_n$. Thus whether Axiom 0 is assumed just affects the wording of theorems.

Axiom 2, that the nonwandering set of $g$ is all of $K$, is of its nature difficult to check. We close this section with a criterion, which though apparently stronger is equivalent in the presence of Axioms 0, 1, and 3 .

We assume (which we can by [3; 3.6]) that $K$ is divided into 1-cells $\{\sigma_i\}_{i=1}^n$ by vertices which include all branch points.

1.2. CRITERION. *If* $g: K \to K$ *satisfies Axioms* 0, 1, *and* 3°, *then* $g$ *satisfies Axiom* 2 *iff there is an integer* $m$ *such that* $g^m$ *maps each one simplex,* $\sigma_i$, *onto* $K$.

PROOF. First, if this criterion holds and $I \subset K$ is a subinterval, then for some $k$, $g^k(I)$ contains a 1-simplex, $\sigma_i$, as $g$ is an expansion. Then $g^{k+m}(I) \subset K \subset I$ which proves Axiom 2. The other way follows from (1.6), below.

1.3. LEMMA. *If* $g: K \to K$ *satisfies Axioms* 1, 2, 3°, *then so also does* $g^n$, $n = 1, 2, \ldots$.

PROOF. That $g^k$ is expanding and is "locally flat" is clear so that we need only verify Axiom 2. Suppose $I \subset K$ is a 1-cell. Then there is a periodic point $a \in I$ [**3**; Theorem A]; say $g^m(a) = a$. Then $g^{nm}(a) = a$ so that $g^{nm}(I) \cap I \neq \varnothing$. Therefore $\Omega(g^n) = K$.

1.4. LEMMA. *If* $g: K \to K$ *satisfies Axioms* 1, 2, 3° *and* $I$ *is an interval in* $K$, *then there is an integer* $m$ *such that* $g^m(I) \supset I$.

PROOF. Let $a_1, \ldots, a_{n-1}$ be the branch points in $I$ and $a_0, a_n$ its end points, labeled in their natural order. Choose periodic points $x_1, \ldots, x_n$ of $I$, one between each two of the $a_i$'s. Let $m$ be a common period of $x_1, \ldots, x_n$, and choose $m$ so large that $g^m([a_i, x_j])$ is longer than $I$, for each $i, j$. It follows that $g^m([a_i, a_{i+1}]) \supset [a_i, a_{i+1}]$ for each $i$ so that $g^m(I) \supset I$.

1.5. LEMMA. *Let* $g: K \to K$ *satisfy Axioms* 0, 1, 2, 3° *and let* $I$ *be an interval in* $K$. *Then* $\bigcup_{i=1}^{\infty} g^i(I)$ *is dense in* $K$.

PROOF. Assume false, let $I^\circ$ be the interior of $I$, let $A$ be the closure of $\bigcup_{i=1}^{\infty} g^i(I)$ and let $B$ be the closure of $K - A$. Note that $A$ is invariant under $g$. Likewise $B$, for if $b \in K - A$, $g(b) \notin \bigcup_{i=1}^{\infty} g^i(I^\circ)$, as then $b$ would be a wandering point. Hence $g(b) \in B$. Then by Axiom 0 there is a point $x \in A \cap B$.

Then there is an interval $J$ about $x$ intersecting both $\bigcup_{i=1}^{\infty} g^i(I^\circ)$ and $K - A$, such that $J$ contains no branch point of $K$ except (possibly) $x$. Choose a periodic point $a \in g^i(I^\circ) \cap J$; for a large period, $m$, of $a$, $g^{m+i}(I^\circ)$ contains an interval $J'$ having $x$ as an interior point. By Axiom 3, there is a neighborhood $N$ of $x$ and an integer $r$ such that $g^r(N) \subset g^r(J')$. But then there is a point $b \in (K - A) \cap N$ such that $g^r(b) \in g^r(J') \subset g^{m+i+r}(I^\circ)$, which is impossible. This contradiction completes the proof of (1.5).

1.6. LEMMA. *If* $g: K \to K$ *satisfies Axioms* 0–2, 3° *and* $I \subset K$ *is an interval, then* $g^m(I) \supset K$ *for some* $m$.

PROOF. By (1.4) there is an $n$ such that $g^n(I) \supset I$. Then $I \subset g^n(I) \subset g^{2n}(I) \subset \ldots$. Now suppose $x \in K$ and let $J$ be a small interval with $x$ an end point, containing no branch point except (possibly) $x$. Then by [**3**; Theorem A], (1.3) and (1.5), there is a periodic point $a \in J \cap g^{nk}(I)$. Hence for $m$ a large multiple of $nk$ and the period of $a$, $J \subset g^m(I)$. As $x$ has a neighborhood $N$ which is the union of finitely many such intervals $J$, it follows that $N \subset g^{ni}$ for some $i$. The lemma follows by the covering theorem.

2. **Solenoids as bundles over $S^1$.** In this section we assume that $K$ is connected and orientable and that $g: K \to K$ satisfies Axioms 1, 2, 3°. (Axiom 0 follows from connectedness.) We let $h: \Sigma \to \Sigma$ be determined by $g$ and $\pi_0: \Sigma \to K$, the projection of $\Sigma$ onto its 0th coordinate. Choose a point $p \in K$ and let $C = \pi_0^{-1}(p)$. Then

2.1. THEOREM. *There is a smooth bundle $C \to \Sigma \to S^1$.*

PROOF. $p$ has a neighborhood which is a union of intervals $[a_i, b_i]$ which we take to be oriented so that $a_i < p < b_i$. Now by [3; Theorem B], each point $x \in C$ has a neighborhood of the form $C_x \times [0, 1]$, which maps under $\pi_0$, like the projection onto $[0, 1]$, into one of the $[a_i, b_i]$. Finitely many of these cover $C$, and their union contains a neighborhood of the type $C \times [-\varepsilon, \varepsilon]$, with $C$ embedded as $C \times 0$. This is so chosen that the various $x \times [-\varepsilon, \varepsilon]$, $x \in C$, are coherently (say) positively oriented.

The characteristic map $f: C \to C$ is defined (as usual) as follows. For $x \in C$ construct a maximal arc $[x, y)$ subject to the condition that $(x, y)$ is positively oriented and does not hit $C$. We claim $y \in C$ and may set $y = fx$.

To prove this, note that we may assume $g$ is orientation preserving, as $g^2$ is in any case. Next, by Axiom 3, there is an $\varepsilon > 0$ such that

(*) if $N$ is a connected neighborhood of diameter $\leq \varepsilon$ then $g(N)$ is a 1-cell.

By (1.6), there is an $m > 0$ such that $g^m(N) \supset K$ for any interval of length $\geq \varepsilon$. Then let $[x_m, a_m] = I_m$ be an interval in $K$, positively oriented and with one end point $= x_m$ (the $m$th coordinate of $x$), of length $\varepsilon$. Then for each $i = 0, 1, \ldots, n - 1$, $g^{n-i}(I_m) = I_i$ is compact and connected. Extend these to define $I_i$, $i > m$, to be the component of $g^{-(i-m)}(I_m)$ which has $x_i$ as an end point. Then the inverse system

$$I_1 \leftarrow I_2 \leftarrow \ldots I_{m-1} \leftarrow I_m \leftarrow I_{m+1} \leftarrow \ldots$$

$$[0, 1] \quad [0, 1]$$

has a connected set as inverse limit, which by (*) is an arc (see [3, p. 480]). Call it $[x, a]$, as its initial point is $x$ by construction, and it is positively oriented. By choice of $m$, $\pi_0([x, z]) = g^m[x_m, a_m] = K$, so that $(x, z] \cap C \neq \emptyset$. The first point of $(x, z]$ in $C$ is $fx$, as required.

This guarantees that $\Sigma$ is an $S^1$-bundle with fiber $C$ and characteristic map $f$. It is not hard to see that the projection $\Sigma \to S^1$ can be chosen to be smooth, and an immersion. (Recall that $\Sigma$ has a tangent bundle—just the inverse limit of the tangent bundles of $K$. To say that $\pi$ is an immersion is obviously equivalent to saying that $\pi$ restricted to each arc in $\Sigma$ is an immersion.)

3. **The Ladder Theorem.**

3.1. LADDER LEMMA. *Suppose $g_i: K_i \to K_i$ satisfies Axioms 1–3 and determines the solenoid and shift map $h_i: \Sigma_i \to \Sigma_i$, $i = 1, 2$. Suppose further that $\Phi: \Sigma_1 \to \Sigma_2$*

is a map such that

$$\begin{array}{ccc} \Sigma_1 & \xrightarrow{h_1} & \Sigma_1 \\ \Phi\downarrow & & \downarrow\Phi \\ \Sigma_2 & \xrightarrow{h_2} & \Sigma_2 \end{array}$$

is commutative. Then there is a map $f: K_1 \to K_2$ and an integer $n$ such that

$$\begin{array}{ccc} K_1 & \xrightarrow{g_1} & K_1 \\ f\downarrow & & \downarrow f \\ K_2 & \xrightarrow{g_2} & K_2 \end{array}$$

is commutative, and

$$\Phi(x_0, x_1, \ldots) = (fx_n, fx_{n+1}, fx_{n+2}, \ldots)$$

for all $(x_0, x_1, \ldots) \in \Sigma_1$.

PROOF. *Case* 1. $g_i$ satisfies Axiom 4, $i = 1, 2$. This allows $K_i$ to be partitioned into 1-cells and vertices by branch points and possibly other points such that $g_i$ (vertex) = a vertex [3; 3.6]. Let $\pi_{ij}: \Sigma_i \to K_i$, $i = 1, 2$, be the projection map onto the $j$th coordinate, $j = 0, 1, 2, \ldots$. Then for each 1-cell $\sigma \in K_2$, $\Phi^{-1}\pi_{21}^{-1}(\sigma)$ has the form $[0, 1] \times C_\sigma$, where $C_\sigma$ is a Cantor set [3; Theorem B].

3.2. LEMMA. *There is an integer* $n = n_\sigma$ *such that if* $\Phi x$, $\Phi y \in \pi_{21}^{-1}(\sigma)$ *and* $\pi_{21}(\Phi x) \neq \pi_{21}(\Phi y)$, *then* $\pi_{1n}(x) \neq \pi_{1n}(y)$.

PROOF.    Assume the lemma is false, say for $\sigma \in K_2$, and let $\Phi^{-1}\pi_{21}^{-1}(\sigma) = [0, 1] \times C_\sigma$. Then either

(a) there are two points $x, y \in [0, 1] \times C_\sigma$ such that $\pi_{1n}(x) = \pi_{1n}(y)$ for all $n$ though $\pi_{21}\Phi(x) \neq \pi_{21}\Phi(y)$, or

(b) there is a point $x_0 \in [0, 1] \times C_\sigma$ and distinct points $x_i, y_i \in [0, 1] \times C_\sigma$ such that $x_i \to x_0$, $y_i \to x_0$, $\pi_{21}\Phi(x_i) \neq \pi_{21}\Phi(y_i)$, but $\pi_{1i}(x_i) = \pi_{1i}(y_i)$, $i = 1, 2, 3, \ldots$.
Case (a) cannot hold as then $x = y$ so that $\pi_{21}\Phi(x) = \pi_{21}\Phi(y)$. Thus (b) holds.

Now as $g_2$ is an expansion, there are constants $\lambda_1 > 1$ and $C > 0$, such that if $x, y \in \Sigma_2$ and $\pi_{21}(x) \neq \pi_{21}(y)$, then $\rho(h_2 x, h_2 y) > \min \{\lambda\rho(\pi_{21}x, \pi_{21}y), C\}$. Next, let $\varepsilon_n$ be the maximum distance between two points $x, y \in \Sigma_1$ such that their coordinates $x_i = y_i$ for $i \leq n$. That is,

$$\varepsilon_n = \max \{\rho(x, y): x, y \in \Sigma_1 \quad \text{and} \quad \pi_{2n}(x) = \pi_{2n}(y)\}.$$

Then $\varepsilon_n \to 0$ as $n \to \infty$.

Choose $n$ so that if $x, y \in \Sigma_1$, and $\rho(x, y) < \varepsilon_n$, then $\rho(\Phi x, \Phi y) < C/2$. There is a 1-cell $I \subset K_1$ such that $\pi_{1n}(x_0) \in \text{Int } I$ and $x_0 \in \pi_{1n}^{-1}(I)$. As $\pi_{1n}^{-1}(I)$ is a neighborhood of $x_0$, it follows that for some $i$ ($>n$) $x_i, y_i \in \pi_{1n}^{-1}(I)$ and $\pi_{1n}(x_i) = \pi_{1n}(y_i)$.

Then $\pi_{1n}(h_1^j x_i) = g_1^j \pi_{1n}(x_i) = g_1^j \pi_{1n}(y_i) = \pi_{1n} h_1^j(y_i)$ for all $j \geq 0$. Hence $\rho(h_1^j x_i, h_1^j y_i) < \varepsilon_n$, by definition of $\varepsilon_n$, for all $j \geq 0$. Then by our choice of $n$, $\rho(\Phi h_1^j(x_i), \Phi h_1^j(y_i)) < C/2$ for $j \geq 0$. Hence $\rho(\Phi h_1^j(x_i), \Phi h_1^j(y_i)) = \rho(h_2^j \Phi x_i, h_2^j \Phi y_i) \geq \lambda^j(\pi_{21}\Phi(x_i), \pi_2\Phi(y_i))$ for all $j \geq 0$, which is impossible. This completes the proof of (3.2).

There is an $n > n_\sigma$ for all $\sigma \in K_1$, where $n_\sigma$ is as in (3.2). Then the diagram

$$
\begin{array}{ccc}
K_1 & \xleftarrow{\pi_{1n}} & \Sigma_1 \\
f \downarrow & & \downarrow \Phi \\
K_2 & \xleftarrow{\pi_{21}} & \Sigma_2
\end{array}
$$

defines $f$, by $f = \pi_{2n}\Phi\pi_{1n}^{-1}$. $f$ is well defined, as otherwise there would be $x, y \in \Phi^{-1}\pi_{21}^{-1}(\sigma)$ with $\pi_{21}(\Phi x) = \pi_{21}(\Phi y)$ but $\pi_{1n}(x) = \pi_{1n}(y)$, contradicting the choice of $n$.

Next, if $x, y \in \Sigma_1$ and $\pi_{22}x \neq \pi_{22}y$ then $\pi_{1n+1}x \neq \pi_{1n+1}y$. For otherwise,

$$\pi_{21}h^{-1}x = \pi_{22}x \neq \pi_{22}y = \pi_{21}h^{-1}y$$

though

$$\pi_{1n}h^{-1}x = \pi_{1n+1}x = \pi_{1n+1}y = \pi_{1n}h^{-1}y$$

contradicting the definition of $n$. Hence the diagram

$$
\begin{array}{ccccc}
K_1 & \xleftarrow{g_1} & K_1 & \xleftarrow{\pi_{1n+1}} & \Sigma_1 \\
\downarrow f & & \downarrow f_1 & & \downarrow \Phi \\
K_2 & \xleftarrow{g_2} & K_2 & \xleftarrow{\pi_{22}} & \Sigma_2
\end{array}
$$

defines $f_1$ by $f_1 = \pi_{22}\Phi\pi_{1n+1}^{-1}$. We claim $f_1 = f$. For let $a \in K_1$, say $a = \pi_{1n+1}x$. Then $f_1 a = \pi_{22}\Phi x$, whereas $a = \pi_{1n}h_1 x$, so that $fa = \pi_{21}\Phi h_1 x = \pi_{21}h_2\Phi x = \pi_{22}\Phi x$.

Iterating this procedure completes the proof of case 1 of (3.1), except that we must show that $f$ is continuous. It is, for otherwise there would be a sequence $x_{in} \in K_1$, $x_{1n}, x_{2n}, \ldots \to x_{0n}$ such that $f(x_{in})$ does not converge to $f(x_{0n})$. But as $g_1$ is onto, for each $i = 0, 1, \ldots$, there is a point $x_i \in \Sigma_1$ such that the $n$th coordinate of $x_i$ is $x_{in}$. This means that $x_i \to x_0$ though $\Phi(x_i)$ does not converge to $\Phi(x_0)$, which is impossible.

*Case* 2. $g_i$ satisfies Axioms 1–3 only. Then by [3; 3.6] there is a commutative diagram

$$
\begin{array}{ccc}
K_i & \xleftarrow{g_i} & K_i \\
r_i \downarrow \uparrow s_i & \searrow & \downarrow r_i \\
K_i' & \xleftarrow{g_i'} & K_i'
\end{array}
$$

where $g_i'$ satisfies Axioms 1–4, $i = 1, 2$. Thus we have a diagram

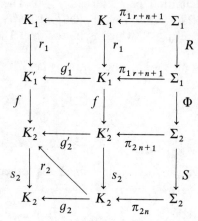

for $n = 0, 1, 2, \ldots$. Here, $R$ is induced by $r_1$, $S$ by $s_2$ and $\Phi: \Sigma_1 \to \Sigma_2$ is a map. Note that $R$ and $S$ are homeomorphisms, so that if we take $\Phi' = S\Phi R$, we get the following "ladder":

$$
\begin{array}{ccccc}
K_1 & \xleftarrow{g_1} & K_1 & \xleftarrow{\pi_{1\,r+n+1}} & \Sigma_1 \\
\downarrow{\scriptstyle s_2 f r_1} & & \downarrow{\scriptstyle s_2 f r_1} & & \downarrow{\scriptstyle \Phi} \\
K_2 & \xleftarrow{g_2} & K_2 & \xleftarrow{\pi_{2n}} & \Sigma_2
\end{array}
$$

for all $n$, which is what was to be proved.

3.3. THEOREM. *If $h_i: \Sigma_i \to \Sigma_i$ is the shift map of a solenoid presented by $g_i: K_i \to K_i$, satisfying Axioms 0–2, $3°$, for $i = 1, 2$, then $h_1$ and $h_2$ are topologically conjugate iff there exist an integer $m$ and maps $f: K_1 \to K_2$, $f': K_2 \to K_1$ such that the following diagrams are commutative.*

$$
\text{(i)}\quad
\begin{array}{ccc}
K_1 & \xrightarrow{g_1} & K_1 \\
\downarrow{\scriptstyle f} & & \downarrow{\scriptstyle f} \\
K_2 & \xrightarrow{g_2} & K_2
\end{array}
\qquad
\text{(ii)}\quad
\begin{array}{ccc}
K_1 & \xrightarrow{g_1} & K_1 \\
\uparrow{\scriptstyle f'} & & \uparrow{\scriptstyle f'} \\
K_2 & \xrightarrow{g_2} & K_2
\end{array}
\qquad
\text{(iii)}\quad
\begin{array}{ccc}
K_1 & \xrightarrow{g_1^m} & K_1 \\
{\scriptstyle f}\searrow & \nearrow{\scriptstyle f'} & \searrow{\scriptstyle f} \\
K_2 & \xrightarrow{g_2^m} & K_2
\end{array}
$$

PROOF. Assume $\Phi: \Sigma_1 \to \Sigma_2$ is a topological conjugacy of $h_1$ and $h_2$. Then applying the ladder lemma to $\Phi$ and $\Phi^{-1}$, we get maps $f: K_1 \to K_2$ and $f': K_2 \to K_1$ such that

$$
\begin{aligned}
\Phi(x_0, x_1, \ldots) &= (f(x_r), f(x_{r+1}), \ldots), \quad x \in \Sigma_1, \\
\Phi^{-1}(y_0, y_1, \ldots) &= (f'(x_s), f'(x_{s+1}), \ldots), \quad y \in \Sigma_2
\end{aligned}
$$

and such that (i) and (ii) are commutative.

Let $m = r + s$. Then $\Phi^{-1}\Phi(x_0, x_1, \ldots) = \Phi^{-1}(f(x_r), f(x_{r+1}), \ldots) = (f'f(x_{r+s}),$ $f'f(x_{r+s+1}), \ldots)$. Hence $g_1^m(x_m) = x_0 = f'f(x_m)$, all $x_m \in K_1$. That is, $f'f = g_1^m$. Similarly $ff' = g_2^m$, which proves (iii) is commutative and completes the proof of (3.3) in this direction.

Now suppose there exist maps $f, f'$ and an $m$ satisfying (i–iii). Then $f$ defines a map $\Phi: \Sigma_1 \to \Sigma_2$ by $\Phi(x_0, x_1, \ldots) = (f(x_0), f(x_1), \ldots)$ and $f'$ defines a map $\Psi: \Sigma_2 \to \Sigma_1$ by

$$\Psi(y_0, y_1, \ldots) = (f'(x_m), f'(x_{m+1}), \ldots).$$

Then $\Psi\,\Phi(x_0, x_1, \ldots) = (f'fx_m, f'fx_{m+1}, \ldots) = (g_1^m x_m, g_1^m x_{m+1}, \ldots) = (x_0, x_1, \ldots)$. Hence $\Psi\,\Phi = 1$. Similarly, $\Phi\Psi = 1$.

### 4. Shift equivalence.

4.0. DEFINITION. Two maps $f: X \to X$, $g: Y \to Y$ are *shift equivalent*, provided there are maps $r: X \to Y$, $r': Y \to X$ and an integer $m$ such that the following diagrams are commutative:

(4.1)
$$
\begin{array}{ccc}
X \xrightarrow{f} X & \qquad X \xrightarrow{f} X & \qquad X \xrightarrow{f^m} X \\
r\downarrow \quad \downarrow r & \qquad r'\uparrow \quad \uparrow r' & \qquad r\downarrow \ \nearrow^{r'} \ \downarrow r \\
Y \xrightarrow{g} Y & \qquad Y \xrightarrow{g} Y & \qquad Y \xrightarrow{\quad} Y
\end{array}
$$

Notation: $f \sim_s g$.

4.2. REMARK. Shift equivalence is an equivalence relation.

PROOF. That shift equivalence is reflexive and symmetric is clear. Suppose $f \sim_s g$, $g \sim_s h$, where $f: X \to X$, $g: Y \to Y$, $h: Z \to Z$. Then there are maps $r: X \to Y$, $r': Y \to X$, $s: Y \to Z$, $s': Z \to Y$ and integers $m, n$ such that (5.1) holds for the two sets of data. Then the following diagrams are commutative.

$$
\begin{array}{ccc}
X \xrightarrow{f} X & \qquad X \xrightarrow{f} X \\
t\downarrow \quad \downarrow t & \qquad r'\uparrow \quad \uparrow r' \\
Y \xrightarrow{g} Y & \qquad Y \xrightarrow{g} Y \\
u\downarrow \quad \downarrow u & \qquad s'\uparrow \quad \uparrow s' \\
Z \xrightarrow{h} Z & \qquad Z \xrightarrow{h} Z
\end{array}
$$

where $t = r(r'r)^k$, $u = s(s's)^l$, for any choice of $k, l$.

If we choose $k = n - 1$, $l = m - 1$, we have in addition

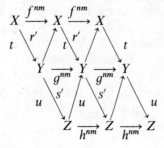

so that $ut: X \to Z$, $r's': Z \to X$ are the required maps and $2nm$ the required integer.

4.3. DEFINITION. For $f: X \to X$, define $X_{\leftarrow}$ to be the inverse limit of the sequence

$$X \underset{f}{\leftarrow} X \underset{f}{\leftarrow} X \leftarrow \ldots$$

and $f_{\leftarrow}: X_{\leftarrow} \to X_{\leftarrow}$ by

$$f_{\leftarrow}(x_0, x_1, \ldots) = (fx_0, fx_1, \ldots).$$

Then $f_{\leftarrow}$ is a homeomorphism. The following is the (easy) converse of the Ladder Theorem (3.3) but it is true quite generally.

4.4. REMARK. If $f \sim_s g$, then $f_{\leftarrow}$ and $g_{\leftarrow}$ are topologically conjugate.

PROOF. We are given $f: X \to X$, $g: Y \to Y$, $r: X \to Y$, $r': Y \to X$ and $m$ satisfying (4.1). Since the ladder

$$\begin{array}{ccccc}
X & \overset{f}{\leftarrow} & X & \overset{f}{\leftarrow} & X \leftarrow \ldots \\
r\downarrow & & r\downarrow & & r\downarrow \\
Y & \underset{g}{\leftarrow} & Y & \underset{g}{\leftarrow} & Y \leftarrow \ldots
\end{array}$$

is commutative, it defines a map $R_i: X_{\leftarrow} \to Y_{\leftarrow}$ by $R_i(x_0, x_1, \ldots) = (rx_i, rx_{i+1}, \ldots)$ for any choice of $i = 0, 1, 2, \ldots$. Similarly $r'$ defines a map $R'_j: Y_{\leftarrow} \to X_{\leftarrow}$ for any choice of $j$. Choose $i + j = m$; then $R'_j R_i(x_0, x_1, \ldots) = (r'rx_m, r'rx_{m+1}, \ldots) = (f^m x_m, f^m x_{m+1}, \ldots) = (x_0, x_1, \ldots)$ so that $R'_j R_i = 1$. Similarly, $R_i R'_j = 1$.

In the special case $m = 1$ of the definition (4.0) of $f \sim_s g$ means

(4.5)            $f = sr$    and    $g = rs$,       $r: X \to Y, s: Y \to X$.

We will use the notation $f \sim_* g$ for this. (It is *not* transitive.)

4.6. LEMMA. *If* $f \sim_s g$, $r, s$ *are the maps and* $m$ *is the integer of the definition, then there are maps* $f_0 = f_1, f_2, \ldots, f_m = g|r(X)$ *such that* $f_{i-1} \sim_* f_i$, $i = 1, \ldots, m$.

PROOF. For $x, y \in X$, say $x \sim_1 y$ iff $fx = fy$ and $rx = ry$. Then $\sim_1$ is an equivalence relation and yields a quotient space $X_1$ roughly as good as $X$ and $Y$. E.g., if they are Hausdorff or metric, or abelian, so is $X_1$. Let $r_1: X \to X_1$ be the quotient map. By definition, both $r$ and $f$ factor through $r_1$, say $f = s_1 r_1$ and $r = t_1 r_1$. Define $f_1: X_1 \to X_1$ by $f_1 = r_1 s_1$.

Next, for $x, y \in X$, let $x \sim_2 y$ iff $f^2 x = f^2 y$ and $rx = ry$. This is again an equivalence relation and yields the quotient space $X_2$. The quotient map has the form $r_2 r_1: X \to X_2$. Also $r$ can be factored as $r = t_2 r_2 r_1$. We claim that $f^2$ can be factored as $f^2 = s_1 s_2$ where $s_2: X_2 \to X_1$. Indeed, we define $s_2 = f_1 r_2^{-1}$ and need only check that $s_2$ is well defined. Thus let $x, y \in X$ determine the same point $r_2 r_1 x = r_2 r_1 y$ in $X_2$. It suffices to show that $f_1 r_1 x = f_1 r_1 y$, or equivalently, that $r_1 fx = r_1 fy$, or that $fx \sim_1 fy$. But $f(fx) = f(fy)$ and $rfx = grx = gry = rfy$ which shows $fx \sim_1 fy$. (The equality $grx = gry$ follows from $rx = ry$.) Again introduce $f_2 = r_2 s_2: X_2 \to X_2$.

So far we have the diagram

(A)

$$X \xrightarrow{f} X \xrightarrow{f} X \longrightarrow \cdots$$

$$r_1 \downarrow \quad s_1 \quad \downarrow r_1 \quad s_1 \quad \downarrow r_1$$

$$X_1 \xrightarrow[f_1]{} X_1 \xrightarrow[f_1]{} X_1 \longrightarrow \cdots$$

$$r_2 \downarrow \quad s_2 \quad \downarrow r_2 \quad s_2 \quad \downarrow r_2$$

$$X_2 \xrightarrow[f_2]{} X_2 \xrightarrow[f_2]{} X_2 \longrightarrow \cdots$$

Clearly this can be continued, getting $X_3, \ldots, X_m$, $f_i : X_i \to X_i$, where $f_{i-1} \sim_* f_i$, $i = 1, \ldots, m$. Note that $X_m$ can be considered as the image $r(X)$ of $X$ in $Y$, because $(f^m x = f^m y$ and $rx = ry) \Leftrightarrow rx = ry$. That is, $r = r_m r_{m-1} \cdots r_1$.

We claim $s' = s_1 s_2 \ldots s_m = s|r(X)$. For, if $rx \in r(X)$, $s(rx) = f^m x = s'(rx)$, by commutativity of the large "scaffold" diagram like (A). Finally, $grx = frx = rf_m x$, i.e., $g|r(X) = f_m$. This completes the proof of (4.6).

4.7. REMARK. Since the definitions and proofs of this section are entirely formal they apply in most categories: spaces and maps, groups and homomorphisms, vector spaces and linear maps. In particular, we have the following corollary to (4.6):

4.8. COROLLARY. *Let $V, W$ be finite dimensional vector spaces and let $f : V \to V$, $g : W \to W$ be shift equivalent linear maps. Then the characteristic polynomials of $f$ and $g$ are equal, modulo a factor of the form $\pm t^j$.*

PROOF. Let $r : V \to W$, $s : W \to V$ be the maps and $m$ the integer given by the definition of shift equivalence and let $W' = $ image of $r$. Then applying (4.6) we have

$$\operatorname{tr} f^i = \operatorname{tr}(g|W')^i$$

for $i = 1, 2, \ldots$, as $\operatorname{tr}(AB) = \operatorname{tr}(BA)$. Therefore the corresponding Weil zeta functions ([2, p. 768] or [3, p. 486]) are equal:

$$(-t)^p \chi_1(1/t) = (-t)^q \chi_2(1/t)$$

where $\chi_1$ and $\chi_2$ are the characteristic polynomials of $f$ and $g|W'$. This proves the proposition for $f$ and $g|W'$. But since $g$ leaves $W'$ invariant $(gr = rf)$ and $g^m(W) \subset W'$, it follows that the characteristic polynomial of $g$ is the same as that of $g|W'$, modulo a factor of the form $\pm t^j$, which proves (4.8).

5. **Elementary presentations.**

5.1. DEFINITION. By an *elementary branched 1-manifold $K$* we mean one which is topologically a wedge of 1-spheres. In detail, $K$ has a single vertex $b$ and a number of 1-cells of which

(a) some (type $O$) leave $b$ to the right and return to $b$ from the left;

(b) some (type $R$) leave and return rightwards;

(c) some (type $L$) leave and return leftwards.

Note that each of these 1-cells forms a 1-cycle, of which type $O$ are orientable, the others not. If $K$ is an elementary branched 1-manifold and $g: K \to K$ satisfies Axioms 1, 2, $3^\circ$, then we say $g$ is an *elementary presentation* of the solenoid

$$K \underset{g}{\leftarrow} K \underset{g}{\leftarrow} K \leftarrow \ldots$$

and shift map $h$.

5.2. THEOREM. *Suppose $\Sigma$ is a connected solenoid with shift map $h$. Then there is an integer $m$ such that $(\Sigma, h^m)$ has an elementary presentation $g: K \to K$.*

PROOF. Let $g_0: K_0 \to K_0$ be a presentation of $(\Sigma, h)$ satisfying Axioms 1–3. Then

5.3. LEMMA. *There is an integer $m$ and a point $x_0$ such that*
(a) $g^m(x_0) = x_0$; *and*
(b) *each (smoothly) embedded one-sphere in $K$ contains a point of $g^{-m}(x_0)$.*

PROOF. There is a finite set of open 1-cells $I_0, I_1, \ldots, I_r$ such that any embedded 1-sphere in $K$ contains some $I_i$, $i = 0, 1, \ldots, r$. Then there is an integer $n$ such that $f^n(I_i) \supset I_0$, $i = 0, 1, \ldots, r$. Choose $x_0 \in I_0$, a periodic point, say of period $s$, and let $m = ns$. Then $g^m(x_0) = x_0$ and for each integer $i = 0, 1, \ldots r$, $x_0 \in g^m(I_i)$ so that $g^{-m}(x_0) \cap I_i \neq \varnothing$. This last clearly implies (b) and proves (5.3).

Next, let $x_1, \ldots, x_t$ be a finite set such that $g_0^m(x_i) = x_0$, $i = 1, \ldots, t$, and each embedded 1-sphere intersects $\{x_0, x_1, \ldots, x_t\}$. If we identify the points $x_0, \ldots, x_t$ to a single point, $b$, we form a new branched 1-manifold $K_1$ with $b$ a point of high ramification. We also identify the tangent space at $x_i$ with that at $x_0$ under the map $Dg^m$, $i = 1, 2, \ldots, r$.

Note that there is the commutative diagram:

$$
\begin{array}{ccc}
K_0 & \xleftarrow{g_0^m} & K_0 \\
{\scriptstyle r}\downarrow & {\scriptstyle s}\searrow & \downarrow{\scriptstyle r} \\
K_1 & \xleftarrow{g_1} & K_1
\end{array}
$$

in which $r$ is the decomposition map, $s$ is guaranteed by the fact that $g_0^m(x_i) = x_0$, and $g_1 = rs$. Note that $g_1$, $r$, and $s$ are immersions. Finally that $g_1$ satisfies Axioms 1, 2, $3^\circ$ and, except at the point $b$, the stronger Axiom 3, is easily checked. Then

(5.4)                    each embedded 1-sphere of $K_1$ contains $b$.

If $K_1$ has no other branch point, then we have found an elementary presentation of $(\Sigma, h^m)$. Thus we need only show that we may "remove" any other branch points, by a reduction process. To this end, we introduce the

5.5. *Four models for reduction.* Let $V$, $Y$, $X$ and $C$ be the branch manifolds and boundaries as in Figure 1.

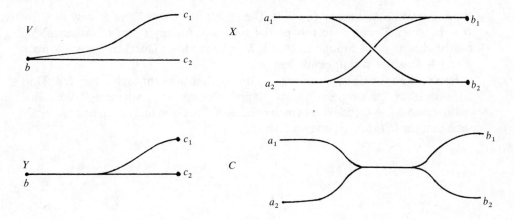

Note that there are "natural" maps $V \to Y$ and $X \to C$. The *stem* of $C$ is $[a_1, b_1] \cap [a_2, b_2]$. The stem of $Y$ is $[b, c_1] \cap [b, c_2]$.

We proceed to remove these branch points by a series of two types of "moves".

(1) Remove a copy of $C$ from $K - b$ and replace it with a copy of $X$.

After all copies of $C$ have been removed we begin the second series.

(2) Remove a copy of $Y$ and replace it with a copy of $V$.

At each move we pass from $g_i: K_i \to K_i$ to a shift equivalent $g_{i+1}: K_{i+1} \to K_{i+1}$. With the exception of $b$, no branch point of ramification higher than three is ever allowed. Property (5.4) is maintained; we call the distinguished branch point $b$ throughout. To show that this process converges, we introduce two weight functions:

$w_1(K)$ is the number of stems of $C$-sets in $K - b$.

$w_2(K)$ is the number of stems of $Y$-sets in $K - b$.

5.6. REMARK. If $C_0$ is a copy of $C$ in $K - b$, where $C_0 - \{a_1, a_2, b_1, b_2\}$ is open in $K$, and $K'$ is formed by replacing $C_0$ with a copy $X_0$ of $X$, then $w_2(K') < w_2(K)$.

PROOF. For if $C_1'$ is a copy of $C$ remaining in $K'$, such that

(a) $C_1' \cap X_0 = \varnothing$. Then $C_1'$ corresponds to $C_1 \subset K$;

(b) $C_1' \cap X_0 = [a_i, b_j]$, then $C_1'$ corresponds to $C_1 \subset K$ where $C_1 \cap C_0 = [a_i, b_j]$; or

(c) $C_1' \cap X_0 = [a_1, b_1] \cup [a_1, b_2]$ (or one of the four other copies of $Y$ in $X_0$), then there is a corresponding $C_1 \subset K$, where $C_1 \cap C_0 = [a_1, b_1] \cup [a_1, b_2]$, etc. Finally, note that the stem of $C_0$ has been removed and not replaced so that $w(K) = w(K') + 1$.

5.7. REMARK. If $Y_0$ is a copy of $Y$ in $K$ with $Y_0 - \{b, c_1, c_2\}$ open in $K$ and $K'$ is formed by replacing $Y_0$ with a copy of $V$, then $w_2(K) = w_2(K') + 1$.

PROOF. Similar to and easier than (5.6).

For $g: K \to K$ satisfying Axioms 1–3, condition (5.4) and $S_1, S_2$ stems of copies of $C$ in $K - b$, we say $S_1 < S_2$ provided $g^n(S_1) \subset S_2$ for some $n > 0$. This is a partial ordering; in particular $S_1 < S_1$ is impossible as $g$ is an expansion.

**5.8. REMARK.** Assume $w_1(K) > 0$. Then there is a stem $S_0$ of a copy of $C$ in $K - b$, minimal relative to this partial ordering. A copy $C_0$ of $C$ with stem $S_0$ can be chosen small enough so that if $K'$ is formed as (5.6), then there is a map $g': K' \to K'$ which is shift equivalent to $g$.

**PROOF.** We take $C_0$ small enough so that $g|C_0$ factors through a one-cell. This is possible, as the stronger Axiom 3 applies except at $b$, where $3°$ holds. The natural map $X_0 \to C_0$ provides a map $r: K' \to K$. We claim there is a map $s: K \to K'$ such that the following is commutative:

$$
\begin{array}{ccc}
K' & \xrightarrow{\ g'\ } & K' \\
{\scriptstyle r}\downarrow & {\scriptstyle s}\nearrow & \downarrow{\scriptstyle r} \\
K & \xrightarrow{\ g\ } & K
\end{array}
$$

For suppose $x \in K$ and consider

*Case* 1. $g(x) \notin C_0$. Then $s(x)$ is the unique point $r^{-1}g(x)$.

*Case* 2. $x \in I_1 \cap I_2$, where $g(I_i) = [a_i, b_i]$, $i = 1, 2$. This is impossible, by the minimality of $S_0$.

*Case* 3. $x \in I_1 \cap I_2$, where $g(I_j) = [a_i, b_j]$, $j = 1, 2$ for some $i = 1, 2$. Then define $g'|I_j$ to map $I_j$ to $[a_i, b_j] \subset K'$.

*Case* 4. For some $i, j$, $x$ lies on one or more 1-cells $I$, each of which maps under $g$ to $[a_i, b_j] \subset C_0$. Then $s$ takes such an $I$ to $[a_i, b_j] \subset X_0$, so that $rs = g$.

Similarly, for $g: K \to K$ satisfying (5.4), Axioms 1, 2, $3°$ and except at $b$ the stronger Axiom 3, $w_1(K) = 0$ and $S_1, S_2$ two stems of copies of $Y \subset K - b$, then $Y_1 < Y_2$ provided $g^n(S_1) \subset S_2$ for some $n$. This is again a partial ordering and just as before one easily proves

**5.9. REMARK.** Assume $w_2(K) > 0$. Then there is a stem $S_0$ of a copy of $Y$ in $K - b$, minimal relative to this partial ordering. A copy $Y_0$ of $Y$ with stem $S_0$ can be chosen small enough so that if $K'$ is formed as in (5.7), then there is a map $g': K' \to K'$ which is shift equivalent to $g$.

This completes the reduction process and establishes Theorem (5.2).

6. **Measure.** The purpose of this section is to derive a natural measure on a branched 1-manifold $K$, given that $g: K \to K$ satisfies Axioms 1–2, $3°$. We use the following part of the Frobenius-Perron theorem [1]:

6.1. Any square matrix with nonnegative (positive) entries has a unique positive eigenvalue $\lambda$ and a unique associated eigenvector with only nonnegative (positve) entries.

6.2. If $g: K \to K$ is an elementary presentation then there is a measure $\mu$ on $K$ and real $\lambda > 1$ such that

(a) $\mu$ (open set) $> 0$,

(b) $\mu(gI) = \lambda\mu(I)$ for any small interval $I \subset K$,

(c) $\mu(K) = 1$,

(d) $\mu$ and $\lambda$ are unique.

PROOF. Let $e_1, ..., e_n$ denote the one cells of $K$, with a chosen orientation. Then as $g|e_i$ is an immersion (and hence doesn't double back) it determines a well defined word in the $e_j$. Let $A_{ji}$ be the number of times this word "covers" $e_j$, regardless of the sign (or direction) of the covering. Then $A$ is an $n \times n$ matrix so that 6.1 applies, yielding an eigenvalue $\lambda > 0$ and an eigenvector $r = (r_1, ..., r_n)$ with nonnegative entries.

The criterion (1.2) says that for some $k$, $A^k$ has only positive entries; thus $A^{2k}$ has entries $\geq n$. Then $A^{2k}r = \lambda^{2k}r$ has entries which are term by term larger than those of $r$, i.e. $\lambda > 1$. We choose $r$ so that $\sum_{i=1}^{n} r_i = 1$, $r_i$ the coordinates of $r$. Then assign the measure $r_i$ to $e_i$, $i = 1, ..., n$.

The total "length" of $g(e_i)$, counting multiplicities, is $l(ge_i)$ where

$$\pm l(ge_i) = \Sigma A_{ji}\mu(e_j) = \Sigma A_{ji}r_j = A(r_i) = \lambda r_i.$$

Now if $I$ is a 1-cell in $K$ such that $g^t(I) = e_i$, covering just once, we define $\mu(I) = \lambda^{-t}r_i$. This is clearly consistent and assigns measures to arbitrarily small intervals in $K$. But then by a well known procedure, this determines a measure on $K$ such that (b) and (c) hold. As each open set in $K$ contains an interval $I$, and $g^t(I) \supset K$ for some $t$ (1.2), (a) follows.

6.3. LEMMA. *If $\mu, \lambda$ satisfy (6.2 a–c) for $g: K \to K$, and $g': K' \to K'$ is shift equivalent, to $g$, then $\mu$ induces a measure $\mu'$ on $K'$ so that $\mu', \lambda$ satisfy (6.2 a–c) for $g'$.*

PROOF. Let $r: K \to K'$, $s: K' \to K$, $m$ be as given in the definition of shift equivalence. Then for a small interval $I \subset K'$, set

$$\mu'(I) = \sum_{i=0}^{m-1} \lambda^{m-i}\mu(sg'^iI).$$

Then

$$\mu'(g'I) = \sum_{i=1}^{m-1} \lambda^{m-i+1}\mu(sg'^iI) + \lambda\mu(sg'^mI)$$

$$= \sum_{i=1}^{m-1} \lambda^{m-i+1}\mu(sg'^iI) + \lambda\mu(g^msI)$$

$$= \lambda \sum_{i=1}^{m-1} \lambda^{m-i}\mu(sg'^iI) + \lambda \cdot \lambda^m\mu(sI)$$

$$= \lambda\mu'(I).$$

This determines a measure which is normalized as usual.

6.4. COROLLARY. *If $K$ is connected and $g: K \to K$ satisfies Axioms 1–3 then there is a measure $\mu$ for $K$ and $\lambda > 1$ such that $\mu, \lambda$ satisfy (6.2 a–c).*

PROOF. Then by 5.2 $g^k: K \to K$ is shift equivalent to an elementary presentation so that by 6.3 we have $\mu_0, \lambda_0 > 1$ for $g^k$. Let $\lambda = \lambda_0^{1/k}$ and for small 1-cells $I \subset K$ define

$$\mu(I) = \sum_{i=0}^{k-1} \lambda^{k-i}\mu_0(g^iI).$$

As above, this is the required measure when normalized.

## 7. The shift class of a shift map and the classification theorem.

7.1. DEFINITION. Suppose $h$ is a shift map of a solenoid $\Sigma$ and assume that $h$ has a fixed point $x$. Choose a presentation $g: K \to K$ and note that $g$ has a fixed point $x_0$ such that $x = (x_0, x_0, x_0, \ldots)$ in this presentation. Then let $S(h)$ be the shift equivalence class of

$$g_*: \pi_1(K, x_0) \to \pi_1(K, x_0).$$

7.2. LEMMA. $S(h)$ does not depend upon the choice of the presentation.

PROOF. Let $g': K' \to K'$ be another presentation and $x_0'$ the corresponding fixed point of $g'$. Then by the Ladder Theorem (3.3) there are diagrams

$$
\begin{array}{ccc}
K \xrightarrow{g} K & \qquad & K \xrightarrow{g} K \\
r \downarrow \quad \downarrow r & \qquad & r' \uparrow \quad \uparrow r' \qquad r'r = g^n \\
K' \xrightarrow{g'} K' & \qquad & K' \xrightarrow{g'} K' \qquad rr' = g'^n
\end{array}
$$

in which $r(x_0) = x_0'$ and $r'(x_0') = x_0$. Then $r_*: \pi_1(K, x_0) \to \pi_1(K', x_0')$ and $r_*': \pi_1(K', x_0') \to \pi_1(K, x_0)$ are the required maps and $n$ the integer to show that $g_*$ is shift equivalent to $g_*'$.

7.3. 1. CLASSIFICATION THEOREM. *Let $h$ and $h'$ be shift maps of solenoids and assume $h$ and $h'$ have elementary presentations. Then $h$ is topologically conjugate to $h'$ if and only if $S(h) = S(h')$.*

The proof is in several steps. First we must consider the fundamental group $\pi_1(K, b)$ of an elementary branched 1-manifold (see 5.1). We see from (5.1) that $\pi_1(K, b)$ is free on a set consisting of three distinguished subsets of generators: (1) $O$ or orientable, (2) $R$ or right nonorientable and (3) $L$ or left nonorientable. In order that a homeomorphism $\alpha: \pi_1(K, b) \to \pi_1(K', b')$ be induced by an immersion $f: (K, b) \to (K', b')$ it must satisfy certain special conditions. This is simplest when no nonorientable generators occur: $f$ is either orientation preserving in which case each word $\alpha(x)$, ($x$ a generator of $\pi_1(K, b)$) is a word with only positive exponents, in the distinguished generators of $\pi_1(K', b')$, or $f$ is orientation reversing, in which case only negative exponents occur in the words $\alpha(x)$.

The general case is somewhat tedious. We write $x \in 0^+$ or $x^{-1} \in O^-$ for $x \in O$, etc.; the corresponding sets of generators for $\pi_1(K', b')$ are written $O'$, $O'^+$, etc.

7.3.1. IMMERSION CONDITION. If $x \in O \cup R \cup L$, then $\alpha(x) = y_1 y_2 \ldots y_r$, where each $y_i$ or its inverse is in $O' \cup R' \cup L'$, and satisfies:

(i) $\qquad y_{i+1} \in \begin{array}{l} O^+ \cup R^{\pm} \quad \text{if } y_i \in O^+ \cup L^{\pm}, \\ O^- \cup L^{\pm} \quad \text{otherwise;} \end{array}$

(ii) $\qquad \alpha$ is locally orientation preserving [reversing] at $b$, i.e.

(a) $\qquad y_1 \in \begin{cases} O^+ \cup R^{\pm} & \text{if } x \in O^+ \cup R^{\pm} [O^- \cup L^{\pm}], \\ O^- \cup L^{\pm} & \text{otherwise;} \end{cases}$

(b) $\qquad y_r \in \begin{cases} O^+ \cup L^{\pm} & \text{if } x \in O^+ \cup R^{\pm} [O^- \cup R^{\pm}], \\ O^- \cup R & \text{otherwise.} \end{cases}$

Note that condition (i) says that the immersion $f$ which is to induce $\alpha$ doesn't double back on any of the 1-cells of $K$. (iia) says (in the orientable case) that if a 1-cell $e$ begins pointing rightward, then so does its image under $f$. (iib) says that such an $e$ will end pointing leftward, as will its image, etc.

7.3.2. LEMMA. *Given two elementary presentations* $g: K \to K$, $g': K' \to K'$ *and a homeomorphism* $\alpha: \pi_1(K, b) \to \pi_1(K', b')$ *satisfying* (7.3.1) *such that*

$$\begin{array}{ccc} \pi_1(K, b) & \overset{g_*}{\to} & \pi_1(K, b) \\ \downarrow \alpha & & \downarrow \alpha \\ \pi_1(K', b') & \underset{g'_*}{\to} & \pi_1(K', b') \end{array}$$

*is commutative, there is a unique immersion* $f: (K, b) \to (K', b')$ *such that* $h_* = \alpha$ *and*

$$\begin{array}{ccc} K & \overset{g}{\to} & K \\ f \downarrow & & \downarrow f \\ K' & \overset{g'}{\to} & K' \end{array}$$

*is commutative.*

PROOF. Note that there is a map $f_0: (K, b) \to (K', b')$ which induces $f_{0*} = \alpha$ on the fundamental group level; we need only choose $f_0(x_i) = \alpha(x_i)$, where this last is interpreted as a path in $K'$. Let $M$ be the space of all such maps; we need to find $f \in M$ such that $fg = g'f$.

To this end let $G: M \to M'$ be the "operator" defined by the diagram

$$\begin{array}{ccc} K & \overset{g}{\to} & K \\ G(f) \uparrow & & \uparrow f \\ K' & \underset{g'}{\to} & K'. \end{array}$$

That is, given $f \in M$, there is a unique map $G(f) \in M$ making this diagram commutative. We claim $G$ has a unique fixed point—which solves our problem—because we can supply $M$ with a metric relative to which $G$ is a contraction.

First, let $\mu, \lambda$ be the measure and "eigenvalue" for $g'$ as given by (6.2). Then, for $f, f' \in M$, and $x \in K$, there is the unique minimal path $\sigma(x)$ connecting $f(x)$ to $f'(x)$ which expresses the fact that $f$ and $f'$ are homotopic and are in $M$. Let

$$\rho(f, f') = \max_{x \in K} l(\sigma(x)).$$

Here $l(\sigma(x))$ means the length of the path $\sigma(x)$, based on the measure $\mu$. Note $\sigma: K \to R$ is continuous.

7.4. LEMMA. $\rho$ *is a metric for* $M$ *in which* $M$ *is complete.*

PROOF. That $\rho$ is symmetric, nonnegative and 0 only if $f = f'$ is clear. The triangle inequality follows from the fact that if $\sigma$ is a path connecting $f$ to $f'$ and $\tau$ is one between $f'$ and $f''$ then $\sigma(x) \circ \tau(x)$ is a path connecting $f$ to $f''$ so that its length is $\geq$ the minimal one.

It is not hard to see that $(M, \rho)$ is complete. E.g., $\rho$ is equivalent topologically to the usual "sup norm" metric.

7.5. LEMMA. $G$ is a contraction.

PROOF. Let $f, f' \in M$ and $\sigma(x)$ be the minimal (continuous) path connecting $Gf(x)$ to $Gf'(x)$. Then $g'\sigma(x)$ connects $f(x)$ to $f'(x)$ and $l(g'\sigma(x)) = \lambda l(\sigma(x))$. Thus $\rho(f, f') \geq \rho(Gf, Gf')$.

Thus there is a unique $f \in M$ such that

$$
\begin{array}{ccc}
K & \xrightarrow{g} & K \\
f\downarrow & & \downarrow f \\
K' & \xrightarrow{g'} & K'
\end{array}
$$

is commutative.

PROOF OF 7.3. If $h$ and $h'$ are topologically conjugate, then (7.2) shows that $S(h) = S(h')$. Now assume $S(h) = S(h')$. Then $h$ and $h'$ have elementary presentations $g : (K, b) \to (K, b)$ and $g' : (K', b') \to (K', b')$ such that $g_* : \pi_1(K, b) \to \pi_1(K, b)$ and $g'_* : \pi_1(K', b') \to \pi_1(K', b')$ are shift equivalent. I.e., there are homomorphisms $\alpha : \pi_1(K, b) \to \pi_1(K', b')$ and $\beta : \pi_1(K', b') \to \pi_1(K, b)$ and an integer $n$ such that the usual diagrams are commutative.

We first show that $\alpha$, $\beta$ satisfies the immersion condition (7.3.1). Let $e_0$ be a one cell of $K$ which begins toward the right. Then $[e_0]$ represents a 1-cycle in $\pi_1(K, b)$ and $\alpha([e_0])$ begins with a cycle $[\varepsilon_0]$, where $\varepsilon_0$ is a 1-cell of $K'$. By choice, we may suppose $\varepsilon_0$ also begins toward the right. Now if $e_1$ is another 1-cell of $K$, and $g([e_1])$ begins with $[\varepsilon_1]$ one easily sees that if $e_1$ begins rightward (leftward), then so does $\varepsilon_1$ because $\alpha g_* = g'_* \alpha$. This proves (iia), and with a little modification, (iib) as well.

To see (i) suppose it fails, say for a generator $[e] \in \pi_1(K, b)$, when $e$ is a 1-cell of $K$. There are two possibilities which are quite similar so we suppose that $\alpha([e]) = \ldots [\varepsilon_1][\varepsilon_2] \ldots$, where $\varepsilon_1, \varepsilon_2$ are 1-cells of $K'$, $\varepsilon_1$ ending to the left and $\varepsilon_2$ beginning to the right. It follows that $g'_* \alpha([e])$ will be a "bent" word, also. But $g_*([e])$ is not "bent", so that by part (iia), $\alpha g_*([e])$ is not either, contradicting the fact that $\alpha g_* = g'_* \alpha$.

Thus 7.3.2 applies yielding maps $r : (K, b) \to (K', b')$ and $s : (K', b') \to (K, b)$ such that $r_* = \alpha, s_* = \beta$ and

$$
\begin{array}{ccc}
K & \xrightarrow{g} & K \\
r\downarrow & & \downarrow r \\
K' & \xrightarrow{g'} & K'
\end{array}
\quad \text{and} \quad
\begin{array}{ccc}
K & \xrightarrow{g} & K \\
s\uparrow & & \uparrow s \\
K' & \xrightarrow{g'} & K'
\end{array}
$$

are commutative. Now consider the problem of finding a map $f$ satisfying $f_* = \beta\alpha$, making

$$
\begin{array}{ccc}
K & \xrightarrow{g} & K \\
f\downarrow & & \downarrow f \\
K & \to & K
\end{array}
$$

commutative. Both $g^n$ and $sr$ solve this problem, so that $sr = g^n$ by uniqueness. Similarly, $rs = g'^m$. Therefore $g$ and $g'$ are shift-equivalent and thus (3.3) $h$ and $h'$ are topologically conjugate, completing the proof of (7.3).

Let $\alpha: F \to F$ be an endomorphism. We say $\alpha$ satisfies the $\Omega$-*condition* provided that for some $m$, the word $\alpha^m(x)$ contains the whole alphabet $O \cup R \cup L$, for each letter $x$. A map $\lambda: X \to X$, $X$ a finite set, is *eventually constant* provided that for some $m \in \mathbf{Z}^+$, $\lambda^m$ is a constant map.

7.6. REALIZATION THEOREM. *In order that a shift class occur as $S(h)$ for an $h$ with elementary presentation it is necessary and sufficient that it contains an endomorphism $\alpha$ of a free group $F$ on generators $O \cup R \cup L$ such that*

(a) *$\alpha$ satisfies the immersion condition (7.3.1)*;

(b) *$\alpha$ satisfies the $\Omega$-condition*;

(c) *for $x \in O$, $R$, $L$, respectively, $\alpha(x)$ begins with $\lambda(x)$, $\lambda(x)$, $\mu(x)$ and ends with $\mu(x)$, $\lambda(x)$, $\lambda(x)$, where $\lambda$ and $\mu$ are eventually constant.*

PROOF. Given $\alpha$ we can clearly define an elementary presentation $g: (K, b) \to (K, b)$ such that $\alpha = g_*: \pi_1(K, b) \to \pi_1(K, b)$. Condition (b) allows us to make $g$ an expansion and at the same time verifies the criterion (1.2). Condition (c) means that Axiom $3°$ is satisfied at the branch point, so that the axioms are verified. The proof in the other direction is similar.

8. **Computation.** We saw in the previous section that the study of conjugacy classes of solenoid-shift maps with elementary presentations is equivalent to the study of shift equivalence classes of certain endomorphisms of finitely generated free groups. These in turn could be fairly easily computed using (4.6) were it not for the fact that the intervening groups (the $X_i$ of the proof of (4.6)) might be of various orders.

Thus let $F$ be free on generators $a, b$ and let $g_1, g_2: F \to F$ be defined by

$$g_1(a) = a^2 b^2 a, \qquad g_2(a) = ababa,$$
$$g_1(b) = a, \qquad g_2(b) = a.$$

It is not known whether $g_1$ is shift equivalent to $g_2$. The characteristic polynomial of $g_i$, abelianized, is $t^2 - 3t - 2$, and one can easily write down the set $\mathscr{G}$ of all endomorphisms $g$ of $F$ satisfying the criteria of (7.6), which have this characteristic polynomial. (Note by [2], [3; Theorem E] or [4], that the corresponding map $K \to K$ would have to have exactly two fixed points. Taking advantage of the symmetries at hand, one need only consider twelve of $\mathscr{G}$'s forty-six maps.)

Though it takes a bit of time, one can write down all factorizations $g = \alpha\beta$, $\alpha, \beta: F \to F$, for each such $g$, and hence discover all "*-equivalences" (see (4.5)), $\alpha\beta \sim_* \beta\alpha$ among the elements of $\mathscr{G}$. Stringing together such *-equivalences would, were it not for the difficulty pointed out above, give all shift equivalences among the elements of $\mathscr{G}$, and incidentally decide whether $g_1 \sim_s g_2$.

Let us write $g \sim_{**} g'$ provided one can proceed from $g$ to $g'$ by a string of equivalences $\alpha\beta \sim_* \beta\alpha$, where $\alpha, \beta$ are endomorphisms of $F$. Then $\mathscr{G}$ has four of these equivalence classes, those containing $g_1, g_2, g_3$ and $g_4$, where

$$g_3(a) = a^2b^4, \qquad g_4(a) = bab,$$
$$g_3(b) = ab, \qquad g_4(b) = ba^2b.$$

Since the geometric counterpart to a $g_i$ has two fixed points, one can find an elementary presentation of it, based at the *other* fixed point. We label the map induced by this other presentation, on the fundamental group level, $g_i'$. One finds that

$$g_1' \sim_{**} g_1, \qquad g_2' = g_2, \qquad g_4' \sim_{**} g_1,$$

but $g_3'$ is a homomorphism of a free group on three letters, to wit,

$$g_3'(a) = ab^4c, \qquad g_3'(b) = abc, \qquad g_3'(c) = ac.$$

It should be emphasized again that it is not known how these maps divide into *shift* equivalence classes. However, it would be very surprising to the author if $g_1$ were shift equivalent to $g_2$.

## 9.  General—not so algebraic—classification theorem.

9.1. DEFINITION. For a group $F$ a subset (not necessarily subgroup) $G$ of $F$, and an endomorphism $\phi: F \to F$, we write $\phi: (F, G) \to (F, G)$ provided $\phi(G) \subset G$. Then $\phi$ and $\phi': (F', G') \to (F', G')$ are shift-equivalent provided there are homeomorphisms $\lambda: (F, G) \to (F', G')$ and $\lambda': (F', G')$ and an integer $m$ such that $\phi'\lambda = \lambda\phi$, $\lambda'\phi' = \phi\lambda'$, $\lambda'\lambda = \phi^m$, and $\lambda\lambda' = \phi'^m$, as before. This is an equivalence relation, and we call the corresponding equivalence classes *shift classes*.

Now let $g: K \to K$ be an immersion satisfying Axioms 1–4 (§1). Label the oriented 1-simplexes of $K$ $x_1, x_2, ..., x_r$, and let $F$ be the free group generated by $x_1, ..., x_r$. Let $G$ consist of the subset of $F$ of all words in $F$ of length at least two which correspond to oriented paths in $K$, that is, immersions of $[0, 1]$ into $K$. (Note that $G$ contains just the information needed to construct $K$.) As oriented paths which can be composed, compose to yield oriented paths, $G$ is a monoid. Note that $g: K \to K$ defines a map $\phi_g: (F, G) \to (F, G)$.

If $g: K \to K$ and $g': K' \to K'$ represent topologically conjugate shift maps $h: \Sigma \to \Sigma$ and $h': \Sigma' \to \Sigma'$, then $\phi_g$ is shift equivalent to $\phi_{g'}$. Thus we may assign to each conjugacy class $[h]$, a shift class $S(h)$ of endomorphisms of $[F(g), G(g)]$, where $g: K \to K$ is a presentation of $h$. Then proceeding just as in the proof of (7.3), we can prove the

9.2. GENERAL CLASSIFICATION THEOREM. *In order that two shift maps $h, h'$ be topologically conjugate it is necessary and sufficient that $S(h) = S(h')$.*

As pointed out in the introduction, the algebraic device $(F, G)$ is a rather thin disguise of the geometric $K$. Nevertheless, (9.2) has some content. Finally, one can easily write down a general realization theorem, corresponding to (7.6).

## BIBLIOGRAPHY

**0.** R. Abraham and S. Smale, *Nongenericity of Ω-stability*, these Proceedings, vol. 14.

**1.** I. N. Herstein, *A note on primitive matrices*, Amer. Math. Monthly, **61** (1954), 18–20.

**2.** S. Smale, *Differential dynamical systems*, Bull. Amer. Math. Soc. **73** (1967), 747–817.

**3.** R. F. Williams, *One dimensional non-wandering sets*, Topology, **6** (1967), 473–487.

**4.** ———, *The zeta function of an attractor*, Conference on Topology of Manifolds (Michigan State Univ.), Prindle, Weber, & Smith, Boston, Mass., pp. 155–161 (to appear).

NORTHWESTERN UNIVERSITY

# AUTHOR INDEX

Roman numbers refer to pages on which a reference is made to an author or a work of an author.
*Italic numbers* refer to pages on which a complete reference to a work by the author is given.
**Boldface numbers** indicate the first page of the articles in the book.

Abraham, Ralph, **1**, *3*, **5**, *8*, 23, *41*, 51, 52, 54, *54*, *92*, *163*, *189*, 191, *202*, 203, 205, 213, *220*, 233, 234, 237, 238, 241, *243*, *272*, *297*, 341, *361*
Adler, R. L., 23, 24, 38, *41*
Anderson, R. D., *123*
Andronov, 278
Anosov, D. V., 9, 61, 62, *92*, *163*, 245, *252*, 283, *326*
Arens, R., *131*
Arnold, V. I., *163*, 179, *183*
Artin, M., *163*, 335, 336, *338*
Auslander, Louis, 9, *15*, 63, 76, 78, *92*, 125, *131*, *184*, 245, 250, 251, 274, 275, *276*, 301, 309, 312, 313, 322, 323, 324, *326*
Avez, A., *131*, *163*, 299

Badé, W., 156
Banchoff, Thomas F., **17**
Bareto, A., 267, *272*
Bialynicki-Birula, A., 312, *326*
Birkhoff, G. D., 253, *264*
Borel, A., 12, *92*, 290, 317, *326*
Bott, R., *3*, 121, *123*
Bourbaki, N., 319, *326*
Bowen, Robert, **23**, *41*, **43**, 290, 336, 338, *338*, *339*
Brouwer, 205
Buchner, Michael A., **51**

Chern, S.-S., 121, *123*
Chevalley, C., 319, 321, *326*
Chow, W. L., 252, *252*
Coddington, E., *183*
Conley, Charles, 55
Conze, J. P., 27, *41*

Dieudonné, *163*
Deprit, A., *189*

Easton, R. W., **55**
Epstein, D. B. A., 274, *276*
Evens, L., 341

Fatou, P., 48, *49*, 95, 97, *123*
Fine, N., 174, *183*
Fomin, S. V., 299, 301, 308, *327*
Franks, John, **61**, 125, 126, *131*, 273, 275, 276, *276*, 283, 299, *327*
Frederickson, Paul, 254, *264*
Furstenburg, H., 23, *41*

Gårding, L., *327*
Gelfand, I. M., 299, 301, 308, *327*
Gottschalk, W., 37, *123*
Green, L., 301, 309, 312, 323, 324, *326*
Gromoll, D., 1, *3*
Guckenheimer, John, *41*, **95**, 203, 204, 215, 216, 217, *220*, 277

Haefliger, A., 81, 82, *92*
Hahn, F., 312, *326*
Harish-Chandra, *326*
Hartman, P., *92*, *163*, *184*, 231
Hedlund, G., *123*
Hedlund, W., *184*
Helgason, S., 315, 316, *327*
Helson, H., 216, *220*
Henrard, J., *189*
Hermann, R., 252, *252*
Herstein, I. N., *361*
Hille, E., *123*
Hirsch, Morris, 8, 27, *41*, 63, 65, 67, 76, *92*, **125**, *131*, **133**, *163*, 167, 192, 199, *202*, 222, 224, 227, 273, 274, 290, 291, 297
Hochschild, G., *92*, 310, *327*
Holmes, *163*

Jacobson, M. B., 122
Jacobson, N., *327*
Julia, C., 95, 97, 103, *123*

Kelley, J., *54*, 133
Keynes, H. B., 27, 37, *41*
Klingenberg, W., 3
Konheim, A. G., 23, 24, *41*
Kopell, Nancy, **165**
Kripke, B., 167, 168
Kupka, I., 167, 170, 233, 239

Lam, P. F., 166
Lanford, O. E. III, 36, *41*, **43**, 336, 338, *339*
Lang, S., *163*, *272*
Lattes, M., 103
Levine, H. I., *189*
Levinson, N., *183*

MacLane, S., *92*
McAndrew, M. H., 23, 24, 38, *41*

Mackey, G. W., 248, *252*, 325, *327*
Maïer, A., 254, *264*

Malcev, A., *15*, 76, *92*, *327*
Malgrange, B., 54, *54*
Margulis, G. A., *327*
Marsden, J., *3*, 51, 52, 54, *54*, 233, 234, *243*, *272*
Mather, John, 61, 125, 148, 155, *163*
Mautner, F., 301, 308, *327*
Mazur, B., 335, 336, *338*
Meyer, Kenneth R., 3, *3*, **185**, 235, *243*, 335, 336, *339*
Meyer, W., 1, 3, *3*
Milnor, J., 92, *93*, 276, *276*, 300, 317, *327*
Montel, 97, 119
Montgomery, D., 127, 128, *131*
Moore, C. C., 301, 322, 325, *327*
Moore, R. L., 167, 263, *264*
Morse, Marsten, 1, 278
Moser, J., 61, 135
Mostow, George D., 301, 312, 324, *327*
Munkres, J. R., *93*
Myers, S. B., 128, *131*

Nachbin, L., 322, *327*
Nagy, *163*
Narasimhan, R., 53, *54*
Nemytskii, V. V., *264*
Newhouse, Sheldon E., **191**, 210, *220*
Nitecki, Zbigniew, **203**
Novikov, S. P., 85

O'Meara, O. T., 320, *327*

Palais, R., 2, *3*, 167, 180
Palis, Jacob, 167, 182, *184*, **221**, *222*, **223**, *231*, 280, 286, 287, 290, 338
Palmore, Julian, **185**, *189*, *243*
Parry, W., 36, *41*, 43
Peixoto, M., 166, 174, *184*, 203, 204, *220*, 235, 239, *243*, 267, *272*, 329, *334*
Poincaré, H., 121, 189, *189*
Pólya, G., *49*
Pontrjagin, L., *93*, 278
Pugh, Charles C., *8*, 27, *41*, 65, *92*, *131*, **133**, *163*, 191, 192, 199, *202*, 204, 205, *220*, *222*, 224, 272, 290, 291, *297*, *327*, *334*

Reeb, G., *93*, 168, *184*
Riesz, *163*
Ritt, J. F., 95, 97, *123*
Robbin, J., *3*, *8*, 52, *54*, *92*, *163*, 203, 205, 213, *220*, 237, 238, 241, *243*, *272*
Robertson, J. B., 27, 37, *41*
Robinson, R. Clark, 3, *3*, **233**, 243, *243*
Rokhlin, V. A., 23, 35, *41*
Rosen, Michael I., **17**
Rosenberg, H., **289**, *297*
Rosenlicht, M., 312, *326*

Sacksteder, Richard, *93*, **245**
Saks, S., 247, *252*

Sarason, D., 216, *220*
Schenkman, E., 275, *276*
Scheuneman, John, **9**, *15*, 251, 313, *326*
Schlessinger, M., 49
Schwartz, A. J., **253**
Schweigert, G. E., 174, *183*
Selberg, A., 317, *327*
Séminaire S. Lie, *327*
Shahshahani, S., **265**
Sherman, 325
Shub, Michael, 27, *41*, 64, 6⁵, 66, 69, 90, 91, *93*, 95, 102, 119, *123*, 125, 127, *131*, 135, 167, 182, 203, 204, 212, *220*, **273**, 274, *276*, 287, 290, 299, 336, *339*, 341
Sinai, Ja. G., 34, *41*, 245, *252*
Smale, Stephen, 2, **5**, *8*, *15*, 17, *21*, 23, 24, 26, 27, 30, 34, 36, 37, *41*, 49, 61, 63, 65, 67, 81, 89, *93*, 95, 96, *123*, 135, 155, *163*, 167, 182, *184*, 191, 192, 197, 202, *202*, *220*, 221, 222, 222, **223**, *231*, 233, 239, 272, *272*, *273*, *276*, **277**, 280, 287, *287*, **289**, *297*, 299, 301, *327*, 329, 330, *334*, 335, 336, 337, 338, *339*, 341, *361*
Smith, P. A., 81
Solovay, R., 167
Spanier, E., *93*
Steenrod, N., 128, *131*
Stepanov, V. V., *264*
Sternberg, S., 121, *123*, *163*, 167, 168, 172, 173, 180, 182, *184*
Szegö, G., *49*

Tamagawa, T., *327*
Tate, *163*
Thom, R., 17, 271, 329
Thomas, E. S., **253**
Tomter, Per, **299**

Van Dantzig, D., 126, *131*
Van der Waerden, B., 126, *131*
Van Kampen, E. R., *184*
von Neumann, J., 248, *252*

Wallach, Nolan R., 301
Walters, P., 27
Walton, Robert, 54
Weil, 335
Weinstein, A., *3*
Weyl, H., *123*
Wilder, R., *123*
Williams, R. F., 34, *41*, 112, 113, *123*, 135, 160, 161, *163*, 191, 192, *202*, 273, *276*, 285, 287, **329**, *334*, **335**, *339*, **341**, *361*
Wolf, Joseph A., 92, *93*, 276, *276*, 300, 301, 305, 308, 316, 317, *327*

Zelinsky, Daniel, 329
Zippin, L., 127, 128, *131*, 258, *264*

# SUBJECT INDEX

Absolutely continuous spectrum, 323
Admissible metric, 157
Adopted metric, 148
Affine algebraic group, 319
Affine transformations, 312, 313
Aggregate, 111
Algebra
    Clifford, 300, 320
    Dual, 14
    Free nilpotent Lie, 10
    Quaternion, 320
Algebraic groups, 310
Algebraic number field, 20
Almost conformal, 130
Almost invariant, 247
Anisotropic, 320, 321
    Over $\mathbf{Q}$, 321
Anosov, 283
    Derived from (DA), 329
Anosov automorphism, 9
Anosov coverings, 62
    Splitting, 67
Anosov diffeomorphism, 17, 61, 159, 302, 312, 313
    $N$-induced, 302, 310, 313, 324, 326
Anosov flow, 299, 302, 305
    $G$-induced, 301
    $(G, \Gamma)$-induced, 299, 301, 305, 313, 317, 319
    Of "Mixed type", 321
Arithmetic groups, 319
Arithmetic subgroups, 301, 319, 320
Attractive, 135
Attractor, 341
    1-dimensional, 160, 341
Axiom A, 5, 191, 341

Baire property, 265
Basic sets, 289
Branched 1-manifold, 341, 343
Bumpy metric, 1
Bundles, 345
    Contracting, 289
    Expanding, 289

Canonical coordinates, 27
Cantor set
    $k$-, 195
    $(k_1, k_2)$-thick, 195, 197
    Two component, 194
Cartan-decomposition, 316
Cartan subalgebra, 314, 315, 321
Cayley hyperbolic plane, 316
Central sequence, 253

Characteristic polynomials, 351
Clifford algebra, 300, 320
Clifford group, 321
Clifford-Klein form, 314, 317
Closed orbit, 277
    Elementary, 234
    Generic, 234, 282
    Periodic, 277
Conformal expanding maps, 130
Conjugate
    $\Omega$, 5, 96, 285, 289
    $\Pi$, 63
    Semi-, 273
    Topologically, 61, 96, 125, 273, 282, 342, 356, 360

Diffeomorphisms
    Anosov, 17, 61, 159, 302, 312, 313
        $N$-induced, 302, 310, 313, 324, 326
    Axiom A, 5, 191, 341
    Future stable, 287
    Local stable, 133
    Mixing, 249
    No cycle property, 289
    Nonwandering, 5, 96, 191, 253
    $\Omega$-conjugate, 5, 96, 285, 289
    $\Omega$-explosion, 296
    $\Omega$-stable, 5, 285, 289
    Periodic point, 17
    Poisson stable, 254
    Structurally stable, 61, 96, 223, 230, 267, 278, 283, 302
    Topologically conjugate, 61, 96, 125, 273, 283, 342, 356, 360
    Toral, 17
Differentiable dynamical system, 277
Differential equation, second order ordinary, 265
Differentiation in a metric space, 247
Defining sequence, 194
Diophantine equation, 320
Dual algebra, 14

Equivalent
    $G$-, 303
    $(G, \Gamma)$-, 304
Ergodic, 301, 322, 323, 324
Ergodic measure, 130
Expanding maps, 60, 90, 125, 273
    Almost conformal, 130
    Conformal, 130
Expansive homeomorphism, 26, 336
Exponential growth, 300, 313, 317, 319

Fiber contraction theorem, 136
Fibonacci number, 17
Finite type, 36, 43, 338
Fixed point, 133, 135
    Attractive, 135
    Homoclinic points, 197
    Lefschitz fixed point formula, 336
Flow, 253
    Anosov, 299, 302, 305
        G-induced, 301
        (G, Γ)-induced, 299, 301, 305, 313, 317, 319
    G-induced, 302, 303
        G-equivalent, 303
    (G, Γ)-induced, 304, 305
        (G, Γ)-equivalence, 304
    Of "Mixed type", 321
    Geodesic, 299, 308, 314
    Recurrence, 253
    Strongly mixing, 322
    Suspended, 301
    Wandering, 253
    Weakly mixing, 301, 322, 323, 324
Frobenius Theorem, 251

Generalized solenoid, 329, 341, 343
    Presentation, 341, 342
        Elementary, 342, 351
Generalized stable and unstable manifolds, 330
Geodesic flows, 299, 308, 314
Gradient, 277
    H-elementary, 51
Graph transform, 134, 141
Group
    Affine algebraic, 319
    Algebraic, 310
    Arithmetic, 319
    Exponential growths, 300, 313, 317, 319
    Isogeny, 319, 321
    Of type I, 323

Hirsch-Plotkin radical, 275
Homoclinic points, 197
Hyperbolic, 96, 133, 279, 286, 302
Hyperbolic affine transformations, 300
Hyperbolic automorphisms, 300, 302, 312, 322
Hyperbolic set, 160, 161
Hyperbolic nilmanifold endomorphisms, 63
Hyperbolic space
    Complex, 316
    Quaternionic, 316
    Real, 316
Hyperbolic structure, 5, 191, 289
Hyperbolic toral endomorphisms, 63

Infrahomogeneous spaces, 125
Infranil-expanding map, 274
Infranil-manifold, 125, 299, 302, 324
Intregral manifolds, 251
Invariant, 248

Inverse limit, 34
    Isogeny, 319, 321
Isolated invariant set, 55
Isolating submanifold, 55
Iwasawa-decomposition, 316, 325

Kupka-Smale theorem, 202, 266

Ladder, 345
Ladder map, 341
Lebesgue spectrum, 301, 325, 326
Lefschitz fixed point formula, 336
Levi complement, 318
Lie algebras, free nilpotent, 10
Lie derivative, 251
Lipschitz inverse function theorem, 137
Lipschitzian, 248

Manifold
    Branched 1-, 341, 343
    Center of, 253
    Generalized stable and unstable, 330
    Infranil-, 125, 299, 302, 324
    Integral, 251
    Isolating sub-, 55
    Nil —, 17
    Semilocal stable, of f, 98
    Stable, 141
    Unstable, 133, 141, 279
Map
    Conformal expanding, 130
    Equivalent, 342
    Eventually constant, 359
    Expanding, 64, 90, 125
    Infranil-expanding, 274
    Inverse limit, 34
    Ladder, 341
    Nil-expanding, 273, 274
    Ω-condition, 359
    Shift, 341
Measure, 343, 354
    Ergodic, 130
    Spectral, 323, 325
Metric
    Admissible, 157
    Adopted, 148
    Bumpy, 1
Metrically splitting, 67
Mixing diffeomorphisms, 249

Nil-expanding map, 273, 274
Nilmanifolds, 17
Nilradical, 322
No cycle property, 289
Nondegenerate critical point, 278
Nonwandering point, 5, 96, 191, 253
Nonwandering set, 285
Number field
    Algebraic, 20
    Totally real, 17

Periodic point, 17
Poincaré map, 269
Poincaré upper half-plane, 314
Poisson stable, 254
Polynomial growth, 91
Presentation, 341, 342
    Elementary, 342, 351
Principal characteristic multipliers, 233
Product neighborhood, 67
Pseudotransverse, 241

Quadratic space, 320
Quaternions, 320, 321
Quaternion algebra, 320

Recurrence, 253
Regular sequence tending to $x$, 247
Riemannian spaces, two-point homogeneous, 316

Semilocal stable manifold of $f$, 98
Semiconjugate, 273
Semidirect product, 302, 307, 310
Sequence
    Central, 253
    Defining, 194
    $i$-gap, 194
    $\omega(\alpha)$-limit point, 254
    Regular, tending to $x$, 247
Set
    Basic, 289
    Hyperbolic, 160
    Isolated invariant, 55
    $k$-Cantor, 195
    $(k_1, k_2)$-thick Cantor, 197
    Stable, 282
    Subbasic, 5
    Two component Cantor, 194
Shift class, 342, 356
    Equivalence, 342
Shift equivalence, 349
Shift map, 281, 341
    Sub-shifts of finite type, 36, 43, 338
    Unilateral, 130
Space
    Infrahomogeneous, 125
    Complex hyperbolic, 316
    Quadratic, 320
    Quaternionic hyperbolic, 316
    Real hyperbolic, 316
    Symmetric, 314, 315, 316

Two-point homogeneous Riemannian, 316
Spectral measure, 323, 325
Spectrum
    Absolutely continuous, 323
    Lebesgue, 301, 325, 326
Stable
    $\Omega$-, 5, 285, 289
    Structurally, 61, 96
Stable manifold, 141
Stable Manifold Theorem, 5
    For a hyperbolic set, 149
    For a point, 146
Stable sets, 282
Strongly mixed modulo, 245, 246, 247, 249
    Structural stability, 329
Structurally stable, 223, 230, 267, 278, 283, 302
Subbasic set, 5
Suspended flow, 301
Suspension, 302, 310
Symmetric space, 314, 315, 316

Tangency, one-sided, 201
Topologically conjugate, 61, 96, 125, 273, 283, 342, 356, 360
Topological entropy, 23
Topologically equivalent, 278
Toral diffeomorphisms, 17
Totally real number field, 17
Transversality condition, 280
    Strong, 287
Transversality theorem, 2
    Density, 53
Tubular family, 224
Tubular families, system of, 224
    Compatible, 224

Unilateral shift, 130
Unstable ideal, 309, 323
Unstable manifolds, 133, 141, 279

Vector field
    Baire Property, 265
    Property H1, 51, 234
    Property H2, 234
    Topologically equivalent, 278

Weil zeta functions, 351
Whitney-$C^r$ topology, 265

$Z$-subalgebra, 9
Zero-dimensional basic sets, 338
Zeta function, 34, 43, 335